Ubiquitous Computing

Ubiquitous Computing

Smart Devices, Environments and Interactions

Stefan Poslad

Queen Mary, University of London, UK

A John Wiley and Sons, Ltd, Publication

This edition first published 2009
© 2009 John Wiley & Sons Ltd.,

Registered office
John Wiley & Sons Ltd, The Atrium, Southern Gate, Chichester, West Sussex, PO19 8SQ, United Kingdom

For details of our global editorial offices, for customer services and for information about how to apply for permission to reuse the copyright material in this book please see our website at www.wiley.com.

Library of Congress Cataloging-in-Publication Data

Poslad, Stefan.
  Ubiquitous computing : smart device, environment, and interactions / Stefan Poslad.
    p. cm.
  Includes index.
  ISBN 978-0-470-03560-3 (cloth)
  1. Ubiquitous computing. 2. Context-aware computing. 3. Human-computer interaction.
  I. Title.
  QA76.5915.P67 2009
  004—dc22

                                                                                        2008052234

A catalogue record for this book is available from the British Library.

ISBN 978-0-470-03560-3 (H/B)

To my family, Ros and Ben here, and to friends and family in three wonderful parts of the world, South Wales (UK), Glandorf and Brisbane.

Contents

List of Figures

List of Tables

Preface

Ubiquitous Computing, often also referred to as Pervasive Computing, is a vision for computer systems to infuse the physical world and human and social environments. It is concerned with making computing more physical, in the sense of developing a wider variety of computer devices can be usefully deployed in more of the physical environment. It is concerned with developing situated and pervasive technology that is highly accessible and usable by humans that can be designed to operate in harmony in human and social environments.

Audience

This book is primarily aimed at computer scientists and technologists in education and industry to enable them to keep abreast of the latest developments, across a diverse field of computing, all in one text. Its aim is to also to promote a much more cross-disciplinary exchange of ideas within the sub-fields of computing and between computer science and other associated fields. It interlinks several sub-fields of computing, distributing computing, communication networks, artificial intelligence and human computer interaction at its core, as well as explaining and extending designs which cover mobile services, service-oriented computing, sensor nets, micro-electromechanical systems, context-aware computing, embedded systems and robotics, and new developments in the Internet and the Web. This is a good text to apply models in these fields.

The main prerequisite needed to understand this book is a basic level of understanding of computer science and technology. Parts of the book should be readily understandable by students towards the middle and end of undergraduate courses in computer science, although parts of it may also be used as an introduction textbook to highlight some of the amazing things that are happening in the world of ICT systems. It is also suitable for students at MSc level and for cross-disciplinary use in courses which include computing as just one of the elements of the course. It is the author's hope that this text will contribute to a renewed interest in some of the advanced ideas of computing by a wider audience and will lead to new advanced courses in computing being developed. An overview of the book is found at the end of the first chapter.

Teaching with this Book

The author's website for the book is available at http://www.elec.qmul.ac.uk/people/stefan/ubicom. The website contains PowerPoint slides for the book, additional exercises and selected solutions to exercises, on-line bibliography for the book, etc. The book site also gives advice about how to use this book in different types of educational courses and training programs.

Acknowledgements

Patricia Charlton, Michael Berger and Robert Patton were involved with this book project at the start. In particular, Patricia Charlton contributed many good ideas particularly in the AI chapters, two of which she co-authored. Several international colleagues gave feedback on specific sections of the book: Barbara Schmidt-Belz, Heimo Laamanen, Jigna Chandaria and Steve Mann; as did several colleagues at QMUL: Athen Ma, Chris Phillips, Karen Shoop and Rob Donnan.

The contents of this book arose in part out of teaching various distributed computing, AI, HCI and other applied computing courses at undergraduate level and at MSc at several universities but in particular through teaching the ELEM038, Mobile Services courses to students at Queen Mary, University of London. Second, this book arose out of research in the following projects: AgentCities, CASBAH, Context-aware Worker (an EPSERC Industrial Case-award project with BT, John Shepherdson), CRUMPET, EDEN-IW, iTrust, My e-Director 2012 and from work with my research assistants: uko Asangansi, Ioannis Barakos, Thierry Delaitre, Xuan (Janny) Huang, Kraisak Kesorn, Zekeng Liang, Dejian Meng, Jim Juan Tan, Leonid Titkov, Zhenchen Wang and Landong Zuo. Several of them helped to review parts of this text.

At Wiley, Birgit Gruber instigated this book project. Sarah Hinton and Anna Smart guided the book through the various stages of drafting to the finished product while Susan Dunsmore and Sunita Jayachandran helped apply the finishing touches. Finally, my family offered a high-level of support throughout, encouraging me onwards, to finish it.

1

Ubiquitous Computing: Basics and Vision

1.1 Living in a Digital World

We inhabit an increasingly digital world, populated by a profusion of digital devices designed to assist and automate more human tasks and activities, to enrich human social interaction and enhance physical world[1] interaction. The physical world environment is being increasingly digitally instrumented and strewn with embedded sensor-based and control devices. These can sense our location and can automatically adapt to it, easing access to localised services, e.g., doors open and lights switch on as we approach them. Positioning systems can determine our current location as we move. They can be linked to other information services, i.e., to propose a map of a route to our destination. Devices such as contactless keys and cards can be used to gain access to protected services, situated in the environment. Epaper[2] and ebooks allow us to download current information onto flexible digital paper, over the air, without going into any physical bookshop. Even electronic circuits may be distributed over the air to special printers, enabling electronic circuits to be printed on a paper-like substrate.

In many parts of the world, there are megabits per second speed wired and wireless networks for transferring multimedia (alpha-numeric text, audio and video) content, at work and at home and for use by mobile users and at fixed locations. The increasing use of wireless networks enables more devices and infrastructure to be added piecemeal and less disruptively into the physical environment. Electronic circuits and devices can be manufactured to be smaller, cheaper and can operate more reliably and with less energy. There is a profusion of multi-purpose smart mobile devices to

---

[1] The physical world is often referred to as the *real-world* or environment in order to distinguish this both from a perceived human view of the world (*imaginary worlds*) not related to reality and basic facts and from computer-generated views of the world (*virtual worlds*).
[2] A distinction needs to be between digital hardware versions of analogue objects, e.g., epaper, versus soft or electronic copies of information held in analogue objects, e.g., etickets, for airlines. The latter type is referred to as vtickets, short for virtual tickets.

Ubiquitous Computing: Smart Devices, Environments and Interactions   Stefan Poslad
© 2009 John Wiley & Sons, Ltd

access local and remote services. Mobile phones can act as multiple audio-video cameras and players, as information appliances and games consoles.[3] Interaction can be personalised and be made user context-aware by sharing personalisation models in our mobile devices with other services as we interact with them, e.g., audio-video devices can be pre-programmed to show only a person's favourite content selections.

Many types of service provision to support everyday human activities concerned with food, energy, water, distribution and transport and health are heavily reliant on computers. Traditionally, service access devices were designed and oriented towards human users who are engaged in activities that access single isolated services, e.g., we access information vs we watch videos vs we speak on the phone. In the past, if we wanted to access and combine multiple services to support multiple activities, we needed to use separate access devices. In contrast, service offerings today can provide more integrated, interoperable and ubiquitous service provision, e.g., use of data networks to also offer video broadcasts and voice services, so-called triple-play service provision. There is great scope to develop these further (Chapter 2).

The term 'ubiquitous', meaning appearing or existing everywhere, combined with computing to form the term Ubiquitous Computing (UbiCom) is used to describe ICT (Information and Communication Technology) systems that enable information and tasks to be made available everywhere, and to support intuitive human usage, appearing invisible to the user.

1.1.1 Chapter Overview

To aid the understanding of Ubiquitous Computing, this introductory chapter continues by describing some illustrative applications of ubiquitous computing. Next the proposed holistic framework at the heart of UbiCom called the Smart DEI (pronounced smart 'day') Framework UbiCom is presented. It is first viewed from the perspective of the core internal properties of UbiCom (Section 1.2). Next UbiCom is viewed from the external interaction of the system across the core system environments (virtual, physical and human) (Section 1.3). Third, UbiCom is viewed in terms of three basic architectural designs or design 'patterns': smart devices, smart environments and smart interaction (Section 1.4). The name of the framework, DEI, derives from the first letters of the terms **D**evices, **E**nvironments and **I**nteraction. The last main section (Section 1.5) of the chapter outlines how the whole book is organised. Each chapter concludes with exercises and references.

1.1.2 Illustrative Ubiquitous Computing Applications

The following applications situated in the human and physical world environments illustrate the range of benefits and challenges for ubiquitous computing. A personal memories scenario focuses on users recording audio-video content, automatically detecting user contexts and annotating the recordings. A twenty-first-century scheduled transport service scenario focuses on the transport schedules, adapting their preset plans to the actual status of the environment and distributing this information more widely. A foodstuff management scenario focuses on how analogue non-electronic objects such as foodstuffs can be digitally interfaced to a computing system in order to monitor their human usage. A fully automated foodstuff management system could involve robots which can move physical objects around and is able to quantify the level of a range of analogue objects. A utility management scenario

---

[3] And of course there is nothing stopping this happening vice versa – games consoles can act as phones, audio-video players and recorders, etc., and cameras can act as phones, etc.

focuses on how to interface electronic analogue devices to an UbiCom system and to manage their usage in a user-centred way by enabling them to cooperate to achieve common goals.

1.1.2.1 Personal Memories

As a first motivating example, consider recording a personal memory of the physical world (see Figure 1.1). Up until about the 1980s, before the advent of the digital camera, photography would entail manually taking a light reading and then manually setting the aperture and shutter speed of the camera in relation to the light reading so that the light exposure on to a light-sensitive chemical film was correct.[4] It involved manually focusing the lens system of the camera. The camera film behaved as a sequential recording media: a new recording requires winding the film to the next empty section. It involved waiting for the whole film of a set of images, typically 12 to 36, to be completed before sending the recorded film to a specialist film processing company with specialist equipment to convert the film into a specialist format that could be viewed. The creation of additional copies would also require the services of a specialist film processing company.

A digital camera automatically captures a visual of part of the physical world scene on an inbuilt display. The use of digital cameras enables photography to be far less intrusive for the subject than using film cameras.[5] The camera can autofocus and auto-expose recorded images and

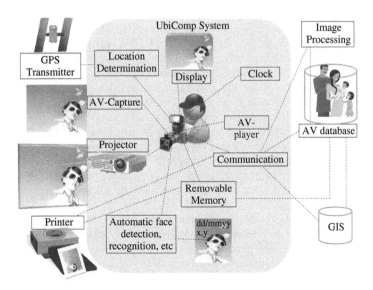

Figure 1.1 Example of a ubiquitous computing application. The AV-recording is person-aware, location-aware (via GPS), time-aware and networked to interact with other ICT devices such as printers and a family-and-friends database

---

[4] There was an easier-to-operate camera called the compact camera that used a fixed focus (set at infinity) and a fixed exposure for the film.

[5] E.g., digital cameras are less obtrusive: they lessen the need for a photographer to ask the subject to say 'cheese' to make the subject focus on the camera because it is easy and cheap to just shoot a whole series of photographs in quick succession and delete the ones that are not considered aesthetic.

video so that recordings are automatically in focus and selected parts of the scene are lit to the optimum degree. The context of the recording such as the location and date/time is also automatically captured using inbuilt location and clock systems. The camera is aware that the person making a recording is perhaps interested in capturing people in a scene, in focus, even if they are off centre. It uses an enhanced user interface to do this which involves automatically overlaying the view of the physical world, whether on an inbuilt display or through a lens or viewfinder, with markers for parts of the face such as the eyes and mouth. It then automatically focuses the lens so faces are in focus in the visual recording.

The recorded content can be immediately viewed, printed and shared among friends and family using removable memory or exchanged across a communications network. It can be archived in an external audio-visual (AV) content database. When the AV content is stored, it is tagged with the time and location (the GIS database is used to convert the position to a location context). Image processing can be used to perform face recognition to automatically tag any people who can be recognised using the friends and family database. Through the use of micro electromechanical systems (MEMS (Section 6.4) what previously needed to be a separate decimetre-sized device, e.g., a projector, can now be inbuilt. The camera is networked and has the capability to discover other specific types of ICT devices, e.g., printers, to allow printing to be initiated from the camera. Network access, music and video player and video camera functions could also be combined into this single device.

Ubiquitous computing (UbiCom) encompasses a wide spectrum of computers, not just devices that are general purpose computers,[6] multi-function ICT devices such as phones, cameras and games consoles, automatic teller machines (ATMs), vehicle control systems, mobile phones, electronic calculators, household appliances, and computer peripherals such as routers and printers. The characteristics of embedded (computer) systems are that they are self-contained and run specific predefined tasks. Hence, design engineers can optimise them as follows. There is less need for full operating system functionality, e.g., multiple process scheduling and memory management and there is less need for a full CPU, e.g., the simple 4-bit microcontrollers used to play a tune in a greeting card or in a children's toy. This reduces the size and cost of the product so that it can be more economically mass-produced, benefiting from economies of scale. Many objects could be designed to be a multi-function device supporting AV capture, an AV player, communicator, etc. Embedded computing systems may be subject to a real-time constraint, real-time embedded systems, e.g., anti-lock brakes on a vehicle may have a real-time constraint that brakes must be released within a short time to prevent the wheels from locking.

ICT Systems are increasing in complexity because we connect a greater diversity and number of individual systems in multiple dynamic ways. For ICT systems to become more useful, they must in some cases become more strongly interlinked to their physical world locale, i.e., they must be context-aware of their local physical world environment. For ICT systems to become more usable by humans, ICT systems must strike the right balance between acting autonomously and acting under the direction of humans. Currently it is not possible to take humans completely out of the loop when designing and maintaining the operation of significantly complex systems. ICT systems need to be designed in such a way that the responsibilities of the automated ICT systems are not clear and the responsibilities of the human designers, operators and maintainers are clear and in such a way that human cognition and behaviour are not overloaded.

---

[6] Of the billions of microprocessors manufactured every year, less than 5% of them find their way into multi-application programmable computers, the other 95% or so are deployed in a range of embedded systems and applications.

1.1.2.2 Adaptive Transport Scheduled Service

In a twentieth-century scheduled transport service, timetables for a scheduled transport service, e.g., taxi, bus, train, plane, etc. to pick up passengers or goods at fixed or scheduled point are only accessible at special terminals and locations. Passengers and controllers have a limited view of the actual time when vehicles arrive at designated way-points on the route. Passengers or goods can arrive and wait long times at designated pick-up points. A manual system enables vehicle drivers to radio in to controllers their actual position when there is a deviation from the timetable. Controllers can often only manually notify passengers of delays at the designated pick-up points.

By contrast, in a twenty-first-century scheduled transport service, the position of transport vehicles is determined using automated positioning technology, e.g., GPS. For each vehicle, the time taken to travel to designated pick-up points, e.g., next stop, final stop, is estimated partly based on current vehicle position, progress and historical data of route users. Up-to-date vehicle arrival times can then be accessed ubiquitously using mobile phones enabling JIT (Just-In-Time) arrival at passenger and goods collection points. Vehicles on the route can tag locations that they anticipate will change the schedule of other vehicles in that vicinity. Anticipated schedule change locations can be reviewed by all subsequent vehicles. Vehicles can then be re-routed and re-scheduled dynamically, based upon 'schedule change' locations, current positions and the demand for services. If the capacity of the transport vehicles was extensible, the volume of passengers waiting on route could determine the capacity of the transport service to meet demand. The transport system may need to deal with conflicting goals such as picking up more passengers and goods to generate more revenue for services rendered versus minimising how late the vehicle arrives at pre-set points along its route.

1.1.2.3 Foodstuff Management

A ubiquitous home environment is designed to support healthy eating and weight regulation for food consumers. A conventional system performs this manually. A next generation system (semi-)automates this task using networked physical devices such as fridges and other storage areas for food and drink items which can monitor the food in and out. Sensors are integrated in the system, e.g., to determine the weight of food and of humans. Scanners can be used to scan the packaging of food and drink items for barcodes, text tables, expiry dates and food ingredients and percentages by weight. Hand-held integrated scanners can also select food for purchase in food stores such as supermarkets that should be avoided on health or personal choice grounds. The system can identify who buys which kind of food in the supermarket.

The system enables meal recipes to be automatically configured to adapt to the ingredients in stock. The food in stock can be periodically monitored to alert users when food will becomes out of date and when the supply of main food items is low. The amount of food, at different levels of granularity in terms of the overall amount of food and in terms of weight in grams of fat, salt and sugar, etc, consumed per unit time and per person can be monitored. The system can incorporate policies about eating a balanced diet, e.g., to consume five pieces of fruit or vegetables a day.

System design includes the following components. Scanners are used to identify the types and quantities of ingredients based upon the packaging. This may include a barcode but perhaps not all food has barcodes and can be identified in this way. The home food store can be designed to check when (selected) food items are running low. Food running low can be defined as there is a quantity of one item remaining but items can be large and partially full. The quantity of a foodstuff remaining needs to be measured using a weight transducer but the container weight overhead is needed in order to calculate the weight of the foodstuff. The home food store could be programmed to detect when food is out of date by reading the expiry date and signalling the food as inedible.

Many exceptions or conditions may need to be specified for the system in order to manage the food store. For example, food may still be edible even if its expiry date has past. Food that is frozen and then thawed in the fridge may be past its sell-by date but is still edible. Selected system events could automatically trigger actions, e.g., low quantities of food could trigger actions to automatically purchase food and have it delivered. Operational policies must be linked to context or situation and to the authorisation to act on behalf of owner, e.g., food is not ordered when consumers are absent or consumers specify that they do not want infinite repeat orders of food that has expired or is low in quantity. There can be limitations to full system automation. Unless the system can act on behalf of the human owner to accept delivery, to allow physical access to the home food store and to the building where consumers live, and has robots to move physical objects and to open and close the home food store to maintain temperature controlled environments there, these scenarios will require some human intervention. An important issue in this scenario is balancing what the system can and should do versus what humans can and should do.

1.1.2.4 Utility Regulation

A ubiquitous home environment is designed to regulate the consumption of a utility (such as water, energy or heating) and to improve usage efficiency. For example, currently utility management, e.g., energy management, products are manually configurable by human users, utilise stand-alone devices and are designed to detect local user context changes. User context-aware energy devices can be designed to switch themselves on in a particular way, e.g., a light switches on, heating switches on when it detects the presence of a user otherwise it switches off. These devices must also be aware of environmental conditions so that artificial lights and heating would not switch on if it determines that the natural lighting and heating levels will suffice.

System design includes the following components and usage patterns. Devices that are configured manually may waste energy because users may forget to switch them off. Devices that are set to be active, according to pre-set user policies, e.g., to control a timer, may waste energy because users cannot always schedule their activities to adhere to the static schedule of the timer. Individually, context-aware devices such as lights, can waste energy because several overlapping devices may be activated and switch on, e.g., when a user's presence is detected.

A ubiquitous system can be designed, using multi-agent system and autonomic system models, to operate as a Smart Grid. Multiple devices can self-manage themselves and cooperate to adhere to users' policies such as minimising energy expenditure. For example, if several overlapping devices are deemed to be redundant, the system will decide which individual one to switch on. Energy usage costs will depend upon multiple factors, not just the time a device is switched on, but also upon the energy rating which varies across devices and the tariff, i.e., the cost of energy usage varies according to the time of day. Advanced utility consumption meters can be used to present the consumption per unit-time and per device and can empower customers to see how they are using energy and to manage its use more efficiently. Demand-response designs can adjust energy use in response to dynamic price signals and policies. For example, during peak periods, when prices are higher, energy-consuming devices could be operated more frugally to save money. A direct load control system, a form of demand-response system, can also be used, in which certain customer energy-consuming devices are controlled remotely by the electricity provider or a third party during peak demand periods. Further examples of ubiquitous computing applications are discussed in Chapter 2.

1.1.3 Holistic Framework for UbiCom: Smart DEI

Three approaches to analyse and design UbiCom Systems to form a holistic framework for ubiquitous computing are proposed called the smart DEI[7] framework based upon:

- *Design architectures to apply UbiCom systems*: Three main types of design for UbiCom systems are proposed: smart device, smart environment and smart interaction. These designs are described in more detail in Section 1.4.
- *An internal model of the UbiCom system properties based upon five fundamental properties*: distributed, iHCI, context-awareness, autonomy, and artificial intelligence. There are many possible sub-types of ubiquitous system design depending on the degree to which these five properties are supported and interlinked. This model and these properties are described in Section 1.2.
- *A model of UbiCom system's interaction with its external environments.* In addition to a conventional distributed ICT system device interaction within a virtual[8] environment (C2C), two other types of interaction are highlighted: (a) between computer systems and humans as systems (HCI); (b) between computers and the physical world (CPI). Environment interaction models are described in Section 1.3.

Smart devices, e.g., mobile smart devices, smart cards, etc. (Chapter 4), focus most on interaction within a virtual (computer) world and are less context-aware of the physical world compared to smart environment devices. Smart devices tend to be less autonomous as they often need to directly access external services and act as personal devices that are manually activated by their owner. There is more emphasis on designing these devices to be aware of the human use context. They may incorporate specific types of artificial intelligence, e.g., machine vision allows cameras to recognise elements of human faces in an image, e.g., based upon eyes and mouth detection.

Smart environments consist of devices, such as sensors, controller and computers that are embedded in, or operate in, the physical environment, e.g., robots (Section 6.7). These devices are strongly context-aware of their physical environment in relation to their tasks, e.g., a robot must sense and model the physical world in order for it to avoid obstacles. Smart environment devices can have an awareness of specific user activities, e.g., doors that open as people walk towards them. They often act autonomously without any manual guidance from users. These incorporate specific types of intelligence, e.g., robots may build complex models of physical behaviour and learn to adapt their movement based upon experience.

Smart interaction focuses on more complex models of interaction of distributed software services and hardware resources, dynamic cooperation and completion between multiple entities in multiple devices in order to achieve the goals of individual entities or to achieve some collective goal. For example, an intelligent camera could cooperate with intelligent lighting in a building to optimise the lighting to record an image. Multiple lighting devices in a physical space may cooperate in order to optimise lighting yet minimise the overall energy consumed. Smart interaction focuses less on physical context-awareness and more on user contexts, e.g., user goals such as the need to reduce the overall

---

[7] Smart DEI stands for the Smart Devices, Environments and Interactions model. It is pronounced 'Smart Day' in order to allude to the fact that the model focuses on the use of systems support for daily activities.

[8] A virtual (computing) environment comprises the distributed shared ICT infrastructure in which individual UbiCom system applications operate. Note also there are other sub-types of virtual environment such as *virtual reality* environments in which humans users can interact with computer simulations of parts of imagined worlds using multimodal sensory interfaces.

energy consumption across devices. Smart interaction often uses distributed artificial intelligence and multi-agent system behaviours, e.g., contract net interaction in order to propose tasks.

The Smart DEI model represents a holistic framework to build diverse UbiCom systems based on smart devices, smart environments and smart interaction. These three types of design can also be combined to support different types of smart spaces, e.g., smart mobile devices may combine an awareness of their changing physical environment location in order to optimise the routing of physical assets or the computer environment from a different location. Each smart device is networked and can exchange data and access information services as a core property. A comparison of a type of smart device, smart environment and smart interaction is also made later (see Table 1.6) with respect to their main UbiCom system properties of distributed, context-aware, obtrusive HCI, autonomy and intelligence and with respect to the types of physical world, human and ICT interactions they support.

1.2 Modelling the Key Ubiquitous Computing Properties

A world in which computers disappear into the background of an environment consisting of smart rooms and buildings was first articulated over fifteen years ago in a vision called ubiquitous computing by Mark Weiser (1991). Ubiquitous computing represents a powerful shift in computation, where people live, work, and play in a seamless computer-enabled environment, interleaved into the world. Ubiquitous computing postulates a world where people are surrounded by computing devices and a computing infrastructure that supports us in everything we do.

Conventional networked computer[9] systems[10] or Information Communication Technology (ICT) systems consider themselves to be situated in a virtual world or environment of other ICT systems, forming a system of ICT systems. Computer systems behave as distributed computer systems that are interlinked using a communications network. In conventional ICT systems, the role of the physical environment is restricted, for example, the physical environment acts as a conduit for electronic communication and power and provides the physical resources to store data and to execute electronic instructions, supporting a virtual ICT environment.

Because of the complexity of distributed computing, systems often project various degrees of transparency for their users and providers in order to hide the complexity of the distributed computing model from users, e.g., anywhere, anytime communication transparency and mobility transparency, so that senders can specify who to send to, what to send rather than where to send it to. Human–computer interaction (HCI) with ICT systems has conventionally been structured using a few relatively expensive access points. This primarily uses input from keyboard and pointing devices which are fairly obtrusive to interact with. Weiser's vision focuses on digital technology that is interactive yet more non-obtrusive and pervasive. His main concern was that computer interfaces are too demanding of human attention. Unlike good tools that become an extension of ourselves, computers often do not allow us to focus on the task at hand but rather divert us into figuring out how to get the tool to work properly.

Weiser used the analogy of writing to explain part of his vision of ubiquitous computing. Writing started out requiring experts such as scribes to create the ink and paper used to present the information. Only additional experts such as scholars could understand and interpret the information. Today, hard-copy text (created and formatted with computers) printed on paper and soft-copy

---

[9] Here the computer is considered to be any device, simple or complex, small or large, that is programmable and has a memory to store data and or code.

[10] A system, at this stage, is defined as a set of interlinked components of interest. Systems are often a system of systems. Everything external to the system's boundary is the system's environment.

text displayed on computer-based devices are very pervasive. Of the two, printed text is still far more pervasive than computer text.[11] In many parts of the world, the majority of people can access and create information without consciously thinking about the processes involved in doing so.[12] Additional visions of Ubiquitous Computing are discussed in Chapter 2 and in the final chapter (Chapter 13).

1.2.1 Core Properties of UbiCom Systems

The features that distinguish UbiCom systems from distributed ICT systems are as follow. First, they are situated in human-centred personalised environments, interacting less obtrusively with humans. Second, UbiCom systems are part of, and used in, physical environments, sensing more of the physical environment. As they are more aware of it, they can adapt to it and are able to act on it and control it. Hence, Weiser's vision for ubiquitous computing can be summarised in three core requirements:

1. Computers need to be networked, distributed and transparently accessible.
2. Human–computer interaction needs to be hidden more.
3. Computers need to be context-aware in order to optimise their operation in their environment.
   It is proposed that there are two additional core types of requirements for UbiCom systems:
4. Computers can operate autonomously, without human intervention, be self-governed, in contrast to pure human–computer interaction (point 2).
5. Computers can handle a multiplicity of dynamic actions and interactions, governed by intelligent decision-making and intelligent organisational interaction. This may entail some form of artificial intelligence in order to handle:

   (a) incomplete and non-deterministic interactions;
   (b) cooperation and competition between members of organisations;
   (c) richer interaction through sharing of context, semantics and goals.

Hence, an extended model of ubiquitous system is proposed. These two additional behaviours enable ubiquitous systems to work in additional environments. These environments are clustered into two groups: (a) human-centred, personal social and economic environments; and (b) physical environments of living things (ecologies) and inanimate physical phenomena. These five UbiCom requirements and three types of environment (ICT, physical and human) are not mutually exclusive, they overlap and they will need to be combined.

1.2.2 Distributed ICT Systems

ICT systems are naturally distributed and interlinked. Multiple systems often behave as and appear as a single system to the user, i.e., multiple systems are transparent or hidden from the user. Individual systems may be heterogeneous and may be able to be attached and detached from the ICT system infrastructure at any time – openness.

---

[11] Many people thought that the rise of computers would lead to a paperless world but this has not happened yet, see Exercises.
[12] Note the ability to read and write text and understand an average vocabulary currently requires several years of training. Hand-writing is inherently dependent on natural language interfaces which have fundamental limitations.

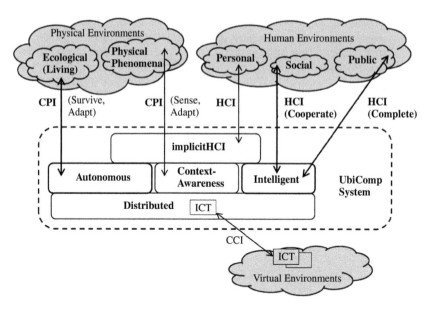

Figure 1.2 A UbiCom system model. The dotted line indicates the UbiCom system boundary

1.2.2.1 Networked ICT Devices

Pervasive computers are networked computers. They offer services that can be locally and remotely accessed. In 1991, Weiser considered that ubiquitous access via 'transparent linking of wired and wireless networks, to be an unsolved problem'. However, since then both the Internet and wireless mobile phones networks have developed to offer seemingly pervasive network access. A range of communication networks exists to support UbiCom interaction with respect to range, power, content, topology and design (Chapter 11).

1.2.2.2 Transparency and Openness

Buxton (1995) considered the core focus of Weiser's vision of ubiquitous computing to be ubiquity (access is everywhere through diverse devices) and transparency (access is hidden, integrated into environments) but that these appear to present an apparent paradox in, how can something be everywhere yet be invisible? The point here is not that one cannot see (hear or touch) the technology but rather that its presence does not intrude into the workplace environment, either in terms of the physical space or the activities being performed. This description of transparency is strongly linked to the notion that devices and functions are embedded and hidden within larger interactive systems. Note also that the vision seems to be associated with a binary classification of system transparency, moving from no transparency to complete transparency. In practice system transparency is often more fuzzy. Systems can have partial connectivity and a limited ability to interoperate with their environment, making transparency more difficult to support. The properties of ubiquity and transparency are core characteristics of types of distributed systems.

A final key property of distributed systems is openness – open distributed systems. Openness allows systems to avoid having to support all their functions at the design time, avoiding closed implementation. Distributed systems can be designed to support different degrees of openness to dynamically discover new external services and to access them. For example, a UbiCom camera can be set to discover printing services and to notify users that these are available. The camera can then transmit its data to the printer for printing.

Openness often introduces complexity and reduces availability. When one function is active, others may need to be deactivated, e.g., some devices cannot record one input while displaying another output. Openness can introduce heterogeneous functions into a system that are incompatible and make the complete system unavailable. Openness can reduce availability because operations can be interrupted when new services and functions are set up. Note many systems are still designed to restrict openness and interoperability even when there appears to be strong benefits not to do so. For example, messages stored in most home answering machines cannot easily be exported, for auditing purposes or as part of a discourse with others. It would be very easy to design phones to share their content via plug and play removable media and a wireless network and to make them more configurable to allow users to customise the amount of message storage they need. Vendors may deliberately and selectively reduce openness, e.g., transparently ignore the presence of another competitor's services, in order to preserve their market share.

Distributed ICT systems are typically designed in terms of a layered model comprising: (1) a hardware resource layer at the bottom, e.g., data source, storage and communication; (2) middleware and operating system services in the middle, e.g., to support data processing and data manipulation; and (3) a human–computer interaction layer at the top. Such a layered ICT model oversimplifies the UbiCom system model because it does not model heterogeneous patterns of systems' interaction. This ICT model typically incorporates only a simple explicit human interaction and simple physical world interaction model. Distributed computer systems are covered in most chapters but in particular in Chapters 3, 4, and 12. Their communications infrastructure is covered in Chapter 11.

1.2.3 Implicit Human–Computer Interaction (iHCI)

Much human–device interaction is designed to support explicit human–computer interaction which is expressed at a syntactical low level, e.g., to activate particular controls in this particular order. In addition, as more tasks are automated, the variety of devices increases and more devices need to interoperate to achieve tasks. The sheer amount of explicit interaction can easily disrupt, distract and overwhelm users. Interactive systems need to be designed to support greater degrees of implicit human–computer interaction or iHCI (Chapter 5).

1.2.3.1 The Calm Computer

The concept of the calm or disappearing computer model has several dimensions. It can mean that programmable computers as we know them today are replaced by something else, e.g., human brain implants, that they are no longer physically visible. It can mean that computers are present but they are hidden, e.g., they are implants or miniature systems. Alternatively, the focus of the disappearing computer can mean that computers are not really hidden; they are visible but are not noticeable as they form part of the peripheral senses. They are not noticeable because of the effective use of implicit human–computer interaction. The forms and modes of interaction to enable computers to disappear will depend in part on the target audience because social and cultural boundaries in relation to technology drivers may have different profile-clustering attributes. For some groups of people, ubiquitous computing is already here. Applications and technologies, such as mobile phones, email and chat messaging systems, are considered as a necessity by some people in order to function on a daily basis.

The promise of ubiquitous computing as technology dissolving into behaviour, invisibly permeating the natural world, is regarded as being unattainable by some researchers, e.g., Rogers (2006). Several reasons are given to support this view. The general use of calm computing removes humans from being proactive – systems are proactive instead of humans. Calm computing is a computationally intractable problem if used generally and ubiquitously. Because technology by its very nature is artificial, it separates the artificial from the natural. What is considered natural is subjective and cultural and to an extent technological. This is blurring the distinction between the means to directly re-engineer nature at the molecular level and the means to influence nature at the macro-level, e.g., pollution and global warming (Chapter 13).

The obtrusiveness of technology depends in part on the user's familiarity and experience with it. Alan Kay[13] is attributed as saying that 'Technology is anything that was invented after you were born.' Everyone considers the technology to be something invented before they were born. If calm computing is used in a more bounded sense in deterministic environments, in limited applications environments and is supported at multiple levels depending on the application requirements, it becomes second nature[14] – calm computing models can then succeed.

1.2.3.2 Implicit Versus Explicit Human–Computer Interaction

The original UbiCom vision focused on making computation and digital information access more seamless and less obtrusive. To achieve this requires in part that systems do not need users to explicitly specify each detail of an interaction to complete a task. For example, using many electronic devices for the first time requires users to explicitly configure some proprietary controls of a timer interface. It should be implicit that if devices use absolute times for scheduling actions, then the first time the device is used, the time should be set. This type of implied computer interaction is referred to as implicit human–computer interaction (iHCI). Schmidt (2000) defines iHCI as 'an action, performed by the user that is not primarily aimed to interact with a computerised system but which such a system understands as input'. Reducing the degree of explicit interaction with computers requires striking a careful balance between several factors. It requires users to become comfortable with giving up increasing control to automated systems that further intrude into their lives, perhaps without the user being aware of it. It requires systems to be able to reliably and accurately detect the user and usage context and to be able to adapt their operation accordingly.

1.2.3.3 Embodied Reality versus Virtual, Augmented and Mediated Reality

Reality refers to the state of actual existence of things in the physical world. This means that things exist in time and space, as experienced by a conscious sense of presence of human beings, and are situated and embodied in the physical world. Human perception of reality can be altered by technology in several ways such as virtual reality, augmented reality, mediated reality and by the hyperreal and telepresence (Section 5.4.4).

Virtual reality (VR) immerses people in a seamless, non-embodied, computer-generated world. VR is often generated by a single system, where time and space are collapsed and exists as a

---

[13] Kay worked at the Xerox Corporation's Palo Alto Research Center (PARC) in the 1970s and was one of the key researchers who developed early prototypes of networked workstations that were later commercialised by Apple, i.e., the Apple Macintosh.

[14] Second nature is acquired behaviour that has been practised for so long that it seems innate.

separate reality from the physical world. Augmented reality (AR) is characterised as being immersed in a physical environment in which physical objects can be linked to a virtual environment. AR can enhance physical reality by adding virtual views to it e.g., using various techniques such as see-through displays and homographic views. Augmented reality can be considered from both an HCI perspective (Section 5.3.3) and from the perspective of physical world interaction (Section 6.2).

Whereas in augmented reality, computer information is added to augment real world experiences, in the more generic type of mediated reality[15] environment, reality may be reduced or otherwise altered as desired. An example of altering reality rather than augmenting it is, rather than use lenses to correct personal visual deficiencies, is to use them to mask far field vision in order to focus on near field tasks.

Weiser drew a comparison between VR and UbiCom, regarding UbiCom to be the opposite of VR. In contrast to VR, ubiquitous computing puts the use of computing in the physical world with people. Indeed, the contrary notion of ubiquitous, invisible computing compared to virtual reality is so strong that Weiser coined the term 'embodied virtuality'. He used this term to refer to the process of 'drawing computers out of their electronic shells'. Throughout this text, the term 'device' is used to focus on the concept of embodied virtuality rather than the more general term of a virtual service. Multiple devices may also form systems of devices and systems of systems. In very open virtual systems, data and processes can exist anywhere and can be accessed anywhere, leading to a loss of (access) control. The potential for privacy violations increases. In physical and virtual embodied systems, such effects are reduced via the implicit restrictions of the embodiment.

Embodied virtuality has several connotations. In order for computers to be more effectively used in the physical world, they can no longer remain embodied in limited electronic forms such as the personal computer but must exist in a wider range of forms which must be more pervasive, flexible and situated. Hence, the emphasis by Weiser of explicitly depicting a larger range of everyday computer devices in the form of tabs, pad and boards (Section 1.4.1.1). Distributed computing works through its increasing ability to interoperate seamlessly to form a virtual computer out of a group of individual computers; it hides the detailed interaction with the individual computers and hides the embodiment within individual forms forming a virtual embodiment for computing.

The use of many different types of physical (including chemical and biological) mechanisms and virtual assembly and reassembly of nature at different levels, can also change the essence of what is human nature and natural (Sections 5.4, 13.7). Through increasing dependence on seamless virtual computers, UbiCom, humans may also risk the erasure of embodiment (Hayles, 1999).

1.2.4 Context-Awareness

The aim of UbiCom systems is not to support global ubiquity, to interlink all systems to form one omnipresent service domain, but rather to support context-based ubiquity, e.g., situated access versus mass access. The benefits of context-based ubiquity include: (1) limiting the resources needed to deliver ubiquitous services because delivering omnipresent services would be cost-prohibitive; (2) limiting the choice of access from all possible services to only the useful services; (3) avoiding overburdening the user with too much information and decision-making; and (4) supporting a natural locus of attention and calm decision-making by users.

---

[15] Reality can also be modified by many other mechanisms, not just virtual computer ones, e.g., chemical, biological, psychological, etc.

1.2.4.1 Three Main Types of Environment Context: Physical, User, Virtual

There are three main types of external environment context-awareness[16] supported in UbiCom:

- *Physical environment context*: pertaining to some physical world dimension or phenomena such as location, time, temperature, rainfall, light level, etc.
- *Human context (or user context or person context)*: interaction is usefully constrained by users: in terms of identity; preferences; task requirements; social context and other activities; user experience and prior knowledge and types of user.[17]
- *ICT context or virtual environment context*: a particular component in a distributed system is aware of the services that are available internally and externally, locally and remotely, in the distributed system.

Generally, the context-aware focus of UbiCom systems is on physical world awareness, often in relation to user models and tasks (Section 5.6). Ubiquitous computers can utilise where they are and their physical situation or context in order to optimise their services on behalf of users. This is sometimes referred to as context-awareness in general but more accurately refers to physical context-awareness. A greater awareness of the immediate physical environment could reduce the energy and other costs of physical resource access – making systems more eco-friendly.

Consider the use of the digital camera in the personal visual memories application. It can be aware of its location and time so that it can record where and when a recording is made. Rather than just expressing the location in terms of a set of coordinates, it can also use a Geographical Information System to map these to meaningful physical objects at that location. It can also be aware of its locality so that it can print on the nearest accessible computer.

1.2.4.2 User-Awareness

A camera can be person-aware in a number of ways in order to detect and make sure people are being recorded in focus, so that it configures itself to a person's preferences and interests. These are all specific examples of physical context-awareness.

User context-awareness, also known as person-awareness, refers to ubiquitous services, resources and devices being used to support user-centred tasks and goals. For example, a photographer may be primarily interested in capturing digital memories of people (the user activity goal) rather than capturing memories of places or of people situated in places. For this reason, a UbiCom camera can be automatically configured to detect faces and to put people in focus when taking pictures. In addition, in such a scenario, people in images may be automatically recognised and annotated with names and named human relationships.

Note that the user context-awareness property of a UbiCom system, i.e., being aware of the context of the user, overlaps with the iHCI property. User context-awareness represents one specific sub-type of context-awareness. A context-aware system may be aware of the physical

---

[16] UbiCom systems may also have an internal system context because a system reflects on its own internal system operation. The internal context may affect adaptation to the external context.

[17] It is not only users who fully determine a system context but other stakeholders such as providers and mediators.

world context, e.g., the location within and the temperature of the environment, and aware of the virtual world or ICT context, e.g., the network bandwidth being consumed for communication (Section 7.6).

In practice, many current devices have little idea of their physical context such as their location and surroundings. The physical context may not be able to be accurately determined or even determined at all, e.g., the camera uses a particular location determination system that does not work indoors. The user context is even harder to determine because the users' goals may not be published and are often weakly defined. For this reason, the user context is often derived from users' actions but these in turn may also be ambiguous and non-deterministic.

1.2.4.3 Active Versus Passive Context-Awareness

A key design issue for context-aware systems is to balance the degree of user control and awareness of their environment (Section 7.2). At one extreme, in a (pure) active context-aware system, the UbiCom system is aware of the environment context on behalf of the user, automatically adjusting the system to the context without the user being aware of it. This may be useful in applications where there are strict time constraints and the user would not otherwise be able to adapt to the context quickly enough. An example of this is a collision avoidance system built into a vehicle to automatically brake when it detects an obstacle in front of it. In contrast, in a (pure) passive context-aware system, the UbiCom system is aware of the environment context on behalf of the user. It just reports the current context to the user without any adaptation, e.g., a positioning system reports the location of a moving object on a map. A passive context-aware system can also be configured to report deviations from a pre-planned context path, e.g., deviations from a pre-planned transport route to a destination. Design issues include how much control or privacy a human subject has over his or her context in terms of whether the subject knows: if his or her context is being acquired, where the context is being kept and to who and what the context is distributed to. Context-awareness is discussed in detail in Chapter 7.

1.2.5 Autonomy

Autonomy refers to the property of a system that enables a system to control its own actions independently. An autonomous system may still be interfaced with other systems and environments. However, it controls its own actions. Autonomous systems are defined as systems that are self-governing and are capable of their own independent decisions and actions. Autonomous systems may be goal- or policy-oriented: they operate primarily to adhere to a policy or to achieve a goal.

There are several different types of autonomous system. On the Internet, an autonomous system is a system which is governed by a router policy for one or more networks, controlled by a common network administrator on behalf of a single administrative entity. A software agent system is often characterised as an autonomous system. Autonomous systems can be designed so that these goals can be assigned to them dynamically, perhaps by users. Thus, rather than users needing to interact and control each low-level task interaction, users only need to interact to specify high-level tasks or goals. The system itself will then automatically plan the set of low-level tasks needed and schedule them automatically, reducing the complexity for the user. The system can also replan in case a particular plan or schedule of tasks to achieve goals cannot be reached. Note the planning problem is often solved using artificial intelligence (AI).

1.2.5.1 Reducing Human Interaction

Much of the ubiquitous system interaction cannot be entirely human-centred even if computers become less obtrusive to interact with, because:

- Human interaction can quickly become a bottleneck to operate a complex system. Systems can be designed to rely on humans being in the control loop. The bottleneck can happen at each step, if the user is required to validate or understand that task step.
- It may not be feasible to make some or much machine interaction intelligible to some humans in specific situations.
- This may overload the cognitive and haptic (touch) capabilities of humans, in part because of the sheer number of decisions and amount of information that occur.
- This original vision needs to be revisited and extended to cover networks of devices that can interact intelligently, for the benefit of people, but without human intervention. These types of systems are called automated systems.

1.2.5.2 Easing System Maintenance Versus Self-Maintaining Systems

Building, maintaining and interlinking individual systems to be larger, more open, more heterogeneous and complex systems is more challenging.[18] Some systems can be relatively simply interlinked at the network layer. However, this does not mean that these can be so easily interlinked at the service layer, e.g., interlinking two independent heterogeneous data sources, defined using different data schemas, so that data from both can be aggregated. Such maintenance requires a lot of additional design in order to develop mapping and mediating data models. Complex system interaction, even for automated systems, reintroduces humans in order to manage and maintain the system.

Rather than design systems to focus on pure automation but which end up requiring manual intervention, systems need to be designed to operate more autonomously, to operate in a self-governed way to achieve operational goals. Autonomous systems are related to both context aware systems and intelligence as follows. System autonomy can improve when a system can determine the state of its environment, when it can create and maintain an intelligent behavioural model of its environment and itself, and when it can adapt its actions to this model and to the context. For example, a printer can estimate the expected time before the printer toner runs out based upon current usage patterns and notify someone to replace the toner.

Note that autonomous behaviour may not necessarily always act in ways that human users expect and understand, e.g., self-upgrading may make some services unresponsive while these management processes are occurring. Users may require further explanation and mediated support because of perceived differences between the system image (how the system actually works) and users' mental model of the system (how users understand the system to work, see Section 5.5.5).

From a software engineering system perspective, autonomous systems are similar to functionally independent systems in which systems are designed to be self-contained, single-minded, functional, systems with high cohesion[19] and that are relatively independent of other systems (low-coupling)

---

[18] The operating system software alone can contain over 30 million lines of code and require 4000 programmers for development (Horn, 1999). The operating system is just one part of the software for a complex distributed system. There is also the set of operating system utilities to consider.

[19] Cohesion means the ability of multiple systems or system components to behave as a single unit with respect to specific functions.

(Pressman, 1997). Such systems are easier to design to support composition, defined as atomic modules that can be combined into larger, more complex, composite modules. Autonomous system design is covered in part in Chapter 10.

1.2.6 Intelligence

It is possible for UbiCom systems to be context-aware, to be autonomous and for systems to adapt their behaviour in dynamic environments in significant ways, without using any artificial intelligence in the system. Systems could simply use a directory service and simple event condition action rules to identify available resources and to select from them, e.g., to discover local resources such as the nearest printer. There are several ways to characterise intelligent systems (Chapter 8). Intelligence can enable systems to act more proactively and dynamically in order to support the following behaviours in UbiCom systems:

- *Modelling of its physical environment*: an intelligent system (IS) can attune its behaviour to act more effectively by taking into account a model of how its environment changes when deciding how it should act.
- *Modelling and mimicking its human environment*: it is useful for a IS to have a model of a human in order to better support iHCI. IS could enable humans to be able to delegate high-level goals to the system rather than interact with it through specifying the low-level tasks needed to complete the goal.
- *Handling incompleteness*: Systems may also be incomplete because environments are open to change and because system components may fail. AI planning can support re-planning to present alternative plans. Part of the system may only be partially observable. Incomplete knowledge of a system's environment can be supplemented by AI type reasoning about the model of its environment in order to deduce what it cannot see is happening.
- *Handling non-deterministic behaviour*: UbiCom systems can operate in open, service dynamic environments. Actions and goals of users may not be completely determined. System design may need to assume that their environment is a semi-deterministic environment (also referred to as a volatile system environment) and be designed to handle this. Intelligent systems use explicit models to handle uncertainty.
- *Semantic and knowledge-based behaviour*: UbiCom systems are also likely to operate in open and heterogeneous environments. Types of intelligent systems define powerful models to support interoperability between heterogeneous systems and their components, e.g., semantic-based interaction.

Types of intelligence can be divided into individual properties versus multiple entity intelligence properties (see Table 1.5).

1.2.7 Taxonomy of UbiCom Properties

There are many different examples of defining and classifying ubiquitous computing. Weiser (1991) referred to UbiCom by function in terms of being distributed, non-obtrusive to access and context-aware. The concept of UbiCom is related to, and overlaps with, many other concepts, such as pervasive computing, sentient computing, context-aware computing, augmented reality and ambient intelligence. Sentient computing is regarded as a type of UbiCom which uses sensors to perceive its environment and to react accordingly. Chen and Kotz (2000) considers context-awareness use as more specifically applied to mobile computing in which applications can discover and take advantage of contextual information (such as user location, time of day, nearby people and devices, and user activity). Context-aware computing is also similar to sentient computing, as is agent-based computing in which agents construct and maintain a model of their environment to more effectively act in it. Ambient intelligence (ISTAG, 2003) characterises systems in terms of supporting the properties of intelligence using ambience and iHCI. Aarts and Roovers (2003) define the five key features of ambient intelligence to be embedded, context-aware, personalised, adaptive and anticipatory.

Buxton (1995) considers ubiquity and transparency to be the two main properties of UbiCom. Aarts and Roovers (2003) classify ubiquitous systems in terms of disposables (low power, low bandwidth, embedded devices), mobiles (carried by humans, medium bandwidth) and statics (larger, stationary devices with high-speed wired connections. Endres *et al.* (2005) classify three types of UbiCom System: (1) distributed mobile systems; (2) intelligent systems (but their focus here is more on sensor and embedded systems rather than on intelligence *per se*); and (3) augmented reality. Milner (2006) considers the three main characteristics of UbiCom as follows: (1) they are capable of making decisions without humans being aware of them, i.e., they are autonomous systems and support iHCI; (2) as systems increase in size and complexity, systems must adapt their services, and (3) more complex unplanned interaction will arise out of interactions between simple independent components, i.e., emergent behaviour.

Rather than debate the merits or select particular definitions of UbiCom, the main properties are classified into five main types or groups of system properties to support the five main requirements for ubiquitous computing (see Figure 1.2). These groups of properties are not exclusive. Some of these sub-types could appear in multiple types of group. Here are some examples. Affective or emotive computing can be regarded as sub-types of IHCI and as sub-types of human intelligence. There is often a strong notion of autonomy associated with intelligence as well as being a more distributed system notion. Goal-oriented systems can be regarded as a design for intelligence and as a design for iHCI. Orchestrated and choreographed can be regarded as a way to compose distributed services and as a way to support collective rational intelligence. Personalised can be regarded as sub-type of context-awareness and as a sub-type of iHCI.

Different notions and visions for ubiquitous computing overlap. There are often different compositions of more basic types of properties. Ambient intelligence, for example, combines embedded autonomous computing systems, iHCI and social type intelligent system. Asynchronous communication enables the components in distributed systems to be spatially and temporally separated but it also enables automatic systems to do more than simply react to incoming events, to support anytime interaction.

Some properties are similar but are referred to by different terms. The terms pervasive computing and ambient computing are considered to be synonymous with the term ubiquitous computing. Systems are available anywhere and anytime, to anyone, where and when needed. UbiCom is not intended to mean all physical world resources, devices and users are omnipresent, available everywhere, at all times, to everybody, irrespective of whether it is needed or not. Ubiquity to be useful is often context-driven, i.e., local ubiquity or application domain bounded ubiquity.

The taxonomy proposed in this text is defined at three levels of granularity. At the top level five core properties for UbiCom systems are proposed. Each of these core properties is defined in terms of over 70 sub-properties give in Tables 1.1–1.5. These tables describe more finely grained properties of UbiCom systems and similar ones.[20] Thus a type of distributed UbiCom can be defined in terms of being networked and mobile. Several of these sub-properties defined are themselves such rich concepts that they themselves can be considered in terms of sub-sub-properties. For example, communication networks (Chapter 11) include sub-properties such as wired or wireless, service-oriented or network oriented, etc. Mobility (Chapter 4) can be defined

---

[20] Without formal definitions of terms at this stage, it is not possible to say that terms are equivalent and synonyms.

Table 1.1 Distributed system properties

Distributed System, middleware, set of generic services	
Universal, seamless, heterogeneous	Able to operate across different homogeneous environments, seamless integration of devices and environments, taking on new contexts when new resources become available (Sections 3.2, 3.3)
Networked	UbiCom devices are interlinked using a network which is often wireless (Chapter 11)
Synchronised, coordinated	Multiple entity interaction can be coordinated synchronously or asynchronously over time and space interactions (Section 3.3.3.2)
open	New components can be introduced and accessed, old ones can be modified or retired. Components can be dynamically discovered (Section 3.3.2)
Transparent, virtual	Reduces the operational complexity of computing, acting as a single virtual system even although it is physically distributed (Section 3.4.1)
Mobile, nomadic	Users, services, data, code and devices may be mobile (Sections 4.2, 11.7.5)

Table 1.2 iHCI system properties

Implicit Human–Device Interaction (iHCI)	
Non-intrusive, hidden, invisible, calm computing	ICT is nonintrusive and invisible to the user. It is integrated into the general ecology of the home or workplace and can be used intuitively by users (Section 5.7)
Tangible, natural	Interaction is via natural user interfaces and physical artefact interaction that can involve gestures, touch, voice control, eye gaze control, etc. (Section 5.3)
Anticipatory, speculative, proactive	Improving performance and user experience through anticipated actions and user goals in relation to current context, past user context and group context. This overlaps with user context-awareness (Sections 5.6 and 7.2)
Affective, emotive	Computing that relates to, arises from, or influences human emotions. This is also considered to be a sub-type of human intelligence (Section 5.7.4)
User-aware	ICT is aware of presence of user, user ID, user characteristics, current user tasks in relation to users' goals (as part of iHCI and context-awareness)
Post-human	Sense of being in a world that exists outside ourselves, extending a person's normal experience across space and time (Section 5.4.1)
Sense of presence immersed, virtual, mediated reality	A person is in a real-time interactive environment which experiences an extended sense of presence that combines the virtual and the real, often by overlaying virtual views on real views (Section 5.4.4)

in terms of sub-sub-sub-properties of mobile services mobile code, and mobile hardware resources and devices and in terms of being accompanied, wearable and implanted or embedded into mobile hosts. Over 20 different sub-sub-properties for autonomic and self-star computing are described (Section 10.4).

These groups of properties act to provide a higher level of abstraction of the important characteristics for analyzing and designing ubiquitous systems. It is assumed that generic distributed system services such as directory services and security would also be needed and these may be need to be designed and adapted for ubiquitous computing use.

Table 1.3   Context-aware system properties

Context-aware	
Sentient, unique, localized, situated	Systems can discover and take advantage of the situation or context such as: location, time and user activity. There are three main sub-types of context-awareness; physical-world, user and virtual (ICT) device awareness
Adaptive, active context-aware	Systems actively adapt to context changes in a dynamic environment rather than just present context changes to the user (Section 7.2.4)
Person-aware, user-aware, personalised, tailored,	Tailored to an individual user or type of user, based on personal details or characteristics that a user provides or is gathered about a user. This may trigger system adaptation (Sections 5.6, 5.7)
Environment-aware, context-aware, physical context-aware	Sometimes physical world context-aware awareness is taken by some researchers to mean general context-awareness or general environment awareness. Physical context-awareness includes spatial and temporal awareness (Sections 7.4, and 7.5)
ICT awareness	Awareness of ICT infrastructure in which an UbiCom system exists, e.g., awareness of network QoS when transmitting messages (Section 7.6)

Table 1.4   Autonomous system properties

Autonomous	
Automatic	Operates without human intervention (Section 10.2.1.1)
Embedded, encapsulated embodied	System input-output and computation is completely encapsulated by, or contained in, the device it controls, e.g., a system that acts as a self-contained appliance (Section 6.5)
Resource-constrained	Systems are designed to be constrained in size to be portable or embeddable; to use constrained computation, data storage, input and output and energy (Section 13.5.2)
Untethered, amorphous	Able to operate independently and proactively, free from external authority, external dependencies are minimized (Sections 2.2.3.2, 6.4.4)
Autonomic, self-managing, self-star	Able to support various self-star properties such as self-configuring, self-healing, self-optimising and self-protecting behaviour (Section 10.4)
Emergent, self-organising	More complex behaviour can arise out of multiple simple behaviours (Section 10.5)

Table 1.5   Intelligent system properties

Individual Intelligent Systems	
Reactive, reflex[1]	Environment events are sensed. Events then trigger action selection that may lead to actuators changing their environments (Section 8.3.2)
Model-based,Rule/ policy-based logic/reasoning	Systems use a model of how itself operates and the how the world works (Section 8.3.3), There are many types of model representation such as rule-based, different types of logic-based, etc.
goal-oriented, planned, proactive	User goals can be used to plan actions dynamically rather than pre-programmed actions (Section 8.3.4)
Utility-based, game theoretic	Systems can be designed to handle multiple concurrent goals (Section 8.3.5)
Learning, adaptive	Systems can be designed to improve their own performance (Section 8.3.6)

Table 1.5 (*continued*)

Individual Intelligent Systems

Multiple Intelligent System, Collective or Social Intelligence	
Cooperative collaborative, benevolent	Multiple agents can share tasks and information in order to achieve shared goals (Section 9.2.3)
Competitive, self-interested, antagonistic, adversarial	Individual agents and organizations have private goals and utility functions that they seek to achieve in a multi-entity setting without requiring collaboration Entities could also act malevolently (Section 9.2.4)
Orchestrated, choreographed, mediated	Multiple interactions can: be controlled and ordered by designating some leader (orchestrated) who acts as a central-planer; allow some freedom of interaction by participants (choreographed) or constrained by the use of some common entity or resource (mediated) (Sections 3.3.2, 9.2.2)
Task-sharing	
Communal, shared meaning	System interaction is sharable, commonly understood within a limited or well-defined domain (Section 8.4)
Shared knowledge	
Speech-act based,[2] intentional, mentalistic.	Multiple agents interact based upon propositional attitudes, i.e., relationships based upon beliefs, desires or wants and intentions[3] (Section 9.3.3.4)
Emergent	Organizations lead to levels of interaction that are not level of the individual interactions (Section 9.2.3.3)

[1] Note a reflex system is different from a reflective system, whereas the former type of system is designed to react to environment stimuli, the latter type of system is designed to think about what it is doing (Section 10.3).

[2] A speech act-based system is different from a speech-based system – whereas the former using a particular linguistic theory to form sentence like structures, the latter is a system that can process human speech input and or convert its output to human-like speech.

[3] Although mental and intentional computing seems like a form of human intelligence, it is usually deployed in terms of a rational model such as such as a **BDI** type logic (Chapter 7).

Each individual property has its own domain of a more finely grained set of discrete values, rather than being seen as a property that is present or absent. Here are some examples:

- from wireless to wired, ad hoc to fixed and from client-server to P2P communication;
- from full local access only, to partial remote access, to full remote access;
- from asynchronous, to synchronous, to coordinated, to organisational to conventions;
- from mobility ranging from: being static at place of manufacture; moved to the place of installation, e.g., embedded then static; mobile between sessions but static during sessions; mobile (roaming from home) during sessions; to being free roaming without a home, untethered;
- from transportable to portable to hand-held to wearable to implants;
- from macro to micro to nano;
- from fully integrated, to embedded inside, to surface-mounted, to various forms of loose attachments such as amorphous computing;
- from total physical reality, to augmenting reality with virtual reality, to mediated reality, to pure virtual reality to synthetic reality;
- from operating as individuals, to operating as part of societal groups, to globally interacting.

In Section 13.2.2, a multi-lateral model that offers different degrees of support for Ubiquitous computing properties from minimal support to full support is proposed.

There are a few closing remarks about the terminology and meaning of the system properties and concepts. Different fields of computer science may use the same term differently. For example, when HCI refers to the adaptive system, it means the focus of the adaptation is the front end of the system or the UI which is adapting to human behaviour, and the adaptation is driven fully by external concerns. In artificial intelligence, an adaptive system often refers to a system which incorporates machine learning so that the system can improve or adapt its performance over time. There are also many nuances and different contexts of the use of the terms. The grouping of terms in the left column of Tables 1.1–1.5 indicates that these terms have a strong overlap and similarity. This does not necessarily mean that terms within a grouping are fully equivalent to each other.

1.3 Ubiquitous System Environment Interaction

At a high level of abstraction, we can distinguish three types of system environment[21] for each particular UbiCom system: (1) other UbiCom systems which form the ICT infrastructure, supporting services and act as middleware for that particular ICT system applications (virtual worlds);[22] (2) human individuals and human organisations; and (3) physical world[23] systems including ecological and biologic systems. Together, the virtual (computer) environment humans and the physical world can be considered as forming an external environment for UbiCom systems. Note that each of these three main environments appear to have quite different design models and processes. Physical world phenomena are governed by well-known laws of physical forces such as electromagnetism and gravity. Living entities in the physical world are governed by ecological models of their habitat. Human living entities are often governed by economic and social models.

A UbiCom system is often organised conventionally as a layered information system stack with a bottom layer of information resources, a middle layer of processing and a top layer of user information abstractions to view and interact with the information. A common communications pipe allows these to be distributed in different ICT systems.

Humans who own and operate the UbiCom systems and are situated in the physical world regard the physical world and ICT devices as their human environment. Humans perceive and act on their environment, often through visual and touch senses. Their actions can be driven by a world model that guides their actions, consisting of prior experiences that are learnt. The physical environment can be represented using multiple models. In a local physical control model, e.g., lighting controls can sense the existing natural lighting and switch on artificial lighting when the natural light is below a certain threshold. More sophisticated control systems can use feedback control. A second type of physical world model is an ecology system, a self-sustaining system inhabited by multiple autonomous organisms that self-regulate their behaviour in the face of different driving pressures and events in the system.

There are three basic types of environment for UbiCom systems: (1) the infrastructure of other ICT systems; (2) the physical world environment; and (3) the human environment. Several basic types of system environment interaction occur: between humans and ICT systems, HCI (see Figure 1.3); between ICT systems and the physical world, CPI (see Figure 1.4); between ICT systems, C2C or CCI. In addition, interactions can occur between the non-ICT systems such as between different physical world entities and between humans (H2H or HHI), also called social interaction. These types of interaction all coexist. The interrelation and simplification of these interactions is

---

[21] These three types of environment are also collectively known as *physical space, cyber space and mental space* or as the world of *atoms, bits and minds*.

[22] The ICT system infrastructure is also referred to as a virtual environment or as cyberspace.

[23] Anything that exists in a physical space, natural or artificial, inherently occupies physical space and consumes physical resources, e.g., a desktop computer often rests on a hard physical surface and consumes energy generated by other physical world resources. The nature of the physical world in our model is as an external environment to an UbiCom system which the UbiCom system may sense and control in order to support some specific application or use.

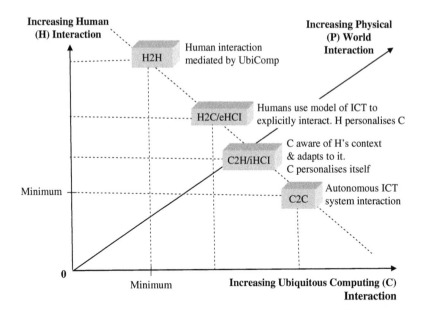

Figure 1.3 Human–ICT device interaction (HCI) is divided into four sub-types of interaction H2H, H2C, C2H and C2C

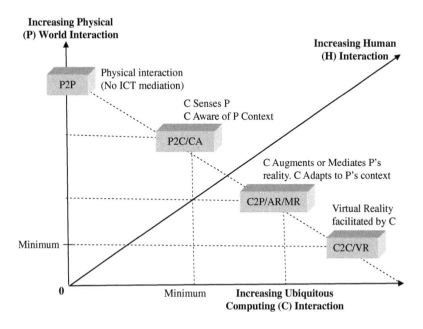

Figure 1.4 ICT device and Physical World Interaction (CPI) is divided into four sub-types of interaction: P2P, P2C, C2P and C2C

discussed further in Section 9.1. Of these three environments, humans have the highest intelligence overall and can act the most autonomously. This needs to be taken into account. Humans are an embodiment of parts of the physical world but also cause the most changes to the physical world. ICT devices are manufactured from the physical world and act in it. ICT devices have a profound effect on humans leading to changes in societal values and norms.[24]

Each of the types of interaction of HCI and CPI is illustrated by describing four degrees of interactions that span their interaction domain. This division of interactions into human–physical, physical–computer and human–computer interaction is a framework to analyse the range of UbiCom systems. Some interactions may span and combine interactions, e.g., mediated reality interaction combines human-physical and human-computer interaction.

1.3.1 Human–ICT Device Interaction (HCI)

For the interaction between two systems, e.g., Humans (H) and ICT systems (C), four characteristic points across the interaction domain are considered, maximum H interaction (H2H or HHI) with minimal C interaction, more H interaction facilitated by C interaction (H2C), more C interaction that leads to human interaction (C2H) and maximum ICT interaction (C2C). Whereas in H2C, Humans have a model of computers, e.g., H has a mental model of C's tasks and goals in order to interact more effectively with the use of C. With C2H, C also has a partial model of H in order to reduce H's interaction (Section 1.3.3). ICT device to physical world interaction and human to physical world interaction are described in a similar way.

ICT device to ICT device (C2C), also called distributed computer systems, is the main focus of computer science, telecoms and networks but we take a much wider perspective here. C2C facilitates all the other types of interaction below, e.g., H2H is often mediated by C2C. C2C is often used to automate tasks and is used for pre-processing to filter out unneeded resources or to filter needed resources transparently to the user.

CCI, in turn, depending on how this is defined, interlinks and requires human interaction. Human interaction is required in different parts of the life-cycle of a CCI system. Humans are involved in the design phase of the UbiCom system, often performing some inspection phase during operations and are involved in the maintenance phase when changes to the design are needed to maintain the system operation (Horn, 2001). Kindberg and Fox (2002) state that in the short term, UbiCom systems will involve humans and that system designers should make clear the system boundary[25] between the ICT system and the human, making clear the responsibilities of both, i.e., what ICT system cannot do and what humans will do.

Humans use multiple devices, explicitly personalised by the user, that are situated in a human's personal space and social space.[26] Humans explicitly access non-interactive and interactive multimedia information services for entertainment and leisure. Humans explicitly access business (enterprise)-

---

[24] For example, in the 1990s, if two strangers were walking down the street and one starting talking, saying hello, one would think the other was talking to him or her and reply. In the 2000s, if this happens, there is a higher probability that the person is talking on a mobile phone to someone else remotely, so it is more more likely that the other will stay silent.

[25] Kindberg and Fox (2002) refer to this ICT system–human boundary as the *semantic Rubicon*, named after the River Rubicon in Italy in ancient times that marked the boundary of part of what was Italy, where Julius Caesar hesitated before crossing it with his troops into Italy.

[26] See Hall (1996), personal space is the region surrounding each person, or that area which a person considers their domain or territory. It is determined to be up to about 120 cm. This personal space travels with us as we move. These spaces are fluid and multi-functional. Personal space also overlaps with social space (used for interacting with acquaintances, up to about 360 cm) and with public space. The specific dimensions and use of these spaces vary, e.g., with culture and age.

related information away from the office and various other supporting virtual ICT services for education, personal productivity, etc. (Explicit Human to ICT Device Interaction H2C or eHCI).

ICT applications can use a model of the person, perhaps created and maintained based upon observed user interaction, and their activities (Implicit ICT Device to Human Interaction, C2H or iHCI). The C model of H can be used to inform users of timely activities, to automatically filter and adapt information and tasks, and to anticipate human actions and interactions and adapt to them.

Social and organisational interaction can be mediated by ICT devices (Human to Human, Social, Interaction, H2H). Two humans may interact, one to one, e.g., unicast voice calls between two people. Computers may facilitate basic information and task-sharing but computers can also be used to facilitate richer sharing of language, knowledge, experiences and emotions. Humans may interact within social spaces and within enterprise organisational spaces, e.g., to support intra or inter-organisational work-flows and to complete as well as cooperate together to attain resources, e.g., interact within auctions to accrue goods.

1.3.2 ICT Device to Physical World Interaction (CPI)

Physical World to Physical World Interaction (P2P) refers to interactions within nature that are (as yet) not mediated by any significant ICT system. There are a variety of simple animal life interactions used in nature, in contrast to the more complex human to human interaction. These involve shared chemical scents, visual signage and different types of audio signals such as drumming, buzzing and vocal calls. While this type of biological interaction appears to be quite esoteric, models of this interaction can be mimicked in CCI and can be surprisingly effective at solving some interaction problems. In addition, the ways that organisms interact within their natural habitat to maintain a balanced ecosystem are quite effective models for self-regulation of autonomous systems (Chapter 10).

Physical Environment to Computer Device Interaction (P2C) covers context-aware ICT systems. These can be designed to be aware of changes in specific physical world phenomena and to react to this in simple ways, e.g., if the temperature is too low, turn up the heating. ICT systems can also be designed to act on the physical world, changing the state of part of the physical world, according to human goals.

Computer Device to Physical Environment Interaction (C2P) refers to augmented and mediated reality systems. ICT systems are used to augment, to add to, physical reality, e.g., physical world views can be annotated with virtual markers. In the more general mediated reality cases, reality may be diminished and filtered not just enhanced. The interplay between physical world and virtual world reality is a strong theme in electronic games. The term hyperreality is used to characterise the inability of human consciousness to distinguish reality from fantasy, a level of consciousness that can be achieved by some electronic games players.

In pure virtual reality interaction, Computer to Computer Interaction (CCI), the physical world may be used as a conceptual space for virtual interaction. In a virtual reality ICT system, humans can use sensory interfaces such as gloves and goggles to be interfaced to support more natural interaction. Humans may also contain implants for medical conditions that can transmit digital data streams into ICT medical monitoring services. Humans can be represented virtually as avatars in order to explore and interact more richly in virtual ICT worlds.

It is also noted that there is some minimal physical world interaction even with maximum CCI as computers consume physical world resources, e.g., energy. CCI is affected by physical world phenomena, wireless and wired signals will become attenuated to different degrees, partially dependent on their frequency. There is in addition increasing awareness of UbiCom systems operating as part of the physical world ecology, in harmony with it. This must occur throughout the full life-cycle of the UbiCom system including operation (optimising energy use) and destruction (through remanufacturing and recycling).

1.4 Architectural Design for UbiCom Systems: Smart DEI Model

Three basic architectural design patterns for ubiquitous ICT system: smart devices, smart environment[27] and smart interaction are proposed (Figure 1.5). Here the concept smart simply means that the entity is active, digital, networked, can operate to some extent autonomously, is reconfigurable and has local control of the resources it needs such as energy, data storage, etc. It follows that these three main types of system design may themselves contain sub-systems and components at a lower level of granularity that may also be considered smart, e.g., a smart environment device may consist of smart sensors and a smart controller, etc. There is even smart dust (Section 2.2.3.2). An illustrative example of how these three types of models can be deployed is given in Figure 1.5.

These are many examples of sub-types[28] of smarts for each of the three basic types of smarts which are discussed in detail in the later chapters of this book. The three main types of smart design also overlap, they are not mutually exclusive. Smart devices may also support smart interaction. Smart mobile start devices can be used for control in addition to the use of static embedded environment devices. Smart devices may be used to support the virtual viewpoints of smart personal (physical environment) spaces in a personal space that accompanies the user wherever they are.

Satyanarayanan (2001) has also postulated different architectures and paths for developing UbiCom systems, first, to evolve from distributed systems, mobile distributed systems into ubiquitous computing and, second, to develop UbiCom systems from smart spaces characterised by

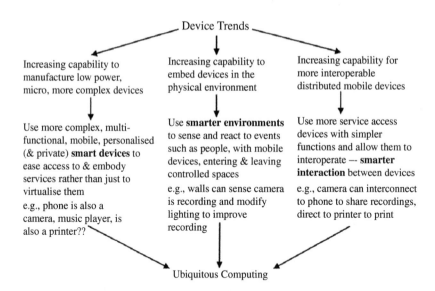

Figure 1.5 Three different models of ubiquitous computing: smart terminal, smart interaction, and smart infrastructure

---

[27] Note: some people just consider the smart environment model to comprise ubiquitous computing but here ubiquitous computing is also considered to comprise the smart device model, e.g., mobile communicators, and smart interaction model.

[28] Further levels of granularity of the sub-types of smarts could be added, e.g., sub-types of smart embedded environments devices such as implants but these are not indicated in order to simplify Figure 1.6.

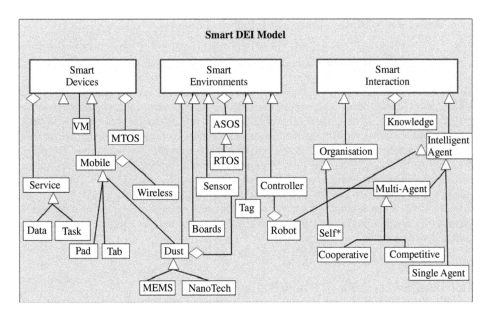

Figure 1.6 Some of the main subtypes (triangle relationships) of smart devices, environments and interactions and some of their main aggregations (diamond relationships) where MTOS is a Multi-Tasking Operating System, VM is a Virtual Machine, ASOS is an Application Specific or embedded system OS, RTOS is a Real-Time OS and MEMS is a Micro ElectroMechanical System

invisibility, localised scalability and uneven conditioning. The Smart DEI model is similar to Satyanarayanan's, except it also incorporates smart interaction. Smart DEI also refers to hybrid models that combine the designs of smart device, smart environments and smart interaction (Figure 1.6). Gillett *et al.* (2000) speculate that general purpose end-user equipment will endure but evolve into a more modular form, driven by user frustration with a proliferation of devices with overlapping functionality and the desire for consistency across multiple environments (such as home, car and office). This is motivation for smart interaction rather the smart device model. However, in practice, users appear to be very tolerant of the vast majority of devices with overlapping functions that are inconsistent and often non interoperable.

1.4.1 Smart Devices

Smart devices, e.g., personal computer, mobile phone, tend to be multi-purpose ICT devices, operating as a single portal to access sets of popular multiple application services that may reside locally on the device or remotely on servers. There is a range of forms for smart devices. Smart devices tend to be personal devices, having a specified owner or user. In the smart device model, the locus of control and user interface reside in the smart device. The main characteristics of smart devices are as follows: mobility, dynamic service discovery and intermittent resource access (concurrency, upgrading, etc.). Devices are often designed to be multi-functional because these ease access to, and simplify the interoperability of, multi-functions at run-time. However, the trade-off is in a decreased openness of the system to maintain (upgrade) hardware components and to support more dynamic flexible run-time interoperability.

1.4.1.1 Weiser's ICT Device Forms: Tabs, Pads and Boards

We often think of computers primarily in terms of the multi-application personal or server computers, as devices with some type of screen display for data output and a keyboard and some sort of pointing devices for data input. As humans, we routinely interact with many more devices that have single embedded computers in them, such as household appliances, and with complex machines[29] that have multiple embedded computers in them. Weiser noted that there was a trend away from many people per computer,[30] to one computer per person,[31] through to many computers per person.

Computer-based devices tend to become smaller and lighter in weight, cheaper to produce. Thus devices can become prevalent, made more portable and can appear less obtrusive. Weiser considered a range of device sizes in his early work from wearable centimetre-sized devices (tabs), to hand-held decimetre-sized devices (pads) to metre-sized (boards) displays. ICT Pads to enable people to access mobile services and ICT tabs to track goods are in widespread use. Wall displays are useful for viewing by multiple people, for collaborative working and for viewing large complex structures such as maps. Board devices may also be used horizontally as surface computers as well used in a vertical position.

1.4.1.2 Extended Forms for ICT Devices: Dust, Skin and Clay

The three forms proposed by Weiser (1991) for devices, tabs, pads and boards, are characterised by: being macro-sized, having a planar form and by incorporating visual output displays. If we relax each of these three characteristics, we can expand this range into a much more diverse and potential more useful range of ubiquitous computing devices.

First, ICT devices can be miniaturised without visual output displays, e.g., Micro Electro-Mechanical Systems (MEMS), ranging from nanometres through micrometers to millimetres (Section 6.4). This form is called Smart Dust. Some of these can combine multiple tiny mechanical and electronic components, enabling an increasing set of functions to be embedded into ICT devices, the physical environment and humans. Today MEMS, such as accelerometers, are incorporated into many devices such as laptops to sense falling and to park moving components such as disk arms, are being increasingly embedded into widely accessed systems. They are also used in many devices to support gesture-based interaction. Miniaturisation accompanied by cheap manufacturing is a core enabler for the vision of ubiquitous computing (Section 6.4).

Second, fabrics based upon light-emitting and conductive polymers, organic computer devices, can be formed into more flexible non-planar display surfaces and products such as clothes and curtains (Section 5.3.4.3). MEMS devices can also be painted onto various surfaces so that a variety of physical world structures can act as networked surfaces of MEMS (Section 6.4.4). This form is called Smart Skins.

Third, ensembles of MEMS can be formed into arbitrary three-dimensional shapes as artefacts resembling many different kinds of physical object (Section 6.4.4). This form is called Smart Clay.

---

[29] For example, new cars have several tens of embedded computers and sensors to support assisted braking, airbag inflation, etc.

[30] Thomas J. Watson, who led the world's first and largest computer company, IBM, from the 1920s to the 1950s, is alleged to have made the statement in 1943 that: 'I think there is a world market for maybe five computers.' This would mean a ratio of one computer to about a billion people.

[31] The one-computer-to-one-person phase may not have existed for any significant period depending on the definition of a computer. Certainly, people who had personal computers, also had many embedded digital devices at that time too.

1.4.1.3 Mobility

Mobile devices usually refer to communicators, multimedia entertainment and business processing devices designed to be transported by their human owners, e.g., mobile phone, games consoles, etc. There is a range of different types of mobiles as follows:

- *Accompanied*: these are devices that are not worn or implanted. They can either be portable or hand-held, separate from, but carried in clothes or fashion accessories.
- *Portable*: such as laptop computers which are oriented to two-handed operation while seated. These are generally the highest resource devices.
- *Hand-held*: devices are usually operated one handed and on occasion hands-free, combining multiple applications such as communication, audio-video recording and playback and mobile office. These are low resource devices.
- *Wearable*: devices such as accessories and jewellery are usually operated hands-free and operate autonomously, e.g., watches that act as personal information managers, earpieces that act as audio transceivers, glasses that act as visual transceivers and contact lenses. These are low resource devices (Sections 2.2.4.5, 5.4.3).
- *Implanted or embedded*: these are often used for medical reasons to augment human functions, e.g., a heart pacemaker. They may also be used to enhance the abilities of physically and mentally able humans. Implants may be silicon-based macro- or micro-sized integrated circuits or they may be carbon-based, e.g., nanotechnology (Section 6.4).

Static can be regarded as an antonym for mobile. Static devices tend to be moved before installation to a fixed location and then reside there for their full operational life-cycle. They tend to use a continuous network connection (wired or wireless) and fixed energy source. They can incorporate high levels of local computation resources, e.g., personal computer, AV recorders and players, various home and office appliances, etc. The division between statics and mobiles can be more finely grained. For example, statics could move between sessions of usage, e.g., a mobile circus containing different leisure rides in contrast to the rides in a fixed leisure park. Mobile ICT is discussed in detail in Chapter 4.

1.4.1.4 Volatile Service Access

Mobiles tend to use wireless networks. However, mobiles may be intermittently connected to either wireless networks (WAN is not always available) or to wired communications networks (moving from LAN to LAN) or to both. Service access by smart mobile devices is characterised as follows.

Intermittent (service access) devices access software services and hardware intermittently. This may be because resources are finite and demand exceeds supply, e.g., a device runs out of energy and needs to wait for it to be replenished. This may be because resources are not continually accessible.

Service discovery: devices can dynamically discover available services or even changes in the service context. Devices can discover the availability of local access networks and link via core networks to remote network home services. They can discover local resources and balance the cost and availability of local access versus remote access to services. Devices can be designed to access services that they discover on an intermittent basis. Context-aware discovery can improve basic discovery by limiting discovery to the services to the ones of interest, rather than needing to be notified of many services that do not match the context.

With asymmetric remote service access, more downloads than uploads, tends to occur. This is in part due to the limited local resources. For example, because of the greater power needed to transmit rather than receive communication and the limited power capacity, high power consumption results in more received than sent calls. Apart from the ability to create and transmit voice signals, earlier phones were designed to be transreceivers and players. More recently, because of miniaturisation, mobile devices not only act as multimedia players, they can also act as multimedia recorders and as content sources.

1.4.1.5 Situated and Self-Aware

Smart devices although they are capable of remote access to any Internet services, tend to use various contexts to filter information and service access. For examples, devices may operate to focus on local views of the physical environments, maps, and to access local services such as restaurants and hotels.

Mobiles are often designed to work with a reference location in the physical environment called a home location, e.g., mobile network nodes report their temporary location addresses to a home server which is used to help coordinate the mobility. Service providers often charge access to services for mobile service access based upon how remote they are with respect to a reference ICT location, a home ICT location. During transit, mobiles tend to reference a route from a start location to a destination location.

Mobile devices support limited local hardware, physical, and software resources in terms of power, screen, CPU, memory, etc. They are ICT resource constrained. Services that are accessed or pushed to use such devices must be aware of these limitations, otherwise the resource utilisation by services will not be optimal and may be wasted, e.g., receiving content in a format that cannot be played. In the latter case, the mobile device could act as an intermediary to output this content to another device where it can be played.

Mobile devices tend to use a finite internal energy cache in contrast to an external energy supply, enhancing mobility. The internal energy supply may be replenished from a natural renewal external source, e.g., solar power or from an artificial energy gird: energy self-sufficiency. This is particularly important for low-maintenance, tetherless devices. Devices can automatically configure themselves to support different functions based upon the energy available. Without an internal energy cache, the mobility of devices may be limited by the length of a power cable it is connected to.

There is usually a one-to-one relationship between mobiles and their owners. Devices' configuration and operation tends to be personalised, to support the concept of a personal information and service space which accompanies people where ever they are.

1.4.2 Smart Environments

In a smart environment, computation is seamlessly used to enhance ordinary activities (Coen, 1998). Cook and Das (2007) refer to a smart environment as 'one that is able to acquire and apply knowledge about the environment and its inhabitants in order to improve their experience in that environment'. A smart environment consists of a set of networked devices that have some connection to the physical world. Unlike smart devices, the devices that comprise a smart environment usually execute a single predefined task, e.g., motion or body heat sensors coupled to a door release and lock control. Embedded environment components can be designed to automatically respond to or to anticipate users' interaction using iHCI (implicit human–computer interaction), e.g., a person walks towards a closed door, so the door automatically opens. Hence, smart environments support a bounded, local context of user interaction.

Smart environment devices may also be fixed in the physical world at a location or mobile, e.g., air-born. Smart environments could necessitate novel and revolutionary upgrades to be incorporated into the environment in order to support less obtrusive interaction, e.g., pressure sensors can be incorporated into surfaces to detect when people sit down or walk. A more evolutionary approach could impart minimal modifications to the environment through embedding devices such as surface mounted wireless sensor devices, cameras and microphones.

1.4.2.1 Tagging, Sensing and Controlling Environments

Smart environment devices support several types of interaction with environments such as the physical environment (Chapter 6) as follows:

- *Tagging and annotating the physical environment*: tags, e.g., RFID[32] tags, can be attached to physical objects. Tag readers can be used to find the location of tags and to track them. Virtual tags can be attached to virtual views of the environment, e.g., a tag can be attached to a location in a virtual map.
- *Sensing or monitoring the physical environment*: Transducers take inputs from the physical environment to convert some phenomena in the physical world into electrical signals that can be digitised, e.g., how much ink is in a printer's cartridges. Sensors provide the raw information about the state of the physical environment as input to help determine the context in a context-aware system. Sensing is often a pre-stage to filtering and adapting.
- *Filtering*: a system forms an abstract or virtual view of part of its environment such as the physical world. This reduces the number of features in the view and enables viewers to focus on the features of interest.
- *Adapting*: system behaviour can adapt to the features of interest in the environment of adapt to changes in the environment, e.g., a physical environment route is based upon the relation of the current location to a destination location.
- *Controlling the physical world*. Controllers normally require sensors to determine the state of the physical phenomena e.g., heating or cooling systems that sense the temperature in an environment. Controlling can involve actions to modify the state of environment, to cause it to transition to another state. Control may involve changing the order (assembly) of artefacts in the environment or may involve regulation of the physical environment.
- *Assembling*: robots are used to act on a part of the physical world. There is a variety of robots. They may be pre-programmed to schedule a series of actions in the world to achieve some goal, e.g., a robot can incorporate sensors to detect objects in a source location, move them and stack them in a destination location (palletisation). Robots may be stationary, e.g., a robot arm, or be mobile.
- *Regulating*: Regulators tend to work in a fixed location, e.g., a heating system uses feedback control to regulate the temperature in an environment within a selected range.

1.4.2.2 Embedded Versus Untethered

Smart environments contain components that have different degrees of dependence from their physical and ICT environments. Smart environments may use components that are embedded or untethered. Embedded devices are statics that are embodied in a larger system that may be static or mobile. Embedded systems typically provide control and sensing support to a larger system. Devices may be embedded in: (1) parts of physical environments, e.g., a passenger- or vehicle-controlled area entry system; (2) parts of the human environment, e.g., heart pacemakers; and (3) parts of larger ICT devices, e.g., a location device may be embedded in a phone or camera as opposed to externally connected to it.

Untethererd or amorphous or spray devices are types of environment devices that can be mixed with other particles and spread onto surfaces or scattered into gases and fluids, e.g., smart dust (Sections 2.2.3.2, 6.4.4). They are nomadic or untethered devices that do not need to operate using a home (base) location. They can self-organise themselves to optimise their operation (Section 10.5.1).

---

[32] RFID tags are also often referred to as smart labels or smart tags, however, smart tags in this text include a much wider range of tags and more specific set of properties (Section 6.2).

1.4.2.3 Device Sizes

Smart environment devices can vary in size. This affects their mobility. Macro-sized devices incorporate a range of device sizes from tab-sized (centimetre-sized) devices, through pad-sized (decimetre-sized) devices, to board-sized (metre-sized) devices (Section 1.2.2.2). Micro Electro Mechanical Systems (MEMS) are fabricated using integrated chip technology. This enables the large-scale cheap manufacture (thousands to millions) production of integrated circuit type devices, to be spread-on surfaces or to be airborne. Nanotechnology is 1 to 100 nanometre-sized devices that are built from molecular components. These are either constructed from larger molecules and materials, not controlled at the atomic level (more feasible) or assemble themselves chemically by principles of molecular recognition (less feasible).

1.4.3 Smart Interaction

In order for smart devices and smart environments to support the core properties of UbiCom, an additional type of design is needed to knit together their many individual activity interactions. Smart interaction is needed to promote a unified and continuous interaction model between UbiCom applications and their UbiCom infrastructure, physical world and human environments.

In the smart interaction design model, system components dynamically organise and interact to achieve shared goals. This organisation may occur internally without any external influence, a self-organising system, or this may be driven in part by external events. Components interact to achieve goals jointly because they are deliberately not designed to execute and complete sets of tasks to achieve goals all by themselves – they are not monolithic system components. There are several benefits to designs based upon sets of interacting components.

A range of levels of interaction between UbiCom system components exists from basic to smart. A distinction is made between (basic) interaction that uses fixed interaction protocols between two statically linked dependent parties versus (smart) interaction that uses richer interaction protocols between multiple dynamic independent parties or entities.

1.4.3.1 Basic Interaction

Basic interaction typically involves two dependent parties: a sender and a receiver. The sender knows the address of the receiver in advance; the structure and meaning of the messages exchanged are agreed in advance, the control of flow, i.e., the sequencing of the individual messages, is known in advance. However, the content, the instances of the message that adhere to the accepted structure and meaning, can vary. There are two main types of basic interaction, synchronous versus asynchronous (Section 3.3.3):

- *Synchronous interaction*: the interaction protocol consists of a flow of control of two messages, a request then a reply or response. The sender sends a request message to the specified receiver and waits for a reply to be received,[33] e.g., a client component makes a request to a server component and gets a response.
- *Asynchronous interaction*: The interaction protocol consists of single messages that have no control of flow, a sender sends a message to a receiver without knowing necessarily if the receivers will receive the message or if there will be a subsequent reply, e.g., an error message is generated but it is not clear if the error will be handled leading to a response message.

---

[33] This quotes synchronisation at the interaction level, the sender waits for a reply. Synchronisation may also occur at the message level rather than the interaction level. In the case the control of flow of the individual messages is synchronised, the sender waits for some acknowledgement that the receiver has received.

1.4.3.2 Smart Interaction

Asynchronous and synchronous interaction is considered part of the distributed system communication functions (Section 3.3.3.2). In contrast, interactions that are coordinated, conventions-based, semantics and linguistic-based and whose interactions are driven by dynamic organisations are considered to be smart interaction (Section 9.2.3). Hence, smart interaction extends basic interactions as follows:

- *Coordinated interactions*: different components act together to achieve a common goal using explicit communication, e.g., a sender requests a receiver to handle a request to complete a sub-task on the sender's behalf and the interaction is synchronised to achieve this. There are different types of coordination such as orchestration (use of a central coordinator) versus choreography (use of a distributed coordinator).
- *Policy and convention-based interaction*: different components act together to achieve a common organisational goal but it is based upon agreed rules or contractual policies without necessarily requiring significant explicit communication protocols between them. This is based upon previously understood rules to define norms and abnormal behaviour and the use of commitments by members of organisations to adhere to policies or norms, e.g., movement of herds or flocks of animals are coordinated based upon rules such as keeping a minimum distance away from each other and moving with the centre of gravity, etc.
- *Dynamic organisational interaction*: organisations are systems which are an arrangement of relationships (interactions) between individuals so that they produce a system with qualities not present at the level of individuals. Rich types of mediations can be used to engage others in organisations to complete tasks. There are many types of organisational interactional protocol such as auctions, brokers, contract-nets, subscriptions, etc.
- *Semantic and linguistic interactions*: communication, interoperability (shared definitions about the use of the communication) and coordination are enhanced if the components concerned share common meanings of the terms exchanged and share a common language to express basic structures for the semantic terms exchanged.

Consider a scenario in which light resources are designed to be context-aware in order to save energy. They are designed to be actuated by human presence. If they detect a human is present, they automatically switch on. If they detect no one is present, they switch themselves off to save energy. However, if there were several lights in a semi-dark room and they were merely context-sensitive, they would all switch on when someone enters, but this wastes energy unnecessarily. If instead they were designed to support smart interaction, they could decide among themselves which lights to switch on in order to best support particular human activities and goals. Smart interaction requires devices to interact to share resource descriptions (e.g., desk-light, wall light, main ceiling light) and goals (e.g., reading, watching a video, retrieving something). This example is more complex in practice as it may need to support several users and possibly conflicting user-goals. Smart interaction also requires some smart orchestrator (central planner) entity or choreographer (distributed planning) entities to establish goals and be able to plan tasks with the participation of others, directed towards achieving those goals.

Resources and users could compete against each other and participate in market-places in which the use of a resource is assigned a utility value and users are required to make the best bid to acquire the use of a resource (auction interaction). Resources may interact and self-organise themselves to offer a combined service (Chapter 9).

1.5 Discussion

1.5.1 Interlinking System Properties, Environments and Designs

In Table 1.6 (and Figure 1.7) a further comparison is made of the Smart DEI models with respect to the internal system and system environment models. There are different ways to model how smart devices, smart environments and smart interaction interlink

There are relations between each of these smart designs to the three main types of environment, human, physical and virtual. For example, smart environment designs focus on distributing multiple devices in the physical environment which are context aware of their human users and their physical locality. Smart environment devices tend to operate more autonomously, using minimal human operational management, compared to smart devices. Smart devices may also support some

Table 1.6 Comparison of smart device, smart environment and smart interaction

Type	Smart Device	Smart Environment	Smart Interaction
Characteristics	Active multi-function devices based in a virtual computing environment	Active single function devices embedded or scattered in a physical environment	Individual components that must cooperate or compete to achieve their goals
System environment interaction	Weak CPI, strong H2C, weaker C2H and strong C2C	Strong C2P and C2H	Rich H2H, P2P models that apply to HCI and CPI
Distributed system: openness, Dynamic services, Volatile ICT Access mobility	Dynamic ICT service, resource discovery	Dynamic physical resource discovery	Dynamic composition of entities and services
Context-awareness: of physical world, user & ICT infrastructure	Low-medium	High	Low-medium
	High, Personalized, 1.1[1] interaction	Shared between users 1-M Interaction	M-M interaction Coordinate, orchestrate
	High	Medium	Low
HCI: locus of control	Some innovative iHCI in smart ICT devices	Very Diverse UIs. iHCI focuses on context-awareness	Language-based interaction and iHCI
	Localized in ICT device	Localized in part(s) of Physical World	Distributed in physical and virtual world
Autonomy	Autonomous control of local ICT resources, less autonomous control of remote services	Autonomous control of local ICT resources	High autonomy of actions and interaction
Intelligence	Low to medium individual rational intelligence	Low to medium individual rational intelligence	High collective intelligence: semantic sharing, social cooperation and competition

[1] 1.1, 1-M and M-M refer to the cardinality of interaction, representing one-to-one, one-to-many and many-to-many interaction respectively.

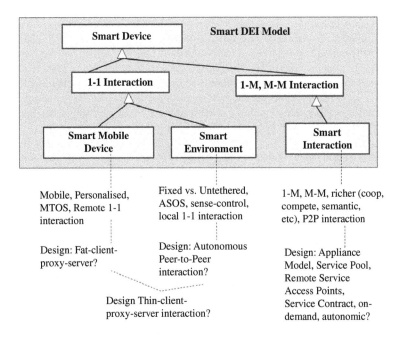

Figure 1.7 Alternate viewpoints and organisations for the device, environment and interaction entities in the Smart DEI model

awareness of the physical environment such as spatial awareness. Smart interaction uses the strongest notion of intelligence compared to the other smart designs in terms of intelligent selection, mediation and composition when multiple cooperating and competing entities interact. One important division[34] the Smart DEI model makes is the separation of the interaction between individual users and UbiCom systems which is modelled as part of smart device design, and the interaction between groups such as social and economic groups of users which is modelled as part of the smart interaction design. Social and economic models of humans may also be used for interaction between multiple ICT devices and services in virtual environments.

1.5.2 Common Myths about Ubiquitous Computing

Ubiquitous computing is quite a broad vision. There is a danger that it becomes too encompassing. Here a few unrealistic expectations about ubiquitous computing are discussed:

- *There is a single definition which accurately characterises ubiquitous computing*: rather there is a range of properties and types for ubiquitous computing which vary according to the application.

---

[34] This division may seem artificial but is done to separate design concerns. Although many-to-many interactions naturally occur between services, devices and human users, this is often modelled as one-to-one interaction through the use of mediators which act to serialise interactions. In practice, individual interactions naturally overlap or need to be combined with group interactions, e.g., using a phone in a meeting. The smart device and smart environment design may also be combined with the smart interaction design depending on the application.

- *The ideal type of ubiquitous computing is where all the properties of ubiquitous must be fully supported*: it may not be required, useful or usable in many cases in practice, to support the full set of these properties.
- *Ubiquitous computing means making computing services accessible everywhere*: this is unnecessary, too costly and makes smart environments become too cluttered, overloading the user with too many choices and contravening the hidden computer idea. Ubiquitous computing is also about computing being localised within a context and being available only when needed. Hence it is more appropriate to speak of context-aware ubiquity.
- *Ubiquitous computing is boundless computing*: this means that the virtual ICT world can extend fully into the physical world and into the human environment, replacing human and physical world systems and their interactions with computer interaction. But there limits to what computer systems can achieve, at least in the short term, e.g., UbiCom systems are not (yet) capable of completely supplanting human cognition and behaviour. Hence, UbiCom must strike a careful balance between supporting being human and living in harmony and experiencing the physical world, between being designed to give humans more fulfilled control of the their environment and taking away the less fulfilling control of the environment.
- *Ubiquitous computing is just about HCI*: automatic interaction and decisions are also needed in order to reduce human task and cognition overload and to enable tasks to be performed more safely, quicker, repeatedly and accurately. It is also less practical for humans to interact with micro-sized devices in the same way as interacting with macro-sized devices. Human interaction with compositions of multiple devices spatially distributed in shared physical spaces and time cannot be controlled centrally in the same way that humans can control a single device.
- *Calm computing should be used as a model for all HCI*. Calm computing is where the system is active, reducing some decision-making by humans. There are many applications and situations, where human users should clearly lead and control the interaction. Calm computing needs to be selectively used. Degrees of calm computing are needed from weak to strong.
- *Ubiquitous computing is just about augmenting reality*: UbiCom systems may not only enhance human–physical world interaction but it may also change it in wider ways. It may even diminish reality in some ways in order to aid the user in focusing on particular contexts. UbiCom is more about mediated reality.
- *Ubiquitous computing is just distributed or virtual computing*: UbiCom is more than being distributed in terms of interlinked, transparent and open ICT systems. UbiCom also focuses on particular models of human and physical world interaction involving context-awareness of the physical world and human and on supporting implicit human computing interaction.
- *Ubiquitous computing is just mobile wireless computing*: The ability to carry around higher resourced, multi-functional wireless mobile devices is useful but is also limited. Too many functions can cause clutter. Increasing numbers of functions can interfere with each other. It can be complex to make mobile devices strongly locally situated and adapt to the physical world. Ubiquitous computing also concerns being situated and embedded in the physical world.
- *Ubiquitous computing is just about smart environments*: while smarter physical world interaction can be facilitated through embedding active computing in the real world, UbiCom also involves interactions of smart, flexible, mobile devices which are human-centred and which support personal and social interaction spaces.
- *Ubiquitous computing need to be highly autonomous systems*: systems' autonomy is often limited in practice as computers are not able to design themselves, to completely adapt to new environments and user requirements and to maintain themselves in the face of changing requirements.
- *Ubiquitous computing is just about physical world context-awareness*: many types of context-aware systems are episodic, considering only the current physical environment state in order to determine their next actions. This is not effective in a partially observable and

non-deterministic world. In addition, the physical world context needs to be considered as part of the user context.

- *Ubiquitous computing is just distributed intelligence*: action selection and many operations can become overly complex and computationally intractable, requiring substantial computation to enable intelligent deliberation to reach an outcome. Interaction is more effective and easier to compute and execute if it has minimal intelligence, e.g., it is based upon reactive system design, rule-based behaviour as used in self-organising and self-creating systems. However, intelligence is very useful when systems have to deal with uncertainty and to handle autonomous systems that are themselves complex and intelligent.
- *Ubiquitous computing systems can operate effectively in all kinds of environments*: It is unrealistic to expect that ubiquitous computing systems can behave deterministically in non-deterministic, partially observable, etc., human and physical environments. Current systems cannot reliably actively adapt to user contexts where users act in an ad hoc manner. A weather context-aware system cannot reliably and accurately predict which clothes users should wear when the weather itself is unpredictable.

1.5.3 Organisation of the Smart DEI Approach

The book chapters are organised as follows:

- *Basics*: Basic and Vision (Chapter 1); Applications and Requirements (Chapter 2);
- *Smart Devices*: Smart Devices and Services (Section 3); Smart Mobiles, Cards and Device Networks (Chapter 4); Human–computer Interaction (Chapter 5);
- *Smart Environments*: Tagging, Sensing and Controlling (Chapter 6); Context-aware Systems (Chapter 7);
- *Smart Interaction*: Intelligent Systems (IS) (Chapter 8); Intelligent System Interaction (Chapter 9); Autonomous Systems and Artificial Life (Chapter 10).
- *Middleware and Outlook*: Ubiquitous Communication (Chapter 11); Management of Smart Devices (Chapter 12); Ubiquitous System Challenges and Outlook (Chapter 13).

This book can be studied in several ways. The traditional way by starting with the basics part and then reading on to understand more advanced topics that build on these. This tells a story from the more abstract and the underlying concepts to more technology-driven approaches. UbiCom is multi-disciplinary with some common themes across disciplines including: Basics and Vision (Chapter 1); Applications and Requirements (Chapter 2), Ubiquitous System Challenges and Outlook (Chapter 13) (see Table 1.7).

Table 1.7 Book chapters and their relation to the Smart DEI Model (Smart Device, Environment, Interaction), UbiCom system properties, and to system to environment interaction

No	Chapter Title	DEI	UbiCom Property	Environment Interactions
1	Basics and Vision	DEI	All	All
2	Applications and Requirements	DEI	Distributed, iHCI, Context-aware	All
3	Smart Devices and Services	Devices	Distributed	C2C

(*continued overleaf*)

Table 1.7   (*continued*)

No	Chapter Title	DEI	UbiCom Property	Environment Interactions
4	Smart Mobile Devices, Device Networks and Smart Cards	Devices	Distributed	C2C
5	Human−computer interaction	Devices	iHCI	HCI
6	Tagging, Sensing and Controlling	Environment	Context-aware	CPI
7	Context-Awareness	Environment	Context-aware	CPI
8	Intelligent Systems	Interaction	Intelligent	C2C, HCI
9	Intelligent Interaction	Interaction	Intelligent, iHCI	H2H, C2C
10	Autonomous Systems and Artificial Life	Interaction	Autonomy, Intelligence,	C2C
11	Communication Networks	Devices	Distributed	C2C
12	Smart Device Management	Devices	Distributed, iHCI, Context-aware	C2C, HCI, CPI,
13	Ubiquitous System Challenges and Outlook	DEI	All	All

These disciplines in the book can be studied separately. This book is a good way of illustrating their applied use. These discipline specific topics studied in addition to the basic part (Chapters 1 and 2) as follows:

- *UbiCom*: all chapters but in particular HCI (Chapter 5), Tagging, sensing, controlling the physical world (Chapter 6), context-awareness (Chapter 7).
- *ICT*: Smart devices and services (Chapter 3), smart mobile devices, device networks and smart cards (Chapter 4), ubiquitous communication (Chapter 11), smart device management (Chapter 12).
- *HCI*: HCI (Chapter 5), context-awareness (Chapter 6), managing smart devices in human-centred environments (Section 12.3), ubiquitous system challenges and outlook (Sections 13.6, 13.7,13.8).
- *AI*: intelligent systems (Chapter 8), intelligent interaction (Chapter 9). autonomous systems and artificial life (Chapter 10).

EXERCISES

1. Suggest some new ways to advance the personal visual memory application introduced in Section 1.1.1. (Hint: what if the camera was wearable? What if the camera supported other forms of image processing such as text recognition?)
2. Discuss why a paperless information environment in the world has not occurred but why a film-less photography world seems to be occurring. Compare and contrast the use of paper and photographs versus computer pads as information systems with respect to their support for the UbiCom requirements.
3. Give some further examples of ubiquitous computing applications today and propose future ones.
4. Analyse three different definitions of ubiquitous computing and distinguish between them.
5. Debate the benefits for ubiquitous systems: to support a strong notion of autonomy, to support a strong notion of intelligence.

EXERCISES   (continued)

6. Debate whether ubiquitous systems must fully support all of the five Ubiquitous system properties and whether or not the five Ubiquitous system properties are independent or are highly interlinked.
7. Debate the need for UbiCom systems to be intelligent, Argue for and against this.
8. Debate the point that whereas mainstream computer science focuses on computer to computer interaction, Ubiquitous computing in addition focuses strongly on ICT device–physical world interaction and on, ICT device–human interaction.
9. Describe the range of interactions between humans and computers, computers and the physical world and humans and the physical world. Illustrate your answer with specific UbiCom system applications. (Additional exercises are available on the book's website.)

References

Aarts, E. and Roovers, R. (2003) Ambient intelligence: challenges in embedded system design. In *Proceedings of Design Automation and Test in Europe*, Munich, pp. 2–7.

Buxton, W. (1995) Ubiquitous media and the active office. Published in Japanese (only) as Buxton, W. (1995). Ubiquitous video, *Nikkei Electronics*, 3.27 (632), 187–195. English translation accessed from http://www.bill-buxton.com/UbiCom.html, on April 2007.

Chen, G. and Kotz, D. (2000) A Survey of Context-Aware Mobile Computing Research. Technical Report TR2000-381. Available from http://citeseer.ist.psu.edu/chen00survey.html. Accessed November 2006.

Coen, M.H. (1998) Design principles for intelligent environments. In *Proceedings of 15th National/10th Conference on Artificial Intelligence/Innovative Applications of Artificial Intelligence*, pp. 547–554.

Cook, D.J. and Das, S.K. (2007) How smart are our environments? An updated look at the state of the art. *Pervasive and Mobile Computing*, 3(2): 53–73.

Endres, C., Butz, A., and MacWilliams, A. (2005) A survey of software infrastructures and frameworks for ubiquitous computing. *Mobile Information Systems*, 1(1): 41–80.

Gillett, S.E., Lehr, W., Wroclawski, J. and Clark, D. (2000) *A Taxonomy of Internet Appliances*. TPRC 2000, Alexandria, VA. Retrieved from http://ebusiness.mit.edu on 2006-09.

Greenfield, A. (2006) *Everyware: The Dawning Age of Ubiquitous Computing*. London: Pearson Education.

Hayles, N.K. (1999) *How We Became Posthuman: Virtual Bodies in Cybernetics, Literature, and Informatics*. Chicago: University of Chicago Press.

Horn, P. (2001) Autonomic Computing: IBM's Perspective on the State of Information Technology, also known as IBM's Autonomic Computing Manifesto. Retrieved from http://www.research.ibm.com/autonomic/manifesto/autonomic_computing.pdf, accessed Nov. 2007.

ISTAG. (2003) IST Advisory Group, Advisory Group to the European Community's Information Society Technology Program. Ambient Intelligence: from Vision to Reality. Retrieved from http://www.cordis.lu/ist/istag.htm. Article on May 2005.

Kindberg, T. and Fox, A. (2002) System software for ubiquitous computing. *IEEE Pervasive Computing*, 1(1): 70–81.

Milner, R. (2006) Ubiquitous computing: shall we understand it? *The Computer Journal*, 49(4): 383–389.

Pressman, R.S. (1997) Design for real-time systems. In *Software Engineering: A Practitioner's Approach*. 4th edn. Maidenhead: McGraw-Hill, pp. 373–377.

Rogers, Y. (2006) Moving on from Weiser's vision of calm computing: engaging UbiComp experiences. In P. Dourish and A. Friday (eds) *Proceedings of Ubicomp 2006, Lecture Notes on Computing Science*, 4206: 404–421.

Russell, S. and Norvig, P. (2003) *Artificial Intelligence: A Modern Approach.* 2nd edn. Englewood Cliffs, NJ: Prentice Hall.

Satyanarayanan, M. (2001) Pervasive computing: vision and challenges. *IEEE Personal Communications*, 8: 10–17.

Schmidt, A. (2000) Implicit human–computer interaction through context. *Personal Technologies*, 4(2&3): 191–199.

Warneke, B., Last, M., Liebowitz, B. and Pister, K.S.J. (2001) Smart dust: communicating with a cubic-millimeter. *Computer*, 34(1): 44–51.

Weiser, M. (1991) The computer for the twenty-first century. *Scientific American*, 265(3): 94–104.

2

Applications and Requirements

2.1 Introduction

Ubiquitous computing postulates a world where people are surrounded by computing devices and a computing infrastructure that supports us in everything we do. People will live, work, and play in a seamless computer-enabled environment that is interleaved into the world. The focus is on developing ubiquitous computer systems to support people in their daily activities in the physical world tasks to simplify these and to make these less obtrusive. Bushnell (1996) has coined variations of the term 'ware' such as deskware, couchware, kitchenware, autoware, bedroomware and bathware[1] to reflect the use of ubiquitous computing for routine tasks. More recently, Greenfield (2006) has revisited this idea and coined the term everyware to encompass the many different types of ware.

2.1.1 Overview

Many ICT companies and research institute have undertaken UbiCom initiatives. This chapter first surveys a sample of the early noteworthy projects (Section 2.2). The survey excludes intelligent and autonomous system projects as these are discussed in detail later in the book. Second, the current UbiCom ICT infrastructure and sets of everyware UbiCom applications are discussed (Section 2.3). Third, an analysis of these projects is undertaken and the main design challenges highlighted (Section 2.4).

2.2 Example Early UbiCom Research Projects

Some[2] of the early noteworthy research on UbiCom projects is presented below. These are also considered with respect to their current technological (ICT) context at the time.

---

[1] See http://wearcam.org/safebath_leonardo/safebath_leonardo207s.txt.htm, for an article about intelligent bathroom fixtures and systems by Steve Mann, accessed February 2008.
[2] There are many innovative UbiCom projects, space limitations in this section means that only a selection of these can be given here. Two of the main conferences that cover a greater range of UbiCom projects can be found on the IEEE www.UbiCom.org Web site and on the ACM www.percom.org website.

2.2.1 Smart Devices: CCI

According to Weiser *et al.* (1999), work on ubiquitous computing began at Xerox Palo Alto Research Center or PARC[3] in 1987 when Bob Sprague, Richard Bruce, and others proposed developing wall-sized displays. At the same time, anthropologists at PARC, led by Lucy Suchman, were observing actual work practices and technology leading researchers at PARC to think how computers were embedded within the complex social framework of daily activity. Weiser's early work was on distributing computing which used wireless communication in novel forms. His later work focused more on iHCI. From these converging forces, from atoms to culture, emerged the UbiCom program at PARC in 1988.

2.2.1.1 Smart Boards, Pads and Tabs

In the first UbiCom program, three main intertwined devices and applications were proposed: a large wall-display program called LiveBoard which migrated from amorphous silicon to rear screen projection; smaller computers, the book-sized MPad and the palm-sized ParcTab computer. These became known as tabs, pads and boards. The initial systems were designed to support distributed information access and use by mobile users (tabs and pads). Note, in the late 1980s, wireless networks were not pervasive or integrated with wired networks. In particular, the tabs and pads were designed for context-awareness not just of the device ID but also of their location, situation, usage, connectivity, and owner.

The ParcTab system consisted of palm-sized mobile computers that could communicate wirelessly through infrared transceivers to workstation-based applications. The ParcTab was designed for portability and could also be carried or worn, e.g., clipped on a belt. It was intended to promote on-demand computing, constant connectivity and supported location-awareness. All pad, tab, and badge prototypes were fully functional and used in everyday use by PARC experimenters. At its peak in 1994, about 40 lab members at PARC used ParcTabs during their daily activities. The aim of the LiveBoard was to support collaborative group design and work. It later became a commercial product.

2.2.1.2 Active Badge, Bat and Floor

The Active Badge system of Want *et al.* (1992) and its associated Location System, was developed at Olivetti Research Labs/University of Cambridge in 1989. This was perhaps the first context-aware computing application, designed to enhance user mobility and to support location-awareness. It was intended to be an aid for a telephone receptionist before mobile phone networks became widespread so that employees could be contacted when they were away from their desk or home location. Once a person was located, phone calls could be forwarded to a desk-phone closest to where the person was located (Section 7.3.1). An Active Badge periodically sends infrared signals to sensors embedded in rooms throughout the building. This became the forerunner that led to the development of the ParcTab at PARC when Roy Want left and later joined PARC. The location determination accuracy of the Active Badge system was limited to identifying which room a person was in. In 1995, Ward *et al.* (1997) developed a new active tag system, called the Active Bat, based on ultrasound[4] that could locate people up to an accuracy of 3 cm. The system used a base station to ask the Bat for a signal that

---

[3] See http://www.parc.com, retrieved June 2007.
[4] Hence the name Bat, because ultrasound is used by bats in nature to determine location.

is then measured in multiple ceiling receivers, the position being determined using trilateration (Section 7.4.1). Unlike the active badge and active bat, the active floor or smart floor did not require someone to explicitly carry some identifying token. Instead, this type of device is designed to identify people indirectly, e.g., in this case by their type of walk or gait (Addlesee *et al.*, 1997). The design of an active floor requires a careful analysis to specify an appropriate spatial resolution and robustness to allow users to walk on sensors without damaging them.

2.2.2 Smart Environments: CPI and CCI

2.2.2.1 Classroom 2000

One potentially useful feature of future computing environments will be the ability to capture the live experiences of its inhabitants and to provide records to users for later access and review. In 1995, a group at the Georgia Institute of Technology undertook a three-year project called Classroom 2000 in an attempt to support both teaching and learning in a university through the introduction of automated support for lecture capture (Abowd, 1999). Whereas most work in courseware development focused on the development of multimedia-enhanced materials, Classroom 2000 attempted to improve content generation by instrumenting[5] a room with the capabilities to record a lecture rather than students having to take their own notes manually and to perhaps transcribe them later. The aims of the project were twofold: (1) to understand the issues in designing a ubiquitous computing application that provides effective capture and access capabilities for rich live experiences; and (2) to understand what it takes to produce a robust, ubiquitous computing application whose impact in its targeted domain can be evaluated over a long period of time.

The initial prototype used a large electronic whiteboard system that allowed the lecturer to display and annotate slides. Students could also use a tablet-type computer to annotate their own copy of the slides. These were not initially networked. In January 1997, a second prototype classroom opened that was specially instrumented for Classroom 2000 use. Microphones and video cameras were embedded in the ceiling and the signals from the microphones and cameras were stored. The electronic whiteboard was used again but this time networked. Two ceiling-mounted projectors attached to networked computers were also for display.

2.2.2.2 Smart Space and Meeting Room

The NIST Smart Space and Meeting Room Projects, 1998–2003[6] focused on the use of pervasive devices, sensors, and networks to provide an infrastructure for context-aware smart meeting rooms that sense ongoing human activities and respond to them (Stanford, 2003). It was split into two phases. In phase 1, experimental smart spaces were prototyped focusing on: advanced forms of human–computer interaction, integrating pico-cellular wireless networks with dynamic service discovery, automatic device configuration, and the software infrastructures required to successfully program pervasive computing applications. Phase 2, focused on developing metrics, test methods,

---

[5] Instrumentation in this context refers to the process of adding sensors, acting as information sources in the physical world. These sources can then be configured and recorded for on-line and off-line analysis. Other systems such as software-based systems may also be instrumented in order to monitor their operation and to reflect on and even modify their behaviour.

[6] Estimated from published papers.

and standard reference data sets to progress the technology and to provide reference implementations to serve as models for possible commercial implementations.

The Meeting Room digitised signals from two hundred microphones, five video cameras, i.e. direct sensor data, and had a smart whiteboard. Two sets of tools are used to manage this sensory data. The sensor streams are managed using the NIST SmartFlow system, a data flow middleware layer that provides a data transport abstraction, and offers consistent formats for the data streams. Metadata or annotations of the data stream are addressed with semantic descriptions using the Architecture and Tools for Linguistic Analysis Systems (ATLAS). When people in the meeting room are talking to each other, the system can take dictation, record a transcript of the meeting, track individual speakers, follow what the conversation is about and trigger associated services from the Internet. This supports an iHCI model for taking notes and to assist speakers by intuitively providing further information.

2.2.2.3 Interactive Workspaces and iRoom

The Interactive Workspaces project started at Stanford University in 1999, and investigated the design and use of interactive rooms called iRooms that contained one or more large displays, with the ability to integrate portable devices and to create applications integrating the use of multiple devices in the user space (Johanson *et al.*, 2002). There were several design principles:

- *Emphasise co-location* (location awareness): take advantage of the shared physical space for orientation and interaction, e.g., team meetings in single spaces. This is in contrast to teleconferencing and videoconferencing which interlink multiple physical spaces.
- *Rely on social conventions to help make systems intelligible*: the room is designed to provide affordances necessary for a group to adjust the environment as they proceed with their task, i.e., users and social conventions take responsibility for actions. The system infrastructure is responsible for providing a fluid means to execute those actions.
- *Wide applicability (interoperability)*: create standard abstraction and application design methodologies that apply to any interactive workspace.
- *Keep it simple (intelligible) for user*: the system must remain accessible to non-expert intermittent users and on the software development side, try to keep APIs as simple as possible both to make the client-side libraries easier to port and to minimise the barrier to entry for application developers.

The second version of the iRoom contained three touch-sensitive whiteboard-sized displays along the side wall, a large display supporting pen interaction called the interactive mural built into the front wall and a custom-designed display to look like a standard conference room table. The room also had cameras, microphones, wireless LAN support, and a variety of wireless buttons and other interaction devices. The room supported three main interactive tasks: moving data, moving control and dynamic coordination of multiple applications.

2.2.2.4 Cooltown

The Cooltown project by HP, 2000–2003, developed a vision of UbiCom to support mobile users providing access to information via wireless hand-held devices based upon Web technology and linking the virtual ICT world to the physical world (Kindberg *et al.*, 2000). A key feature of the Cooltown approach is that a physical world resource can have a Web presence. Physical resources are associated with a simple standard resource identifier, a URL or Universal Resource Locator. URLs of physical and virtual resources can be discovered and exchanged in very simple local device–physical world transfers. Thus, when entering a new room, a PDA device can receive a message that contains the URL of that room via an IR transmission from a RF beacon in the room, by reading a barcode in the room, etc. The PDA will then be able to access the website for the room to view the available facilities and functionalities. Kindberg *et al.* state that there are three important benefits of using the Web for mobile users situated in the physical world:

- *Ubiquitous access*: the Web offers accessibility that supports mobility in two senses. Resources on the Web can be accessed from any device that supports the standard HTTP protocol and that supports transparent access to resources anywhere in the Web, providing the ephemeral criterion (see Section 2.2.4.5) is supported.
- *Just enough middleware (thin client access)*: Web standards are used and simple URLs are exchanged. No application programming-specific language middleware is needed to run on the device, just a Web browser. (Section 3.3.3.4).
- *Keep it local where possible*: Web services can be delivered to mobile users using a local Web-server without requiring a global wireless connection like a cell-phone or mobile IP. This minimises how much of the infrastructure is needed for users to interact with local services.

2.2.2.5 EasyLiving and SPOT

Microsoft has undertaken several UbiCom research projects, developing several prototype products, this work is ongoing. These include: the EasyLiving project, 1997–2002, Smart Personal Object Technology (SPOT) initiative that started in 2003 and the MyLifebits project (Section 12.2.9.2). The focus of the EasyLiving Project, 1997–2003, was on developing intelligent environments which allow the dynamic aggregation of diverse I/O devices into a single coherent user experience (Brumitt *et al.*, 2003). In UbiCom applications, there is often the need for the separation of hardware device control, internal computational logic and user interface presentation rather than tightly coupling input/output devices to applications. For example, in order to allow content created in one device to be output to a different device and controlled by yet another device, or because the input control device is not conveniently collocated with a remote display device, such as when remotely controlling a large screen display. EasyLiving enables this kind of flexible interaction by providing abstract descriptions of their capabilities, geometric modelling of the location of devices in relation to other devices and through sensing capabilities. In contrast, the desktop PC model assumes that computer peripheral devices such as display, mouse, and keyboard are connected to a single machine and are all appropriately physically located. When working in a distributed environment, it is no longer viable to assume this static fixed device configuration, both in terms of device presence and physical configuration.

In the SPOT initiative, SPOT devices were designed to listen for digitally encoded data such as news stories, weather forecasts, personal messages, traffic updates, and retail directories transmitted on frequency sidebands leased from commercial FM radio stations (Krumm *et al.*, 2003). Such devices could provide valuable broadcast notifications and alerts to millions of people.[7]

---

[7] The SPOT initiative seems to have stagnated, although there are many potential applications for this. For example, electronic devices that contains a timer chip could also be designed as a VHF receiver and automatically synchronise their time to a radio broadcast rather than requiring users to manually set the time via a proprietary control interface. A local radio receiver could also acts as a store and forward gateway, supporting a local cost short range network, e.g., Bluetooth, to distribute this data. The use of VHF radio to date has been underexploited as a data channel to support UbiCom. Maybe the increasing availability of digital radio broadcasts will act as an enabler.

2.2.2.6 HomeLab and Ambient Intelligence

Ambient Intelligence or AmI was proposed by Philips in the late 1990s as a novel paradigm for consumer electronics that are sensitive to, and responsive to, the presence of people, e.g., person-awareness. This was regarded as somewhat different in focus from the original ubiquitous computing vision. AmI was developed within Philips in a company-wide strategy called healthcare-lifestyle technology (Weber, 2003). In 2002, after two years of design and construction, the Philips HomeLab was opened, an advanced laboratory that could be used to conduct feasibility and usability studies in Ambient Intelligence. Philips HomeLab looked and felt like a regular home with modern furniture in every room and even a fully stocked kitchen. Prototypes of Ambient Intelligence were installed that ranged from electronics that could recognise voice and movement to digital displays within a bathroom mirror to new 'toys' that were designed to help children expand their creativity.

While no one lives at Philips HomeLab, temporary 'residents' can stay at the facility for as long as needed, depending on the type of research being conducted. Researchers are able to carefully watch how their tenants are living with these technologies 24 hours a day through tiny cameras and microphones that are hidden unobtrusively throughout HomeLab. During the first year of HomeLab, three consumer needs were explored:

- *Need to belong and to share experiences*: research focused on connectivity as an enabler to support the sharing of content and experiences and create the feeling of being together. For example, the presence of one remote group was electronically mediated to another group in different ways while sharing a common experience such as watching a football match (de Ruyter *et al.*, 2005).
- *Need for thrills, excitement and relaxation*: Research focussed on enhancing experiences by adding intelligence to remote control light and sound environments.
- *Need to balance and organise our lives*: Research focused on developing intuitive navigation concepts that put users in control through developing a context-aware personal remote control (PRC). This used context-awareness to filter events. Relying on a number of sensors, the remote control could detect contexts such as being in the user's hand, lying in a drawer, the user being around, and so on. The remote control also changed the way it issued reminders for upcoming programs depending on the personal importance to users.

In parallel, the vision was exported and developed as part of an EU-wide research framework (ISTAG, 2003). The two key properties of ambient intelligent systems are the following. Networked devices are integrated into the environment, including sensor networks and embedded control systems. User-aware[8] systems can tailor themselves by adapting to changes in the personal context with respect to a person's goals. The system has a model of a person's goals and behaviours. It anticipates a person's goals and self-adapts, or automatically adapts to them. It is this person-aware property that mainly implies a particular, social, type of (ambient) intelligence in the sense that systems are designed to closely cooperate with their human users.

2.2.3 Smart Devices: CPI

2.2.3.1 Unimate and MH-1 Robots

Machines are used to perform physical tasks that are very labour-intensive and repetitive or are too dangerous or difficult for humans to implement directly. Automated machines that just do one thing

---

[8] User-aware combines four overlapping properties described by Aarts (2003): context-aware, personalised, adaptive and anticipatory.

are not robots. Robots have the capability to handle a range of programmable jobs, e.g., in a factory, perhaps using different end effectors, i.e., hands or grippers. In 1961, Heinrich Ernst developed the MH-1, a computer-operated mechanical hand at MIT (Ernst, 1961). The first industrial computer-controlled robot, the Unimate, designed by Joseph Engelberger, followed an earlier patent he had filed in 1956. It had one powerful arm with five articulations. It could be programmed to load and unload machine tools, palletise parts, handle welding guns and operate die casting machines and forging presses. The very first application was die casting at a General Motors car factory in 1962.

2.2.3.2 Smart Dust and TinyOS

Micro fabrication and integration of low-cost sensors, actuators and computer controllers, MEMS (Micro-Electro-Mechanical Systems) enable devices or motes to be small enough to be sprayed, or scattered untethered into the air, to become embedded throughout a digital environment, creating a digital skin that senses a variety of physical and chemical phenomena of interest (Section 6.4.4). This has been termed Smart Dust[9] (Warneke *et al.*, 2001). Unlike the Internet where data is often generic and can get stale, information from this digital skin can be localised, current and directly accessible by end-users and applications.

The Smart Dust project at the University of California, Berkeley (UCB), circa 2001, led by Kris Pister, explored whether or not an autonomous sensing, computing, and communication system can be packed into a cubic-millimetre-sized mote (a small particle or speck) to form the basis of

Figure 2.1 Example of Smart Dust, Golem Dust, solar-powered mote with bi-directional communications and sensing, acceleration and ambient light, about 10 mm³ total circumscribed volume and 5 mm³ total displaced volume. Reproduced by Permission from Warneke, B.A., Pister, K.S.J. (2004) An Ultra-Low Energy Microcontroller for Smart Dust Wireless Sensor Networks. Int'l Solid-State Circuits Conf. 2004, (ISSCC 2004): 316–317. © 2004 IEEE

---

[9] The name 'smart dust' apparently started out as a joke according to Kris Pister (Frost, 2003) 'Everyone was talking about smart houses, smart buildings, smart bombs, and I thought that it was funny to talk about smart dust.'

integrated, massively distributed sensor networks (Figure 2.1). The Smart Dust mote consisted of a thick-film battery, a solar cell with a charge-integrating capacitor for periods of darkness, or both. Depending on its application, it integrates various sensors. An integrated circuit provides sensor-signal processing, communication, control, data storage, and energy management. A photodiode allows optical data communication.

TinyOS started out as a collaboration between UCB and Intel Research in 1999 (Hill *et al.*, 2000) and has since grown to be an international consortium in 2007. TinyOS is an embedded open-source operating system and platform aimed at wireless sensor network (WSNs) applications. It was intended to be incorporated into smart dust and into its follow-on projects.

2.2.4 Smart Devices: iHCI and HPI

These projects here are to an extent combinations of types of smart device, smart environment and smart interaction that focus on supporting iHCI.

2.2.4.1 Calm Computing

In the mid to late 1990s, Weiser became more interested in a vision of UbiCom he called Calm Technology or Calm Computing (Weiser and Brown, 1997). Weiser noted whereas computers and games for personal use have focused on the excitement of interaction, when computers are all around, we interact with them differently. We often want to compute while doing something else. The term 'periphery' refers to what we are attuned to without attending to explicitly. This is akin to foreground processing and attention versus background processing and attention. Things in the periphery are attuned to by the large portion of our brains devoted to peripheral (sensory) processing. Calm technologies are said to calm us as they can empower our periphery in three ways:

- to engage both the centre of our locus of attention and the periphery of our attention, which moves back and forth between the two;[10]
- to enhance our peripheral reach by bringing more details into the periphery. The periphery informs without overburdening us. The periphery has a higher capacity for storage than at the centre of attention, e.g., a video conference in contrast to a telephone conference allows people to focus on facial expressions and body posture that would otherwise be inaccessible.
- to offer locatedness (location-awareness): when our periphery is functioning well, we are tuned into what is happening around us, and so also to what is going to happen, and what has just happened. We are connected effortlessly to a myriad of familiar details.

An example of calm technology was the 'Dangling String' created by artist Natalie Jeremijenko. This is an 8-foot piece of plastic spaghetti that hangs from a small electric motor mounted in the ceiling. The motor is electrically connected to a nearby Ethernet network cable so that each bit of information that goes past causes a tiny twitch of the motor. Hence the degree of twitching indicates the degree of network traffic in that Ethernet segment.

2.2.4.2 Things That Think and Tangible Bits

The Things That Think[11] (TTT) research consortium was established at MIT Media Lab in 1995 to look at how the physical world meets the logical world or virtual computer world. It can be grouped

---

[10] It is not clear how exactly this can be controlled or manipulated.
[11] Things That Think, http://ttt.media.mit.edu/, accessed January 2008.

into bits, people and atoms, i.e., virtual computing environment, human environment and physical world environment (Gershenfeld, 1999). There are a whole host of projects which have taken place at MIT such as Oxygen and Tangible Bits (see below). Gershenfeld leads the physics and media group, the bits layer of TTT gives an overview of some of the projects which members of his group were involved in such as e-ink (Section 5.3.4), wearable computers (Sections 2.2.1.4, 5.4.3), use of digital computers to enhance stringed musical instruments, body area networks (Section 11.7.4), and e-cash.

The Tangible Bits project at MIT led by Ishii (Ishii and Ullmer, 1997), began in 1995 as part of the TTT program. This explored a future where intelligence, sensing, and computation move off the desktop into 'things'. Conventional computer systems are largely characterised by exporting a GUI-style interaction metaphor to large and small computer terminals situated in a virtual environment. In contrast, the Tangible Bits project aimed to change 'painted bits' of a GUI, into 'tangible bits' by taking more advantage of multiple senses and multimodal human interactions with the real world and is similar to Weiser's Calm Computing vision in his later work. Tangible User Interfaces emphasise both visually intensive, hands-on foreground interactions, and background perception of ambient light, sound, airflow, and water flow at the periphery of our senses.

Three specific applications were developed called metaDESK, transBOARD and AmbientROOM. The metaDESK consists of a nearly horizontal back-projected graphical surface. Physical objects called onto this surface were sensed by an array of embedded optical, mechanical and electromagnetic field sensors. Users could use these physical objects to interact with the virtual world views of satellite-images (augmented reality). The transBOARD is a networked digitally-enhanced physical whiteboard designed to explore the concept of interactive surfaces which absorbs information from the physical world, transforming this data into bits and distributing it into cyberspace. The ambientROOM used ambient media, ambient light, shadow, sound, airflow, water flow, as a means of communicating information at the periphery of human perception. For example, the sound of heavy rain indicated many visits to the web page and success in attracting customer attention, while no rain might indicate poor marketing or a potential breakdown of a web server.

2.2.4.3 DataTiles

The focus of the Sony DataTiles project (Rekimoto *et al.*, 2001) was an interactive user interface that uses task-specific physical objects as alternatives to manipulating virtual information systems, rather than using general-purpose input devices such as a mouse and keyboard (see Figure 2.2). There are several potential advantages to this approach. Physical objects can offer stronger affordances[12] than purely visual and virtual ones. This enables people to use additional haptic skills to manipulate objects, not only pointing and clicking, but also rotating, grasping, attaching, etc. Interactions may involve two hands. It allows several people to interact cooperatively in the same physical interaction space. Unlike virtual GUI objects, physical objects do not suddenly disappear or reappear when the system changes modes or become unresponsive when the system is busy with other tasks. Another motivation behind this work is the increasing complexity of orchestrating digital devices: users need to be able to focus on the task itself rather than on the underlying ICT system. The abundance of task-specific devices, information appliances, will be the major interfaces to the digital world that interact with each other to support our daily lives.

The DataTiles system consisted of an acrylic transparent tiles with embedded RFID tags; a flat display that also acts as a tray for the tiles, an electromagnetic pen tablet behind the display, RFID readers (sensor coils) mounted behind the display's cover glass; an electronic circuit for sensing

---

[12] Things that suggest obvious actions based upon shape and other attributes, e.g., a knob or dial can be rotated.

Figure 2.2   The DataTiles system integrates the benefits of two major interaction paradigms, graphical and physical user interfaces. DataTiles, reproduced by permission of © Sony Computer Science Laboratories, Inc.

multiple sensor coils using a single RFID reader (Section 6.2.4). Three key types of interaction were embodied in the system. DataTiles can act both as physical windows for information and can trigger specific actions when tiles are placed on a sensor-enhanced display surface. Several combinations of physical and graphical interfaces are possible, e.g., grooves can be engraved upon the tile surfaces and also act as passive haptic guides for pen operations. A simple physical language for combining multiple tiles is naturally implied, e.g., stacking several tiles in a particular order could imply a sequence of actions.

2.2.4.4 Ambient Wood

In the Ambient Wood project, which started in 2002, a field trip with a difference was created, where children had to discover, hypothesise about and experiment with biological processes taking place within an outdoors physical environment (Rogers *et al.*, 2004). As well as being able to explore the environment itself, pupils used tools that digitally augmented the environment, and enabled them to take their own readings of the area. Their positional information triggered a variety of further details about the environment and its inhabitants. A digitally enhanced probe tool was available to children to enable them collect information about moisture and light in their habitats. Readings could be displayed on a hand-held computer as an image showing relative rather than numerical values. Information about the children's position in the wood could be recorded and location-relevant information about living organisms in the wood could be transmitted to their hand-held computer.

2.2.4.5 WearComp and WearCam

Wearable computing is considered here as a more specific case of surface mounting on a mobile host. When the mobile host is a human, this is referred to as a wearable ICT device rather than a surface-mounted device. In turn, surface-mounted devices are regarded as a sub-type of mobile ICT devices

Evolution of Steve Mann's "wearable computer" invention

1980 | Mid 1980s | Early 1990s | Mid 1990s | Late 1990s

Figure 2.3 Type of wearable computer devices prototyped by Mann
Source: Reproduced, with permission, from http://en.wikipedia.org/wiki/Wearable_computing, © Steve Mann

(Section 4.2). One of the pioneers of Wearable Computing[13] is Steve Mann whose first experiments with wearable computers started in the late 1970s. His first major application focused on recording personal visual memories that could be shared with others via the Internet. In a review of his work, Mann (1997) differentiated three (overlapping) generations of wearable computing (see Figure 2.3). These all used a head-mounted display that was permanently available in the user's field of vision to support augmented or mediated reality, a wireless transceiver to exchange information with external entities and nodes and a programmable computer for local processing. Generation three, mid-1980s to 1997, enabled computer-assisted forms of interaction in ordinary day-to-day situations, while walking, shopping and meeting people. Uptime was measured in days and the body network circuit used threads of the clothing as conductors. This new generation supported three key features: hidden computing, e.g., customised glasses were used as the head-mounted display; special conductive fabric was used as a body area network; mediated reality and homographic modelling was used.

Homographic[14] modelling enables the view to be annotated with text or simple graphics. It was noted that by calculating and matching homographies of the plane, an illusory rigid planar patch appeared to hover upon objects in the real-world, giving rise to a form of computer-mediated reality, e.g., a person might leave a virtual grocery list on a view of a refrigerator that would be destined for a particular individual. Although the message is sent right away, it remains dormant in the recipient's WearComp memory until a corresponding homographic view triggers the message. This has the advantage that information can be hidden, reducing the overload on the user, until an appropriate context, e.g., visiting a grocery store in this case, triggers the message.[15] This represents a form of context-aware stigmergy (Section 10.5.1) and prospective memory (Section 5.7.6).

The applications for wearable computers include: monitoring the human body's physiological functions, the distance walked, route taken, and projecting hands-free information for viewing.

---

[13] A history of wearable computing can be found at http://www.media.mit.edu/wearables/lizzy/timeline.html, retrieved on November 2007. However, the definition of computer here is vague and includes some analogue devices which were later replaced by digital versions. There is also the issue of what defines a computer. In UbiCom, MTOS computers (Section 3.4.3), embedded computing and microprocessor applications are included (Section 6.5).

[14] Homography is defined as a relation between two figures, such that any given point in one figure corresponds to one and only one point in the other, and vice versa.

[15] It is easy to think up many applications of this, e.g., someone may leave a note for someone in advance to eat more healthily which is only triggered when the Webcam views some food being eaten.

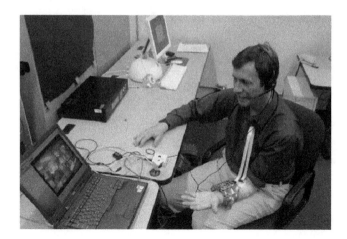

Figure 2.4   An electrode array surgically implanted into Warwick's left arm and interlinked into median nerve fibres is being monitored. Reproduced by permission of © University of Reading

2.2.4.6 Cyborg 1.0 and 2.0

Implanted ICT devices into human mobile hosts is regarded here as a more general case of embedded devices, i.e., devices which are embedded into inanimate or animate devices (Section 4.2.1) which in turn are regarded as a specific kind of mobile ICT device. In 1998, Kevin Warwick[16] underwent an operation to surgically implant a silicon chip transponder in his forearm which remained in place for 9 days (see Figure 2.4). This experiment, Cyborg 1.0, enabled radio-frequency readers embedded in the environment to react to Warwick as he moved through halls and offices of the Department of Cybernetics at the University of Reading (Warwick, 2003). A unique identifying signal emitted by the implanted chip could be used to trigger the doors to open, lights, heaters and other computers to be activated without him lifting a finger. However, this implant, although in his body, was not part of his body in terms of being interlinked to it. Carrying around a wearable transponder or a card transponder would also trigger the same events in the physical environment. A second experiment, Cyborg 2.0, got underway in 2002 that examined how a new implant in his lower arm could send signals back and forth between Warwick's nervous system and a computer (Warwick, 2003). He was able to remotely control an electric wheelchair and an intelligent artificial hand using his arm. This required a training phase to condition his brain to recognise signals and his brain could only attune to these signals for about an hour. He was also able to create artificial sensations by stimulating individual electrodes within the array and to share these with another human, his wife, who had a second, less complex implant connected to her nervous system. The implant was removed after about 90 days when the implanted sensor was about to fail.

2.2.5 Other UbiCom Projects

There are many other examples of smart UbiCom projects[17] and applications given throughout this text. Mobile devices and applications are discussed further in Chapter 4. Additional information

---

[16] See also http://www.kevinwarwick.com, accessed 22 December 2007.

[17] Some oft quoted projects in the literature are not included because no detailed peer-reviewed scientific publications could be accessed to ascertain how they worked.

about HCI projects and applications are covered in Chapter 5. Micro and nano-sized devices and robot projects are covered in Chapter 6. Context-aware projects and applications are covered in detail in Chapter 7.

Here are also some brief outlines of a few further smart environment projects. The Aura project (Garlan *et al.*, 2002) at Carnegie Mellon University, aimed to introduce the concept of a personal information aura: an invisible halo of computing and information services that persists regardless of location and that spans wearable, hand-held, desktop and infrastructure computers. It investigated how to create a home environment that is aware of its occupants' whereabouts and activities.

The Aware Home Research Initiative at the Georgia Institute of Technology, led by Abowd, started in 1998 and is still ongoing. It simulates and evaluates user experiences with off-the-shelf and state-of-the-art technologies applied to the home. It researches how to provide services to residents that enhance their quality of life or help them to maintain their independence as they age (Kidd *et al.*, 1999).

The Cognitive Lever, or CLever, project, started in the early 2000s to investigate computationally enhanced environments was designed to assist not only people with a wide range of cognitive disabilities, but also their support community focusing on Intelligence Augmentation (IA) approaches with the aim of complementing, empowering and augmenting human capabilities (Kintsch and dePaula, 2002). Further UbiCom projects and applications are also considered in the next section.

2.3 Everyday Applications in the Virtual, Human and Physical World

The UbiCom applications that are the focus of this book are the everyware applications in which UbiCom systems facilitate people's routine activities. UbiCom applications can be grouped according to the type of interaction they facilitate: human–computer interaction, human–physical world interaction and computer– physical world interaction and human–human interaction.

2.3.1 Ubiquitous Networks of Devices: CCI

Information and communication technologies (ICT) have become a critical component of the global infrastructure. Three types of wide-area ICT networks are quite ubiquitous at this time: (1) GSM and other wireless telecoms networks; (2) TCP/IP-based wireless access networks attached to a wired Internet backbone; and (3) satellite networks including Global Positioning System (GPS) networks. The TCP/IP-based Internet, a network of heterogeneous networks, is increasingly being used as a universal backbone network to deliver many different logical media applications, e.g., email, Web, video and audio data over a variety of physical media networks (Chapter 11).

Video broadcasts, voice unicasts and data can often be clustered, by various content, service and network providers, as triple-play services, delivered over a single wide area network. These may still be split into single media content and accessed by non-interoperable individual devices. If we add mobile information services to this bundle, these are called quad-play services. There are many further service bundles which could be considered. If we add radio audio broadcasts we could call these a pentad-play service bundle. Mobile devices such as phones, PDAs and mobile phones already offer pentad-play service bundles.[18] We can also create further service provision bundles, e.g., sextet-play and septet-play to combine interactive audio-video games and power distribution respectively and so on (see Section 2.3.2.1).

---

[18] Currently, many other types of devices such as games-consoles, various audio-video broadcast receivers could also offer such pentad-play services but they are not designed to (see Section 2.5).

Within specific domains, such as the home domain, TCP/IP-based WLANs, fixed local area networks can easily be established to distribute multimedia content within that domain. However, it is also common to use a computer as a hub to access application services and to use removal media such as memory sticks, CDs and DVDs and short serial line wire links to exchange content. The means to access and distribute AV and data content received within a local domain may easily be extended to enable it to be distributed further, outside the home. Video content received under licence in the home, could be place-shifted to be accessed outside the home using various store and stream-forward multimedia services. This has been referred to as place-shifting and is in contrast to time-shifting, the playing of multimedia streams in order to pause or delay them.

Some countries have adopted specific strategies for targeting ubiquitous services and applications. For example, since 2001, the Nomura Research Institute in Japan has reported a more limited vision of ubiquitous computing for a ubiquitous networked society in form of the u-Japan plan (Murakami, 2004). Murakami considers two main strategies for ICT evolution: development of an improved ICT infrastructure, and promotion of improved ICT utilisation. This vision concentrates on the evolutionary development of current mainstream networked ICT rather than on more futuristic ICT computing such as wearable computing, digital paper and ink, which are referred to as esoteric computing. The ubiquitous network development is envisaged along three dimensions:

- *any place*: at the PC, in other rooms, outdoors, in moving vehicles.
- *any time*: while indoors at the PC, indoors not at the PC, out and about (at a destination away from home, e.g., shops, cinema, and friends, etc.) and on the move.
- *any object*: PC to PC, person to person, person to object and object to object.[19]

2.3.2 Human–Computer Interaction

2.3.2.1 Ubiquitous Audio-Video Content Access

Multi-media content should be available over every network and accessed by any suitable device in order to provide the greatest flexibility for users. Much broadcast multimedia content created by third parties that is downloaded is generally non-interactive, read-only content and is often stored and manipulated in the access device. Even if content is read-only, users would often like to add value to content by selectively splitting content, annotating content and through integrating multiple heterogeneous content. Locally created content by the user is modifiable. For content to be used in this way requires it to be annotatable and for multimedia sources to be interoperable.

Individual voice, video and audio services are often not aware of each other and sometimes are not user configurable to enable users to dynamically orchestrate more advanced services. For example, when a voice call arrives, TV and radio are automatically paused or muted. Voice calls can be recorded in answerphone devices but they cannot easily be exported to other systems or converted into different formats, e.g., text for a document or emailed to someone. To support such dynamic service composition requires the use of a pervasive network infrastructure, standard multimedia data exchange formats, dynamic service discovery and the use of metadata. Metadata describes the content, the characteristics of the network and the characteristics of associated mediated services in order to interlink and to compose heterogeneous content and to enable anywhere, anytime access on any device in any composition.

---

[19] In our model, a PC can be substituted using any other ICT system including embedded computer systems. Objects are physical world objects.

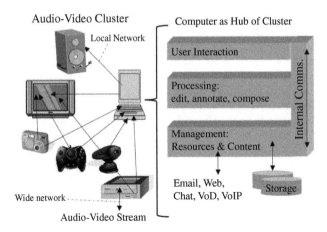

Figure 2.5   Audio-video cluster distributed over a local home network with a PC as the hub

A typical audio-video device cluster is given in Figure 2.5. Audio-video devices are connected in a simple Web star local network often with a PC as the hub of the device network. Other home clusters could exist for home security, lighting and heating. The majority of electronic devices that have embedded computers have no network interface, e.g., washing machines, answer-phones, vacuum cleaners, etc. Thus the user can only check the status or configure devices, manually, using the local control interface. Smart devices and services are discussed further in Section 3.1.

2.3.2.2 Ubiquitous Information Access and Ebooks

A personal digital calendar is a good example of a distributed information application. It can be accessed via a variety of devices such as PCs, set-top boxes, smart phones and PDAs and used to provide a partial user context for synchronising user activities in a variety of situations. Pull-type interaction allows users to initiate the information exchange such as searching the Web for information and services. Push-type notification services are used for customers to be notified of events, e.g., news, lottery numbers.

Surveys in 2004 confirmed that the PC remained the dominant interactive information access device, however, the beginnings of the uptake of digital TV and mobile and multi-platform access can be seen (EU, 2005). Generally, PCs and various AV device displays and controls are not available or conveniently positioned throughout buildings and can be very cumbersome to use in daily tasks. For example, positioning an ICT system in a kitchen can provide instructions for cooking meals. In this case, ICT systems must be able to tolerate liquid and food spillages and support hands-free interaction because hands may be dirty or involved in other activities. Hence mobile devices are often used to support ubiquitous access to selective content. However, in some cases, even mobile devices are similarly restricted in terms of the HCI.

Electronic information access, compared to paper-based systems such as newspapers, magazines and books, has several advantages. Electronic information access offers more flexible text size adjustment, indexing, searching, interlinking different articles, browsing, annotation, editing, and enough capacity to store thousands of articles and books all in one access device. Electronic information also supports information exchange across applications and media, e.g., text can be translated into a different language, can be converted into speech and emailed.

The PC as a ubiquitous electronic information access device supports both reading and writing and can support relatively free copying of content. However, the PC suffers from a number of limitations compared to its paper counterpart. PC information devices are relatively expensive, power-hungry, heavy, slow to start up, unable to be read in bright sunlight, unable to be controlled in the dark (input keys are not lit), and awkward to control using the input keys. Paper also has the flexibility to be rolled and folded to support larger area formats whereas computers screens are fixed in size and users need to scroll around or zoom in and out to display larger area content. Many current PC soft readers seem also not to be very adept at adapting the layout and column format of the original published format to new formats, i.e., to facilitate reading in different size display devices. Instead viewing often requires users to undertake a heavy degree of awkward scrolling to view continuous text passages.

To this end, more specialised electronic book reading devices such as ebooks and epaper have been developed. Ebook readers[20] are lightweight, thinner, long-lasting powered, pocket-sized devices with touch screens, enabling pages to be turned by touch. A key difference between computer displays and ebooks is the type of display used. An ebook screen is designed to be more like paper, thus epaper,[21] reflecting rather than transmitting light. It is also readable in direct sunlight, is equally viewable from any angle and it uses a static image. There is a zero refresh rate when the image does not change, so it does not drain the battery. Content can be delivered to an eBook over a wireless link, avoiding the need to physically go to collect a copy of a book or today's newspaper. Whereas paper can be recycled to be reused, electronic paper can be reused in an even more eco-friendly process.[22] A good candidate technology for epaper and ebooks is an Electrophoretic Display (Section 5.3.4).

2.3.2.3 Universal Local Control of ICT Systems

Computers are normally controlled via input control devices that are integrated into the computer and used while the person is seated at the computer. These may also have an inbuilt wireless remote control interface that allows them to be controlled at a distance, for example, use of a wireless mouse and keyboard, but these still tend to be used by a user seated at the computer. Many appliances have device interfaces such as infrared ones allowing them to be controlled via short-range local control devices,[23] within a distance of a few metres. Well-known conventions are used to label buttons associated with common functions such as changing the volume and switching the device power on and off. However, controller device designers activate uncommon tasks using a

---

[20] Two eBook readers available in 2008 are Sony's Reader (www.sonystyle.com) and Amazon's Kindle (www.amazon.com). Interoperable content formats such as .epub, an XML extension defined by http://www.idpf.org/ are being supported but different digital rights schemes can still make these non-interoperable. According to Harrison (2000), the concept of the ebook was first envisaged by 1895, prior to electricity, television, and aviation, by Frenchman Albert Robida.

[21] There are several different technologies to build epaper, some of which can use plastic substrate, organic materials and electronics, so that the display is flexible. Not only can epaper be designed for reading and writing, paper can also be used as a substrate for electronic circuits (Berggren et al., 2001), paving the way to transfer electronic circuits over a network into special printers – electronic circuit reproduction.

[22] Hard-copy paper reuse involves recycling. Recycling involves physical transport of paper to collection centres, sorting and then transport to recycled paper mills. Paper fibres are pulped before being washed and screened to remove unwanted smaller contaminates like staples, glues, adhesives, ink and plastics. The pulp is then pumped, pressed and dried into paper rolls for reuse, etc. Epaper reuse simply requires transmission of the new content over the air and a recharge to change the molecular structure of the eink on the epaper.

[23] These control devices are referred to as remote control devices but they are really better described as local control devices as they are able to operate only within the local vicinity, i.e. over 1–10 metres in many cases.

combination of buttons and modes with the corresponding instructions for these combinations, listed in a complementary manual. These instructions may be hard to follow because they are very terse and often assume no errors. There is a profusion of local control devices but each of these is configured to control only a single device, leading to clutter and searches for misplaced controllers.

Two types of hand-held universal local control device have been proposed that can be configured for multiple local devices: hardware and software devices. Commercial versions of both hard and soft universal controller devices[24] have been launched. With both these types of universal controller, some non-trivial manual configuration and knowledge of a manufacturer's control code for the device is needed. Simple universal hardware-based controllers have a fixed number of buttons, can control a fixed number of (often up to five) devices and often contain no display to return the status.

Software-based universal device controllers contain both hardware buttons and a display for soft buttons giving users more flexibility for associated buttons with actions and to display status information (see Figure 2.6). Key design issues are how a controller can discover which devices are situated locally, what features the device supports and how to describe them. One way to do this is to enable the device to notify users of various Web URLs that specify the presentation, model and manufacturer URL. The universal device must then download the device information over an Internet connection and configure the device UI for use. This device and UI discovery add complexity and a delay to device access.

Software-based universal device controllers can be customised by the user. They can be configured to provide a simplified interface for casual infrequent users, or a more functional one for regular users. They can adapt to reflect only those devices that are locally accessible. They can be personalised by users. It is also suggested that the use of such universal soft controllers can be

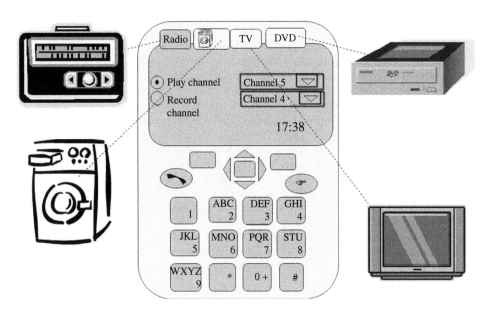

Figure 2.6   Use of a soft universal local controller to interact with washing machine, TV, DVD recorder and radio

---

[24] See, for example, the Philips Pronto programmable touch screen remote controls, http://www.pronto. philips.com, accessed November 2007.

economically attractive for device manufacturers because improving the traditional user interface of appliances can be very costly. Instead the user interface for such complex tasks could be offloaded to a universal local controller (Roduner, 2006).

The use of mobile phones and PDAs as universal local controllers seems attractive because of the inherent features in mobile devices, powerful microprocessors, local and remote wireless network support and a reprogrammable display and soft keys. Studies by Roduner *et al.* (2007) have shown that using soft universal controllers can make simple tasks harder to perform than in the proprietary mobile device, however, they can make more complex tasks easier to perform. The reason for the latter is the mobile device can provide additional information regarding the state of the appliance, either by providing a display to appliances without one, or by extending an already existing, but smaller embedded display. Roduner (2006) has considered the limitations of the mobile phone as a universal interaction device. The main limitation is that in many everyday situations, direct manipulation of the appliance is easier, faster, and more convenient than hand-held-mediated interaction. Some appliances may also require a user presence at the device for safety reasons such as controlling a cooker or microwave. Universal controllers also introduce concurrent control issues when multiple users try to control the same device using multiple controllers.

2.3.2.4 User-Awareness and Personal Spaces

Personalisation enables content and services to be tailored to individuals based upon knowledge of their preferences and behaviour (Section 5.7.4). Personalised services have several benefits such as greater convenience and more relevant filtered information, but perhaps with some trade-off against a loss in privacy. Users can personalise the configuration of services, e.g., each home occupant can access content filtered to their preferences. Users can personalise the annotation of content, enabling this to be organised according to their personal preferences. The same configuration and personal preferences could follow the user and be shared among devices and services freeing the user from having to manually reconfigure similar preferences in similar services.

Mobile devices are often designed for use on a personal scale, for discrete use within our personal space. Mobile devices provide an obvious means for users to personalise their environment, however, mobile devices do not necessarily allow the personal model with the mobile device to be shared. In Section 2.3.2.3, one approach to personalisation has been proposed based upon devices exporting their presentation functions and these being personalised in the mobile device. Nakajima and Satoh (2006) propose a model to coordinate and configure multiple distributed home appliances using a pervasive personal model based upon everyone having their own (centralised) personal home server. The personal home server can seamlessly discover and configure appliances at any location. There are three general requirements for the personal home server: (1) discovering services when devices spontaneously interact; (2) simplicity of interoperability; and (3) maintaining the privacy of personal profiles.

Multiple personal spaces may also expand into shared social spaces. Some concurrent control may be needed when multiple users try to personally interact with devices in conflicting or incompatible ways, e.g., one user wants the music volume turned up while another user wants the volume turned down.

2.3.3 Human-to-Human Interaction (HHI) Applications

People belong to multiple organisational groupings such as family, friends, colleagues, acquaintances, neighbourhoods and communities, at work, at home, and away from work and home. Whereas in the past, social interaction was more physically situated, wide area electronically mediated HHI enables social links to persist by voice, video and text when people are not so co-located.

2.3.3.1 Transaction-based M-Commerce and U-Commerce Services

E-commerce enables local and remote services and products to be provided and exchanged for money from customers. While e-commerce could be completely automated, it is typically only partially automated. Human interaction is involved in sales negotiations, payment interaction, goods delivery and after-sales services. More automated processes could be involved in product selection, product dispatch and payment. E-commerce payment typically involves: a payment transfer from the payer to the payee or merchant in exchange for goods or services. It involves a funds transfer from the payer's bank indirectly to the payee's bank. It involves divulging payment authorisation over a secure channel in the form of authorisation codes, e.g., payment card number, payment card expiry date or personal identification number (PIN) that are private to the payer and quoted as the authorisation code to transfer funds to the payee. It also involves some audit trail of receipts for the e-commerce transaction to the parties concerned.

M-commerce, short for mobile commerce services, is e-commerce services accessed via mobile devices. There are two types of m-commerce in terms of whether or not remote services are accessed locally by mobile users or if local services are accessed locally by mobile users. Remote services and data such as music, ring-tones and video clips can be downloaded and purchased via the mobile phone. Payment for smaller amounts can be charged to the customer's phone bill, based upon the type and size of content that is downloaded and a settlement made from the phone company to the merchant at a later date. Not only can ICT services be accessed and paid for in this way but so can various physical world situated services, see Section 2.3.4.2. Payment for larger transactions can occur in the same way as for e-commerce.

An extension of m-commerce for UbiCom use, called u-commerce, has been proposed by Watson *et al.* (2002). This is the use of ubiquitous networks to support personalised and uninterrupted communications and transactions between an organisation and its various stakeholders, to provide a level of value over, above, and beyond traditional commerce. It is characterised by *Ubiquity*, being available everywhere; *Universality*, being able to operate (everywhere) in heterogeneous environments; *Uniqueness*, relating services to a context such as a location and *Unison*, service orchestration, allowing multiple parties to work together.

There are some obvious differences between u-commerce and m-commerce. First, u-commerce services can use a wider variety of smart devices for payment authorisation such as the various contactless smart cards (Section 4.4) that can contain encrypted payment authorisation details. These can be swiped on or near a secure payment card reader in order to authorise transfer funds to a trader, transmitting the same sort of authorisation information that is used for e-commerce. Second, u-commerce provides more focus for dynamic pricing through context-aware service provision (Section 2.3.3.2). U-commerce may also modify social behaviour when purchasing items in that the speed and ease with which items can be purchased may exacerbate impulse buying of items that can be bought without sufficient consideration.

2.3.3.2 Enhancing the Productivity of Mobile Humans

Productivity can suffer from a bottleneck that occurs when people do not have the right information where they need it and when they need it. Mobile users may need to be better informed in moving from A to B. Location-awareness enables mobile devices to determine their position and for content such as local services to adapt to the position, see Section 2.3.4.1. The most common location-aware devices are stand-alone map-based devices such as SatNav (satellite GPS navigation) devices that display the positions, routes and location-specific services over map views. Location-aware services can also be embedded into devices such as mobile devices or even umbrellas that are connected to the Internet and designed to make walking in rainy days more fun.

The Pileus Smart Umbrella[25] uses the top underneath surface of the umbrella as the screen for the display and has a built-in camera, a motion sensor, GPS, and a digital compass. The Pileus umbrella supports two main functions: social photo-sharing and a 3D map navigation powered by Google Earth. Matsumoto *et al*. (2008) considered two types of technology for the display, using the screen itself as the display, e.g., using OLED displays (Section 5.3.4.3), versus using a light-weight video projector. The video projector was chosen to enable low cost standard umbrellas to be used and then to add the interfaces to the umbrella to support projection, positioning, etc. If the umbrella gets torn or broken during the storm, only a cheap umbrella needs to be replaced rather than more expensive OLED displays. Further, in theory, the development of MEMS micro-projectors (Section 6.4.2) should lead to much lighter-weight, smaller, low-energy projectors.

Corporate business applications for mobile users include access to document and sales data to databases, automation, email and collaborative workflows. Mobile users can access applications based on mobile devices such as a diary, calendar and notepads. Follow-on services can be configured to support automatic call-forwarding and the transmission of a remote workspace based on a mobile device.

However, much interaction tends to be unilateral. Supporting greater two-way interaction across space may be beneficial, e.g., field engineers and sales staff, leading to faster decision-making based on local evaluation and knowledge being returned back to managers. There is an issue as to whether mobile service access empowers workers or enslaves them because they are always contactable by others such as bosses and customers.

The term 'communities of practice' describes the use of more informal information and task exchange that can take place among peers. Communities of practice can be set up for home-based activities e.g., gardening, home maintenance and ICT home maintenance. Assimilating, organising and harmonising such tacit knowledge will also be very challenging. (Shepherdson *et al.*, 2003) have developed and applied a workflow management framework that supports more decentralised worker-oriented teamwork coordination, enabling workers to schedule work requests, work schedules, to trade work requests and work-shifts, to make collective decisions, to extend or reduce work hours and to call on additional expertise. More flexible and utility-based travel can also be planned and re-planned.

2.3.3.3 Care in the Community

'Vulnerable' individuals at home can be monitored by friends, family and health professionals situated elsewhere. These are subsequently notified of events such as abnormal immobility or lack of activity by a subject. Non-invasive, low-cost technologies, at home or in an assisted living or residential care facilities, can enable the elderly, disabled, and others with chronic ill-health, to remain in the community longer, to be independent longer, hence, to reduce their use of professional healthcare services and their demands on and cost to society.

There are two basic kinds of approaches in terms of whether the subject explicitly asks for help from others or whether the subject is monitored and modelled so others can anticipate when the subject requires help. An example of the former, user-requested care approach is that approximately one million people in the UK have access to pendant or pull-cord systems (Edwards *et al.*, 2000). An example of the latter approach is the BT CAREnet trial that finished in 1999 in which

---

[25] See http://www.pileus.net/, accessed January 2008. *Pileus* is the Latin word for skullcap. It is also now used to describe a cloud that appears above a cumulus in meteorology. The project, a spin-off from Keio University Okude Lab, uses Pileus for both metaphors, for a physical umbrella (mushroom cap) and for cybernetwork services (overlaid cloud).

sensors such as passive infrared (PIR) movement sensors and magnetic proximity switches in rooms and appliance doors in strategic locations around the home allow a third party such as a family or professional caregiver to remotely monitor the activities and status of a subject. First, a normal activity profile of users is built up. Next, new data can be analysed automatically for anomalies. For example, alert situations in the CAREnet trial were triggered for the user not getting out of bed, being less active than usual, using kitchen appliances less than usual and the room temperature being too low. There are many different variants of projects depending on the type of sensing used. It can, however, be challenging to differentiate true positive and true negative events from false positive and false negative events and to balance the intrusion into an individual's life to monitor a subject versus maintaining their privacy.

2.3.4 Human-Physical World-Computer Interaction (HPI) and (CPI)

Innovative ways of supporting human-to-physical space interaction mediated have already been given (Section 2.2.4) and include smart tables, smart floors, smart rooms and smart tiles. Additional everyware applications are described below.

2.3.4.1 Physical Environment Awareness

Services can be slanted towards specific physical environment contexts such as location awareness, temperature and rainfall awareness. We can distinguish two types of location-awareness: longer-range mobility-based location-awareness and static, short-range location-awareness. Location awareness is considered by its proponents to be one of the main drivers for mobile communication, e.g., person or business asset tracking and navigation. Here the focus is on long-range tracking of the position of a moving asset in relation to a destination position along a preset route to the destination. Static location awareness involves dynamically discovering services such as a meeting place or tagged personal items such as keys and library books, etc. within a locality. Context-aware systems and applications are dealt with in detail in Chapter 7.

Sensors for specific physical world phenomena are statically embedded into specific devices and services, rather than being network-enabled. For example, a heating system can switch heating on if the temperature drops, light sensors can switch lights on when it gets dark and sprinkler systems can water the garden periodically if it does not rain.

2.3.4.2 (Physical) Environment Control

A mobile phone or other hand-held device can use a wireless link to issue simple control instructions to start or stop network-enabled devices embedded in the environment or to control access to physical resources in a similar way to locally controlling ICT devices (Section 2.3.2.3). Resources may be privately owned, e.g., garage door or car door, or may be provided as pay per use services such as a drinks dispenser or shoe-cleaning machine. In the former case, successful authorisation leads directly to limited free access. In the latter case, authorisation to pay for use of the service is needed before access is granted. This could involve adding a special network billing charge for dialling a number associated with a product and then sending a special code to confirm the intent to purchase. The network provider then passes on some of the payment to the service provider (Section 2.3.3.1). However, the control and reconfiguration of many current electronic devices are not designed for wide area remote access. These are designed only to be enacted manually via a front-panel or via a proprietary short-range IR remote controller.

2.3.4.3 Smart Utilities

Energy is required to power embedded computer and analogue systems to support various physical world activities such as heating, and lighting and to support various human activities such as travelling and eating. These energy production networks are in turn computer-controlled. Energy appears pervasive in many affluent countries, supplied from continuously operating energy grids and from a mix of non-renewable and renewable energy sources. Some devices have no internal energy stores and must remain connected to the energy grid to function. Other devices can use refillable, rechargeable or renewable energy supplies to enable them to be mobile and to be operated for finite periods of time while not being connected to an energy grid. Mobile devices may also take advantage of self-powering systems, powered by walking, by arm movement, by vibration, etc.

The demand for energy is increasing. There is an incentive to increase the energy efficiency of devices and how they are operated to reduce wastage and to reduce costs, e.g., efficient energy use in mobile devices enables them operate disconnected from external energy source for longer periods. It is estimated that modern power supplies (internal and external) of electronic devices are often only 50% efficient so large energy losses occur. Hence, there is much room for further improvement in terms of energy efficiency.

Smart energy utility meters are networked and allow the real-time energy consumption to be tracked enabling customers to be better informed to self-regulate their own consumption. In a demand-response system, customers can choose to save money by adjusting energy use in response to dynamic price signals and policies. For example, during peak periods, when prices are higher, energy-consuming devices could be operated more frugally to save money. In Direct Load Control systems, certain customer energy-consuming devices are controlled remotely by the electricity provider or a third party during peak demand periods.

Context-aware energy devices can switch themselves on in a particular way or off when not in use, e.g., the heating system could also be aware of the presence (or not) of the inhabitants in the building when regulating the heat. Smart Grid multi-agent technology applications allow these types of products to function together as resources within the electricity delivery system, e.g., not all the lights switch on when someone is near, just selected ones that are deemed to best support that activity.

2.3.4.4 Smart Buildings and Home Automation

Automation is increasingly used in building automation, such as light and climate control, control of doors and window shutters, security and surveillance systems, etc. It can also be used to control multi-media home entertainment systems, automatic plant watering and pet feeders. Some control devices can be surface-mounted, e.g., building lighting, etc. They can be designed to switch on automatically when it gets dark or when movement or body heat is detected. In other cases, it may be best to design systems as part of the building, e.g., doors that open as people walk towards them and close afterwards. During construction of a new building, control wires are usually added before the drywall is installed. These wires run to a controller, which will then control the environment. In a retrofit situation, or when installing control wiring which may be too expensive or simply not possible, there are alternatives such as powerline protocols, e.g., X10, Universal powerline bus, and wireless protocols such as ZigBee or using standard PC wired interfaces.

Home automation seems to be more common in the USA than in Europe. Spinellis (2003) provides a neat summary of the underused potential of current home control systems such as heating, lighting, garden watering, media players and recorders that are often used in stand-alone device mode and controlled by independent local, closed system controllers, each with its own powerline or wireless interface.

There is a profusion of IR remote controls that have non-uniform interfaces for performing the same control operations. For the appliances described, it is thus difficult to determine what actions are possible at any moment, the system's conceptual model and current state are hidden from the user, and there are no natural mappings between a user's intentions, the required actions, and the resulting effect. Spinellis (2003) proposed the use of a PC hub to integrate existing consumer home-control, 'infotainment' (replaying stored music files), security (serial-line connection to security alarm devices), and communication technologies (phone interfaces used to connect to a door entry system). Arens *et al.* (2005) report how wireless sensor network technology could affect future building design and operation. Flexible location of sensors and increased density of multiple sensor types can make significant improvements to building energy efficiency. The Future_Home project enabled people at home to move, control, communicate and enjoy AV entertainment without noticing the underlying technologies or networks (Alves *et al.*, 2004).

An important motivation is that the house of today is not well suited to keep pace with rapid technological changes and with recent sustainability concerns. It is an inflexible space in a dynamic world, unable to respond to the continuously changing requirements of its inhabitants. In part, this is due to the characteristics of the meta-sector of a habitat, where traditionalism, fragmentation, lack of innovation capacity and failure to communicate characterise many of the actors involved. Alves *et al.* noted the need to specify a methodology for the construction and maintenance of an experimental future house. They also noted that the technology needed for the modernisation of the habitat meta-sector requires action on many fronts such as building processes, production technologies, technical solutions for structural components and systems as well as investments in cooperation actions with research organisms, universities and firms. They note:

> The project-based nature of work in the construction sector implies that firms have to manage networks of highly complex innovation interfaces. As such, construction can be viewed as a complex systems industry in which there are many interconnected and customised elements organised in a hierarchical way, with small changes to one element of the system leading to large changes elsewhere.

2.3.4.5 Smart Living Environments and Smart Furniture

Several smart environment devices can adapt to human activities. Doors, lighting, taps and air ventilation can be designed to detect the presence of humans, to be activated by them and to adapt to them. An oft quoted example of an intelligent environment is a fridge that is aware of the stored ingredients it has in stock. It may also be aware of the ones needed for meals scheduled later in the week via external system applications such as diaries. The smart fridge behaves as a stock control system which automatically detects low quantities or out-of-date food items such as fresh vege-tables and milk. This is simple to operate if it just re-orders when only one container remains but containers can be large. There is the complication of how the quantity remaining is detected, e.g., using some transducer to weigh incoming food, but there is a need to know the container weight overhead too. The system is empowered to act on context to order new food if authorised. Policies could include that the 'order new food action is not triggered if consumers are absent' or that it cannot order more food than the store size, even if there is a bulk discount. Ordering could take into account preferences for organic or skimmed (low fat) milk. Ordering does not result in fully automatic delivery of food to storage. This needs human assistance and human coordination for delivery. Food ordered could be tagged in a way so that the customer can track their orders, so if they knew the delivery schedule, they could estimate when it would be delivered to them to better plan to physically be there when deliveries arrive.

A ubiquitous home environment could be designed to support healthy eating and weight regulation by its inhabitants. This could be supported using the following networked components: a distributed core ICT system that supports multi-sized and flexible positioned displays and mobile terminals; fridge(s); multiple storage areas that sense what food and drink items they contain; weighing scales for food and humans and hand-held scanners. Scanners are used to identify food and drink from packaging, this may include a barcode but not all food may have barcodes. Scanners can be used to identify and select healthy food for purchase in the home and in food stores such as supermarkets. The system may also scan for text labels, expiry dates, food ingredients and the percentages of ingredients by weight.

The system could monitor the amount of food consumed per day and per week, the amount of calories consumed and the weight in grams of fat, salt and sugar daily. It could proactively propose and support eating habits to eat a balanced diet, e.g., by aiming to consume five pieces of fruit or vegetables a day. It could maintain a list of unhealthy or undesired food ingredients that cannot be selected: hydrogenated oils or trans-fats, mono-sodium glutamate, etc. and food items not allowed on religious grounds. It could support food stock control to minimise food wastage and to procure new supplies when food is out of date.

Smart objects in the home environment such as cups can be designed to sense their physical state and to map sensor readings autonomously to a virtual computer model. Simple electronic tags such as RFID tags could be simply attached to objects. More complex embedded networked sensing and controller devices can also be added, e.g., the MediaCup (Beigl et al., 2001). MediaCup was not designed for any specific application, however, several applications have since arisen, for example, door sensors can be used to determine the aggregation of hot cups in a room to infer and indicate meetings. Users can be tracked via their cups. Users can be warned if they pick up a mug whose contents are still too hot to drink.

Smart chairs such as SenseChair (Forlizzi et al., 2005) can be designed to take information about a sitter's behaviour and to adapt to it. The need for direct user input to this chair is designed to be limited to turning the chair on and off. The chair can determine the distribution of body weight across the surface of the chair measured by pressure sensors and relate this to the length of time since the last substantial body movement and the time of day, e.g., to suggest that the participant has fallen asleep in the chair after dark, prompting the chair to respond accordingly.

Smart clocks can be used not only to indicate what the current time is but to give context information such as where people are that time or what the weather is at that time. For example, the Whereabouts Clock[26] (Sellen et al., 2006) is essentially a situated display. It is a persistent 'at-a-glance' display of information that can be public or privately shared. This application is location-aware, through determining a person's cell phone location, each person's location from a fixed group of static locations is indicated by the position of a clock hand.

Smart mirrors can automatically optimise the field of vision, e.g., near-side wing mirrors of cars can automatically face downwards to see the kerb when reversing into the kerb. Mirrors can be smart because they can be linked to actuators and sensors in the physical world, e.g., mirrors can be linked to toothbrushes that contain cameras and sensors so that they automatically guide themselves or show the view to assist users in which parts of the teeth need to be especially cleaned. Glass and mirrors can be used as augmented displays, e.g., a vehicle glass windscreen can show the speed of the vehicle so that driver does not need to glance away from the road.

---

[26] The idea of a clock displaying people's location rather than time was inspired by J.K. Rowling's *Harry Potter* stories. The Weasley family has a magic clock, first introduced in book 2, *Harry Potter and the Chamber of Secrets*, with hands for each member of the family, indicating their location or state.

Park *et al.* (2003) discuss a range ideas for smart homes as follows. The smart wardrobe digitally looks up the weather forecast for the user so that they can comfortably and adequately coordinate what they wear with the outside environment before they leave the house. A smart bed can be programmed to remember your preferred sound, smell, light and temperature settings to gently wake up all your senses and give you a good start every morning. A smart pillow can read any books of your choice to you at bedtime and can play your favourite music to drift off to when you start to get sleepy. Once your body goes into deep sleep, it will automatically check the condition and quality of your sleep, gradually reducing the volume of the music accordingly and, eventually, turning it off completely. A smart mat situated at the entrance of every home can be used to sense the body weight and footprint of the users, enabling the smart mat to perhaps differentiate and recognise who is stepping on the mat. A smart sofa can enhance your experience when watching the television or playing video games. Depending on the visuals and the sounds on the screen, it uses vibrations to enhance the viewing experience in action scenes.

2.3.4.6 Smart Street Furniture

Street furniture refers to equipment installed on streets and roads such as benches, bollards, lighting, traffic lights, traffic signs, public transport stops, waste-bins, taxi stands, public lavatories and fountains. For example, in some cities, bollards can be sunk and raised by remote control from buses, thus providing secure access control to reserved lanes for buses but not for other vehicles. Street furniture typically utilises solar panels to supplement power. They are networked and may house other devices such as video cameras and sensors. Traffic lights may also adapt to the traffic patterns rather than be timer-driven.

2.3.4.7 Smart Vehicles, Transport and Travel

Embedded computer systems are increasingly being used within vehicles. This helps to improve their operation such as automatically controlling or providing assisted control, in which the driver still has some control, of the antilock brakes, air-bag inflation, cruise speed, in-vehicle climate, collision avoidance via automatic braking, etc. Location-determination computer systems also enable vehicles and goods to be remotely tracked and interested parties to be informed of their schedule. Some transport systems can be automatically guided along tracks and controlled, with no driver.[27] In some cases the stations themselves are unmanned too. Train, flights and bus services can be designed to inform waiting passengers of the status of arriving and departing vehicles. Current transport information can now be distributed much more and accessed much more conveniently, e.g., by mobile device, and at transport way-points such as bus-stops. Tickets to provide authorisation to travel are also smarter. In some cases, electronic tickets can be requested, paid for and issued remotely. Smart tickets can also be designed to be pay-before and to support contactless verification at passenger entry and exit points.[28]

Passengers and drivers increasingly have access to the Internet in moving smart vehicles such as trains, boats and planes. Once connected to the Internet, not only can Internet data be accessed from

[27] For example, the Docklands Light Railway in East London, opened in 1987.

[28] Smart tickets often contain RFID chips that can be read from a short distance away by swiping the card over the reader, e.g., the Phillips MiFare ticket, is used in London, where it is called the Oyster Card, and in other cities world-wide. This system uses the ISO 14443A industry standard for contactless radio-frequency, smart cards. The card is also swiped on exit too to set the fare in proportion to the length of journey.

the vehicle, but vehicles can also be monitored and various environmental information generated within the vehicle by embedded computers and sensors[29] can be uploaded to the Internet. Since 1995, the InternetCAR (Internet Connected Automobiles Research) project at Keio University, Japan, has been investigating how vehicles can be connected to the Internet in a transparent manner (Ernst *et al.*, 2003). Three separate networks of data sources are identified for vehicles: (1) information services such as audio broadcasts and navigation services; (2) devices controlled by passengers (some lights, power windows, etc.); (3) devices that operate the vehicle (engine and brakes, sensors that monitor the air pressure in the tyres or amount of fuel left in the tank, etc.). The focus of the InternetCar project has been on the mobility of cars modelled as mobile computer nodes within a wireless network and to consider the ability of some cars not just to act as service providers or users but to also to act as mobile routers as part of a large-scale ad hoc network.[30]

2.3.4.8 Pervasive Games and Social Physical Spaces

On detecting friends within a local vicinity, a 'friends meeting service' invites the different parties to meet at the trader's local premises, such as a coffee shop. Providers such as the coffee shop may even sponsor the service to make it free for users because presumably there is an increased likelihood that trade will increase as a result. The types of trader locations suggested as meeting places could depend on the dynamic personal preferences of the parties concerned.

Local traders that require customers to be present in person can advertise their offers locally. This requires services to strike a careful balance between providing timely, situated and useful notification services versus providing unwanted floods of notifications from many competing providers, causing users to be overloaded with notifications. Notification services could be message-based or context-driven when a user searches for a particular service. There are many interesting social and economic issues here apart from privacy issues, e.g., dynamic market equilibriums could be reached as customers situated at one business receive a better offer from a competing business nearby and threaten to leave unless the current or situated business makes them a similar or better offer.

Games are an important type of entertainment and social, interactive, application. Traditional or pre-electronic games consist of two types of interaction, human-to-physical world interaction (HPI) and human-to-human interaction (HHI). HPI concerns moving humans and the interaction between them, e.g., sports, or humans moving various physical objects, tokens, across a board, taking cards, etc. HHI includes competing with others to win a game. In contrast, electronic games create the illusion of being immersed in an imaginative virtual world. Computer games can be more interactive than traditional games and can be designed with an optimal level of information complexity to provoke and engage players.

Electronic games often tend to focus more on HCI than on CPI and HHI. Electronic games tend to virtualise humans so that they become immersed in the game, interacting in a virtual ICT environment. There appears to be a sharp social divide between young gamers and older non-gamers. An analysis of the requirements of existing games consoles indicates an important lack of appeal of video games to a larger audience. The game control interface (the d-pad interface, q double joystick and four-way toggle switch) is often considered to be hard to use by newbies or non-gamers. Games console innovation often focuses on developing more powerful and stimulating visual virtual

---

[29] A typical car currently contains about 120 sensors and 50 different embedded computer systems.

[30] In theory, the set of all road vehicles could be configured to form the planet's largest ad hoc computer network with millions of computer nodes being interconnected in a region.

environments (Grossman, 2006). This has led to one games-console manufacturer, Nintendo, to discard the typical games console interface, replacing it by a simpler interface called a Wii that resembles a wand. This is part laser pointer and part motion sensor so it can monitor where it's being aimed at and how fast it's moving. It is also designed to sense gestures.

In pervasive gaming, various social activities and games are created that seek to exploit the potential of combining the physical space, the use of physical objects and a sense of the physical space, enhanced with a virtual space (Magerkurth et al., 2005). Kampmann Walther (2005) gives a good overview of the variety of different types of pervasive games. These include:[31]

- *mobile games*: that take changes in the relative or absolute positions of players and objects into account in the game rules;
- *location-based games*: include relative or absolute locations in the game rules, e.g., treasure hunts;
- *augmented reality games and mixed reality games*: these are an interesting approach to the creation of game spaces that seek to integrate virtual and physical elements within a comprehensibly experienced perceptual game world, e.g., electronic wands with gyroscopes can detect orientation and movement and merge these players' movements into a virtual landscape;
- *adaptronic games*: applications and information systems simulate life processes observed in nature. These games are embedded, flexible, and usually made up of 'tangible bits' that oscillate between virtual and real space.

Pervasive games support the UbiCom properties of being distributed. Players are also often mobile. Players' actions are situated in the physical world and they may be location-aware of themselves and of other players. In addition, games are designed to be able to persist between player sessions and to support transmediality. Transmediality concerns the use of concepts that transcend individual media. The interplay of multiple media spread out over huge networks and accessible through a range of devices is rather a nice instance of how media communicate in circular, not linear, forms. These media carry information, entertainment, games, role-play, and character sketches in a nonstop circuit of jointly coupled citations and codes of utilisation that can be promptly attuned and functionally altered (Kampmann Walther, 2005).

2.4 Discussion

2.4.1 Achievements from Early Projects and Status Today

The early projects focused most on the three basic UbiCom requirements of distributed system support particularly for mobile users, a variety of iHCI, and context-awareness. These particular early projects did not tend to focus on UbiCom systems designed as intelligent systems and as autonomous systems.

2.4.1.1 Smart Devices

The focus of early projects at PARC and by Olivetti, starting in the late 1980s was more towards basic smart device model design for tabs and pads. They could support

---

[31] Pure virtual reality games have been excluded as a sub-type of pervasive game, as the focus of UbiCom is more on mediated rather than virtual reality. Ubiquitous games is regarded as a synonym for pervasive games.

communication and location-awareness for mobile users. In the late 1980s, there were no commercial mobile ICT devices and widely available wireless networks. Apple Inc. produced and marketed the first hand-held computer or Personal Digital Assistant (PDA) in 1993 called the Newton MessagePad, often referred to as the Apple Newton. Although 200,000 Newtons were produced, the system had a number of cost and design problems and Apple discontinued the product in 1998. In the late 2000s, mobile devices such as phones (tabs) and laptops (pads) are widely used and wireless networks are widely available that support mobile data communication routing to users wherever they are. Current tab-sized mobile devices use touch screens and support gesture recognition, e.g., to allow users to move along a photo album and to zoom in and out of images. An even more common application of smart tabs than their use for mobile personal communication devices, is their use as smart cards[32] to support a variety of applications such as a driver's licence, insurance information, chip and pin credit or debit bank card and travel card and tickets (Section 4.4).

Electronic boards, which allow users to collaboratively edit text and graphics, were prototyped at PARC in the early 1990s. These have since become commercial products. Their use in the Classroom 2000 project in 1995–1998, was first reported by Abowd (1999), and they are now routinely used in many educational establishments to support distance learning. There are also some interesting prototypes of tab-sized smart mobile devices that allow them to act as boards, e.g., micro-projectors (Section 6.4.). Electronic boards are designed to be used as large vertical displays for information dissemination to larger groups, led by a single presenter. Much of the group is at a distance and has no easily accessible method to interact with the display. The DataTiles electronic table is similar to the electronic board in that it is designed for collaboration. However, electronic tables are horizontal, designed for use by smaller groups that sit around the table.[33] Any group member can interact with the table. There may not necessarily be a leader or arbitrator, so group members may need some kind of floor-control to prevent multiple conflicting actions being taken. This particular type of electronic table also uses physical objects as affordances to interact with the table. Weiser (Section 1.4.1.1) originally characterised devices by their form factor in terms of tabs (centimetre-inch-sized), pads (decimetre-foot-sized) and boards (metre/yard-sized). This characterisation of tab, pad and board-sized UbiCom devices needs to be expanded to include the use of micro and nano-sized (dust-sized) devices and non-planar surfaces and volumes.

Today, robots have a variety of uses. These uses are mainly industrially but there are increasingly entertainment and social applications too. Robots may be used to assist humans in the domestic environment by acting as intrusion detectors or digital pets.[34]

2.4.1.2 Smart Physical World Environments

Although physical environments are getting smarter, this is happening gradually. Smart home environments are gradually acquiring more wireless networked devices such as sound speaker

---

[32] Whereas there were about 70 million smart phones in use in 2006, there were about 10 billion smart cards in use.

[33] In 2007, Microsoft announced a Surface Computer initiative. Surface computers can turn an ordinary tabletop into a dynamic surface that supports interaction with different forms of digital content through natural gestures, touch and physical objects that can recognise physical objects from a paintbrush to a cell phone, and allows hands-on, direct control of content such as photos, music and maps. It is proposed that these are situated in hotels and homes where social interaction routinely takes place.

[34] The International Federation of Robotics IFR, statistics division, predicts that 3.6 million robots will be sold for domestic and leisure use from 2006–2010, see http://www.worldrobotics.org/index.php, retrieved July 2007.

systems, door bells, movement detectors for security systems, etc. A physical environment in which there are islands of interoperability rather than seamless interoperability will exist in the medium term (over the next five years) if not in the longer term (over the next ten years). Supporting seamless interoperability in the physical environment faces many technical challenges. Physical environments still tend to be quite passive and dumb overall. Cost, reliability, low maintenance and security will be important factors for embedding devices in physical environments in addition to the utility of the applications themselves. There are two main types of smart physical environments: embedded computer systems and amorphous computer systems. Advances in micro-electro mechanical systems (MEMS) is one important enabler for smart environments, resulting in much smaller, low-cost, low-resource and low-maintenance, amorphous computing devices becoming available (Section 6.4.4).

2.4.1.3 Context-Awareness and Service Discovery

Location determinism tends to be supported mainly as stand-alone devices and services that are not readily interoperable. Location determinism is also integrated into some mobile devices such as phones, cameras and cars. However, these are mainly for outdoor use. Systems for indoor use are available today, e.g., based upon trilateration using WLAN transceivers but there is no single standard in routine use to support indoor location awareness. Many services support some degree of user-awareness. Users can configure devices and services to support persistent personal preferences. Users can opt to let services keep their personal details, such as residential or business addresses and payment credentials so that they do not need to keep repeating them to the same service every session. Services can also be interoperable to share personal preferences and information.

Service discovery of local network resources is weak and the discovery of other local environment resources is still virtually non-existent, hence, much of the vision of Cooltown is not routinely available. This is in part because smart environments with a wider spectrum of available environments are not yet mature or widely available. There are a variety of reasons for this, such as the diversity of support needed, cost and the ability to secure fixings.

2.4.1.4 Wearable Smart Devices and Implants

Wearable smart devices are still in their infancy. Devices can be worn as watches, earpieces, glasses, belts and clothes. An early focus of Mann's work was to allow a Cyborglogger or glogger[35] to continuously record, process, computationally interpret, and share personal day-to-day life.[36] Earlier versions in the form of larger head-mounted audio-visual equipment were more obtrusive but later versions in the form of smart glasses makes them less obtrusive. Several prototypes of novel types of smart glasses are available. Glasses can consist of an extremely thin layer of liquid crystal sandwiched between two pieces of glass enabling smart lenses to be able to switch between near and far vision and to support auto-focusing enabling people who currently use bifocals and multifocals. Smart glasses can also be designed as dual LCD displays to display video rather than to see through.

---

[35] Mann's community of cyborgs has grown to more than 30,000 members, see http://glogger.mobi, accessed Feb. 2008.

[36] As observers of the current trend of reality TV shows know, a real challenge is the sheer amount of (manual) editing needed to filter out all the mundane stuff, leaving the noteworthy events.

A second main application of smart glasses is to act as a Heads Up Display or HUD that allows pilots and drivers to keep their attention on what is going on around them and not have to look down at their instruments for critical information too often. HUDS, in the form of helmets, contain a glass with a special coating that reflects the monochromatic light representing digital data and annotations while allowing all other wavelengths of light to pass through, creating a superimposed image. These first demonstrated by Sutherland in 1968 are currently used in some military aircraft. Fixed HUDs which require the user to look through a display element attached to the airframe or vehicle chassis are also available in some current road vehicles, e.g., they can display the speed on a part of the windscreen. Wearable computers are based upon smart head-sets, earpieces, gloves, etc.

Earpieces are also available that act as directional amplifiers, useful for aiding people with hearing impairments. Various watches have been marketed[37] to act as a personal communication and informational manager but they seemed to have a heavy build and have not yet caught on with users. Unlike a mobile phone, which can be operated hands-free, a smart watch is worn on one hand and can only be operated one-handed by the other hand. It may need to be moved closer to the eye to see detailed information. As such, they do not support the full set of requirements for wearable computers as proposed by Mann; they do not fit the interactional constancy criteria (see Section 2.2.4.5).

A wireless, e.g., Bluetooth, earpiece, paired with a mobile phone, can be seen as a type of wearable computer – this system fulfils Mann's three criteria (Section 2.2.4.5) that define wearable computers. Smart earpieces rather than smart glasses may succeed more, precisely because they augment rather than mediate reality. As sight is the dominant sense in that it is the one we use to aid moving about in the physical world and to capture information about the physical world, electronically mediating sight, may have significant side effects on our natural use of this facility. Humans may find it difficult and may tire quickly when they need constantly to switch their focus of attention from a virtual sight view to a normal physical world view.

Different tinted contact lenses can augment human vision even for those with normal eyesight. An amber tint can improve the visibility of moving targets such as baseballs and tennis balls. A grey-green tint can enhance the dips and curves in a distant putting green for golfers. An important application of wearable computers is in virtual reality electronic games in which players can use gloves, headsets, earpieces and body-suits to support multi-modal interaction in the environment.

Implants are routinely placed in some animals to support authentication and traceability. Implants are routinely used in humans for medical reasons such as: cardiac pacemakers to assist and regulate the electrical activity of the heart, hearing aids; micro-pumps for pain suppression, neurological dysfunctions and even for weight control. Key features of implantable medical electronics are their unique combination of extreme low power and high reliability (Gerrish et al., 2005) and bio-compatibility requirements. This requires substantial iterative design to lead to implants that meet these requirements. Hence it is not surprising that the nerve prototype implant investigated by Warwick eventually failed. A second use of implants in humans is to give ourselves abilities over and above those of other humans. As Warwick (2003) points out, this presents ethical problems with regard to how far the research should be taken and whether it is a good thing or bad thing to 'evolve' humans in a technical, rather than biological, way (the Posthuman model, Section 5.4.1). The trade-off between the benefits of using such implants in humans for technical reasons such as for implicit authentication and authorisation, versus the detriments such as the long-term increasing risk of malfunction and rejection by the body and the possible tracking and loss of privacy to the individual, need to be carefully evaluated.

---

[37] For example, 'Swatch the Beep' in the mid-1990s, the Timex DataLink, backed by Microsoft, Fossil Palm WristPDA, Casio Databank and the Microsoft SPOT wristwatch in 2003.

EXERCISES

1. In terms of the six forms for UbiCom: tabs, pads, boards, dust, skin and clay, how would you classify the vibrating string form Calm Computing artefact?
2. Define Mann's three criteria for wearable computers. Discuss whether or not devices such as laptops, mobile phones and wrist-watch information devices fulfil these criteria.
3. Discuss how to design a UbiCom system to support Abowd and Mynatt's (2000) design principles to support daily informal activities.
4. What is Ambient Intelligence? Give some examples. Does smart environment interaction really need intelligence?
5. Consider how the Cooltown model could be deployed beneficially at work or home.
6. Compare and contrast Abowd and Mynatt's (2000) design principles with Johanson *et al.*'s (2002) design principles for interactive work-spaces.
7. Compare and contrast Weiser's idea of Calm Computing with the Philips Homelab ideas of supporting the need to belong and share experiences, the need for thrills, excitement and relaxation and the need to balance and organise our lives.
8. Think up your own tangible UbiCom devices and discuss their benefits and design challenges.
9. Define your own triple-play, quad-play, pentad-play, etc. service bundles along with suitable UbiCom access devices and justify their utility. (Further exercises are available on the book's website.)

References

Aarts, E. (2003) 365 days' Ambient Intelligence research in HomeLab. Retrieved from http://www.research. philips.com/technologies/syst_softw/ami/vision.html; accessed December 2007.

Abowd, G.D. (1999) Classroom 2000: an experiment with the instrumentation of a living educational environment. *IBM Systems Journal*, 38(4): 508–530.

Addlesee, M.D., Jones, A., Livesey, F. and Samaria, F. (1997) The ORL Active Floor. *IEEE Personal Communications*, 4(5): 35–44.

Alves, J., Saur, I and Marques, M.J. (2004) Envisioning the house of the future: a multi-sectorial and interdisciplinary approach to innovation. Paper presented at E-Core Conference, Maastricht, Holland.

Arens, E., Federspiel, C.C., Wang, D. and Huizenga, C. (2005) Ambient Intelligence research in HomeLab: engineering the user experience. In W. Weber, J.M. Rabaey and E. Aarts (eds) Ambient Intelligence. Berlin: Springer Verlag, pp. 63–80.

Beigl, M., Gellersen, H.W. and Schmidt, A. (2001) Mediacups: experience with design and use of computer-augmented everyday objects. *Computer Networks*, 35(4): 401–409.

Berggren, M., Kugler, T., Remonen, T., Nilsson, D., Miaoxiang Chen and Norberg, P. (2001) Paper electronics and electronic paper. In *Proceedings of 1st International IEEE Conference on Polymers and Adhesives in Microelectronics and Photonics*, pp. 300–303.

Brumitt, B., Meyers, B., Krumm, J., Kern, A., and Shafer, S. (2000) EasyLiving: technologies for intelligent environments. In *Proceedings of 2nd International Symposium on Handheld and Ubiquitous Computing, HUC-2000*, Bristol. In *Lecture Notes in Computer Science*, 1927:97–119.

Bushnell, N. (1996) Relationships between fun and the computer business. *Communications of the ACM*, 39(8): 31–37.

Edwards, N., Barnes, N., Garner, P. *et al.* (2000) Life-style monitoring for supported independence. *BT Technology Journal*, 18(1): 64–65.

Ernst, H.A. (1961) MH-1: a computer-operated mechanical hand. PhD thesis, Massachusetts Institute of Technology. Available from http://hdl.handle.net/1721.1/15735; accessed December 2007.

Ernst, T., Uehara, K. and Mitsuya, K. (2003) Network mobility from the InternetCAR perspective. In *Proceedings of 17th International Conference on Advanced Information Networking and Applications (AINA'03)*, pp. 19–26.

EU (2005) Information Society Benchmarking Report. Available from http://europa.eu.int/ information_society/activities/atwork/hot_news/publications/index_en.htm, accessed 1 September 2006.

Forlizzi, J., DiSalvo, C., Zimmerman, J., Mutlu, B. and Hurst, A. (2005) The SenseChair: the lounge chair as an intelligent assistive device for elders. *Proceedings Conference on Designing for User Experience*, DUX05, No.31.

Frost, G.P. (2003) Sizing up smart dust. *Computing in Science and Engineering*, 5(6): 6–9.

Garlan, D., Siewiorek, D.P., Smailagic, A., *et al.* (2002) Project Aura: towards distraction-free pervasive computing. *IEEE Pervasive Computing*, 1(2): 22–31.

Gerrish, P., Herrmann, E., Tyler, L. and Walsh, K. (2005) Challenges and constraints in designing implantable medical ICs. *IEEE Transactions on Device and Materials Reliability*, 5(3): 435–444.

Gershenfeld, N. (1999) *When Things Start to Think*. London: Henry Holt & Co.

Greenfield, A. (2006) *Everyware: The Dawning Age of Ubiquitous Computing*. Harlow: Pearson Education.

Grossman, L. (2006) A game for all ages. *Time Magazine*, 167(20): 32–35.

Harrison, B.L. (2000) E-Books and the future of reading. *IEEE Computer Graphics and Applications*, 20(3): 32–39.

Hill, J., Szewczyk, R., Woo, A. *et al.* (2000) System architecture directions for networked sensors. In *Proceedings of 9th International Conference on Architectural Support for Programming Languages and Operating Systems*, Boston, MA, pp. 93–104.

IFR (2007) 2006 World robot market. IFR (International Federation of Robots) Statistical Department. Retrieved from http://www.worldrobotics.org/index.php, Nov. 2007.

Ishii, H. and Ullmer, B. (1997) Tangible Bits: towards seamless interfaces between people, bits and atoms. In *Proceedings of Conference on Human Factors in Computing Systems (CHI '97)*, Atlanta, USA, pp. 234–241.

ISTAG (IST Advisory Group) (2003) Advisory Group to the European Community's Information Society Technology Program. Ambient Intelligence: from vision to reality. Retrieved from http://www.cordis.lu/ist/ istag.htm. Article on May 2005.

Johanson, B., Fox, A. and Winograd, T. (2002) The Interactive Workspaces Project: experiences with ubiquitous computing rooms. *Pervasive Computing*, 1(2): 67–74.

Kampmann Walther, B. (2005) Atomic actions – molecular experience: theory of pervasive gaming. *ACM Computers in Entertainment*, 3(3): Article 4B.

Kidd, C.D., Orr, R. and Abowd, G.D. (1999) The aware home: a living laboratory for ubiquitous computing research. In *Proceedings of 2nd International Workshop on Cooperative Buildings, Integrating Information, Organisation, and Architecture. Lecture Notes in Computer Science*, 1670: 191–198.

Kindberg, T., Barton, J., Morgan, J., *et al.* (2000) People, places, things: Web presence for the real world. In *Proceedings of 3rd IEEE Workshop on Mobile Computing Systems and Applications*, pp. 19–28.

Kintsch, A. and dePaula, R. (2002) A framework for the adoption of assistive technology. In *SWAAAC 2002: Supporting Learning through Assistive Technology*, pp. 1–10.

Krumm, J., Cermak, G. and Horvitz, E. (2003) RightSPOT: a novel sense of location for a smart personal object. *Lecture Notes in Computer Science*, 2864: 36–43.

Magerkurth, C., Cheok, A.D., Mandryk, R.L. and Nilsen, T. (2005) Pervasive games: bringing computer entertainment back to the real world. *ACM Computers in Entertainment*, 3(3): Article 4A.

Mann, S. (1997) An historical account of the 'WearComp' and 'WearCam' inventions developed for applications in personal imaging. In *1st International Symposium on Wearable Computers*, pp. 66–73.

Matsumoto, T., Hashimoto, S. and Okude, N. (2008) The embodied Web: embodied Web-services interaction with an umbrella for augmented city experiences. *Computer Animation and Virtual Worlds*, 19(1): 49–66.

Maxwell, C. (2000) The future of work – understanding the role of technology. *BT Technology Journal*, 18(1): 55–56.

Murakami, T. (2004) Ubiquitous networking: business opportunities and strategic issues. Available via Nomura Research Institute home page, http://www.nri.co.jp/english/. Accessed 15 September 2006.

Nakajima, T. and Satoh, I. (2006) A software infrastructure for supporting spontaneous and personalized interaction in home computing environments. *Personal and Ubiquitous Computing*, 10(6): 379–391.

Park, S.H., Won, S.H., Lee, J.B. and Kim, S.W. (2003) Smart home – digitally engineered domestic life. *Personal and Ubiquitous Computing*, 7 (3-4): 189–196.

Rekimoto, J., Ullmer, B., and Oba, H. (2001) DataTiles: a modular platform for mixed physical and graphical interactions. In *Conference on Human Factors in Computing Systems, CHI2001*, Seattle, USA, pp. 269–276.

Roduner, C. (2006) The mobile phone as universal interaction device – are there limits? In *Proceedings of MobileHCI '06 Workshop on Mobile Interaction with the Real World*, Espoo, Finland, pp. 30–33.

Roduner, C., Langheinrich, M. and Floerkemeier, C. (2007) Operating appliances with mobile phones – strengths and limits of a universal interaction device. *Lecture Notes in Computer Science*, 4480: 198–215.

Rogers, Y., Price, S., Fitzpatrick, G. *et al.* (2004) Ambient wood: designing new forms of digital augmentation for learning outdoors. *Proceedings of Conference on Interaction Design and Children: Building a Community*, pp. 3–10.

Ruyter, B. de, Aarts, E., Markopoulos, P. and IJselstein, W. (2005) Ambient intelligence research in HomeLab. In W. Weber, J.M. Rabaey and E. Aarts (eds) *Ambient Intelligence*. Berlin: Springer Verlag, pp. 50–61.

Sellen, A., Eardley, R., Izadi, S. and Harper, R. (2006) The Whereabouts Clock: early testing of a situated awareness device. In *Conference on Human Factors and Computing Systems, CHI '06*. Montreal, Canada, pp. 1307–1312.

Shepherdson, J.W., Lee, H. and Mihailescu, P. (2003) mPower: a component-based development framework for multi-agent systems to support business processes. *BT Technology Journal*, 21(4): 92–103.

Spinellis, D.D. (2003) The information furnace: consolidated home control. *Personal and Ubiquitous Computing*, 7(1): 53–69.

Stanford, V., Garofolo, J. Galibert, O., Michel, M and Laprun, C. (2003) The NIST Smart Space and Meeting Room Projects: signals, acquisition, annotation, and metrics.

Want, R., Hopper, A., Falcão, V., and Gibbons, J. (1992) The Active Badge Location system. *ACM Transactions on Information Systems*, 10(1): 91–102.

Ward, A., Jones, A. and Hopper, A. (1997) A new location technique for the active office. *IEEE Personal Communications*, 4(5): 35–41.

Warneke, B., Last, M., Liebowitz, B. and Pister, K.S.J. (2001) Smart dust: communicating with a cubic-millimeter. *Computer*, 34(1): 44–51.

Warwick, K. (2003) A study in cyborgs. *Ingenia, Journal of the Royal Academy of Engineering*, 16: 15–22.

Watson, R.T., Pitt, L.F., Berthon, P. and Zinkhan, G.M. (2002) U-commerce expanding the universe of marketing. *Journal of the Academy of Marketing Science*, 30(4): 329–343.

Weber, W. (2003) Ambient intelligence – industrial research on a visionary concept. *Proceedings of 2003 International Symposium on Low Power Electronics and Design, ISLPED '03*, pp. 247–256.

Weiser, M. and Brown, J.S. (1997) The coming age of calm technology. In P. Denning and R. Metcalfe (eds) *Beyond Calculation: The Next Fifty Years of Computing*. Berlin: Springer-Verlag, pp. 75–93.

Weiser, M., Gold, R. and Brown, J.S. (1999) The origins of ubiquitous computing research at PARC in the late 1980s. *IBM Systems Journal*, 38(4): 693–696.

3

Smart Devices and Services

3.1 Introduction

Smart user devices are driven by the increasing capability and cost to benefit ratio to put powerful integrated resources with significant data processing and storage, network bandwidth access into a variety of static and mobile devices that have a variety of forms such as dust, tabs, cards, pads and boards (Section 1.4.1). Smart devices in this chapter focus on tab, card and pad-sized devices in which the locus of control for the user interaction resides in the device.

3.1.1 Chapter Overview

This chapter is structured as follows. First, the main characteristics of smart devices as a means to provide an embodiment for smart services and viewpoints, abstraction and virtualisation for these are considered. A range of architectural models for UbiCom systems are analysed (Section 3.2). Next the service provision and access life-cycle described (Section 3.3). Finally, operating system support for service execution is discussed (Section 3.4).

3.1.2 Smart Device and Service Characteristics

Smart devices are characterised by the ability to execute multiple, possibly concurrent, applications, supporting different degrees of mobility and customisation and by supporting intermittent remote service access and operating according to local resource constraints (see Table 3.1). Smart devices tend be owned, operated, configured and under the control of individual human users, e.g., personal computers (PC), phones, cameras, games consoles, set-top boxes and other computer peripheral devices such as printers, external disk drives, etc. Some of the more complex devices such as laptops can themselves be considered as aggregates of several smart devices.[1] As the ability to

---

[1] A laptop computer is a complex device consisting of multiple embedded computers, e.g., laptop battery with integrated charger (contains a microprocessor to self-test the charger, sense temperature, regulate fast recharging, etc.), USB flash memory drive or memory stick (contains a microprocessor to control the data transfer to and from the on-chip ROM and RAM types of memory), etc.

Ubiquitous Computing: Smart Devices, Environments and Interactions   Stefan Poslad
© 2009 John Wiley & Sons, Ltd

Table 3.1 Characteristics of service access used by smart devices

Feature	Requirements	Design
General Characteristics for smart devices, environments and interaction		
Multiple applications accessed via a range of devices	Devices execute multiple local processes; act as portals to access preconfigured sets of multiple remote services	Support different mixes of local versus remote processing (Section 3.2.1)
Simple service access	Services should be simple to initiate and to operate	Modularisation and abstraction of ICT resources (Section 3.2.1)
Minimum configuration and maintenance	Device access to local resources and remote services.	Self-discovery networks, services, etc. (Section 3.3) Use of appliance, utility self-management, etc models (Section 12.3.2)
Shared resources & services	Support access to resources by multiple users	Concurrency control (Section 12.3.3)
Internal Smart Device Properties		
Open, Interoperable, Heterogeneous	Users prefer a choice of service instances, access devices and networks	Use virtualisation and mediation combined with abstraction to support these (Section 3.2.1)
Dynamic service provision	Discover, select, compose & invoke services anywhere, anytime from anywhere relevant	Discover, compose, invoke & maintain services (Section 3.3)
Mobility	Range of mobile services and users depending on the application	Mobility support to: route to mobile receivers; to discover mobile access nodes (Section 4.2)
Volatile remote service access	Intermittent access is due to network failures, dynamic service access, concurrency, upgrading, etc.	Design to execute self–sufficiently and in an intermittent off-line mode (section 3.2.3)
Local resource constraints	Some types of mobile & embedded access devices have energy, data storage, display and input constraints	Adapt operation to limited ICT resource context (Section 4.3.3), limited local energy self-sufficiency (section 4.3.4)
External System Interaction		
ICT Context aware	Awareness of local resource availablity Volatile remote resource availability	OS versus application support
User context aware (Personalized)	Smart devices are often personalized but users may interact with many other impersonal services. User Context is dynamic – varies with activity, time, etc.	Devices act as a source for a personal space that can expand into other devices. Support for dynamic direct, indirect user profiling (Section 5.6)
Physical Context aware	Devices are active, situated in a relatively passive physical world environment	Limited awareness of the physical world (Section 7)

manufacture micro-sized devices, supported by powerful service infrastructures increases, other smart devices will emerge such as pens, badges, calculators, cards, glasses, paper and watches.

Thus, considering the personal memories application (Section 1.1.2.1), functions such as audio-video (AV) capture, replay, and context-aware annotation can be accessed via a range of devices such as digital cameras, mobile phones, games consoles, key rings, pens, etc, as the device electronics decreases in size. These devices should be simple to use and self-configurable to detect the network and relevant services. Displays and printers can be dynamically discovered in the locality by mobile users and users can select which ones to interact with based upon personal preferences. Different resources from different vendors can be mixed and matched.

It is worth briefly commenting on the difference between services and devices. Devices represent an execution environment for (service) processes, comprising a device-specific and limited set of local (to the device) ICT system resources and physical environment resources. In device models, it cannot be assumed that a ubiquitous virtual computing environment exists to enable services to be accessed and to be executed ubiquitously in any device because the access by devices to the virtual computing environment may be volatile and non-interoperable. In Chapter 1, devices are referred to as embodied virtual computing systems.

3.1.3 Distributed System Viewpoints

Smart devices embody user access to distributed system components such as information and task-based services, e.g., resource management and control, within a user-centred access device to a distributed ICT service. Distributed ICT systems can be modelled from multiple complementary viewpoints[2] with respect to different stakeholders in the system such as network infrastructure providers, computer device and service infrastructure providers, individual users and enterprise users (see Figure 3.1). These viewpoints can be regarded as architectural patterns, conceptual models that capture the essential elements of an ICT system architecture and its interrelationships.

From the (individual) user view, the system provides information or tasks. From the enterprise user viewpoint, user access is controlled via organisational roles and policies. From the information system, service or computation platform viewpoint, the system is split into three[3] interdependent components that can be distributed and interlinked using a fourth communication component. These three components are: one or more managed information resources such as stored data and information resources, such as sensors and databases; processes to manipulate information; user interfaces in order to interact with a system's resources and processes. The information viewpoint takes an abstract view of communication, viewing the network and resources as passive components and the users and processing as active components. The network providers' view of ICT systems defines two kinds of nodes, computer or service nodes and network or communication nodes, special purpose computing nodes that enable computer nodes to share the use of physical network links (see Chapter 11). The focus in this chapter is the service viewpoint. The user,

---

[2] The Open Distributed Processing Reference Model (RM-ODP) Architecture (ISO/IEC 10746−3, 1996) defines viewpoints for enterprise users, computation, information (distributed, service), Engineering and Technology. The IEEE 1471 architectural model (ISO/IEC 42010, 2000) is a more generic architectural model for software-intensive systems that is less specific about defining viewpoints than the earlier RM-ODP model.

[3] Other models of ICT systems have different numbers of components. The term *ICT* (Information Communication Technology), currently is in widespread use, Garlan and Shaw (1993) define two components: Information or Computation and Communication. Perry and Wolfe (1992) and Fielding (2000) define three general system components: processing, connectors and data components. Alonso *et al.* (2004) define components for data-intensive systems to be resource management (e.g., data storage), business logic (data processing) and presentation logic (data access).

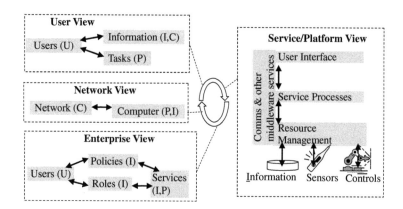

Figure 3.1   Different viewpoints of distributed ICT system components. Components are User Access (U),[1] Service Processing (P),[2] Communication (C[3]) and Resources (R)[4] such as Information (I), sensors, controllers and other device hardware

Notes: [1] The User Interface (UI) is also called the *presentation layer* in business information models.

[2] Processing or Computation is referred to as Application Logic or *Business Logic* layer in business information models.

[3] Note C refers to Communication in the acronym ICT and to Computer in the acronym HCI, CPI etc.

[4] Resources, data sources, data storage, sensors, controllers, etc are encapsulated in the *Resource Management* layer in business information models. Resources are bound to hardware, are passive and activated by users and services.

enterprise viewpoints and network viewpoints are hidden via the service viewpoint and are dealt with in later chapters.

In a monolithic service model, services, user access and resources are all on one device or platform, whereas in a distributed service model, these can be distributed. The distributed service model has several benefits. It supports applications and business that are inherently distributed, e.g., a company with multiple offices, factories and employees at different locations that can be supported by messaging services such as email and via information and knowledge portals. Distributed access increases the use of services that would only be accessed locally. Expensive resources can be shared. Better performance can be achieved through separating the computations so that they can be executed on multiple processes in parallel. Reliability can be increased by replicating tasks and data over several computer nodes so that if one set fails, access to another copy can be configured. Some services are best distributed, even when they needn't be, into collections of co-operating specialised services, in order to give flexibility for maintenance so that parts can be modified without suspending the operation of the whole system.

3.1.4 Abstraction Versus Virtualisation

System architectures focus on the idea of reducing complexity both through a separation of concerns using modularisation and through using abstraction. Meyer (1998) uses five criteria to characterise system modules or components: (1) decomposability to divide a system into modules; (2) composability to assemble particular existing modules to form a new system; (3) understandability of each stand-alone module to make it easier to change and to assemble a system out of them for users; (4) continuity so that changes to one module will reduce side-effects on other modules; and (5) protection so that errors within a module will seek to limit the effects within that module.

To support the properties of distributed systems, modules should support high cohesion performing well-defined functions, loose coupling and few side-effects. Pressman (1997) describes a range of module cohesion from coincidental (low) to functional (high) and a range of module coupling from no direct coupling to content coupling.

The second key concept to reduce complexity is to use *abstractions* which define those things that are important in a system and to hide or make transparent those properties that are not. Important types of transparency for distributed services include the following. *Access transparency* specifies resources that can be accessed from anywhere. Users just define what resources they require, not where the resources are. *Concurrency transparency* facilitates multiple users or (application) programs operating on shared data without interference between users. *Failure transparency (Fault Tolerance)* enables systems to mask partial failures of a system and availability increases. *Replication transparency* enables users or programs to be unaware that a system uses multiple instances of resources to increase reliability or performance. *Migration transparency* permits resources to move during use. *Scaling transparency* facilitates dynamic resource supply so that it can expand (or contract) to meet demand. In practice, the ideal transparency of a single image for all resources, all the time, under all conditions is hard to achieve and is usually only when the distributed system is operating normally.

Using standard homogeneous service interfaces eases interoperability. Using standard homogeneous service implementations eases system management. In practice, heterogeneous service implementations exist for the same service interfaces, produced by different service vendors. The advantage for consumers is that they have more choice and they perceive a bigger critical mass of products when these are compatible. Either the service interfaces or implementations or both can be standardised.

The availability of standard interface specifications does not necessarily guarantee interoperability. Specifications can be interpreted differently by different vendors, for example, parameters may have different default values or may be used as option versus being mandatory. Vendors may find it beneficial to extend standards but want to do it quicker, going alone rather than going through an often protracted standardisation process, hence compliance with standard service specifications can vary. Providers often promote the benefits of upgrades to the latest version of products because they offer improved functionality. The latest version may be less stable and incurs an additional outlay and maintenance cost, hence users may often delay upgrades. Variations and specialist versions of service may be needed, e.g., small portable displays are used for personalised mobile use versus large displays used more for viewing at a distance by groups of users.

Abstraction alone often focuses on simplifying access to resources, e.g., the operating system and application services abstract the details of hard disk addressing such as moving a disk-head over tracks and sectors of multiple rotating hard disks, into random-access files and data records (Figure 3.2). Abstractions are defined using a well-defined *interface*,[4] e.g., a file system and database system that hides the details of the lower-level disk access. Abstractions alone do not necessarily support interoperability between different instances types of related interfaces, e.g., resources designed to support one interface need to be redesigned to support another related one, leading to a lack of interoperability.

Virtualisation provides a way to solve this limitation of abstraction: it supports the ability to map components in one interface at a given level of abstraction into different interfaces and different resources at different levels of abstraction (Smith and Nair, 2005).[5] Virtualisation is particularly

---

[4] An abstraction that is a simplification of functionality an entity provides of itself to the outside is called an *interface*.

[5] *Virtualisation technologies* hide the physical characteristics of computing resources from users, e.g., to allow applications or the OS of one host system to run under another one. There are three requirements (Popek and Goldberg, 1974): (1) equivalence so that an application running under the virtual host is compared to the native host; (2) resource control so that the virtual host is in complete control of the virtualised resources; and (3) efficiency: a statistically dominant fraction of machine instructions must be executed without virtual host intervention.

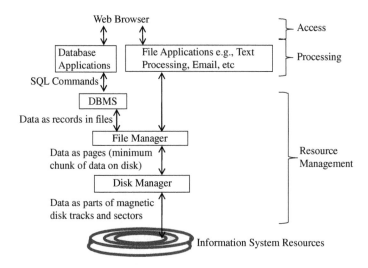

Figure 3.2 Abstract view of user access to database and file applications which in turn see an abstract view of resource managers for the file system which in turn see an abstract view of the disk data storage system

useful for ubiquitous systems as it enables different components to be used in multiple services and at multiple levels of abstraction. A second important difference between abstraction and virtualisation is that virtualisation does not necessarily aim to hide and simplify all the details of accessing services, although it can. The combination of virtualisation and public specifications of interfaces characterises an open system eases the addition, modification and removal of system components. Important uses of virtualisation include operating systems and virtual machines (Section 3.4).

If resources and services in a distributed system are replicated in some way, then services can be designed so that if access to one or more specific components fails, alternate instances of these can be accessed instead. Any component for which at least one alternative is offered is referred to as a redundant component. Components may be redundant because independent services providers are adhering to the same interface specifications and offering similar or the same services. Components may be redundant by design. This is particularly necessary for critical components such as networks paths and directory services. In addition, distributed systems could be designed so that after a failed component is detected, the system could either attempt to restart it, returning it to normal operation or it could automatically switch to make use of an alternate component.

3.2 Service Architecture Models

3.2.1 Partitioning and Distribution of Service Components

There is a range of designs for dividing and distributing services that depend on: (1) the application; (2) the type of communication service; and (3) the type of access device used. Figure 3.3 illustrates two different ways in which a chess application can be designed to run on smart devices. In a high resource device, the application can execute locally without using the network. In low resource devices, a chess application executes remotely because there is not enough CPU power to execute the application locally.

Relatively high resource access devices can act self-sufficiently, operating in an offline mode as monolithic or stand-alone devices – the appliance model. Relatively resource-poor access devices,

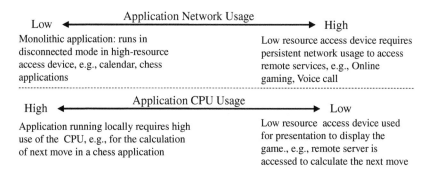

Figure 3.3 Balancing the use of local processing against the amount of communication needed depends upon the application and how it is designed

such as lightweight mobile devices, are often designed so that service execution largely occurs over the network, the utility model. However, the need to cope with unreliable and low-performance networks, as well as the need to adapt power consumption to the power reserves available in mobile applications supports the case for some degree of self-reliance and local processing support – elements of an appliance model. There is a need to balance the degree of local information storage and local processing against the degree of remote processing and communication bandwidth required. The balance may need to shift dynamically depending on the type of ICT infrastructure available and on the type of applications. In the middleware service model, generic services such as communication and data management are factored out of specific application servers and hosted in computer nodes so that can be shared across multiple application services (Section 3.2.3). Service Oriented Computing (SOC), e.g., instantiated as XML Web Service (WS) based SOC, defines an explicit notion of service (Section 3.2.4). An important variant of the client–server model is the mobile client model in which the network addresses of client devices can move during a service invocation session and service access is optimised for low resource mobile devices. The peer-to-peer[6] service model supports more flexible service availability over more dynamic service infrastructures such as service access over mobile ad hoc networks or MANETs (Section 11.6.5.3).

3.2.2 Multi-tier Client Service Models

In Figure 3.4, five different designs for information resource UbiCom systems are given based upon how their A, P and I components are distributed. These functions can be distributed over multiple different computer nodes or tiers. From top to bottom and left to right the designs given in Figure 3.4 are as follows. In a single-tier, monolithic system, the whole application service resides locally, when it is operating. The system may be networked so that under special conditions it can go online to seek help when its operation is interrupted or because of local failures. In a two-tier, thin-client server, the access device (or client device or terminal) supports data access or presentation, service processes execute remotely and the information associated with these services is stored remotely. In a two-tier, fat-client server model, the access device can support some local processing and some local use of services but can also invoke remote services.

---

[6] SOC has also been referred to as a P2P computing model, however, most SOC designs seem to be more accurately characterised by a client–server model.

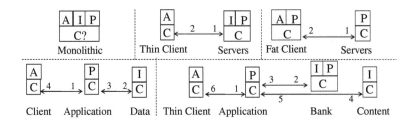

Figure 3.4 Different designs for partitioning and distributing Information (I), Processing (P) and Service Access (A) using communication (C).
Note: The numbers on the arrow indicate the ordering of the interaction.

In multi-tier (3, 4 ... N-tier) systems, rather than access devices being directly connected to the end service nodes, different numbers of intermediate nodes can be used. Examples of the use of single intermediate nodes, leading to three-tier systems, are designs that decouple services access from service provision via the use of discovery services (Section 3.2.2) and services that act on behalf of the access node to simplify operations (Section 3.2.2.4). Often the application processing and application data are put on separate nodes leading to a four-tier system. Multiple application and generic application support services called middleware services (Section 3.2.3) can be defined. A typical ecommerce application consists of four tiers: a user makes a request for a service to an application to download some content, the application checks with an authentication and banking service if this is OK; if this is OK the content is then retrieved from the content server and delivered to the access device. Application services can also be designed to be distributed over five or more tiers depending on the application, e.g., the banking and authentication services could be separated, leading to a five-tier system.

3.2.2.1 Distributed Data Storage

Some of the components such as (information) resources, processing and access can be further split to support different types of application. Types of systems in which information resources are divided and distributed (see Figure 3.5) include: transaction monitors where data transactions are created by distributed data sources such as point-of-sale terminals where data warehouses; centralised analysis of centrally stored data sub-sets that are periodically extracted from distributed data resources is supported; distributed databases where queries are distributed to multiple heterogeneous databases, each individual database is mapped into a common form using a database wrapper.

3.2.2.2 Distributed Processing

Although a single CPU client–server type architecture is the dominant processing model used in distributed systems, sometimes more processing power is needed for a short time. One way to achieve this is by dividing the processing, distributing it among multiple remote processors, each executing part of the processing in parallel and then reassembling the results from the individual pieces to form the whole. For this to be worthwhile, the time gained in increasing the processing must be significantly more than the time taken to partition and distribute the tasks, collect the individual results and reassemble them.

There are several different computing architectures to achieve parallel processing. Previously, support for parallel processing was found only in supercomputers which have hundreds to tens of thousands of independent CPUs and memory such as massively parallel computers and specialised

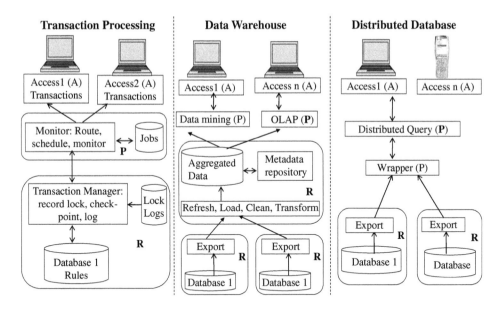

Figure 3.5 Information Resources (R) can be divided within an Information System

computer clusters. These were used to undertake high performance processing applications such as weather prediction (Hord, 1999). Today, some limited yet substantial parallel processing can be achieved using general purpose computers by networking general purpose computers into clusters, such as computer grids. Multiple processing capabilities can be built into general purpose computers, e.g., a multi-core processor that contains multiple CPUs within a single IC chip (Kumar *et al.*, 2003). It has the advantage of reduced latency and lower power consumption compared to the networked single-cored CPU model. Other models to support massively parallel computation included the use of P2P computing (Section 3.2.6) and cellular computing (Section 10.6.1).

3.2.2.3 Client–Server Design

The client–server model has the advantage of more centralised control of distribution but has the disadvantage that the distribution and configuration of servers are fixed. The client–server model is an asymmetric distributed computing model with respect to the resources and the direction of the interaction. Servers are usually resource-rich, e.g., have a higher storage capacity, more powerful CPUs to support intensive processing tasks, a high bandwidth, always on-network connection in order to service multiple service requests, and act as a shared data repository. In contrast to servers, clients are relatively resource poor. Client–server interaction is also asymmetric: client processes on access devices initiate the interaction, making requests to application service processes on servers that wait for client requests. This asymmetry simplifies the synchronisation between clients which start requesting while servers which start waiting for client requests.

The system configuration in terms of partitioning and distribution of service components depends upon: (1) the available ICT infrastructure such as the performance and reliability of network links; (2) the resource constraints of the local device; (3) the remote services that are available; and (4) the type of application and the service maintenance model. The relative resource poverty of some types of computer nodes such as mobile devices argues for reliance on external servers, providing that the communications network supports remote service access on demand – a thin-client server model. An example of a thin-client server model is a mobile terminal that just supports service access using a mini Web browser. All the processing on behalf of clients, see Section 3.2.2.4, is performed in the server requiring substantial server resources to handle many (possibly heterogeneous) clients. A thin-client server model is often considered to be easier to maintain as maintenance can be performed at a centralised server location remotely rather than having to be performed in each distributed local access (client) node. However, Web browser-based thin-clients support a very limited application platform in terms of supporting protocols other than HTTP, support for rich interactive UIs, or support for sophisticated application logic.

The need to cope with unreliable and low performance networks argues for some degree of self-reliance and some use of local processing and data resources, i.e., the use of a fat-client server model. The design of the interaction to tolerate intermittent network access is discussed further in Section 3.3.3.9. The fat-client model is suitable when the access device has a higher system specification than thin-client devices and when a limited network connection is available. It has the advantage of offloading some of the processing and storage from the server. The type of processing needed in the access device depends on the type of application. Location-aware services may determine the location of the access device and transform its location coordinates into a form for use to interoperate with other applications such as map applications, thus avoiding transmitting the data remotely to be transformed. Access devices such as point-of-sale terminals can be used to collect data locally, to perform some local processing and then to upload it later.

3.2.2.4 Proxy-based Service Access

Some applications, in order to mask the complexity of communication from being supported in client access process, use a client proxy. Proxy-based service access can:

- *offload presentation processing and network processing* from the low resource client access device to the third proxy node, e.g., the client proxy could speak XML to a Web server but use a simpler message protocol when communicating with the mobile device, thus avoiding having to parse complex XML data structures in the access terminal;
- *hide the heterogeneity* of different terminal types and different types of networks from the access applications;
- *simplify and compose access to multiple service providers*. A proxy may request a default connection, reducing the processing and communication with wireless end devices. It can support a mobile portal model by aggregating content from multiple service providers.
- *reduce the complexity of communication used in access devices*, e.g., often complex hierarchical data structures need to be encoded and decoded into more efficient serialised data structure for transmission (see Figure 3.6), e.g. by compressing data or transcoding data at the server side thus reducing the amount of air wireless bandwidth consumed (Chapter 11);
- *enable devices to operate intermittently in a disconnected state*. Devices may power off, move out of range of the wireless network, or simply choose to operate in a disconnected mode (Section 3.3.3.9);
- *shield network-based applications from the mobility of the access devices* (Section 4.2).

Mobile applications on resource-constrained access devices tend to use at least a three-tier design model. A thin client is supported on the access device node, the application services are on another node and some mediator or proxy-client resides on another node.

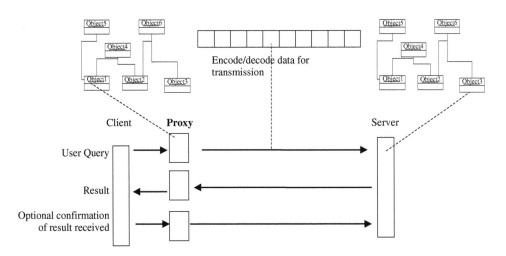

Figure 3.6 Use of proxies to simplify network access by transparently encoding and decoding the transmitted data on behalf of clients and servers

The use of proxies in this way overcomes problems associated with limited network coverage and can support reliable communication. The proxy client also has several disadvantages that must be considered in the design of the proxy client. Disadvantages include: being a single point of failure, the use of additional network hop can increase the latency. In some cases, proxy clients can be positioned far from the application Web servers and access devices, further increasing the latency. Some of these issues can be addressed by having replicated distributed proxy servers, positioned at optimum points in the network.

3.2.3 Middleware

The variety and heterogeneity of services access add more complexity to the design of applications to access this increased range of services. Hence, middleware was introduced in between applications and the operating system, to enable applications to hide and simplify access to the heterogeneous and distributed resources of multiple networked computing systems (Bernstein, 1996).

Middleware essentially factors out a set of generic services, e.g., database access, file system access, messaging, time service, directory service, etc., out of the application services and out of the operating so that they can be application- and operating system-independent. This in theory makes the operating system (OS) much more compact and more flexible. The OS does not need to be rebuilt or rebooted every time new types of hardware are added. The middleware itself is distributed but this is transparent to the application. The middleware model also makes applications simpler to define because the middleware can handle the complexity of dealing with the detailed use of system services.

Sometimes, however, it may be useful for some applications to have some awareness of lower level interactions and not for access to resources to be completely hidden by the middleware (Figure 3.7). For example, application awareness can better handle message latency, service activation and replacement of components that require synchronisation to better cope with deadlocks (when two or more processes are each waiting for another to release a resource) and livelocks (processes constantly change with regard to one another, none progressing).

Alonso *et al.* (2004) view systems of (heterogeneous) models such as EAI (Enterprise Application Integration) and SOA (Service Oriented Architecture) to model business supply-chains, as distinct

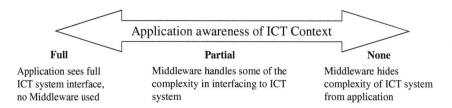

Figure 3.7 The trade-off in using middleware to hide the complexity of the ICT system access from applications and types of middleware service

from middleware models. This is because each type of system has its own type of middleware. In contrast, systems of systems require heterogeneous middleware to interoperate. This requires significant extensions to the middleware model because this type of systems of systems interaction may require harmony between different policies such as security policies, requirements, data formats. This requires more sophisticated ways to model services and service interaction (Section 3.2.4).

3.2.4 Service Oriented Computing (SOC)

Service Oriented Computing (SOC) or Service Oriented Architecture (SOA) focuses on services such as computational or information processing components that are autonomous and heterogeneous, running on different platforms and possibly owned by different organisations. XML-based Web Services and Computer Grids are common examples of a SOA. The OASIS standards forum focuses its SOA model[7] on the concept of ownership: a SOA is 'a paradigm for organizing and utilizing distributed capabilities that may be under the control of different ownership domains'. Its reference model defines the core concepts of Visibility, Service Description, Execution Context, Real World Effect, Interaction, Contract and Policy. Each of these is defined in terms of a more detailed conceptualisation. Similar initiatives to specify standard SOA models are also being undertaken by other bodies such as the Open Group SOA Working Group,[8] which defines an SOA as a style of IT architecture that delivers enterprise agility and 'Boundaryless Information Flow', and the OMG.[9] The notion of a service can be characterised in terms of:

- *Descriptions* (or specification) of some task (a set of one or more actions) that are offered by providers to users. It is assumed that descriptions are discoverable. A complication is that the provider and user may not share a common understanding of the specification or how to specify a service.
- *Outcomes*: the service is the means to achieve a defined outcome for a task, e.g., a repair service enables normal operation to be resumed within a certain time frame.
- *Offers*: (or tenders) to perform a task on behalf of another. It is assumed that if an offer is made that the provider is available.
- *Competency*: to undertake the task, e.g., a provider may publicise qualifications that have been validated by an independent regulatory authority.

---

[7] OASIS SOA RM (2006) OASIS Reference Model for Service Oriented Architecture V 1.0. Available from http://www.oasis-open.org/committees/. Accessed August 2007.
[8] Open Group (2007) The Open Group SOA Home page. http://www.opengroup.org/projects/soa/ Accessed August 2007.
[9] OMG SOA (2007) The OMG SOA Special Interest Group Home Page. Available from http://soa.omg.org/. Accessed August 2007.

- *Execution*: actually performing the service on behalf of someone.
- *Composition*: Multiple services may need to be composed before they can be executed with respect to an outcome and time constraints.
- *Constraints or policies*: for a service, which may be specified either by the user, e.g., for a taxi service 'don't drive too fast', or by the provider 'not exceeding the speed limit'.

Service design may not explicitly define all of the features. In the simple informal case, only the service description may be defined according to the viewpoint of the provider. All the other features are implicitly assumed under some informal agreement. Often, for business services, each of these features needs to be explicitly defined. Some service design models may make a distinction between information processing tasks and information retrieval tasks. An SOA is an architectural paradigm that promotes development of possibly ad hoc applications out of a set of loosely coupled, self-dependent and mutually collaborating services. Services in a SOA can be separated into three layers of functions: basic (lower), composition (middle) and management (higher layer) (Papazoglou *et al.*, 2007). These functions are described as follows:

- *Service discovery* (Basic function): Service descriptions are exposed, published and represented using metadata that can be advertised and discovered via third-party mediating services such as service directories (Section 3.3.2);
- *Enterprise service bus* (Basic function): this supports service, message, and event-based interactions with appropriate service levels and manageability;
- *Service invocation* (Basic function): Services are invoked via public interfaces over an open service infrastructure. Services are accessible via a public network and defined using standards-based representations (Section 3.3.3);
- *Service composition*: services can be combined from simpler components into more complex, composite, executable service processes at run-time (Section 3.3.3.9);
- *Service management*: Services are managed by third-parties, between the user and provider, based upon policies, by exchanging schema-based contracts and Service Level Agreements (SLA), e.g., Dan *et al.* (2004).

The SOC is a design model independent of any specific technology, e.g., Web services or event-driven. This independence can be achieved by limiting the number of implementation restrictions at a level of abstraction in the service interface. SOC only requires that functions, or services, are explicitly defined by a service description language, e.g., using WSDL, and have interfaces to use these descriptions to perform useful business processes (Papazoglou *et al.*, 2007). Not all the service characteristics and not all SOC characteristics given above may be supported in each SOC implementation, e.g., consumer device-based SOCs (Section 4.5) tend not to define service offers, competency and outcome and service composition and service management are often quite simple and avoid policies and contracts.

3.2.5 Grid Computing

Grid computing refers to distributed systems that enable the large-scale coordinated use and sharing of geographically distributed resources,[10] based on persistent, standards-based service infrastructures, often with a high-performance orientation (Foster *et al.*, 2001). Grid computing

---

[10] Requests to use and share resources such as computer resourcing and computer storage are referred to as *Jobs*.

specifies standards for a high performance computing infrastructure rather than support for fault-tolerance and support for highly dynamic ad hoc interaction, which is more the focus of P2P systems (Foster and Iamnitchi, 2003). Three main types of grid system occur in practice: (1) computational grids that have higher aggregate computational capacity available for single applications than the capacity of any constituent machine in the system; (2) data grids that provide an infrastructure for synthesising new information from data repositories such as digital libraries or data warehouses that are distributed in a wide area network; and (3) service grids that provide services that are not provided by any single machine. This category is further subdivided into on-demand, collaborative, and multimedia grid systems (Krauter et al., 2002). Each of these types of grids currently uses a Web-based SOA model. Much effort has gone into developing open models and specifications for grid computing.[11]

Typically, grids focus on providing a single virtual computer view of distributed systems made up of heavyweight servers and fat-client computers that communicate on high-bandwidth highly available fixed networks rather than on lightweight, thin-client devices that can be connected over more volatile, low-bandwidth networks including wireless networks. The resource model used in the grid focuses on shared use of data processing and data storage resources. In contrast, the resource models used in UbiCom focus on a wider variety of ICT and non-ICT resources including energy and environment control.

3.2.6 Peer-to-Peer Systems

Peer-to-peer systems or P2P are service infrastructures.[12] They can be defined as distributed systems consisting of interconnected nodes able to self-organise into network topologies with the purpose of sharing resources such as content, CPU cycles, storage and bandwidth, capable of adapting to failures and accommodating transient populations of nodes while maintaining acceptable connectivity and performance without requiring the intermediation or support of a global centralised servers or authorities (Androutsellis-Theotokis and Spinellis, 2004).

Rather than concentrating sophisticated server processing and resource management in a relatively low number of specialist nodes or servers, for some high-resourced client, service access devices can also themselves act as on-demand servers. Computer nodes in a P2P system can act as both clients and servers. Client and server are considered more as dynamic organisational roles for peers that can be changed in an ad hoc way. P2P interaction often uses an ad hoc application router or service overlay network (Section 11.7.8.4). Ad hoc is generally applied to physical networks in which communication takes place without any pre-existing infrastructure set up between the communicating computers, e.g., Mobile Ad hoc Networks or MANETs. P2P service infrastructures can overlay ad hoc networks.

A P2P service infrastructure seems a very suitable system design to support many of the smart device characteristics given in Section 3.1.2. P2P applications include: content sharing in which

---

[11] See the Open Grid Forum (OGF) at http://www.ogf.org/, accessed Jan. 2008. Note, however, others do not refer to this specific OGF model of the grid but refer to it in a much looser sense as some interconnected network (grid) of distributed resources such as sensor grids, e.g., Hingne et al. (2003).

[12] From the onset, the original Internet was designed to operate as a P2P network (Chapter 11). Many applications focus P2P models as a network model rather than as a service infrastructure model because computer nodes act as ad hoc communication middleware and message routers in order to distribute information in a distributed manner. However, this message routing still occurs at the application level, as an 'overlay service oriented network'. Nodes really behave more as application-level message routers, service proxies and service gateways rather than as network-level routers, hence P2P is referred to as a service rather than a network infrastructure.

anyone can share and publish; spontaneous user collaboration in real time such as VoIP; ad hoc wireless device interaction such as home devices discovering each other and sharing traffic reports in an ad hoc network of cars as network nodes; distributed computation such as sharing processing power to solve complex problems, e.g., the SETI@home[13] project (Anderson *et al.*, 2002), distributed database systems, sensor net applications (Section 6.3) and various middleware services such as privacy protection via anonymous distribution of information.

A P2P computing model offers several important benefits:

- *lower cost of ownership for content sharing*: by eliminating specialised server costs and by distributing the maintenance costs through multiple low cost peers that can interact;
- *performance enhancements*: the resources of all the nodes can be used for storage, computation and data exchange rather than focusing resources mostly in the server type nodes;
- *ad hoc resource utilisation and sharing*: as demand for particular services peaks, more nodes can act as servers, for example, as file servers, to meet demand;
- *autonomous control and ownership*: peers can have a greater degree of and exercise more decentralised, autonomous control over their data and resources. No reliance on a central server to collect and relay information.
- *anonymity and privacy*: session-based IDs and addresses can be assigned and can be hidden and masked as data gets routed; one cannot tell who has created data, who is querying data, who is storing data etc.;
- *fault-tolerance*: there are no central servers that can be attacked or can cause complete system failure, instead, alternative paths and servers can be used when one fails (Section 12.2.6).

The challenges in designing and using P2P systems are:

- *more complex coordination is often needed.* In contrast, with client–server interaction, clients initiate content downloads and uploads while servers react to clients to satisfy these requests or notify clients when service updates are available. In contrast, it is not so clear how to initiate sessions, e.g. two interacting peers can both decide to wait or to send to each other.
- *Nodes can act as freeloaders*: nodes may be happy to play a role of service requesters but are always configured to refuse the requests of others to use their resources.
- *More complex security* may be needed as identification can be masked so access control is harder. *Network address discovery*: peers need to be dynamically assigned network addresses and to discover addresses of destination nodes. The use of broadcasting to discover the network addresses of others can flood networks.
- *Ad hoc network routes*: need to create and discover ad hoc routes between nodes, inefficient multi-hop routes can form to link source nodes to destination nodes.
- *Service discovery*: how to discover the selective nodes where services can be invoked from versus the inefficiency of distributing services to all nodes.

There are three main variations of P2P system depending on the types of computer nodes: pure, decentralised, P2P; partially decentralised (SuperNode) P2P; and hybrid decentralised P2P (see Figure 3.8). These can be grouped into two main types of topologies for P2P systems that overlay

---

[13] In 1995, David Gedye proposed the *SETI@home*, Search for Extraterrestrial Intelligence, project to use a virtual supercomputer composed of large numbers of Internet-connected computers to detect intelligent life outside Earth. To users, this application acts as a screen saver that periodically downloads and processes chunks of astronomical data and notifies someone of any detected patterns, see http://setiathome.berkeley.edu, accessed Dec. 2007.

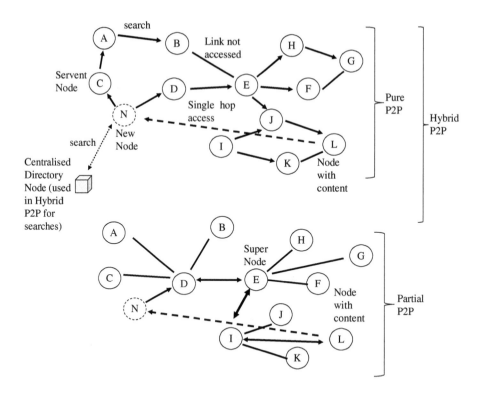

Figure 3.8   Three types of P2P system, pure, hybrid and partial decentralised

the underlying physical network. Unstructured overlay networks, e.g., ad hoc networks are independent of any physical network topology and use decentralised and partially decentralised nodes. Structured overlay networks are dependent on the physical network topology of nodes and use hybrid decentralised nodes (Chapter 11). Rather than connect each node to any other node to form a mesh, nodes are connected only to their nearest neighbours.

A pure P2P service infrastructure uses no notion of fixed clients or servers, only of equal peer nodes that simultaneously function as both dynamic servers and clients (called servents) for other nodes, depending on supply and demand. Service infrastructures such as Gnutella or Freenet (Androutsellis-Theotokis and Spinellis, 2004) use a (pure) P2P organisation for all purposes, and are sometimes referred to as true peer-to-peer networks.

In partially decentralised P2P systems, all nodes are not equal, a few superpeers or supernodes are elected to operate as middleware servers, acting as network relays for other nodes in a VoIP application or to cache indices to locate the distributed content for a cluster of nodes, e.g., KaZaa and Skype VoIP (Montresor, 2004). Peers are automatically elected to become supernodes if they have sufficient bandwidth and processing power. They are also fault-tolerant in that new super-nodes can be elected to replace old ones that fail.

In hybrid P2P networks, e.g., Napster, a client–server organisation is used for specific tasks and interactions, such as searching for services and a P2P organisation is used for others such as service invocation. There are three basic content access processes in distributed systems: (1) to identify nodes; (2) to register nodes that provide content; and (3) to search for content and to retrieve it. How these work depends on the types of nodes and on the topologies for structuring the nodes (Milojicic *et al.*, 2002). Hybrid P2P systems tend to use client–server interaction to register and search for content in a centralised directory and use P2P interaction to retrieve content.

P2P systems tend to use a decentralised approach to search for unknown receiver nodes. They use message broadcasts or message floods to ensure all parts of the system are searched. There are several challenges with message broadcasts: it scales poorly with increasing network size. It can utilise resources and computation power on all nodes, even those that may not answer the request. It consumes significant network bandwidth as it may traverse all nodes through all paths possible. Message loops and partitioned networks can either make broadcasts last overly long or terminate prematurely. Overlay networks of nodes can be used to avoid loops and to overcome partitions in the underlying physical networks and the time to live (TTL) for messages can be limited to make flooding more manageable. This still leaves the challenge of efficiently identifying and annotating nodes that contain certain services and avoiding requests having to be routed through non-relevant nodes and routes.

An important solution to this challenge used in pure and partial P2P is the Distributed Hash Table (DHT). Here, each node is assigned a random ID and each peer also knows about a given number of neighbouring peers. When a document is published (shared), an ID is assigned to the document based on a hash of the document's metadata, which summarises its content, and its name. Each node will then route the document towards another node with the ID that is most similar to the document ID. This process is repeated until the nearest peer ID is the current peer's ID. When a peer requests the document from the P2P system, the request will get routed to the peer with the ID most similar to the document ID. This process is repeated until a copy of the document is found. Then the document is transferred back to the request originator, while each peer participating in the routing will keep a local copy (Milojicic *et al.*, 2002).

3.2.7 Device Models

Devices embody services in fixed systems to support task-specific functions. Devices tend to be very heterogeneous with different degrees of ICT resource limitations. In addition to MTOS-type devices (Section 3.3.3.9), devices include: embedded control devices (Section 6.5), mobile devices (Section 4.2), smart cards (Section 4.3) and micro-electromechanical devices and sensors (Section 6.4). In contrast to information-based systems, e.g., WS-SOC, UbiCom devices focus more on the context-awareness (Chapter 7), iHCI (Chapter 5) and autonomy (Chapter 10) UbiCom system properties. They inherently support more automated network configuration for more dynamic, fragmented, volatile, less structured networks. They support more automated service configuration of more dynamic and less seamless service and resource spaces. Specific designs for dynamic self-managed networks and service discovery are discussed further in Sections 3.3.1 and 3.3.2.

3.3 Service Provision Life-Cycle

The provision of application services for smart devices entails the management of distributed services throughout the whole of their life-cycle (see Figure 3.9) and not just in specific phases such as service discovery. For example, smart devices that operate in dynamic environments, such as smart mobile devices, cannot assume that services will remain static during service operation. The design of service provision must contend with intermittent service access and handovers between different service instances, e.g., wireless communication handovers, There are two separate aspects to this, first, defining a generic life-cycle model for service provision and, second, to manage this life-cycle. In a simple service provision lifecycle model, only two of the five service

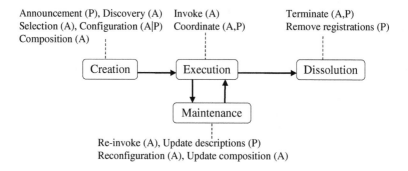

Figure 3.9 The service life-cycle: smart services entail operation and management throughout the whole life-cycle. P and A indicate that service processes and service access are active during each phase

model components are active, the processing services or service provision and service access or clients, the other three components, communication, stored information and information sources, are treated as passive components.

In the service creation phase, service processes register themselves in service directories. Service requesters in access nodes search for services (information processes and repositories). Services get selected, configured, and multiple services need to be composed, e.g., multiple services need to be composed to capture, annotate, transmit, edit, store, configure and print images. In the operational or service execution phase, services are invoked and multiple interlinked services may need to be coordinated. In the service maintenance phase, service processes, access configurations and service compositions can be updated. In the service dissolution phase, services may be put off-line or terminated temporarily by the processes themselves or by requesters. Services may also be terminated permanently and removed.

The design for the service lifecycle depends on application requirements such as the type of mobility needed (Chapter 1). For example, a static device such as a set-top audio-video receiver can support both dynamic service initiation and execution. This enables the device to be preconfigured using default factory settings and then shipped to be used in different regions in which it must detect and tune itself to the variable regional RF broadcast signal sources. This can also enable a static smart device to switch to an alternative service provider when a fault occurs, providing the user has permission to access it, possibly via another service contract.

3.3.1 Network Discovery

Generally, network discovery must precede service registration and service discovery.[14] Dynamic network discovery is used by mobile nodes and when new nodes are introduced into a network. A Domain Name Service or DNS is used to map an IP address to a name of some network node and vice versa. A common approach to discover the network is to use DHCP (Dynamic Host Control Protocol) to ask a DHCP server for an IP address that is leased for a given time. Leasing enables a limited set of resources, in this case, network addresses, to be periodically renewed by active nodes and to be reused and freed from inactive computer nodes. Some nodes that offer long-term services

---

[14] Discovery services may also be sub-classed into *white page* look-up (name, address, etc.), *yellow page* lookup (lookups by type and attribute) and *green page look-up* (information about how to invoke the service).

such as printers may be assigned static IP addresses. The complexity in using DHCP is in setting up and managing DHCP servers and in detecting and resolving duplicate IP addresses being used because: multiple DHCP servers may issues overlapping addresses; permanent IP addresses can conflict with dynamically assigned ones; inactive clients may attempt to use an address that has been reassigned.

Zeroconf or Zero Configuration Networking is a set of techniques that automatically creates a usable IP network without configuration or special servers. This allows inexpert users to connect computers, networked printers, and other items together and expect them to work automatically. Without Zeroconf or something similar, a knowledgeable user must either set up special services, like DHCP and DNS, or set up each computer's network settings by hand, which is a tedious and challenging task for non-technical people. Zeroconf currently solves automating three tasks: choosing network addresses, giving oneself an address, discovering names and discovering service addresses.

Both IPv4 and IPv6 have standard ways of automatically choosing IP addresses. IPv4 uses the 169.254.any, link-local set of addresses, see RFC 3927.[15] For IPv6, zeroconf, see RFC 2462, can be used. There are two similar ways of figuring out which network node has a certain name. Apple's Multicast DNS (mDNS) allows a network device to choose a domain name in a local namespace and announce it using a special multicast IP address. Microsoft's Link-local Multicast Name Resolution (LLMNR) is used less and was not on the IETF standards track.

3.3.2 Service Announcement, Discovery, Selection and Configuration

If service providers and requesters are static, then there is little need for dynamic service discovery. Dynamic service discovery is needed to allow service requesters to change providers when requesters or providers are mobile (Section 4.2), when network access is intermittent (Section 3.3.3.9) and when requesters or providers fail. Dynamic Service discovery[16] involves decoupling service provision from service requests and supporting dynamic announcements and dynamic discovery of service providers and service requesters.

There are two main approaches to dynamic service announcements and discovery: push and pull. Push uses broadcasts or multicasts to announce[17] the available service requests or service capabilities[18] to a number of unknown parties, e.g., Bluetooth. Broadcasting service requests or service descriptions are a sub-type of message broadcasts to unknown message receivers. The requester or provider does the matching. Pull uses lookups to search or browse lists of requests or capabilities previously announced to a directory held by some known third party, e.g., Jini, UPnP, UDDI, etc. The third party does the matching. The advantage of directories over broadcasts is that this minimises network traffic concerning service discovery. The disadvantage of directories is that

---

[15] IETF RFC (Request For Comments) specifications are available from www.ietf.org/rfc.html, accessed Aug. 2007.

[16] The scope of the term service discovery varies depending on the specific design. It could just involve asking for the list of available service providers that match a request. It may or may not include service selection, service configuration, service name to address resolution and even service invocation.

[17] Announcements can be designed to occur at certain times: periodically irrespective of whether any audience exists or not; only when any kind of audience is available; only when a specific type of audience is detected – multicast versus broadcast.

[18] Note service (provider) capabilities are often regarded as being synonymous with service advertisements and service descriptions by many researchers but others, SOA, differentiate capabilities which refer to competency to provide the service from the service description.

this requires third-party administration of the directory, the directory to be available and the directory to have a well-known location for clients and servers to find it.

There are several design dimensions to specifying service selection: request-based versus capability-based versus goal-based, exact versus inexact and syntactical versus semantic. It is more common to match requests from a single user against the service descriptions of multiple providers in directories. The opposite of this is to use a blackboard, a third party service, in which service requests instead of service provision descriptions are announced and service providers search for service requesters with which they want to do business (see Figure 3.10). Whereas discovery identifies the set of possible matches, selection sets the actual service match to be used.

Request-based service selection (and invocation) often imply that services to satisfy the request exist in the (external) virtual environment to the ICT system, rather than internally within the system itself. This begs the question of how a requester of a service knows whether or not it can perform the service itself (because it has or has not the necessary resources, expertise, etc.). Depending on the utility or cost model (Section 8.3.5), in some cases it may be more advantageous to source services internally rather than externally or vice versa. This is also true of business services as well as ICT services. The process of establishing whether or not a service exists internally involves self-descriptions, self-awareness and reflection (Section 10.3).

External service involves matching specific requests by one system with available service advertisements by providers in another system. Providers are selected either by the directory on behalf of the requester or all matches are given to the requester and the requester must decide on the selection of the provider. This involves getting the contact details of providers that match service requests in terms of preferences and constraints and making a selection from multiple service providers to then invoke the service (Figure 3.10). Service matching[19] can be designed to support exact matches or inexact wild-card and conditional matches. Goal-based service matching involves determining which of the available services can be used and how they can be composed to meet a user goal.[20]

A particular service provider selected can be invoked directly, providing the requester is configured to invoke the service. A variant of this is the service broker which differs from the

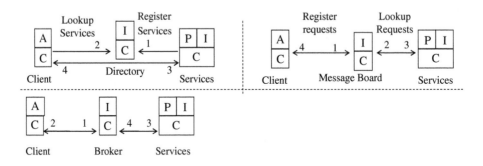

Figure 3.10 Service discovery driven by providers publishing service descriptions

---

[19] Exact matches are cheapest to implement in terms of computation resources. Wild-card matches match any character at specific character positions. Conditional matches match on a condition, e.g., a speed value being equal or greater than a threshold value. Semantic matches involve semantics-based partial matching, e.g., a fast speed and the use of semantic similarity, e.g., fast is similar to quick.
[20] User goals are high-level outcomes or end-points to be achieved from the use context for specific sets of one or more low-level user tasks.

service directory in that the broker not only handles service discovery but is also used to mediate service invocation. Service discovery is usually asymmetric, service providers do not need to look for service requesters as their details are normally supplied in the initial request. Sophisticated directory services could regularly poll servers to check whether the services they register are still alive; they could perform load-balancing of clients by distributing clients over network; or could perform client authentication and only allow specific registered and authenticated clients to use its services. There is a wide range of models for service mediation (Section 9.2.2).

3.3.2.1 Web Service Discovery

A Web service supports interoperable machine-to-machine interaction over a network. It has an interface described in a machine-processable syntactical format. Web service SOAs consist of many possible Web service protocols depending on the application and service requirements. SOAP[21] is used as a lightweight XML-based transport independent (however, usually over HTTP) protocol for exchanging structured information between peers in a distributed environment. SOAP defines a standard message format of an envelope with headers, the message body and possible attachments. WSDL[22] is used to describe services in terms of actions, input data, output data, constraints and service processes (sequences of service actions). UDDI[23] is a directory-based infrastructure that uses WSDL.

3.3.2.2 Semantic Web and Semantic Resource Discovery

Syntactic-level matching and discovery, e.g., for Web services, devices and resource, are challenging in pervasive environments due to the autonomy of service providers and the resulting heterogeneity of their implementations and interfaces devices. The semantic Web represents resources using RDFS and OWL (Section 8.4). Semantic service descriptions can be defined using OWL-S and WSMO. There is as yet little specific semantic middleware available. Because Semantic Web defines much richer XML-based data structures and relationships, heavier computation resources are needed to process these. The Semantic Web can also use WS directory protocols to store and access semantic service capabilities.

Semantic matching of service requests can enable services to be classified and grouped. This promotes bounded advertising of services and service group-based selective, forwarding the discovery requests which, when coupled with peer-to-peer dynamic caching of service advertisements, leads to a service discovery performance that can give better response time and reduces network load compared to syntactic service discovery (Chakraborty et al., 2006). However, semantic-based processing requires very heavyweight computation resources that are not yet present in many lightweight devices. This is in addition to the many other complexities of developing and using semantic services that are discussed later (Section 8.4).

3.3.3 Service Invocation

Specifying an application protocol in terms of a set of service descriptions of service actions is often insufficient to invoke a service. First, requesters need to have the know-how to invoke the service

---

[21] Simple Object Access Protocol (SOAP), http://www.w3.org/TR/soap/, accessed Oct. 2007.

[22] Web Service Description Language (WSDL), http://www.w3.org/TR/wsdl, accessed Oct. 2007.

[23] Universal Description, Discovery, and Integration (UDDI), http://uddi.xml.org/, accessed Oct. 2007.

described. This may entail manually or automatically downloading specific service access software, mobile codes, in order to be able to send and make requests, e.g., invoking hardware resource services such as printers may involve downloading hardware drivers into the access device.

Second, requesters may not know in which order to invoke service actions or how to handle out-of-order message sequences in a process without terminating service processes. The interaction in the process needs to be coordinated. Often the coordination may be hard-coded into each service API and under the control of the provider. This makes the coordination of multiple services inflexible. Clients often need to invoke not just individual service actions in isolation but to invoke a whole series of service interactions as part of a business process (a particular pattern of service actions to achieve a client goal and plan). Multiple heterogeneous processes often need to be interleaved: e.g., select item, order item, receive acknowledgement and receive item, need to be interleaved with a separate pay for item process.

There are different approaches to service invocation depending on application interaction patterns and on the characteristics of the service infrastructure. The main service invocation designs considered include: volatile service access, on-demand service requests; delayed reads and mail-boxes; event-based notification; caches, write ahead and delayed writes. Each of these is considered in turn.

3.3.3.1 Distributed Processes

A simple type of system design is to specify actions that are executed as fixed sequences, or are fully ordered, to complete a process. Earlier actions in the sequence of actions generate data that is used by later actions. The flow control may contain some flexibility in terms of branches, conditions and

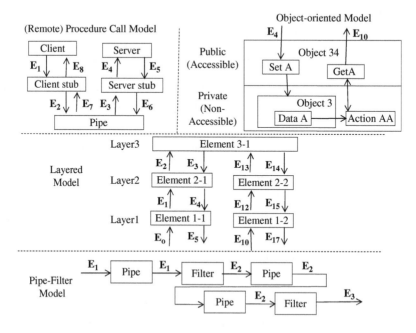

Figure 3.11 Different designs for supporting distributed interaction: (remote) procedure call, object-oriented interaction, layered network interaction and pipes and filters

loops. Systems can be designed so that actions can be distributed but yet support a common process. Example designs for distributed processes are based upon procedure calls, layers, object-oriented and pipe-filter models (Section 3.2.3). Although fixed process systems are a suitable design for closed systems such as embedded systems, they are not suitable for open dynamic environments where environment events can be generated at any time and may require changes to sequences of actions.

The design of remote interaction across different computer nodes via middleware is different from the design of local process interaction within the same computer node because it occurs across the network rather than across local shared memory and because different computer nodes are autonomous and heterogeneous. Designs for distributed interaction (Figure 3.11) include (remote) procedure calls and object-oriented interaction which hide the distribution of local versus remote interaction. The layered model uses high-level interfaces to hide the details of lower-level interaction which is often used as a design to mask and combine the use of multiple network protocols. The pipe-filter model is often used for streaming and combining multiple media to different applications that use different kinds of content filtering.

3.3.3.2 Asynchronous (MOM) Versus Synchronous (RPC) Communication Models

A common problem when a sender issues a request to one or more receivers, e.g., client–server computing (Section 3.2.2.3) is that servers need to be ready before clients start to make requests to them. If a client makes synchronous requests to servers including discovery services that are not ready, the client must block and wait. Asynchronous messaging can solve this issue. Asynchronous messaging applications such as email over the Internet, or SMS over mobile voice networks are often regarded as the first important[24] data applications over these networks respectively.

Two basic variants of asynchronous messaging exist: sender-side versus receiver-side asynchronous requests. Asynchronous requests initiated by the sender can use polling in which the sender periodically repeats the asynchronous request to a receiver.[25] The use of asynchronous communication often results in more responsive user interfaces and easier error handling. Process in senders do not need to block, waiting for a response from message receiver processes (Figure 3.12) but can continue with other interactions and processing for the sender. In contrast, during synchronous communication, process threads of execution must block waiting in the sender, waiting for a response. A common way to design asynchronous communication is to use event-driven interaction which decouples message sources from message consumers, to use message buffering and to use connectionless communication, supported by a third-party mediator, e.g., asynchronous Message Oriented Middleware (MOM[26]). Examples of MOM applications are mailboxes and mobile phone text messaging services such as the Simple Messaging Service (SMS).

Message buffering can occur in the sender or receiver or in some mediating node such as a proxy. The message buffer has its own separate thread of execution, it stores received messages temporarily in some area and keeps trying to send messages, to empty the buffer. Buffer design concerns

---

[24] Also referred to as Killer-Apps because it intentionally or unintentionally gets you to make the decision to buy the system the application runs on. Often Killer Apps cannot be foreseen by the original designers but emerge during usage, e.g., the first main application of the Internet was to increase the utility of expensive processing across users in different time-zones but email quickly emerged as the first main application.

[25] The sender-initiated type of asynchronous interaction has also been called deferred synchronous interaction (Emmerich, 2004).

[26] Note some authors refer to both asynchronous and synchronous designs as MOM (Emmerich, 2000) whereas others assume that MOM is asynchronous and refer to remote synchronous MOM interaction as RPC or Remote Procedure Calls (Menascé, 2005).

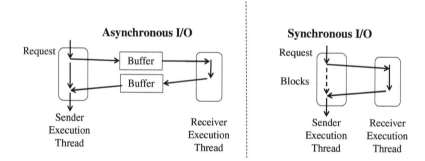

Figure 3.12 Asynchronous versus synchronous I/O: the use of buffering when sending or receiving, either at the sender or receiver, enables senders and receivers to be temporally decoupled

configuring the size of the buffer and the input and output message order, e.g., First-In-Last-Out (FILO) buffer versus First-in-First-Out (FIFO) buffer (Coulouris *et al.*, 1994).

In connectionless communication, no communication channel or connection needs to be pre-configured before any communication takes place, nor do connections need to be dismantled after the communication session has ended, whereas in connection-oriented communication, resources are needed to configure a connection or communication channel and this must be done before any messages can be exchanged. Messages (headers) tend to be shorter with connection-oriented communication as some of the communication context such as the sender and receiver addresses are set only once when the channel is set and assumed to be fixed during the communication session. Message headers tend to be longer when connectionless communication is used as the full communication context must be specified in each message (header) sent and received. Connection-oriented communication is advantageous over low-bandwidth links providing the communication link can be maintained. However, if the communication link intermittently breaks, a common occurrence in low-bandwidth communication, a terminated connection must be cleared up and restarted and this consumes ICT resources and time. Often rather than design communication to be synchronous versus asynchronous at the application level, it is handled at a lower level in the network stack, in the network layer (Section 11.4).

MOM,[27] mailboxes and SMS are examples of asynchronous messaging systems which use third-party mediators. These have facilities to store, route and transform messages. There is a lack of agreed standards for MOM, hence MOM tend to be vendor-specific. In addition, it could be argued that some of the facilities of MOM systems such as application-level message-routing are performed more efficiently at lower network protocol layers and in more specialised hardware. Menascé (2005) provides a useful comparison of the advantages and disadvantages of asynchronous (MOM) versus synchronous (RPC) messaging. For example, MOM solutions tend to be more robust to failures than RPC as MOM enables service requesters to continue to process other requests and not block. Both are complex to design. MOM-based applications are complex to design because distribution is not as transparent to the application as with Remote Procedure Calls

---

[27] There are many vendor and application domain-specific MOM systems available, e.g., IBM WebSphere MQ, MSMQ, WebMethodsEnterprise, etc. (Alonso *et al.*, 2004). One of the first uses of asynchronous messaging in the early 1990s was for Transaction Processing Systems (TPS) used in ecommerce. TPS today can handle very high messaging throughputs, e.g., major credit card transactions typically generate about 100 billion transactions yearly or about 30 million transactions daily.

(RPCs). RPCs are complex to design in order to make remote communication look like local communication, e.g., parameter marshalling (Birrell and Nelson, 1984). For this reason, RPC has evolved into object-oriented style of Object Request Brokers where components interact using Remote Method Invocation (RMI) rather than RPCs (Emmerich, 2000).

3.3.3.3 Reliable versus Unreliable Communication

An additional design issue is to consider the reliability of the network to deliver messages without loss or delay and in order. Service access over wireless networks is often more unreliable than wired networks. Applications can assume no network guarantee about delivery and need to detect message corruption, message loss and to handle these problems. Message corruption can be dealt with using various message integrity checks or can be dealt with at the network layer (Chapter 11).

At the application level, a message protocol can use additional acknowledge-event messages to detect any delayed or lost messages sent and received. In order to handle lost messages, senders and receivers can be aware of states (stateful), retaining some intermediate states about messages sent or to buffer sent messages. This way, replacement messages do not have to be created from scratch, which is a big overhead for data that requires substantial computation to generate. Message senders that do not retain any state about sent messages are called stateless. To an extent, stateful communication is more complex to synchronise than stateless communication because the equivalence of intermediate states may need to be compared. However, this can consume less ICT resources when generating complex messages as there is no need to completely generate these from scratch.

A further issue before repeating message transmission is to consider the consequences of doing this. Messages that can be repeated, at least once, without side-effects are called idempotent messages, e.g., pressing an elevator call button again because the response has not yet been completed. Other messages may be non-idempotent, e.g., a message request that withdraws funds from one bank account. In this case, the challenge is because of the application requirements for the action to be performed only once. Due to partial observability at the sender, the sender may not be able to distinguish between a sender crash before the message is sent, a sent message being lost, a remote server crash and the received message being lost. In the cases of a receiver crash and the received message being lost, it is OK to repeat the request but not if a sent message is lost.

3.3.3.4 Caches, Read-Ahead and Delayed Writes

Two design patterns to deal with intermittent server access are read-ahead and delayed write (Figure 3.13). In read-ahead interaction, information is pre-cached in devices when the network is available. When the data required can be retrieved from the cache, it is referred to as a cache-hit. When the data required cannot be retrieved from the cache, it is referred to as a cache-miss. In this case, the cache must send a request to retrieve the data from source. Design decisions include

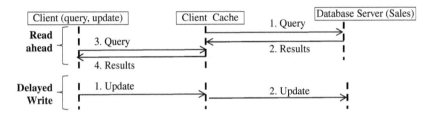

Figure 3.13 Two design patterns to deal with intermittent server access, read-ahead and delayed write

deciding which useful information to cache and to decide the frequency of cache updates. Frequent caching leads to fast location times but eliminates the benefits for reducing control traffic because of the frequent updates. Less frequently cached data can be stale (not up to date), and performance may actually degrade because of the additional data operations to check the cache and synchronise it.

With delayed writes, updates are made to the local cache while services are unreachable which must be later reintegrated upon reconnection. Concurrent local and remote updates may need to be synchronised. Write conflicts need to be detected when the same data has been modified locally and remotely. Satyanarayanan (1996) discusses several techniques to handle cache misses and cache resynchronisation. Model-based cache-miss design can take into account strong relationships between items and to proactively pre-cache strongly associated items with missed cache data. Optimistic replication allows replicated contents in the cache to diverge in the short term in order to support concurrent work practices and to tolerate failures in low-quality communication links. Changes can be propagated in the background, conflicts can be discovered after they happen and agreements on the final contents can be reached incrementally. Cache coherence may also be more usefully maintained at multiple levels of granularity.

3.3.3.5 On-Demand Service Access

On-demand service access implies that services are always available when needed. A common design for this type of interaction involves a pull-type, request-response interaction. One entity initiates requests for application services across an always available network connection, using synchronous communication and block waiting for the response to the request (see Figure 3.12) from the server providers. For on-demand, request-driven, service interaction, a common design is based upon a thin-client, stateless, access node that uses synchronous communication over an always-on network connection. This greatly simplifies the synchronisation between the presentation on the access node and server-based processing in the network. Often multiple request-responses are chained together to complete an application session, for example, clients need to make several requests to prepare to invoke services, i.e., to query catalogues to discover services and to select from multiple providers.

Ecommerce service invocation is an important example of on-demand service access and often involves a sequence of multi-node service interaction by a merchant's customer-facing services such as catalogues and sales that act as a broker or customer proxy to the detailed supporting back-end services of a merchant such as inventory, delivery and customer bank.[28] Commerce and ecommerce service interaction often uses a thin-client access terminal that provides basic UI controls to get an information request and to display simple forms of response. In the simple case, this interaction uses synchronous communication and can be sequenced and driven by the requester. This interaction can be grouped into two distinct customer transactions: the select goods transaction and the purchase goods transaction.

The most important design challenges with on-demand service access are to ensure the right services are available under the right conditions when needed and that service access can tolerate volatile service access.

There are several designs that help support volatile service access. Services interaction can simply be repeated if requests are non-idempotent (Section 3.3.3.3). Local caches, read-ahead and delayed writes can be used to mask volatile service access (Section 3.3.3.4). In addition, when request for

---

[28] Some of the details at the merchant part of the interaction are not shown here, e.g., the funds transfer from the customer bank to the merchant's bank and the audit trail of receipts from the merchant and customer banks to the merchant and customer respectively.

services fail, service discovery and fault-tolerance can be used to search for alternative existing services instances (Section 3.3.2) and service composition can be used to synthesise new service instances (Section 3.3.3.9).

3.3.3.6 Event-Driven Architectures (EDA)

Gelernter and Carriero (1992) assert that coordination mechanisms should be separated from computation mechanisms. This supports several key benefits:

- *portability*: by providing computation language-independent mechanisms for coordination;
- *heterogeneity*: enabling devices and applications to coordinate with each other even when these mechanisms are implemented in different hardware or languages;
- *flexibility*: enabling different coordination mechanisms and computations mechanisms to be mixed and matched.

Garlan and Shaw (1993) have identified and classified several different interaction mechanisms[29] for distributed ICT systems. One common way to decouple coordination from computation is an event-driven system which supports very loose coupled control or coordination between event generators or producers, and event receivers or consumers, e.g., clicking buttons on a User Interface (UI) produces events that trigger associated actions in services. This is also known as publish-and-subscribe interaction. One or more nodes publish events while others subscribe to being notified[30] when specified events occur (Eugster *et al.*, 2003). An event is some input such as a message or procedure call that is of interest. An event may be significant because it may cause a significant change in state, e.g., a flat tyre triggers a vehicle driver to slow down. An event may cause some predefined threshold to be crossed, e.g., after travelling a certain number of miles, a vehicle must be serviced to maintain it in a roadworthy state. An event may be time-based, e.g., at a certain time record a certain audio-video program. External events can trigger services. Services may in turn trigger additional internal events, e.g., the wheel brake pads are too worn and need to be replaced. Event-driven architectures are an important interaction to support service-oriented architectures (Papazoglou *et al.*, 2007). These are referred to as event-driven service oriented architectures.

In an Event-Driven Architecture or EDA,[31] event consumers specify and register events that they are interested in being notified of with some special event dispatcher engine that gathers input, unprocessed events, and buffers them (see Figure 3.14). When events are generated by event producers, they are gathered and distributed by the event dispatcher to the event consumer that wants to be notified of them. The event consumer can perform its own check that events meet certain constraint conditions and if they do, one or more actions or services are triggered.

The events producers are said to be loosely coupled from the event consumers. Event consumers can dynamically register and deregister themselves, e.g., a customer can register itself to be notified for a new vehicle but then deregister itself when it is no longer interested. Event producers can decide not to produce any new events. This loose coupling can lead to complexity and uncertainty. Event producers may not know who will reply and when they will be finished consuming event, e.g.,

---

[29] They classified six main interaction or coordination styles: Pipe and Filter, Object-Oriented Data Abstraction, Event-driven, Layered, Repository, Interpreter and VM.

[30] Event-based notification is similar in some respects to procedure callbacks.

[31] Several standards consortiums standardise event-driven architectures, see http://soa.omg.org/SOA-docs/ EDA-Standards.htm, accessed June 2007.

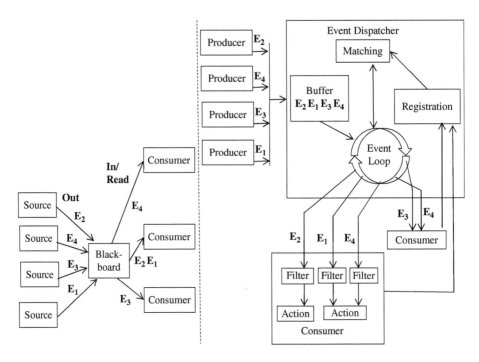

Figure 3.14   Shared Repositories (left) and Event-driven Interaction (right)

users can get frustrated when pushing buttons on a UI as they not sure if anything has happened or what progress has been made, unless feedback is given. In some cases receivers may not be passed the full context or situation in which the event occurred in order to handle the event effectively.

Figure 3.14 shows a basic EDA. Multiple input events are input into an event buffer such as a first in first out (FIFO) or a first in last out (FILO) buffer. Events are matched with consumers that have previously registered an interest in those events. Matching can be simple, e.g., an event is matched with any registered consumers and sent to them, or more complex. In more complex design, events can be consumed using an ECA (Event Condition Action) paradigm, i.e., events are used to trigger actions when certain conditions are met, e.g., if time equals T_1, start recording programme on channel Y. These actions could in turn also trigger new events. Events could be filtered in the event producer rather than in a centralised event dispatcher that is shared between multiple event producers. Examples of the use of an EDA are in asynchronous messaging systems (Section 3.3.3.1) sensor systems, (Section 6.3) and reactive intelligent systems (Section 8.3.2). In a policy-based EDA, policies can be defined for action triggers, for event buffering and for event registration.

There are several design challenges with EDA, such as dealing with event floods and asynchronous coordination. In event floods, a few highly significant events can be lost in a great volume of non-significant events that cannot be processed in time. Event floods can be dealt with by prioritising events, enabling events to expire if not acted upon within a specific time-frame and by using event filters. Event coordination may be needed by applications when events can arrive in any order. EDA generally have no persistence. There is no inherent way to reuse recent events. It is more difficult to keep things running in a failure situation, receivers may also need to track new senders to decide whether or not to subscribe to them. For efficiency, most publish–subscribe systems only broadcast events when receivers are detected (Johanson and Fox, 2002).

3.3.3.7 Shared Data Repository

In a repository style of interaction, two participants communicate by leaving messages for others via some shared intermediary. Hence, a shared repository system consists of two types of components: a central data structure represents the current state, and a collection of independent components operates on the central data store. There are two major sub-types of coordination depending whether transactions in an input stream trigger the selection of executing processes, e.g., a database repository, or if the current state of the central data structure is the main trigger of selecting processes to execute, e.g., a blackboard repository. This represents and stores data that is created and used by other components. Thus we can talk about how data producers input data into the repository and data is output from the repository into data consumers. In this way, repositories are similar to EDA systems in the way they support independence and allow volatile producers and consumers to be tolerated (see Figure 3.14). The main difference between the EDA and shared data repository is the persistent storage of the input data and data management, e.g., consistency management. Examples of shared data repository include electronic bulletin boards, knowledge-based blackboard systems including tuplespaces and relational data type databases.

There are several other sub-types of repository in addition to database versus blackboard: passive versus active, centralised versus distributed, and caches. A repository can be passive: a sender component pushes data to a selected repository component to store it, e.g., relational databases. The repository can be active: it stores and specifies its own actions that are triggered by particular states of the repository, e.g., rule repository and engine, knowledge-based system. A repository does not have be a single centralised repository, it can be highly distributed e.g., the Web. Repositories may also be replicated and distributed to improve availability. A cache is a repository for the replication of results of previous requests so that they can be reused by later requests.

An example of a blackboard repository is a tuplespace in which tuples, ordered typed fields, where each field either contains a value or is undefined, are stored in a persistent shared abstract space called a tuplespace. Three tuplespace operations are supported. 'Out' puts a tuple into a tuplespace. 'In' removes a tuple and 'Read' copies a tuple from the space. In and Read match a tuple to a template tuple where explicit values are used for some fields, and wild cards for others. Of these, the Read is the key operation that differentiates it from event-driven programming as Read events are copied from the store and remain there for later processing.[32] A separate component, a control unit (not shown in Figure 3.14), may be included to support read–write access by multiple concurrent events, data or knowledge producers and consumers. Johanson and Fox (2002) describe the use of a variation of the tuplespace model called an event-heap that they have applied in a distributed shared workspace application called iRoom (Section 2.2.2.3). Examples of tuplespace blackboards can be built using an underlying object-oriented, even-driven design include JavaSpaces[33] and TSpaces (Wyckoff *et al.*, 1998).

3.3.3.8 Enterprise Service Bus (ESB) Model

An Enterprise Service Bus or ESB[34] supports messaging, Web service integration, data transformation and intelligent routing for SOC; it decouples service provision from service access. These

---

[32] Read actually is a misnomer and should be more accurately called 'copy and store'. This operation turns an episodic environment event-driven reactive intelligent system into a model-based intelligent system that captures and takes account of sequential or past events in the environment (Section 8.2.3).

[33] See Sun Microsystems Labs, JavaSpaces, available from http://www.sun.com/jini/specs/js.pdf.

[34] Although the idea of ESB was first reported in about 2004, there is no current standard specification (mid-2008) for ESBs. Several designs and implementations exist, e.g., Keen *et al.* (2004).

functions are distributed. There are two main types of design: message-oriented ESB versus service-oriented ESB. A message-oriented middleware (MOM) design for ESB supports asynchronous messaging, transactions, publish–subscribe interaction styles and application-level routing (Section 3.3.3.1). However, MOMs may mandate a specific application level or transport protocol which may mean gateways are needed to convert from an application which supports another protocol. MOM may itself not be modelled as a direct Web service or first-class service[35] but rather as an API. A service-oriented model for ESB as opposed to a message-oriented model offers fuller support for three types of integration: (1) integrating multiple service access, e.g., behaving as a portal; (2) integrating multiple application service processes, supporting work-flows, brokerage and propagation; and (3) supporting data translation.

3.3.3.9 Volatile Service Invocation

Sometimes service access may be quite intermittent, for example, wireless networks and mobile users tend to suffer higher error rates and more frequent disconnections. This is due to: limited network area coverage and intermittent low bandwidth access via some networks; network hand-offs as mobile users move between different base-stations; intermittent interference and variable signal reception. It may also be due to changing heterogeneity in terms of network bandwidth and coverage. In addition, in open service infrastructures, service access has to contend with variable access as different numbers of requesters try to access variable numbers of services that can go on and off-line and to deal with the natural heterogeneity of open service interaction.

Designs of the application and middleware must take this into account otherwise requests will block or terminate and may need to be repeated and restarted. Basic designs to handle volatile[36] service access include the use of asynchronous communication (Section 3.3.3.1), handling unreliable communication (Section 3.3.3.3) and message caching (Section 3.3.3.4).

Satyanarayanan (1996) has summarised a number of more detailed design mechanisms for handling volatile service access, over possibly low-bandwidth, reliable, network links for applications in which concurrent access to shared data occurs. This concurrency may occur at two different levels: remote versus local and write versus read. Design mechanisms to handle volatility are derived from his group's experiences with developing the Coda file system and Odyssey mobile platform. Application adaptation support includes:

- *caching, logging and synchronisation*: updates that cannot immediately be shared because of disconnections can be pre-cached (read-ahead) or logged (delayed write) for asynchronous exchange. Additional mechanisms are needed to manage data consistency between cached (temporary stores of) data and more permanent data stores.
- *adaptive transport protocols*: for example, 'trickle' reintegration mechanisms for propagating updates over low-bandwidth[37] links;
- *resource reuse*: at any time, the system may revoke resources that it owns and has temporarily delegated to other applications. Resource reuse is vital in low-resource systems.

---

[35] Modelling messaging itself as full service has pros and cons. The advantage is that it can be invoked in a standard way, supporting a generic service model. The disadvantage is that can be quite inefficient for high throughput systems, requiring two service invocations: a service invocation to invoke the messaging service and the messaging service invocation itself that transmits the message.

[36] Volatile services are also sometimes referred to as occasionally connected or intermittently connected or ad hoc connected or spontaneously connected.

[37] Low bandwidth connectivity is also referred to as weak connectivity.

- *partial observability*:[38] access devices may be able to partially sense or estimate (local) changes in its environment, e.g., message round-trip time, and then make inferences about the cause of these changes, e.g., network bandwidth limitations, and react appropriately.

3.3.4 Service Composition

Composition is concerned with synthesising new services and assembling more complex (composite) services from simple (atomic) services to achieve a user or application goal, and then collectively executing them as composite service processes. Service processes are sequences of individual service actions that are scheduled for execution. Service processes may involve one or more entities, one or more actions and involve one or more processes. Statically organising services and actions into the expected preset groups can lead to requests failing when a request is made to a service that does not exist or is unavailable. Dynamic service composition can be triggered to give greater flexibility and to help promote on-demand service access (Section 3.3.3.4).

Composition can occur incrementally over several rounds rather than in a single round, with later rounds perhaps learning from the constraints of and experiences of execution in earlier rounds. This is useful when it may not be possible to pre-plan the composition because it must be derived from the experience of executing the service. Composition can be controlled by a central entity (service orchestration) or controlled by distributed entities on a peer-to-peer basis (service choreography).[39] Generally, the orchestrator tends to hold a global viewpoint of the service actions and constraints of the participants whereas in service choreography, participants are usually more responsive to local viewpoints of the service actions of oneself and the adjacent service processes of others. Service orchestration is simpler to design than service choreography and appears to be more commonly used. Service composition can be specified manually or automatically using various service composition methods.

In terms of automating service composition, four approaches are proposed by different communities: business processes, workflow, Semantic Web and MAS planning. Business collaborations require long-running interactions driven by an explicit WS application process model, e.g., XML-based WS composition and execution standards from the business community such as BPEL4WS, the Business Process Execution Language for Web Services (Bucchiarone and Gnesi, 2005) (van der Aalst *et al.*, 2003). WS models, however, do not inherently offer rich data structures and hierarchies such as class-based hierarchies. They also require devices to contain sufficient computation resources to understand and parse XML.

The Semantic Web community focuses on reasoning about web resources by explicitly declaring their preconditions and effects by means of ontology models (Bucchiarone and Gnesi, 2005). Van der Aalst *et al.* (2003) have compared workflow management systems and Web service composition languages using a set of patterns. The comparison reveals that Web service composition languages adopt most of the functionality present in workflow systems. However, Web service composition languages are more expressive than the traditional workflow products.

---

[38] Partial observability is also referred to as Global Estimation from Local Observations.

[39] The term orchestration is used in music to refer to a conductor who directs a group of individual classical musicians to play together. Choreography is derived from the Greek words for 'dance' and 'write' and refers to the art of scripting dance steps which the individual dancers have some flexibility in providing their own interpretation of dance steps.

3.3.4.1 Service Interoperability

The ability to communicate at the network level is insufficient to interoperate at the service level. First, information and tasks need to be coordinated, e.g., there is no point in transmitting if a receiver is not available. Second, a common syntax and semantics are needed to represent different information and tasks that are exchanged across heterogeneous systems.

There is a difference between integrated services versus interoperable (or federated) services. Integrating services generally means that the individual systems become statically linked, perhaps by combining data and code into a single whole repository, hence losing some of their autonomy. Service interoperability enables services to remain autonomous but to dynamically link to each other to allow them to exchange data in a format that both understand.

Distributed systems that interoperate can exchange data in a variety of data formats, encodings, etc. Two main methods for heterogeneous data exchange can be distinguished: either a common or canonical exchange data format is used, or a receiver or sender makes it right scheme where either party transforms the data so that the other one can understand it. Less mappings need to be maintained when using a common exchange format compared to a receiver or sender makes it right scheme.

Different levels of expressivity for data structures may be used. Depending on the application, hierarchical structures such as XML, object-oriented hierarchies and graph-based structures can be exchanged. Syntactical or semantic structure descriptions can also be exchanged. Application-level communication protocols are specified as linear structures such as byte streams, but this means that hierarchical application data structures need to be serialised for transmission and recomposed in the receiver.

Different kinds of state information can be exchanged as metadata when one process invokes another. There are two main options: RPC versus REST. With an RPC, essentially, information about the state of the sender is transferred to the receiver. However, the RPC is under the control of the receiver. In the Representation State Transfer (REST) model (Fielding, 2000), the receiver is seen as a set of resources identified by URLs, only representations of the resources are exchanged, not any state information.

3.4 Virtual Machines and Operating Systems

3.4.1 Virtual Machines

A Virtual Machine (VM) supports large-scale multi-user concurrent server execution and it enables cross-platform interoperability across a diverse set of hardware resources at multiple levels of abstractions. To understand the concept of a VM, the concept of a computer or machine needs first to be considered from two different viewpoints: from the process and from the (operating) system viewpoint (Smith and Nair, 2005). From an application processing viewpoint, a computer consists of the use of processes that are held in memory in bounded address spaces. Processes consist of a list of instructions defined in a high-level interface, the Application Programmer's Interface (API), that are converted into binary digital instructions at a lower-level Application Binary Interface (ABI), to be executed. The underlying hardware such as the CPU and I/O devices are hidden by the virtualising software or VM Monitor (VMM), the API and ABI. The operating system viewpoint considers the details of how multiple processes can be executed simultaneously on the hardware and when there are more processes than hardware resources available as opposed to dedicated task (embedded) systems. From the operating system's point of view, the hardware interface such as the Instruction Set Architecture (ISA) and System Call interface define the machine and act as the VMM. Processes regard the OS to be the VM while the OS regards the hardware to be the VM.

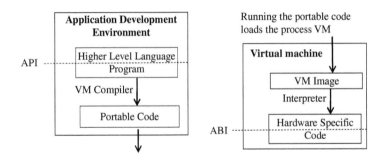

Figure 3.15 a HLL (High-level Language) Program is compiled into intermediate (portable) code (left). When this portable code executes it triggers a Process VM (Virtual Machine) to start up to interpret instructions converting them to executable code that runs on specific hardware (right)

A System Virtual Machine or System VM provides a complete persistent virtual environment for an operating system and application process. The System VM was the original type of VM developed in the 1960s and 1970s for mainframes (Dickman, 1973). A System VM enables multiple OS systems and application to be run on the same hardware. If one system fails, the others are isolated and keep running. This is still a useful technique employed in modern servers and server farms that need to support multiple users, applications and need to share hardware resources.

A Process or Application VM is a virtual platform that executes an individual process. This VM is created when the process starts to execute and persists only as long as the process it is executing. A high-level Language Process VM executes actions specified in some instruction language (see Figure 3.15). Virtual machines are powerful because they can use a hardware-neutral, intermediate language, thus supporting the ability to execute the same intermediate language on a diverse set of hardware, and map this to heterogeneous hardware to actually execute the program. Its central component is an interpreter engine that translates a language instruction-by-instruction, as necessary into binary code.

This is in contrast to a compiler that translates a program in one step into binary code that can be executed on hardware. An interpreter is a program that acts like a CPU with a fetch-and-execute cycle. In order to execute a program, the interpreter runs in a loop in which it repeatedly reads one instruction at a time from the program, it decides how to carry out that instruction, and then performs the appropriate machine-language commands to do so (Smith and Nair, 2005). It maintains a link between the state of the intermediate program being executed and the state of the binary code actually being executed.

An intermediate language interpreter differs from a high-level language interpreter such as a Basic or Web script language interpreter in that the production of the (native) machine-code is optimised to give the interpretation of intermediate code a performance that is comparable to executing a compiled high-level language. It does this by using a Just-In-Time or JIT compilation. This converts code prior to executing it natively at runtime. The performance improvement over interpreters originates from caching the results of translating blocks of code rather than simply re-evaluating each line or operand each time it is met.

3.4.2 BIOS

Often when a computer is started or booted, also called bootstrapped, the software is loaded in stages. First, the BIOS or Basic Input/Output System, a type of firmware, is loaded. This is used to load the operating system kernel. Firmware is a type of low-level software to control hardware. It is a basic computer program that resides in special hardware, typically in ROM in EEPROM or flash memory-type ROM or Read Only Memory. It can be updated over the network, via the OS. The BIOS initialises several motherboard components and peripherals such as the CPU, the system (primary) memory, graphics controller, secondary storage, I/O controllers such the keyboard, mouse, USB ports, and the system clock. Finally, the BIO loads the operating system (OS) and transfers control to it and in modern computers, the OS's own hardware drivers then take over control of the hardware from the BIOS. In addition, many devices attached to MTOS computer systems are actually special-purpose computers themselves, e.g., printers, these may also contain their own firmware in a ROM within the device itself.

3.4.3 Multi-Tasking Operating Systems (MTOS)

Desktop computers and smart-phones (fat-client devices) often require more complex multi-process control and the use of a Multi-Tasking Operating Systems (MTOS), also referred to as General Purpose Operating System (GPOS), to support the execution of multiple users and applications. The operating system can be considered as a VM for processes (Section 3.4.1).

The high-level model of an ICT system, given in Section 3.2, consists of data input, data storage, data output and data processing components that can be distributed and interlinked via communication components. Generally, these components are virtualised by the Operating System or OS for processes. A special part of the OS called the OS kernel has priority control of the computer hardware such as the CPU, memory and I/O ports in order to execute processes. Depending on the design of the OS, some core utilities are part of the kernel, others are outside it. In addition, the operating system also supports software interfaces called device drivers to control hardware devices such as memory, displays and input and devices and it supports software to build applications. The OS can be modelled as a multi-layered architecture (Figure 3.16). The core utilities of an OS are to control their access to data storage such as memory and disk, to control data input and output and to control data communication (Figure 3.17). Computer

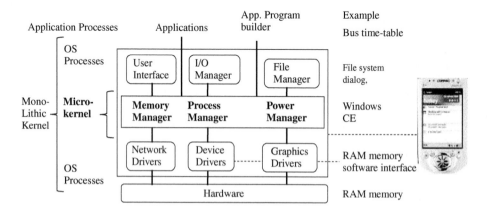

Figure 3.16 The main components of an operating system. There are two basic types of operating system kernel: micro-kernel and monolithic kernel

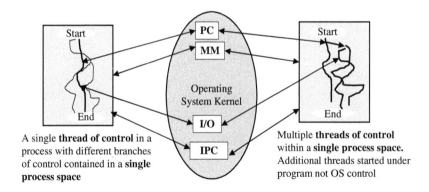

Figure 3.17 Operating System kernel functions: memory management (MM), process control (PC), inter process communication (IPC) and Input/Output Control (IO)

system hardware and computer architectures are considered in more detail in Hennessy *et al.* (2006).

3.4.4 Process Control

The Operating System Kernel is a process which has privileged use of the ICT resources. It has full access rights to all physical resources, it has its own protected address space for its data memory and it runs the CPU in a special mode called the supervisor mode. The kernel controls the access rights to the physical resources for all other processes; it controls the process access to memory and controls process access to input and output devices. It creates an execution environment for processes to run in and sets up an address space for each application process, outside the kernel space, to safely execute in and protects processes from interfering with each other.

Each process starts with a single thread of control, a single-path through different branches and loops of control in a process. Threads of control in process may divide themselves into multiple threads of control so they can have more threads of control than CPU even within a single process space. This allows one thread to be active while others are perhaps temporarily blocked from accessing a resource. The advantage in creating multiple threads within a single process rather than multiple processes with a single thread is that creating a second thread within the same process space is more efficient because a whole new execution space does not need to be created. However, the downside is that the thread control within a process space and inter-thread communication are under the application process control not under the kernel control. This is less robust. The kernel can terminate threads of control and frees up their resources such as memory, file links. It schedules process's threads of control for execution on CPU.

The kernel maintains the process state, the list of processes waiting to run, a list of open file descriptors and the state of each process such as active, terminated and dead-locked. It also coordinates multiple processes and manages inter-process communication. Often there are more runnable (or executable) processes, processes that are waiting and ready to use the CPU, than the number of CPUs available to execute them (Figure 3.18). Executing processes can block waiting for resources. Sometimes processes have not finished but a higher priority one comes along. Some type of multi-task scheduling is needed. In static scheduling, all scheduling decisions are determined before execution and all runnable processes are treated equally. In a round-robin, also called, a pre-emptive task scheduler, each process is allocated a fixed time to use the CPU – too

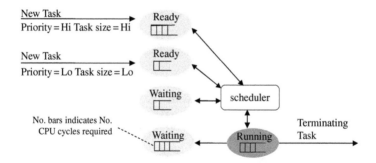

Figure 3.18   Scheduling multiple tasks that exceed the number of CPUs available

short for some processes, too long for others. Short finished processes can waste CPU cycles and interrupt[40] a running process, switching it from a runnable context back to a waiting context. Context switching[41] wastes time and resources. In contrast, in a non pre-emptive process scheduler, processes can run to completion, there is less context switching, no uncompleted tasks need a context switch. However, executing processes can significantly hold up other processes waiting to execute. In dynamic scheduling, run-time decisions are used. For example, in priority scheduling, runnable processes with a higher priority run first. To prevent a high priority task running indefinitely, a scheduler decreases task priority over time.

3.4.5 Memory Management

The second key function of an OS is memory management. Processes are associated with Virtual Memory or an address space that is mapped to main primary physical memory (RAM). The operating system kernel defines a separate region of address space for each process. A process associated with one addressable region cannot directly access another region. Regions are non-overlapping and are separated from each other by buffer memory which is not addressable by the process. A region may have different parts such as read/write (RW), e.g., heap or stack, or read only (RO), e.g., program code regions (RO). The primary memory used by processes may be backed up or mapped yet again to more persistent secondary memory such as ROM or disks: this involves actual copying of data from primary to secondary data. Hardware such as disks, network device interfaces, or graphics card interface appears to a process as an array of bytes of data, i.e., as secondary memory data that is mapped to the primary memory address space.

This secondary memory enables programs and data to access more memory than the amount of primary memory available – this is of particular interest to low-resource devices. A page-fault is generated by a process when it tries to access data not available in primary memory. Demand-paging is commonly used to rectify a page-fault, i.e., to access data not in primary memory. It requires data

---

[40] Interrupts also take time away from the CPU executing processes. The interrupt latency is the time taken between the generation of an interrupt and the servicing of the interrupt by the OS.

[41] (Process) context switching refers to a CPU switching between the state of one active process and that of another waiting process. To do this, a CPU must store the state of the active process, remove the state information from memory, retrieve the state of another process into memory and start it.

to be copied into primary memory from secondary memory. If no free room in primary memory exists, an existing region of primary memory must be swapped out to secondary memory.

3.4.6 Input and Output

I/O devices can be directly referenced in the program as addresses or pointers to memory in the kernel that are mapped to the device interface card memory. Special (operating) system calls can be made to read and write this data. File handles in the C Programming Language or I/O streams in C++ or Java are used to provide a higher-level program API to wrap the lower-level system I/O calls. But this is inflexible, the program would need to be modified every time a different device is introduced to get the output or input, e.g., to output to a printer rather than a video device. I/O support for all devices would need to be supported in each program. A more effective design is to define generic (also called logical or virtual) I/O devices which can be re-mapped by the OS to a specific I/O device.[42] To use a remote I/O device, a standard stream, set with a port number associated with a remote computer and process is created. The network appears to a process as just another I/O channel connected to a network interface called a socket.

EXERCISES

1. Discuss the difference between system designs based upon abstractions versus virtualisation.
2. Compare and contrast the following architectural models for service access: client–server model, application server model, middleware service model, service-oriented computing model and peer-to-peer model.
3. Discuss whether or not the message broadcast or flooding techniques used in P2P networks to locate unknown P2P nodes can also be used more generally for UbiCom service discovery.
4. What is a multi-tier server model? Compare and contrast a thin-client server model versus a fat-client server two-tier model. Give some examples of three-, four-, five- and six-tier models.
5. Describe three different designs for partitioning and distributing: (a) communication; (b) processing; and (c) data resources.
6. Describe the benefits of a proxy-based service access model; outline a proxy-based design for mobile service access.
7. Compare and contrast the following system component interaction paradigms for UbiCom system components: (Remote) Procedure Call, Object-oriented interaction, layered network interaction, pipes and filters, shared repository.
8. Characterise a P2P model The P2P model needs more complex synchronisation compared to the client–server model, why?
9. Give the benefits for system designs that separate coordination or control from computation. Then discuss the pros and cons of object-oriented versus event-driven versus blackboard repository type coordination. (More exercises are available on the book's website.)

---

[42] E.g., in the core Java API, three standard I/O devices are defined as static system objects methods: System.in, System.out and System.err.

References

Allard, J., Chinta, V., Gundala, S. and Richard G.G. (2003) Jini meets UPnP: an architecture for Jini/ UPnP interoperability. In *Proceedings of Symposium on Applications and the Internet (SAINT'03)*, pp. 268–275.

Alonso, G., Casati, F., Kuno, H. and Machiraju V. (2004) *Web Services: Concepts, Architectures and Applications*. Berlin: Springer Verlag.

Anderson, D.P, Cobb, J., Korpela, E., Lebofsky, M. and Werthimer, D. (2002) SETI@home: an experiment in public-resource computing. *Communications of the ACM*, 45(11): 56–61.

Androutsellis-Theotokis, S. and Spinellis, D. (2004) A survey of peer-to-peer content distribution technologies. *ACM Computing Surveys*, 36(4): 335–371.

Bernstein, P.A. (1996) Middleware: a model for distributed system services. *Communications of the ACM*, 39(2): 86–98.

Birrell, A.D. and Nelson, B.J. (1984) Implementing remote procedure calls. *ACM Transactions on Computer Systems*, 2(1): 39–59.

Bucchiarone, A. and Gnesi, S. (2005) A survey on services composition languages and models. In *International Workshop on Web Services Modeling and Testing (WS-MaTe 2006)*, pp. 51–63.

Carman, M., Serafini, L. and Traverso, P. (2003) Web service composition as planning. In *International Conference on Automated Planning & Scheduling, ICAPS'03, Workshop on Planning for Web Services*, pp. 1636–1642.

Chakraborty, D., Joshi, A., Yeshaand Y. and Finin T. (2006) Toward distributed service discovery in pervasive computing environments. *IEEE Transactions of Mobile Computing*, 5(2): 97–112.

Chan, A.T.S. and Wan, D.K.T. (2005) Web services mobility in a pocket. In *Proceedings of IEEE International Conference on Web Services (ICWS'05)*, pp. 159–166.

Coulouris, G., Dollimore, J. and Kindberg, T. (1994) *Distributed Systems, Concepts and Designs*. 2nd edn. Reading, MA: Addison-Wesley.

Dan, A., Davis, D., Kearney, R. *et al.* (2004) Web services on demand: WSLA-driven automated Management. *IBM Systems Journal*, 43(1): 136–158.

DiBona, C., Ockman, S. and Stone M. (eds) (1999) *Open Sources: Voices from the Open Source Revolution*. New York: O'Reilly.

Dickman, L.I. (1973) Small virtual machines: a survey. In *Proceedings of Workshop on Virtual Computer Systems*, Cambridge, MA, pp. 191–202.

Emmerich, W. (2000) Software engineering and middleware: a roadmap. In *Proceedings of Conference on the Future of Software Engineering*, pp. 117–129.

Eugster, P.Th, Felber, P.A., Guerraoui, R. and Kermarrec, A-M. (2003) The many faces of publish/subscribe. *ACM Computing Surveys*, 35(2): 114–131.

Fielding, R.T. (2000) Architectural styles and the design of network-based software architectures. PhD thesis, University of California, Irvine.

Foster, I. and Iamnitchi, A. (2003) On death, taxes, and the convergence of peer-to-peer and grid computing. In F. Kaashoek and I. Stoica (eds) *Proceedings of 2nd International Workshop on Peer-to-Peer Systems (IPTPS'03)*. Berlin: Springer Verlag.

Foster, I., Kesselman C. and Tuecke, S. (2001) The anatomy of the grid: enabling scalable virtual organisations. *International Journal of High Performance Computing Applications*, 15(3): 200–222.

Garlan, D. and Shaw, M. (1993) An introduction to software architecture. In V. Ambriola and G. Tortora (eds) *Advances in Software Engineering and Knowledge Engineering*, Vol. 2. New York: World Scientific Publishing Company, pp. 1–39.

Gelernter, D. and Carriero, N. (1992) Coordination languages and their significance. *Communications of the ACM*, 32(2): 97–107.

Goodman, D.G. (2000) The wireless internet: promises and challenges. *IEEE Computer*, 33(7): 36–41.

Hennessy, J.L., Patterson, D.A. and Arpaci-Dusseau, A.C. (2006) *Computer Architecture: A Quantitative Approach*. 4th edn. New York: Morgan Kaufmann.

Hingne, V., Joshi, A., Finin, T., Kargupta, H. and Houstis, E. (2003) Towards a pervasive grid. In *Proceedings of International Parallel and Distributed Processing Symposium (IPDPS'03)*, pp. 207–335.

Hord, R. M. (1999) *Understanding Parallel Supercomputing*. New York: IEEE Press.

Johanson, B. and Fox, A. (2002) The event heap: a coordination infrastructure for interactive workspaces. In *Proceedings of 4th IEEE Workshop on Mobile Computing Systems and Applications*, pp. 83–93.

Keen, M., Acharya, A., Bishop, S., *et al.* (2004) Patterns: implementing an SOA using an Enterprise Service Bus, IBM Redbook, Retrieved from http://www.redbooks.ibm.com/redpieces/pdfs/sg246346.pdf, accessed Oct. 2007.

Kindberg, T. and Fox, A. (2002) System software for ubiquitous computing. *IEEE Pervasive Computing*, 1(1): 70–81.

Krauter, K., Buyya, R. and Maheswaran, M. (2002) A taxonomy and survey of grid resource management systems for distributed computing. *Software Practice Experience*, 32: 135–164.

Kumar, R., Farkas, K..I., Jouppi, N.P. *et al.* (2003) Single-ISA heterogeneous multi-core architectures: the potential for processor power reduction. In *Proceedings of 36th Annual IEEE/ACM International Symposium on Microarchitecture*, Washington, DC, pp. 81–92.

Menascé, D.A. (2005) MOM vs. RPC: communication models for distributed applications. *IEEE Internet Computing*, 9(2): 90–93.

Meyer, B. (1998) Object-oriented Software Construction. Englewood Cliffs, NJ: Prentice-Hall.

Milojicic, D.S., Kalogeraki, V., Lukose, R., *et al.* (2002) Peer-to-peer computing. HP Lab Technical Report HPL-2002-57. Available from http://www.hpl.hp.com/techreports/, accessed Nov. 2007.

Montresor, A. (2004) A robust protocol for building superpeer overlay topologies. In *Proceedings of 4th International Conference on Peer-to-Peer Computing*, pp. 202–209.

Papazoglou, M.P., Traverso, P., Dustdar, S., *et al.* (2007) Service-oriented computing: state of the art and research challenges. *IEEE Computer*, 40(11): 38–45.

Popek, G.J. and Goldberg, R.P. (1974) Formal requirements for virtualizable third generation architectures. *Communications of the ACM*, 17(7): 412–421.

Pressman, R.S. (1997) *Software Engineering: A Practitioner's Approach*. 4th edn. Maidenhead: McGraw-Hill.

Satyanarayanan, M. (2001) Pervasive computing: vision and challenges. *IEEE Personal Communications*, 8: 10–17.

Smith, J.E. and Nair, R. (2005) The architecture of virtual machines. *Computer*, 38(5): 32–38.

Van der Aalst, W.M.P., Dumas, M. and ter Hofstede, A.H.M. (2003) Web service composition languages: old wine in new bottles? In *Proceedings of 29th Euromicro Conference*, pp. 298–305.

Wyckoff, P., McLaughry, S.W., Lehman, T.J. and Ford, D.A. (1998) TSpaces. *IBM Systems Journal* 37(3): 454–474.

4

Smart Mobiles, Cards and Device Networks

4.1 Introduction

Smart user devices in this chapter focus on tab or card and pad-sized devices in which the locus of control for the user interaction resides in user-customisable systems. Mobility is an important feature of these smart devices.

4.1.1 Chapter Overview

An overview of the different notions of mobility is shown in Table 3.1. This section continues by discussing smart mobile services as an extension of distributed service models, given in Section 4.2. Next (Section 4.3) operating system support for mobile computers and communication devices is discussed, including how to handle resource-constrained ICT resources and power management. Then smart card-type device are discussed, these are much more resource-constrained and application-specific than smart phone-type devices. Finally, some types of device networks are covered.

Some related mobility topics are covered in other chapters such as wearable devices (Section 5.4.3), implanted devices (Section 5.4.4), self-mobile devices, i.e., robots (Section 6.7), smart dust (Section 6.4), location-awareness for mobile devices (Section 7.4) and mobile communication (Section 11.6.5).

4.2 Smart Mobile Devices, Users, Resources and Code

Users are naturally mobile, e.g., users can move in between Internet nodes, to log on and to access Web-based content and email, anywhere, anytime. Users can carry personalised mobile networked devices with them to access services filtered according to their personal preferences and to be aware of their location context and adapt to it (Chapter 7). Other types of inanimate hosts such as transport vehicles can also act as mobile hosts. Each of the main components of a UbiComm system (Sections 3.1.2, Section 3.2) can be mobile:[1] virtual processing and services at the operating system and application level, code, access devices,[2] hardware and data resources.

4.2.1 Mobile Service Design

Mobility service design builds upon the basic design of smart devices and services but is a more specialised variant of it. In order to simplify access by applications and users, mobility, e.g., locating and addressing mobile users and routing data to mobile receivers, should be designed to be more transparent to applications and users. There are three kinds of transparency for mobile services: (1) user virtual environments (UVE); (2) mobile virtual terminals (MVT); and (3) virtual resource management (VRM). UVE provides users with a uniform view of their working environments independent of the location and terminal type, e.g., MobiDesk (Baratto *et al.*, 2004) and Virtual Home Environments or VHE (Moura *et al.*, 2002). MVT preserves the terminal execution state for restoration at new locations, including active processes and subscribed services, e.g., mobile agents (Section 4.2). VRM permits mobile users and terminals to maintain access to resources and services by automatically requalifying the bindings and moving specific resources or services to permit load balancing and replication (Bellavista *et al.*, 2001).

Mobility within a homogeneous network or network of networks can be supported at the Internet or network level of the network protocol stack so that mobility is transparent to applications (Section 4.2). Mobile devices may use a range of heterogeneous wireless communications protocols.[3] Mobility transparency is more challenging across heterogeneous networks. A seamless handover from one type of network to another with a limited delay could be performed by application processes but this is easier if this is OS supported. There must be a smooth transition between being used on the network and being a self-sufficient device. Wireless connectivity is still patchy, with different protocols around the world, fade-outs while moving and incomplete coverage, especially in remote areas, in some buildings or while airborne. Relying on a permanent mobile connection can be very frustrating for some data applications if such a connection is assumed. In short, mobile phone processes should support a volatile connectivity model (Section 3.3.3.9).

Mobile devices can create new local data that may be business-sensitive or personal. A *Denial of Service* (DoS) can occur when a mobile device is stolen or left behind. There are two possible solutions to handle DoS. First, back-ups, occasional wireless-synchronisation or wired synchronisation of data can be used. Second, a remote-access model can be used to support a virtual distributed UVE type desktop on the mobile device so that data that appears to be local is actually managed remotely e.g., MobiDesk (Baratto *et al.*, 2004). MobiDesk adds a virtualisation layer that supports a thin stateless client model but it requires an on-demand, always-on connection. For both these techniques, data synchronisation is needed. In applications, services, devices, computers, it needs to be clear to the user what is being synchronised. In foreground synchronisation, the user initiates synchronisation and selects the data to be synchronised. In background synchronisation, a designated event triggers script-based data selection and subsequent data synchronisation. In

---

[1] Apart from the physical environment which is not normally mobile unless this is some natural disaster.

[2] Although access devices, computation resources and peripheral devices are mainly mobile, middleware devices such as routers can also be mobile and positioned to optimise coverage, etc.

[3] Few mobile devices include support for the complete range of heterogeneous networks, hence network bridges or application gateways are needed, e.g., OSGI model (Section 4.5.2).

addition, concurrency conflicts can occur when changes are made to common data remotely and locally. Database concurrency control models can be used but this is challenging because data is distributed and data storage and media may be heterogeneous unless a UVE mobility model is used. Finally, if mobile devices are no longer accessible, they could be configured to be triggered to delete any important data once they are connected to an unrecognised network.

4.2.1.1 SMS and Mobile Web Services

One of the earliest and but still most widely used data application for many mobile phones users is the *Short Messaging Service* (SMS). SMS is a messaging transport service that supports a reliable two-way, connection-less, messaging protocol. It is very simple protocol that supports send and receive and is often used for notifications. Its main limitations are it is text oriented, that text size is limited to 160 characters or less, message latency is one minute or more, there is poor security at the application level and it is difficult to link messages to interactions and transactions.[4]

Because of the limitations of SMS, a more flexible data protocol, the *Wireless Application Protocol* (WAP)[5] was introduced. The aim of WAP was to become the de facto world standard for the presentation and delivery of wireless information and telephony services – Internet on a mobile phone. WAP supports connection-oriented, interactive communication sessions and multimedia. There is no hard message size limit, it has security, and latency is less than SMS. WAP supports data formatting and navigation for small screens. WAP was designed for sending data over wireless with delays, slow links and low-resource mobile terminals (Goodman, 2000). There are also several design challenges with the *Wireless Markup Language* (WML) used by WAP in addition to it having such significant differences compared to HTML that it requires developers of mobile services to almost learn another mark-up language. WML uses a card deck UI metaphor. The compressed WML deck must not be larger than 1.4K. WAP devices are often not 100% WML compliant or support 100% of its features. Rendering of WML on some micro-browsers makes navigation difficult. Device capabilities sometimes cannot be established. Cache problems are also an issue.

WAP also faces competition from a mobile information service called *i-mode* launched by NTT DoCoMo of Japan in Feb. 1999, based on proprietary technology, the DoCoMo packet communication network protocol (Enoki, 2001). The major advantages over WAP were that i-mode used a content language, compact HTML (cHTML), which was much more similar to HTML than WML. Web-based services were cheaper to use because they were packet-based not time-based. I-mode was also designed for use with higher resolution mobile terminals.

Mobile Web Service design often uses a three-tier thin-client, client-proxy, server architecture (Section 3.2). The thin-client architecture simplifies the features on mobile devices: they are just used for information access and can run a mini-Web browser or *micro-browser* (Lawton, 2001), however, application functionality is limited (see Figure 4.1). A client-proxy (Section 3.2.2.4) is used in order to offload handling the heterogeneity of adapting content to heterogeneous terminals, micro-browsers and content languages. In a passive approach to content adaptation, there is no automatic adaptation to the access terminal and web browser capabilities, users need to select the content format. In order to adapt content to a variety of devices, the following content characteristics must be defined: the access device I/O capabilities, the types of content language supported; the type of presentation, e.g., micro-browser, that is used to render the content and the layout and style of the content presented.

---

[4] Using gateways, it is possible to interchange messages with other systems such as email, Web, etc.
[5] WAP Open forum was founded by Ericsson, Motorola, Nokia, etc. in 1997. In 2002, OMA, the Open Mobile Alliance, was formed by about 200 companies. In 2003, the WAP Forum and other similar forums became integrated into OMA, http://www.openmobilealliance.org/, accessed July 2007.

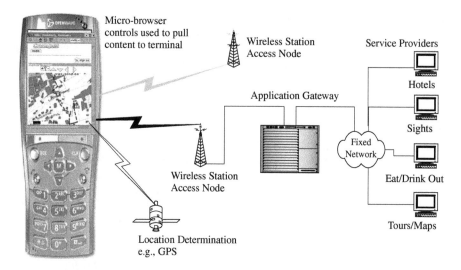

Figure 4.1 Thin client-server architecture example, a micro-browser running on a mobile device is used to retrieve content over a wireless network

Content adaptation can either be static, prepared for the most common sets of content characteristics for a lowest common denominator approach, or dynamically adapted to each set of content characteristics.

Information about the different devices display form-factor capabilities, the number of x–y pixels and colour depth, is needed in order to scale and adapt content to fit the constraints of a particular display. This is especially true when considering multi-publication, author-once-publish-many applications. The Device Independence Group (W3C) has created the *Composite Capabilities/Preferences Profile* (CC/PP) specification[6] in the RDF/XML language. In conjunction with the W3C, OMA has created a mobile device vocabulary for CC/PP called a User Agent Profile (UAProf).[7]

As noted above, there are different content languages for use in micro-browsers, e.g., WML and cHTML. The development of the W3C XHTML/CSS recommendations for the mobile web set expectations that the content languages would become standardised (Pashtan, 2005). In fact, the situation deteriorated because XHTML/CSS had yet more combinations than WML[8] and because experience has shown that it is extremely difficult to develop a XHTML portal for devices without detailed content adaptation solutions to dynamically fix a variety of browser bugs. From an

---

[6] See http://www.w3.org/Mobile/CCPP/, proposed by the W3C Ubiquitous Web Applications Working Group, accessed January 2007. Note other W3C groups have developed terminal definitions, e.g., the Mobile Web Initiative Working Group, http://www.w3.org/2005/MWI/DDWG/, which has proposed the Device Description Repository (DDR). CC/PP and DDR are similar.

[7] Unfortunately device vendors' adoption of CC/PP and UAProfiles to date has been relatively poor. Profiles can be hard to find, often invalid, or just plainly wrong.

[8] To date, there are four versions of XHTML (XHTML 1.0, XHTML 1.1, XHTML-Basic, XHTML 2.0) and OMA has worked on a mobile profile called XHTML-MP. Compatibility between versions is problematic and further amplified in the mobile space where XHTML-MP is no longer compatible with versions developed by the W3C.

operator perspective, the enormous permutations of browser and content language implementations are one key limitation hindering mobile Internet growth or increasing its complexity.

Content layout defines whether or not a full article, e.g., a full news story on a desktop, is
displayed on a mobile or only its title; whether or not to put the menu on the right for desktop
landscape screen and on the top for portrait screen, what should be the type of navigation and how
multiple screen content can be accessed. As mobile devices have limited local information storage, a
strategy is needed to manage local content. Either, the content can automatically expire or content
must be manually deleted.

4.2.1.2 Java VM and J2ME

J2ME, the Java[9] 2 Mobile Environment, supports the development of fat client–server system
and stand-alone system designs in which code is developed. The mobile device can be
networked to servers that are implemented using J2SE (Java Standard Edition) or J2EE
(Java Enterprise Edition). The J2ME platform arose from the need to define a computing
platform that could accommodate consumer electronics and embedded devices. J2EE, J2SE
and J2ME applications are developed to be executed on a Java VM (see Figure 4.2) to support
code portability across heterogeneous devices. Thus providing an appropriate subset of the
Java API is used, the same code can be written once and read and executed on many different
devices.

Rather than being structured as a single, monolithic platform with a large footprint for devices,
J2ME is a multi-layered organisation of software bundles of three types: configurations, profiles,
and optional packages. A configuration, such as CDC or CLDC, provides fundamental services for

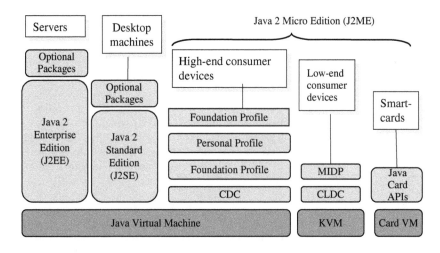

Figure 4.2   J2ME uses a VM to support a variety of devices

---

[9] In the early 1990s, the Green Project at Sun developed the Oak programming language, later renamed Java, for
programming consumer electronic devices but it became used more to develop desktop and server-based
applications. In 2000, the J2ME Platform Specification (JSR 68) was released, thus returning to Java's roots
to support programmable low resource devices.

a broad class of devices and virtual machines. A profile, such as the CLDC Mobile Information Device Profile MIDP, supports higher-level services for more specific class of devices. An optional package adds specialised services that are useful on devices of many kinds, but not necessary on all of them, e.g., those defined in other J2ME JSR Extension specifications.[10]

This organisation promotes both reuse and efficiency by enabling developers to put together a software stack that fits both the capabilities of target devices and the resource needs of applications. J2ME platform delineates[11] devices into two distinct categories or configurations (see Figure 4.2). First, *Connected Device Configuration* (CDC) devices are defined. These are devices with more than 2MB of both RAM and ROM and support constantly connected networks that have a special purpose or are limited in function. They are not general purpose computing machines, e.g., set-top boxes, Internet TVs, Internet-enabled screen phones, high-end communicators, and car entertainment/navigation systems. These devices generally have higher resource UI facilities and have more computing power. Second, *Connected Limited Device Configuration* (CLDC) devices are defined. These are devices with 160–512 KB memory and support personal, intermittently connected mobile information devices with a limited GUI, e.g., mobile phones, two-way pagers, PDAs, and organisers. CLDC also supports card-based ICT devices with limited ICT resources and without a GUI.

4.2.1.3 .NET CF

Windows Mobile[12] is a variation of Microsoft's Windows OS for minimalistic computers and embedded systems. Windows Mobile is now based upon a distinctly different kernel OS, rather than being simply a 'trimmed down' version of desktop Windows. Windows CE is a modular operating system[13] that supports several classes of devices. Some of these modules provide subsets of other components' features, e.g. varying levels of windowing support (DCOM versus COM), others which are mutually exclusive and others which add additional specialist features.

.NET Framework is an integral Microsoft Windows component for building and running software applications and WS (see Figure 4.3). The .NET Framework consists of two main parts: the common language runtime (CLR) and a unified set of class libraries such as ASP.NET for Web applications and services, Windows Forms for smart client applications, and ADO.NET for loosely coupled data access. Code written on the .NET Framework platform is called managed code. This refers to the Common Language Runtime (CLR) providing assurances code that handles certain common errors that plague Win32 programmers. Managed code cannot have bad pointers, cannot create memory leaks and supports strong code type-safety. Managed code is compiled down to a combination of MSIL (Microsoft Intermediate Language) and metadata.

---

[10] There are additional J2ME specifications such as JSR-135 Mobile Media API, JSR-172 Web Services Specification and JSR-82 Bluetooth API. Johnson (2006) reports the experience in using JSR-82, whereas plenty of phones have Bluetooth support, only a few devices specifically support J2ME and JSR 82. JSR specifications are available from http://jcp.org/en/home/index, retrieved Aug. 2007.

[11] This delineation is somewhat fuzzy, because technology continues to enable more and more power to be placed in smaller and smaller devices.

[12] Windows Mobile now represents a unified underlying platform for three kinds of device: Microsoft Smartphones, Pocket PCs, and Pocket PC Phones. There are several other related names. Often Windows CE, Windows Mobile, and Pocket PC are used interchangeably. This practice is not entirely accurate.

[13] WinCE or Windows CE is also a real-time operating system, with deterministic interrupt latency. It supports 256 priority levels and is thread-based to support concurrent programming.

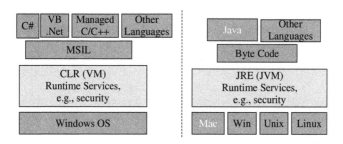

Figure 4.3   .NET VM versus the JRE VM

These are merged into a Pre Execution Environment (PE) file, which can then be executed on any CLR-capable virtual machine.[14] When this is executed, the JIT starts compiling the IL down to native code and executes it.

The Microsoft *.NET Compact Framework* (CF) is a rich subset of .NET that is designed specifically for resource-constrained devices, such as PDAs and smart mobile phones. Its class libraries are a subset of .NET (about 25%). .NET CF also has a few additional libraries that are specific to mobile devices: IrDA support, SQL Server CE and device-specific controls. .NET CF runs on a high performance JIT Compiler for mobile devices. A fat-client server design requires a .NET Framework to be installed on a mobile device.

4.2.2 Mobile Code

Code is usually designed to be downloaded from a remote service point. Installation requires configuring the code installation onto each platform. This is automated using a *Makefile*, e.g., installers can instruct the Makefile in the file system to install the code. Once it is installed, it often remains at one point in the ICT infrastructure, although it can be maintained and updated. Service access devices can also download new operational capabilities at run-time without requiring the capacity to store all possibly needed service support in advance – this reduces the need for resource-rich service access devices in dynamic environments. This paradigm enables providers to maintain, e.g., upgrade and fix, code in consumer devices with a network connection without the provider having to ship physical media to customers.

Unlike mobile computing, in which devices move or users move, and unlike static code where the code resides in one location, mobile code changes the network node where the program is executed (Fuggetta *et al.*, 1998). Code mobility is an important enabler for system extensibility, to support operation in open dynamic environments, in particular using resource-constrained access devices. Commonly used mobile code representations include Java, PostScript which instructs printers how to create images, and many other declarative languages that contain instructions that can be interpreted at a remote service access point. Whereas early mobile

---

[14] The .NET CLR and JVM approaches are similar in that they both use a VM that executes machine independent code. JVM is designed for platform independence (for multiple OSs) using mainly a single programming language (Java). A separate JVM is needed for each OS and type of device. CLR is designed for language independence (development supports multiple languages such as C++, VB, C# etc.) but for a single underlying OS, although multiple versions of Windows OS exist.

devices needed to be activated and to access code updates while attached via a peripheral connection to a PC, many current mobile devices can be activated and updated by code over-the-air (OTA) or via a WWAN.

Brooks (2004) gives an overview of mobile code paradigms based upon on where code executes and who determines when mobility occurs. This includes: client–server and remote evaluation interaction in which the client-side code needs to be (implicitly) downloaded before interacting with the server-side code; code on demand in which clients (explicitly) download and execute code as needed; process migration where processes move from one node to another to balance the processing load; mobile agents where a program code and its state move from one site to another according to its own internal logic; active networks where packets moving through the network reprogram the network infrastructure.

The benefits of mobile code include increased system flexibility, scalability, and reliability but this is tempered by the increased security risks concerned with potentially malicious or malfunctioning code being downloaded onto devices, and frequent disruptions to consumer ICT activities as multiple applications on multiple devices upgrade themselves. There are currently four main approaches to mobile code security (Zachary, 2003): *sandboxes* that limit the local services that code can access; *code signing* which ensures that code originates from a trusted source; *firewalls* that limit the machines that can access the Internet; and *proof-carrying code* (PCC) that carries explicit proof of its security.

4.2.3 Mobile Devices and Mobile Users

Device mobility can be viewed from several dimensions: (1) in terms of physical dimensions); (2) in terms of whether or not the device is mobile or some kind of host to which it is attached to is mobile; (3) in terms of what kind of host, mobile devices can be bound to; (4) in terms of how devices are attached to a host; and (5) in terms of when the mobility occurs. Each of these is discussed in turn.

To some extent, mobility depends upon physical dimensions, the smaller a device is, the less energy is required to move it, increasing the degree of mobility. Several different forms for UbiCom devices discussed in Sections 1.4.1.1 and 1.4.1.2 which could be mobile include dust, tabs, pads and skins. At one extreme, we have the metre-sized board type of UbiComp device which are often centimetres thick and too heavy to be portable by one person. However, if the thickness could be reduced and as a result, the material is made more flexible, then it behaves as a skin and can be made more portable; its size can be shrunk during transit by being rolled or folded. Devices such as organic displays and fabrics (Sections 5.3.4.3) can act as mobile skins. The second largest mobile device is the decimetre-sized pad devices such as a laptop, optical scanner, printer, etc. Some of these run an MTOS, some of these run an ASOS. The second smallest mobile devices are centimetre-sized pad devices such as mobile phones and smart cards. Pad devices and mobile phone-type tab devices tend to be expensive, higher resource ICT devices and tend to run a MTOS (Section 4.3). The card type of tab device tends to be much lower-resource ICT devices and governed by a more specialised CardOS (Section 4.4.1). The smallest type of mobile device is a dust-sized device (Section 6.4).

Devices may be intrinsically mobile using their own rechargeable or renewable power sources to drive them, e.g., a robot (Section 6.7). Devices may be mobile because they are attached to some host which may be mobile. A mobile host may be animate or inanimate. Devices may be mobile because they are bound in some way to some embodied mobile host such as a living thing or some physical world artefact such as a vehicle or they can be attached to non-living things which move, for example, tiny or micro dust-sized devices can be bound to air and fluids that flow around the physical environment driven by physical forces.

There are three basic ways mobile devices can be physically bound to mobile hosts: accompanied, surface-mounted or embedded into the fabric of a host, e.g., an embedded controller embedded in a host device. Accompanied refers to an object being loosely bound and accompanying a mobile host, e.g., a mobile phone can be carried in a bag or pocket but which can easily be misplaced. A device can be surface-mounted onto a mobile host. When the mobile host is a human, we refer to surface mounting as wearing. Smart paint can be sprayed onto a surface (Section 6.4.4). Embedded means a device can become permanently attached inside a mobile host, e.g., a braking control system embedded in a vehicle. When a device is attached in this way to a human or animal, we refer to this as an implant.

The different phases of the device's operational life-cycle when mobility can occur are as follows. A device may move from the place of manufacture to a permanent place of installation; can be mobile between sessions but static during sessions; can be mobile (roam from home) during user sessions but linked to a home location; can roam freely without a home (untethered).

4.3 Operating Systems for Mobile Computers and Communicator Devices

Desktop computers and high ICT resourced mobile devices such as smart phones (fat-client devices) often require more complex multi-process control and the use of a *Multi-Tasking Operating Systems* (MTOS) to support the execution of multiple users and applications.

4.3.1 Microkernel Designs

In a *macro kernel or Monolithic Kernel Operating System* (Section 3.4) all the system utilities such as hardware-related drivers, memory management, process support, process scheduling, network protocol stack and file system are in one, single, large, kernel system (Liedtke, 1996). The main benefit of the monolithic kernel system is that it is more efficient for a single processor system because fewer context switches are needed. Context switches are only needed between processes and the kernel utilities and not generally between processes. The main drawback of a monolithic kernel system for low-resource systems such as mobile systems is that use of the kernel is quite large and requires many system resources. In addition, the monolithic kernel is potentially more complex and has more points of failure and can require more updates, e.g., in order to support adding new hardware.

In contrast, in a microkernel, only the fundamental parts of the operating system such as basic memory management, (limited) process management and inter-process control are supported. The potential benefits of a small kernel are that it is more manageable in a low-resource environment and more robust. It can still function even when system utilities, not in the kernel, fail. The drawbacks are that there is potentially more context switching between application, non-kernel utility and the kernel utility process execution contexts, thus potentially lowering the performance when run on single processor systems.[15]

4.3.2 Mobility Support

Some original designs for OS for low-resource device such as mobile devices were based on creating a cut-down version of PC-type MTOSs, e.g., Windows Mobile, while other designs for an OS were specifically oriented to a lower-resource mobile device and other specialist characteristics from the

---

[15] The debate over macro versus micro-kernel has been discussed further in DiBona *et al.* (1999) with Linus Torvalds, creator of Linux, arguing the case for a macro kernel and Andrew Tanenbaum, creator of Minix, arguing the case for the micro-kernel design.

ground up. Mobile device design is still evolving. Some aspects of mobile OS support are even being proposed to produce low-cost, low-resource, low-power, PC design. In addition, to the normal OS ICT support, mobile devices have several more specialised requirements including HCI for small mobile user interfaces (Section 5.2.3), heterogeneous communication support (Section 11.7.5), intermittent connections (Section 3.3.3.9), data management (Section 12.2.9), mobility support (Section 11.6.5) and power management (Section 4.3.4).

The core OS kernel should be small for a low-resource device. Basic support is needed for memory management, to prevent memory leaks and to release system resources as soon as they are no longer needed. Good strategies for resource reuse are vital as resources are limited. There are different options for multi-tasking support. Multi-tasking is useful to support communications-capable real-time performance in order to talk, to count down alarms that were set and to run and access data and applications on the phone all at the same time, e.g., Symbian OS.[16] Alternatively, a system can schedule one task at a time, wait for it to finish and then switch to another one, i.e., non-pre-emptive task scheduling, e.g., Palm OS. Small mobile devices tend not to use magnetic-disk based secondary storage for persistence, because of the relatively slow access speeds, damage to moving parts and higher power consumption. Permanent storage is used in the form of a Flash ROM to retain files and data, however, flash memory is slower than RAM. Mobile devices tend to boot from ROM. However, flash memory uses less power, so the battery life of devices can be longer. Note in the past, certain types of ROM had a shorter lifetime in terms of number of read writes they supported before they failed.

4.3.3 Resource-Constrained Devices

OSs for resource-constrained devices, e.g., hand-held mobile devices, must cope with a low primary memory, low secondary storage capacity, slow CPU, limited input/output and operating in a low power mode[17] because of a finite energy supply. Strategies to cope with low memory include data compression, offloading data storage from a device (thin-client) to remote networked servers, simply over-writing older data and using larger capacity secondary memory. Strategies to adapt computation to a slow CPU include offloading computation from local to remote networked servers, the use of variable voltage CPUs and the use of energy-based and predictive process OS scheduling (Section 3.4). Strategies to decrease the energy required for communication involve using communication strategies where higher power servers and other devices initiate transmissions[18] and the use of short-range transmissions.

Satyanarayanan (2001) proposes different strategies to adapt to resource constraints. First, the system can automatically guide applications to modify their behaviour so that they use less of a scarce resource. This change usually reduces the user-perceived quality, or fidelity, of an application. Second, a client can ask the environment to guarantee a certain level of a resource. This is the approach typically used by reservation-based quality of service (Section 11.7.1). From the viewpoint of the client, this effectively increases the supply of a scarce resource to meet the client's demand. Third, a client can suggest a corrective action to the human user of the application and it is up to the user to follow this through or not. UbiCom system adaptation in general is considered in more detail as part of context-awareness (Section 7.2.4).

---

[16] The Symbian Developer Network at http://www.symbian.com/, accessed August 2007.

[17] For example, a high-power light bulb typically consumes 100 W/H, a low energy light bulb uses 10W/H, a laptop uses 10W/H and a phone 0.1 W/H.

[18] The energy needed to transmit is greater than the energy needed to receive in part because transmitted signals attenuate with distance.

4.3.4 Power Management

Power management is crucial for many embedded and mobile devices that are not attached to a permanent power supply. It increases their operational time in the field before their energy supply needs to be replenished and it reduces the time and effort needed to periodically replenish or replace nonrenewable energy sources. Power management is also a concern for devices permanently connected to a power source because of environmental and economic concerns. There are various strategies for device power reduction including: manual overrides for users to specify when they no longer need to use devices; the use of simple time-outs to detect long idle states and to power down devices to a sleep mode or off mode; the use of more finely grained techniques than on-sleep-off modes for power consumption (Section 4.3.4.1) and power conversion to store energy for later use that would otherwise be wasted.

4.3.4.1 Low Power CPUs

If devices are fully powered up all the time, power could be being wasted unnecessarily because, for example, the CPU, disk and display are not being used. Devices instead could hibernate. However, they also need to be responsive in all situations. A short boot sequence is needed when the device is turned on. Devices may never be powered down completely since they need to activate timed alarms or handle incoming calls, providing many hours of operation on a single charge or set of batteries. Competing processes and users may need to be scheduled to receive a fair share of battery (power) resources rather than CPU resources, e.g., an application that makes heavy use of DISK I/O may be given lower priority relative to a computer-bound application when energy resources are low. The CPU typically consumes 30–50% of power in a device. The power P of a CMOS processor satisfies Equations 4.1 and 4.2.

$$P = kCV^2 f \tag{4.1}$$

$$f \propto (V - V_t)^2 / V \tag{4.2}$$

where k is a constant, C the capacitance of the circuit, f the CPU frequency, V the voltage, and V_t a threshold voltage. A power consumption approximation is given in Equation 4.3.

$$P \sim (V^3) \tag{4.3}$$

Hence if the voltage of the CPU could be reduced, the power saving would be significant. Note, however, that reducing the CPU also reduces the CPU clock frequency. Dynamic Voltage Scaling (DVS)[19] enables the voltage of a CPU to be dynamically adjusted to save power (Pillai and Shin, 2001) (Figure 4.4). If a CPU runs at a uniform speed, minimum energy is used if the number of CPU cycles can be allocated exactly. However, jobs use cycles statistically, they often finish before using up their allocated time. There is the potential to save even more energy through stochastic, soft real-time, control of the CPU clock. The CPU voltage is determined according to program response time requirement and deadlines. Earliest deadline first (EDF) scheduling allocates CPU cycle budget per task based on profiles. It executes the tasks with the earliest deadlines and a budget based upon the number of cycles consumed by the task. A task is pre-empted when its budget is exhausted (Section 3.4).

---

[19] DVS-enabled processors include AMD's PowerNow! and Intel's SpeedStep and XScale (Pillai and Shin, 2001).

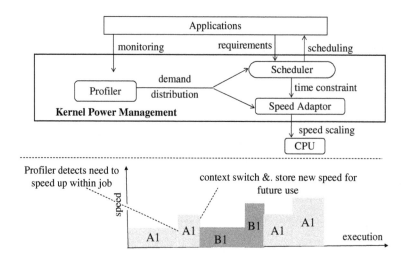

Figure 4.4 Use of Dynamic Voltage Scaling and Soft Real-Time scheduling to reduce CPU usage and power consumption

4.3.4.2 Application Support

Ellis (1999) makes a case that power management by the OS alone will not be optimum, applications need to be involved. He has proposed a wish-list of capabilities for a power-wise OS and what should be exposed to the applications layer. This includes: decoupling the states of various devices so they can be independently specified; use of a notification mechanism for imminent power-related events (e.g. the device is about to enter sleep mode) with an opportunity to respond, e.g., to save data in secondary memory. A key challenge is how to deal with the power management requirements from multiple applications which may conflict. In addition, background task resource and power requirements may differ from foreground tasks which currently control the main device input and output channels.

4.4 Smart Card Devices

A *smart card*[20] is a plastic card embedded with digital memory and (usually) a microprocessor chip, as opposed to cards which store data on magnetic strips. It is reprogrammable, stores and processes data in the card and transacts data between card users and applications. Data can be stored and processed within the card's memory or microprocessor, which is accessed using a card reader (Shelfer and Procaccino, 2002). Smart cards are small and easy to carry around. They provide a secure data container, can be used for authentication purposes, e.g., as a hardware-based digital signature, and can be used for metered services (Husemann, 1999).

---

[20] In 1977, Motorola and Bull, the French computer company, produced the first smart card microchip. The worldwide smart card market was estimated to be about 7 billion units in 2006. Smart cards can be memory only, containing no microprocessor. These are used as prepaid cards to transfer cash to a vendor, e.g., when making a phone call in a public kiosk, and simply discarded when empty.

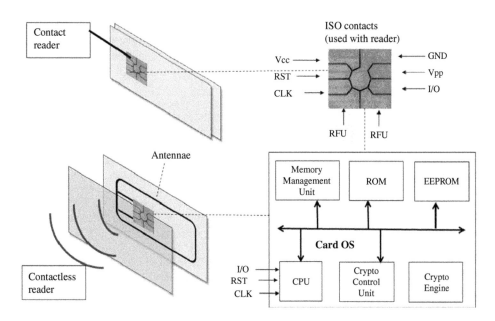

Figure 4.5 Contactless and Contact Smart Cards. Contactless cards include an inbuilt antennae and transceiver to interact with a reader. Contact cards include electrical contacts for a reader. A typical smart card OS is also shown

Many things found in a person's wallet have the potential to be replaced by a smart card, including driver's licence, insurance information, chip and pin credit or debit bank card, travel card and ticket, etc. Smart cards potentially represent a virulent form of Privacy-Invasive Technology or PIT (Section 12.3.4.2).

Smart cards maybe either contact-based or contactless (Figure 4.5). Contact cards have to be inserted in a reader and a physical electrical contact made. This may be chosen for security reasons, e.g., a chip (and enter) PIN (Personal Identification Number) application. Contactless cards also contain an antenna and an RF transceiver or transponder. The card can be waved in the immediate vicinity of a reader or base-station in order to interact, e.g., a smart travel card designed to be contactless reduces the time to gain authorisation compared to the use of a contact card.

Multiple types of plastic cards and applications could be combined into one universal, multi-functional smart card. Open issues include whether the owner of the card owns the private data, how owners can access their personal information, how to manage, differentiate and transfer multiple types of card data across multiple applications on multiple cards and how to manage the life-cycle of smart cards.

4.4.1 Smart Card OS

The primary tasks of a smart card operating system on behalf of applications are transferring data to and from the smart card, controlling the execution of commands, managing files, managing and executing cryptographic algorithms to protect access to stored data, and managing and executing program code. The components of a typical card OS (see Figure 4.5) are combined into a single IC chip to reduce size. To standardise the communication protocol between the on-card application, and off-card programs accessed via the reader, the ISO/IEC 7816 standard[21] has been defined.

4.4.2 Smart Card Development

Java Card refers to the technology that allows small Java-based applications (applets) to be run securely on smart cards and similar small memory footprint devices such as JavaRings. A Java Card is the tiniest version of Java targeted for embedded devices. It is widely used in SIM cards (used in GSM mobile phones) and ATM cards. The first Java Card was introduced in 1997. Java Card applications use Java Card bytecode, a subset of standard Java bytecode, executed in a Java CardVM. The J2ME basic profile and configuration are still too large for a Smart Card, a subset of Java bytecode used by Java KVM is used in the CardVM.[22] A different encoding is also used and optimised for size. This conserves memory, a necessity in resource-constrained devices like smart cards. Techniques exist for overcoming the memory limit in some Smart Cards such as partitioning the application's code.[23]

Java CardVM runs in many smart cards, e.g., the GSM phone Subscriber Identification Module (SIM) card. SIM cards securely store the service-subscriber key (IMSI) that is used to identify a subscriber. SIM cards allow users to change phones by simply removing the SIM card from one mobile phone and inserting it into another mobile phone or broadband telephony device.

Four main steps comprise Java Card applet development using the JavaCard development kit.[24] First, the applet functions are specified, e.g., a security function requires the user to enter a PIN; the card locks after three unsuccessful attempts to enter the PIN, etc. An ID (Applet IDs) is requested and assigned to both the applet and the package containing the applet class. The class structure of the applet programs is defined. The interface between the applet and the terminal application is defined.

Java Card focuses on providing a common programming interface for the development of smart card applications, however, the communication between the card's applet and non-card services is considered to be very basic, non-object-oriented and somewhat ad hoc in nature (Chan and Wan, 2005). Chan and Wan propose a WS SOAP-based wrapper for Java Card communication between the card and reader and a proxy on the reader to map the WS SOAP messages to the lower-level Java Card communication messages. WSDL (and hence UDDI) were not used at the card end to describe the services offered by the Java Card as this was considered too resource-hungry, however, it can be used in a proxy-based design. Their WS architecture, called WSCard, has been applied for use in an online drug-ordering web service.

4.5 Device Networks

The goal of a device network is to enable a wide variety of devices to interoperate. These activities include home automation such as light and climate control, person-aware systems, home security,

---

[21] ISO/IEC 7816 is a series of standards from the International Organisation for Standardisation for Smart Cards: physical characteristics (ISO 7816-1); contact location and dimensions (ISO 7816-2); electrical signals and low-level transport (ISO 7816-3) and high-level application (ISO 7816-4) Available from http://www.iso.org/iso/iso_catalogue/, accessed August 2007.

[22] For example, there is no support for some Java language features (types char, double, float and long; enums; arrays of more than one dimension; finalisation; threads).

[23] Similar techniques were also used in early PCs to overcome the memory limitations.

[24] See http://java.sun.com/products/javacard/dev_kit.html.

care in the community and pervasive AV content access (Section 2.3.2.1). The enablers for this are home network and service infrastructures that are easy to install, to configure and to maintain; low-cost (capital and operational cost) devices and infrastructures, and useful applications.

According to (Vaxevanakis *et al.*, 2003), wireless technologies will dominate in the main home network. In part, this is due to the ease of distribution throughout the home without the main-tenance cost of re-wiring and because the added insecurity of wireless does not yet appear to be a major concern for users. Currently, the home device infrastructure is highly fragmented and far from being a seamless infrastructure. First, many electronic devices are monolithic, may not be digital and are not network-enabled. Second, heterogeneous control devices and networks pre-dominate. Several systems have been proposed to connect and control home appliances in an integrated way including, InfraRed, X10,[25] HAVi and HES (Section 4.5.1), UPnP and Jini (Section 4.5.2) and OSGi (Section 4.5.2). Smart mobile devices such as laptops and mobile phones can use WiFi, DECT and 3G mobile phone networks. Broadcast video typically uses another network (Section 11.6). Bluetooth, supported in increasing numbers of consumer devices, is another type of wireless network. Third, inherent issues with wireless such as the ease of discovery and the finite power supply of many mobile devices are a major concern. For example, consumers may prefer to use humans as networks[26] to move storage media between devices because this reduces the cost of energy for communication transfer and simplifies the pairing of the content source and the content processing application.

4.5.1 HAVi, HES and X10

Home Audio Video Interoperability or *HAVi* is an industrial standard specification that provides standard APIs for typical audio and visual home appliances, for example, televisions. HAVi also provides a discovery service to find appliances connected to an IEEE 1394 network.

The Home Electronic System (HES) is an international standard for home automation under development by experts from North America, Europe and Asia. The Working Group is formally known as ISO/IEC JTC1/SC25/WG1.[27] HES defines the following system components. A Universal Interface is incorporated into an appliance for communicating over a variety of home automation networks. A Command Language for appliance-to-appliance communications is independent of which network carries the messages. A residential gateway, HomeGate, links home control networks with external service provider networks. Standard interfaces are defined to support interoperability among application domains, such as security, lighting, energy manage-ment and for command, control, and communications in commercial and mixed-use buildings such as apartment houses, retail shops and offices.

4.5.2 Device Discovery

If device networks are to support open, dynamic, heterogeneous device access and interoperability, automatic device discovery is necessary to simplify use. Device discovery standards and

---

[25] X10 wireless home networks, http://www.x10.com/

[26] Humans moving data has been used throughout the history of personal computing. In the days of floppy-disk drive personal computers in the 1980s and 1990s, this was referred to as floppy-disk networks or sneaker-net as people with footwear referred to as sneakers would walk or sneak around to transfer data between different multiple computers. Today, USB memory sticks are used instead.

[27] http://hes-standards.org/, accessed January 2008.

technologies include Sun's *Jini*, *UPnP forum* headed by Microsoft, IETF's SLP, DNS Service Discovery and Bluetooth's SDP (Helal, 2002).

Jini, introduced in 1998, consists of three Java language protocols: discovery, join, and lookup. Discovery occurs when a service looks for a lookup service, either using *a priori* information or by using a multicast with which it can register. Join is used by service providers to register service capabilities. Lookup occurs when a client or user locates and invokes a service described by its interface. The service can then be invoked by downloading it to be accessed locally or by using a remote method invocation (RMI) protocol to download the service access proxy software to invoke the service remotely. Jini grants access to services using service leasing for a fixed time period only. Before the lease expires, requesters must ask to renew the service lease. This prevents service connection resources being maintained for service requesters that are no longer active.

Simple Service Discovery Protocol (SSDP) is a Universal Plug And Play (UPnP) protocol that uses HTTP notification announcements to request a service-type URI and a Unique Service Name (USN). SSDP is supported in some firewall appliances so that applications can tunnel through HTTP to get external access and media exchange between host computers and media centres are facilitated using SSDP. The UPnP protocol is based upon HTTP/TCP or UDP/IP and XML. A control point in an UPnP network is similar to a client. It is capable of discovering and controlling other devices.

UPnP uses plain XML rather than WSDL to describe device profiles and supports higher-level descriptions of services in the form of a user interface that can be retrieved from an URL by users. This can be loaded into a Web browser and, depending on the page's capabilities, enables a user to control a device or to view the device's status. Devices can dynamically join a network, obtain an IP address, convey its capabilities on request, and learn about the presence and capabilities of other devices. SSDP is analogous to Jini in terms of supporting three protocols for discovery, join, and lookup. SSDP can work with or without its central directory service, called the Service Directory. When a service wants to join the network, first, it sends an announcement message to notify its presence to the rest of the devices. This announcement may be sent by multicast, so all other devices will see it, and the Service Directory, if present, will record the announcement. Alternatively, the announcement may be sent by unicast directly to the Service Directory. When a client wants to discover a service, it may ask the Service Directory for it or it may send a multicast message asking for it.

IETF Service Location Protocol (SLP) provides a flexible and scalable framework to enable devices to access information about the existence, location and configuration of networked services. SLP provides a dynamic configuration mechanism for applications in local area networks and is designed to scale from small unmanaged networks to large enterprise networks. Applications are modelled as clients that need to find servers attached to any of the available networks within an enterprise. In cases where there are many different clients and services available, Directory Agents that offer a centralised repository for advertising services may be used. Some peripheral devices such as printers use SLP. In order to remain simple and flexible, SLP is designed to support service discovery but not to invoke services, unlike Jini and UPnP.

DNS Service Discovery (DNS-SD) is a way to use DNS to browse for services. It can be used with the multicast DNS (mDNS) protocol to discuss the network. In contrast to Microsoft's SSDP, DNS-SD uses DNS rather than HTTP. The DNS that DNS-SD uses mDNS but unicast DNS can also be used.

Bluetooth Service Discovery Protocol (SDP) is specific to Bluetooth and supports search by service class and by service attributes, and service browsing when a Bluetooth client has no prior knowledge of the services available in its vicinity. If a Bluetooth device is set as discoverable, other devices can find it by broadcasting a query. Devices listed as 'not discoverable' do not respond to the query. If the search is successful, a device replies by sending the service record information, retrieved from its service discovery database, to the requesting device. The requester can then use the service record information to establish a connection with the other device.

Current SDPs are designed more for use in local area networks, e.g., the IP multicast range limits discovery in Jini. This is inadequate for use in mobile devices that require access to services from wide area networks. Device service discovery tends to use a central directory server, e.g., UPnP, Jini and SLP. However, some environments such as home environments cannot be relied upon to have a centralised directory service permanently available as it increases the cost and requires continued maintenance. Decentralised designs often use multicast and broadcast transmissions to locate service directories but this may be power-greedy and network bandwidth-greedy, an important consideration for wireless networks and mobile devices.

The second main type of design for decentralised SDP cuts down on the use of broadcasts. This involves dynamic directories, which initially advertise their presence by sending multicast messages to nodes in their vicinity. Since directories are deployed dynamically, more than one could be present in the same vicinity (Flores-Cortés et al., 2006). Some existing SDPs require computational resources, e.g., Jini requires the use of JVM and RMI which may not be present in many mobile devices. Existing service matching uses simple matching schemas and is done at a syntactic level, e.g., Jini attributes, Bluetooth SDP, etc.

4.5.3 OSGi

The Open Services Gateway Initiative (OSGi)[28] defines and promotes open specifications, such as a core platform specification, for the delivery of managed services into networked environments such as homes and automobiles. The initial market for OSGi was home services gateways, e.g., in video broadcast set-top boxes, broadband modems. These would then act as a gateway, between the end user (and owner) of the devices on a LAN, and the service providers, that could be accessible over the Internet who want to provide (i.e., sell) services for the devices such as home security, home health care monitoring and entertainment services (Hall and Cervantes, 2004). The core OSGi platform specification defines the service framework to include a minimal component model, management services for components, and a service registry. OSGi in turn uses underlying Java VM (Section 4.2.1.2) and OS services. Application services are encapsulated and deployed in bundles which consist of service interfaces along with their implementations and associated resources. Event-driven management mechanisms provided by the framework support the installation, activation, deactivation, update, and removal of bundles. Technically, the essence of an OSGi service framework implementation is a customised, dynamic Java class loader and a service registry that is globally accessible within a single Java virtual machine.

In terms of the basic services supported, OSGI appears similar to the Jini and UPnP frameworks (Section 3.3.2.1), the main difference being the more specialised management device functions. OSGi can also be used to act as a gateway to support interoperability between UPnP and Jini. A bridge between Jini and OSGi and an interface that transforms UPnP services to OSGi services and vice versa can also be defined (Lee et al., 2003). Allard et al. (2003) have built a direct bridge between Jini and UPnP, highlighting the following differences between Jini and UPnP. Jini uses a set of attributes to search for specific services, UPnP does not. Jini is a Java-only development whereas a number of languages are available for UPnP development. Jini events are quite different from UPnP events, where state variable values are transmitted directly to clients: either a remote

---

[28] Open Services Gateway Initiative, http://www.osgi.org, Accessed January 2007. Open source implementation of OSGi is available from http://oscar-osgi.sourceforge.net or Equinox, or available from http://www. eclipse.org/equinox/, accessed January 2008.

callback mechanism or the proxy can use a thread to poll the Jini service periodically and detect state changes. Jini, UPnP and SLP should in theory be simply to be modified to run over Bluetooth because they can run over IP networks and because IP protocols can easily be run over Bluetooth.

UPnP interactions may include multiple devices and services, e.g., when a user wants to set up and control sessions between multiple data sources and sinks. UPnP allows streams and other session-oriented protocols to be added on-demand to specific device or service profiles. There seems to be no compelling need to add stronger partial failure semantics, complex session or transaction semantics to the core UPnP architecture. Even though SOAP on top of HTTP is quite computationally inefficient;[29] neither the cost of its implementation nor its performance is perceived as a significant problem.

EXERCISES

1. Characterise the type of mobility in the following types of devices with respect to the classification of mobility given in Section 4.2.3: WLAN transmitters, computer, phone, smart card, camera, television and printer.
2. In order to simplify access by applications and users, mobility in terms of how to locate and address mobile users, how to route data to mobile receivers, should be transparent.
3. Compare and contrast three kinds of transparency for mobile services: user virtual environments (UVE), mobile virtual terminals (MVT) and Virtual Resource Management (MVT).
4. Discuss the motivation for using mobile code. Discuss some designs for mobile code based upon on where code executes and who determines when mobility occurs. Outline the security challenges and give some solutions to deal with this.
5. Compare and contrast SMS, WAP and i-mode as mobile information service infrastructures; designs based upon three-tier versus two-tier and fat-client versus thin-client, with or without client-proxy, server architectures; technology models for mobile data communication devices such as J2ME and .NET CF.
6. Why are smart cards seen as privacy-invasive technologies?
7. What is OSGI? Discuss how OSGI can be used in a multi-vendor device discovery environment.
8. Compare and contrast a micro-kernel operating system with a monolithic operating system plus middleware service model. Which is better for use in hand-held mobile devices?
9. Discuss the design of an OS for mobile use to deal with the higher prevalence of heterogeneous mobile access terminals; the need to dynamically route messages as the user moves, the need to deal with resource-constrained devices and the need to conserve energy for a mobile terminal with finite energy supply.
10. Outline designs for low power CPUs.

---

[29] Challenges with using SOAP/XML/HTTP for wireless communication include that this is quite verbose and bandwidth-heavy. Some solutions include discarding XML byte encoding, compressing messages either generically or XML-specifically such as binary XML. It is often better to use SAX (process XML events during message reading) rather than DOM (read in all message into tree structure in memory, then parse). HTTP could be discarded and replaced by the use of persistent connections and asynchronous one-way messaging. More efficient HTTP usuage includes negotiating HTTP parameters only once and using more compact protocol headers.

References

Allard, J., Chinta, V., Gundala, S. and Richard G.G. (2003) Jini meets UPnP: an architecture for Jini/UPnP interoperability. In *Proceedings of Symposium on Applications and the Internet (SAINT'03)*, pp. 268–275.

Baratto, R.A., Potter, S. Su, G. and Nieh, J. (2004) MobiDesk: mobile virtual desktop computing. In *Proceedings of MobiCom'04*: 1–15.

Barton, J. (2003) From server room to living room. *ACM Queue*, 1(5): 20–32.

Bellavista, P., Corradi, A. and Stefanelli, C. (2001) Mobile agent middleware for mobile computing. *IEEE Computer*, 34(3): 73–81.

Brooks, R.R. (2004) Mobile code paradigms and security issues. *IEEE Internet Computing*, 8(3): 54–59.

Chan, A.T.S. and Wan, D.K.T. (2005) Web services mobility in a pocket. In *Proceedings of IEEE International Conference on Web Services (ICWS'05)*, pp. 159–166.

DiBona, C., Ockman, S. and Stone M. (eds) (1999) Open Sources: Voices from the Open Source Revolution. New York: O'Reilly.

Ellis, C.S. (1999) The case for higher-level power management. In *Proceedings of 7th Workshop on Hot Topics in Operating Systems*, pp. 162–167.

Enoki, K-I. (2001) i-mode: the mobile internet service of the 21st century. In *IEEE International Solid-State Circuits Conference, ISSCC 2001*, pp. 12–15.

Flores-Cortés, C.A., Blair, G.S., Grace, P. (2006) A multi-protocol framework for ad-hoc service discovery. In *Proceedings of 4th International workshop on Middleware for Pervasive and Ad-Hoc Computing, MPAC '06*, pp. 10–16.

Fuggetta, A., Picco, G.P. and Vigna, G. (1998) Understanding code mobility. *IEEE Trans. Software Engineering*, 24(5): 342–361.

Goodman, D.G. (2000) The wireless internet: promises and challenges. *IEEE Computer*, 33(7): 36–41.

Hall, R.S. and Cervantes, H. (2004) An OSGi implementation and experience report. In *Proceedings of 1st IEEE Consumer Communications and Networking Conference, CCNC 2004*, pp. 394–399.

Helal, S. (2002) Standards for service discovery and delivery. *IEEE Pervasive Computing*: 1(3): 95–100.

Husemann, D. (1999) The smart card: don't leave home without it. *IEEE Concurrency*, 7(2): 24–27.

Johnson, S. (2006) Java in a teacup. *ACM Queue*, 4(3): 36–41.

Lawton, G. (2001) Browsing the mobile internet. *IEEE Computer*, 34(12): 18–21.

Lee, C., Nordstedt, D. and Helal, S. (2003) Enabling smart spaces with OSGi. *Pervasive Computing*, 2(3): 89–94.

Liedtke, J. (1996). Towards real microkernels. *Communications of the ACM*, 39(9): 70–77.

Moura, J.A., Oliveira, J.M., Carrapatoso, E. and Roque, R. (2002) Service provision and resource discovery in the VESPER. In *Proceedings of the VHE.IEEE International Conference on Communications, ICC 2002*, 4: 1991–1999.

Pashtan A. (2005) User mobility and location management. In Mobile Web Services. Cambridge: Cambridge University Press.

Pillai, P. and Shin, K.G. (2001) Real-time dynamic voltage scaling for low-power embedded operating systems. In *Proceedings of 18th ACM Symposium on Operating Systems Principles*, Banff, Canada, pp. 89–102.

Satyanarayanan, M. (2001) Pervasive computing: vision and challenges. *IEEE Personal Communications*, 8: 10–7.

Shelfer, K.M. and Procaccino, J.D. (2002) Smart card evolution. *Communications of the ACM* 45(7): 83–88.

Tao, L. (2001) Shifting paradigms with the application service provider model. *IEEE Computer*, 34(10): 32–39.

Vaxevanakis, K., Zahariadis, Th. and Vogiatzis N. (2003) A review of wireless home network technologies. *ACM SIGMOBILE Mobile Computing and Communications Review*, 7(2), 59–68.

Zachary, J.M. (2003) Protecting mobile code in the wild. *IEEE Internet Computing*, 7(2): 78–82.

5

Human–Computer Interaction

5.1 Introduction

The term Human–Computer Interaction (HCI) has been in widespread use since the advent of the IBM computer for personal use in the mid-1980s. However, the groundwork for the field of HCI certainly started earlier, at the onset of the Industrial Revolution. Tasks became automated and power-assisted, primarily to save labour, but also motivated by the need to overcome some limitations in human abilities and to perform tasks at a reduced cost. This triggered an interest in studying the interaction between humans and machines in order to make the interaction between them more effective. To enable humans to effectively interact with devices to perform tasks and to support activities, systems need to be designed to be useful and to be usable.

5.1.1 Chapter Overview

The chapter is structured as follows. First, HCI is considered from the perspective of the explicitness (eHCI) versus the implicitness (iHCI) of the interaction. Next, the diversity of ICT device interfaces and interaction is considered starting with the interaction of four commonly used ICT devices: desktop and laptop PCs, mobile phones and games consoles (Section 5.2). It is noted that several designs include types of natural interaction such as gesture input, voice input, etc. Then types of interaction and interfaces are considered further in a wider range of computer devices that support much more natural human–computer interaction (Section 5.3). Types of interaction and interfaces for human-embedded devices are considered next (Section 5.4). The design process for human-centred interaction, oriented towards interaction involving explicit human input is considered next with respect to the personal visual memories (PVM) scenario described in Section 1.1.1 (Section 5.4.5). This section also considers how this design process can be enhanced to incorporate elements of implicit HCI to supplement explicit HCI.

Note also that some details of HCI and iHCI are incorporated in other chapters of the book: tangible UIs based upon MEMS devices and amorphous computing (Section 6.4); context-awareness of humans to adapt ICT services to people (Chapter 7); user models based upon human intelligence and artificial intelligence (Chapter 8); group-based and social interaction (Chapter 9); management of smart devices in human-centred environments (Section 12.3); key challenges and outlook for HCI and HPI (Sections 13.5, 13.6).

Ubiquitous Computing: Smart Devices, Environments and Interactions   Stefan Poslad
© 2009 John Wiley & Sons, Ltd

5.1.2 Explicit HCI: Motivation and Characteristics

The basic concepts of HCI are:

- *humans*: single or multiple users, with diverse physical and mental abilities, interacting cooperatively or competitively;
- *computers*: not just PCs but also a range of embedded computing devices and a range of device
  sizes such as dust, tabs, pads and boards (Section 1.4.1);
- *interaction*: may be directed via a command or by manipulating virtual objects (windows,
  desktop) but it can also involve more natural interaction such as speech interaction, gestures, etc.

HCI refers to the processes and the models for design and the operational interface for some
products. For many users, the User Interface (UI) part of the system is the product. Explicit HCI
puts the user at the centre of the interactive systems, so that the control of the system, responds to
and is driven externally by the user, rather than the system being driven internally.

Poorly designed UIs can lead to both higher training costs and higher usage costs and of course
leads to lower product sales. The reasons for higher training costs include users spending time
working out what is happening, trying out inappropriate computer services and impaired task
quality. Users may feel that a particular machine forces them to do tasks in ways they prefer not to
(no control). Users may have to re-learn how to perform tasks: starting work again lowers
productivity. Poor UIs can lead to higher error rates not acceptable in safety-critical systems that
require some human interaction.

The motivation for HCI is clear; to support more effective use (Dix *et al.*, 2004) in three ways to be:

- useful: accomplish a user task that the user requires to be done;
- usable: do the task easily, naturally, safely (without danger of error), etc.;
- used: enrich the user experience by making it attractive, engaging, fun, etc.

The success of a product depends largely on both the user's experience with how usable a product is
and how useful it is in terms of the value it brings to them. This combination has been neatly
summarised as Heckel's law and Heckel's inverse law (Derrett, 2004). Heckel's law states that the
quality of the user interface of an appliance is relatively unimportant in determining its adoption by
users if the perceived value of the appliance is high. Heckel's inverse law states the importance of the
user interface design in the adoption of an appliance is inversely proportional to the perceived value
of the appliance. What these laws express is that although the usability of the UI is important, the
overriding concern is the usefulness of the device itself. If a difficult user interface acts as an
inhibitor to the uptake of an appliance, then the appliance probably does not have enough
perceived value to be useful.

5.1.3 Complexity of Ubiquitous Explicit HCI

Explicit HCI is complex for UbiCom scenarios even if it is well designed for individual devices
because we need to use tasks as part of activities that require access to services across multiple
devices, because devices can be used by different types of people, because users are engaged in
multiple concurrent activities, because users are engaged in activities which may occur across
multiple physical environments, because activities may be shared between participants, and
because activities on occasion need to be suspended and resumed.

The amount of explicit HCI used can overwhelm users as more and more computer devices
require inputs, even if individually, computers are designed to be usable. Individual devices may be

simple to use but sometimes multiple devices may need to be used concurrently to support multiple concurrent user activities such as receiving a phone call while performing some other activity. Some tasks may require multiple individual devices to be configured and their individual behaviour to be orchestrated, e.g., to record live audio–video content while playing already recorded content.

5.1.4 Implicit HCI: Motivation and Characteristics

Explicit HCI (eHCI) design supports direct human intervention at set points during a device's normal operation. Pure explicit interaction is context-free which means that users must repeat and reconfigure the same application access every session even if every session repeats itself. It is also more about H2C (Human-to-Computer) Interaction (Section 1.3.1.2) in which the focus is on the human having a model of the system (a mental model) rather than the system having a model of the individual user.

For example, consider a person entering a very dark room to retrieve an object, he or she will explicitly switch on the lights. A key part of the usability of the interaction design is to position the switch to be conveniently reachable and possibly to illuminate the switch so that it can be activated in the dark. Alternatively, human interaction may be implicitly modelled, e.g., rather than requiring humans to switch on a light when someone enters the room, the environment is smart. It contains devices which monitor people entering the room and switches on the light for authorised people automatically – this is an example of implicit interaction. The shift from explicit interaction design to also include implicit interaction design will be a key enabler of effective ubiquitous computing systems. UbiCom systems need be designed to support both explicit HCI and implicit HCI.

The concept of implicit HCI (iHCI), proposed by Schmidt (2000) was introduced in Section 1.2.3. It is defined as 'an action, performed by the user that is not primarily aimed to interact with a computerised system but which such a system understands as input'. To support iHCI interaction is more about C2H (Computer-to-Human) Interaction (Section 1.3.1.3). Implicit interaction is based on the assumption that the computer has a certain understanding of users' behaviour in a given situation: it is a user-aware type of context-aware system (Section 5.7.2). This knowledge of users' behaviours in given situations is then considered an additional input to the computer while doing a task (Schmidt, 2000). Implicit interaction can allow some of the interaction to be hidden from users and hence for the device to become invisible.

Implicit interaction is also introduced as we seek to design systems with which we can interact with in a more natural[1] way. For example, if we use hand gestures, such as a clap, to control a device to switch it on and off, we may also clap our hands for other reasons such as to express an emotion. Unless the context of the gesture is also defined and shared, clapping cannot be unambiguously used to imply that it is a command to switch a device on or off.

There are some obvious challenges in supporting implicit interaction. It can sometimes be complex to accurately and reliably determine the user context because of: the non-determinism of the subject, the user has not clearly decided what to do, or because of non-determinism in the subject's environment. Implicit interaction, in contrast to explicit interaction, reduces or even removes explicit user control. The user context determination may however invade and distract users' attention in order to directly interact with them or the determination may be inaccurate

---

[1] Natural means as humans we can directly use senses to interface with artefacts in the physical world rather than some tool which causes us to shift our locus of attention from the task to the tool. However, in some cases, our familiarity and expertise with a tool are so great, it appears to become part of us, interaction via these tool interfaces also seems natural, e.g., some humans may find 'writing text' using a keyboard more natural than using a pen. Familiarity is cultural and subjective.

because it can only indirectly model users, e.g., via observing user interaction and inferring user behaviour. Systems may also require time in order to learn and build an accurate model of a user. Partial iHCI systems can be very useful in practice if designed appropriately. For example, they can be used to anticipate actions, to prioritise choices and to lessen the overload on, and the interaction by, users.

5.2 User Interfaces and Interaction for Four Widely Used Devices

Four of the most commonly used networked ICT devices are the personal computer in its various forms such as desktop and laptop, hand-held mobile devices used for communication, games consoles and remote-controlled AV displays, players and recorders. Each of these can be designed to perform a common set of functions such as AV player, record, output, data and AV communication (Section 2.3.2.1) and to act as a hub or portal for multiple service access and interoperability.

5.2.1 Diversity of ICT Device Interaction

The term computer in the acronym HCI has a much more diverse meaning within the field of UbiCom. It refers to any device with a programmable IC chip inside, including a range of multi-task operating system (MTOS) devices. Common multi-task devices include desktop and laptops PCs, mobile phones, games consoles, AV recorders and players such as televisions, radios and cameras. Each of these supports keypad or keyboard haptic inputs and audio outputs and output display interfaces. The user interfaces for these devices are primarily visual. Note the design for a universal visual interface interaction model will not work equally effectively across the wide variety of display and input device types and sizes.

Even more numerous than MTOS computers are computers embedded in devices that perform specialised tasks such as various household appliances, vehicle control systems, travel ticket machines, cash dispensers, building controls, etc. Their interfaces are far more diverse and in many cases have less input controls, typically consisting of a set of control buttons and knobs, and outputs such as one or more LEDs (Light Emitting Diodes) and LCDs (Liquid Crystal Displays).

In Section 1.4, six different form factors for ICT devices were considered. Of these, decimetre-sized, from about 5 to 20 centimetres are the most common form for ICT devices today (Figure 5.1). There are several dimensions devices could be characterised according to:

- *size*: hand-sized, centimetre-sized, decimetre-sized versus micro-sized versus body-sized or larger;
- *haptic input*: two-handed versus one-handed versus hands-free operation;
- *interaction modalities*: single versus multiple;
- *single user versus shared interaction*: in personal space, friends' space or public space;.
- *posture for human operator*: lying, sitting, standing, walking, running, etc.;
- *distance of output display to input control*: centimetres to metres;
- *position during operation*: fixed versus mobile;
- *connectivity*: stand-alone versus networked, wired versus wireless;
- *tasking*: single task devices versus multi-task devices;
- *multimedia content access*: voice and text communication-oriented, alpha-numeric data or text-oriented, AV-content access;
- *integrated*: embedded integrated devices versus dynamically interoperable devices.

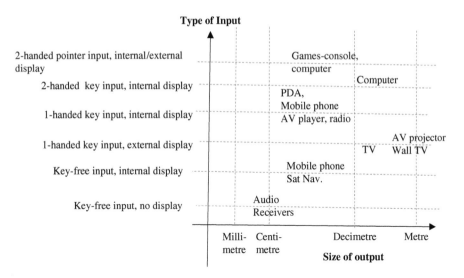

Figure 5.1   The range of ICT device sizes in common use in the 2000s

Current ICT device interaction is dominated by explicit human–device interaction, the use of MTOS systems, and interfaces with visual output.

5.2.2 Personal Computer Interface

Although Windows systems were demonstrated as early as 1965, the basic idea was envisaged as early in 1945 by Bush (1945) in his MEMEX system: 'It consists of a desk, and while it can presumably be operated from a distance, it is primarily the piece of furniture at which he works. On the top are slanting translucent screens, on which material can be projected for convenient reading. There is a keyboard, and sets of buttons and levers. Otherwise it looks like an ordinary desk.' Section 13.6 reviews the changes in UIs and considers UIs in the future. From the early 1980s, computers started to become more widely used by non-specialist users. The keyboard and visual command interface which allowed text to be entered and displayed one line at a time dominated computer interfaces. Users needed to be able to remember the command name and the syntax of any parameters used to qualify the command such as data files for the command to act on. The command is executed when a delimiting character such as the return key is typed. Commands within an application can only be issued sequentially. It is a quite an efficient interface when the same command needs to be repeated, using a loop, on different data sets.

From the early 1990s, the WIMPS interface which supports direct manipulation of visual objects prevailed. Users can interact with computers by typing text commands but for many tasks direct interaction can be used by activating and moving active window sub-areas called icons and menus, using mouse clicks and mouse drag-and-drop respectively. Such an interface is commonly called *WIMPS* (Windows, Icons, Menu and Pointer device) and direct manipulation. It is the dominant interface for desktop and laptop computers at this time. The mouse as a pointer device was first demonstrated in 1965 as a replacement for the light-pen. This did not achieve its first commercial success until the mid-1980s with the advent of the Apple Macintosh (Myers *et al.*, 1996). The WIMPS interface is associated with a desktop metaphor. Its documents relate to windowed areas of the screen. Just like documents, windows can be arranged in stacks or tiled (placed side by side), created, discarded, moved, organised and modified on the display screen using the pointer device (and keyboard) in a technique called *Direct Manipulation Interfaces* (Shneiderman, 1983) because these interfaces allow direct manipulation of the visual objects. The two main advantages of the WIMPS UI over the command UI are that the order of multiple commands can be much more ad hoc and the use of direct manipulation means users do not need to remember command names. A good overview of the history of windows and direct manipulation type PC interaction is given in Myers *et al.* (1996).

Shneiderman and Plaisant (2004, pp. 213–264) describe several challenges with direct manipulation: it is not necessarily an improvement for visually impaired users[2]; it consumes screen space which is more critical in lower resolution displays; the meaning of visual representations may be unclear or misleading to specific users; mouse pointer control and input require good hand–eye coordination and can be slow. Designs to address these challenges include reducing the number of virtual objects which need to be displayed, abstracting and highlighting the essence of object features and browsing and navigating between different interlinked sub-sets of objects that are semantically organised.

Dialogues are mechanisms in which users are informed about pertinent information that they must acknowledge receipt of or they ask for input to constrain a query. Typically, this type of interface is displayed in the form of a pop-up window called a dialog box. Query dialogues may also use a language that requires a specific syntax to constrain the queries, e.g., the SQL relational database query language. Form filling dialog interfaces are used by many applications for alphanumeric data input, e.g., into an information system, or for data output, e.g., a spreadsheet. These enable applications to receive data input in a structured way, reducing the processing used by a computer.

5.2.3 Mobile Hand-Held Device Interfaces

PC-style WIMPS interaction will not work as effectively on mobile devices because the display area is smaller. It is impractical to have several windows open at a time. It can be difficult to locate windows and icons if they are deeply stacked one on another and to resize them. Screen navigation using fingers on a touch-pad or an external device may be too big and unwieldy for small devices. In addition, the keyboard is smaller for user input and there is a greater variety of input devices.

Instead of using the inbuilt device interface, the device can be attached to different kinds of external input interface which are available in the environment. It could become common to have displays, keyboards and Internet work connections at fixed hotspots and to allow users to plug in their own mobile access device. Single hand-held mobile devices such as PDAs and smart phones have used a variety of types of interfaces to overcome their resource-constrained input and output devices (Jones and Marsden, 2006) as follows.

5.2.3.1 Handling Limited Key Input: Multi-Tap, T9, Fastap, Soft keys and Soft Keyboard

The majority of key input techniques introduce modes because of the limited number of keys and the minimum key-size that can be consistently clicked, so that the same interface

---

[2] Note, users may have permanent visual impairment or the environment conditions, e.g., bright sunlight, or user activities such as, running whilst looking at a device, may lead to transient visual impairment for normally unimpaired users.

interaction, e.g., pressing the same key, leads to different kinds of input actions (*multi-tap keypad*). A typical numeric keypad used in current standard mobile devices has 12 keys. Eight of the numeric keys have three alphabetic characters associated with them that can be triggered by multi-tapping keys, e.g., for the 2 key, 1 tap gives 'a', 2 taps gives 'b', etc. Multi-tapping can be enhanced with the introduction of a dictionary-based predictive text method known as T9 and can produce a text entry speed up to 40 words per minute. T9 is an example of implicit interaction complementing explicit interaction – the user types the start of words, the system tries to automatically complete the rest of the words based upon a simple text context

The *Fastap keypad* involves two keypads, one with smaller keys raised at the corners above the other keypad keys. The upper one is used for alphabetic input, the lower one for number input, both laid out over the size of a business card. If several keys are hit at once, a technique called passive chording allows the system to work out what the user intended to enter.

Soft keys enable the meaning of the two left and right keys at the top of keypad to be determined by information on the screen; this allows the same keys to be reused to support application and task specific choices. Instead of having two soft keys, a whole mini keyboard, a soft keyboard, could also be displayed if there is sufficient screen space.

Internal pointer devices such as tracker pad, roller pads, mini-joysticks or even keyboard arrow keys can be used to move the pointer on the screen. Screens may also be designed to be touch-screens whose areas can be activated using some physical stick-like pointer, pen or a finger (Section 5.3.4.1).

Auditory interfaces with speech recognition enable the use of voice activation to accept incoming calls and to initiate outgoing calls. However, there are additional challenges for mobile users when: operating in variable physical environments such as a noisy background; users have limited attention; users are less able to attend to the phone sometimes when they are engaged in other activities.

5.2.3.2 Handling Limited Output

There are multiple techniques for overcoming limited output. This was an issue for many earlier desktop systems and not just an issue for current mobile devices. For example, the UNIX MTOS system used a single line text editor program, 'ed', in the 1980s/1990s in order to maximise use of a low resolution screen. Mobile devices need to be smarter at filtering information to minimise the information available for display. Mobile displays tend not to use a Single Window WIMPS system because of the lack of control in using a pointer to select parts of a screen on a small display. The single window or screen can be switched to other content using a common navigation bar on each page, or hypertext links can be used to support between-page navigation while scrolling can be used to support within-page navigation.

There are several well-known approaches to deal with size differences between the size of visual content and the size of the display. If content is too large, it can simply be cropped or the content resolution can be reduced or a zooming interface can be used. Zooming (in and out) coupled with the use of scrolling (up and down) and panning (side-to-side) control enable users to view content piecemeal in an interactive way. Marking which part of the whole view that is currently zoomed in another miniature view is also for useful for orientation. Another development here is a peephole display (Yee, 2003) which uses sensors to act as a tangible UI. Sensors determine the position of the device in relation to the user. As the device moves towards or away from the user, its position changes and its display can be updated to display different information. Two further approaches to enable small form sizes to display large visual content are to use MEM-type micro projectors (Section 6.4.2) or to use organic displays which can be folded during transit and unfolded for operation.

Audible outputs can also be useful. This is beneficial when the user's visual locus of attention is already engaged. For example, a vehicle navigation system can give audible directions as drivers focus on steering the vehicle while driving. *Haptic outputs* can also be used as an output device. These have the advantage that touch-sensitive signals can be received more silently and privately. For example, vibrations can be used to signal incoming calls in mobile devices carried on a person. The vibrations can vary in intensity and rhythm so that the incoming signal, the phone call, can convey more information, such as the urgency of the call and the type of user.

5.2.4 Games Console Interfaces and Interaction

Games consoles are an important driver and can contribute to UbiCom in a number of ways. They can introduce innovative ways of interacting with physical environments into the mainstream and they can help to advance the use of augmented and the more general mediated reality environments. Bushnell[3] has highlighted that computer games have often acted as an incubator for many innovations driving computing (Bushnell, 1996). For example, in 1971, at the start of computer games development, computers had no monitors but used paper, punched cards and paper tape for input and output. To create a supply of monitors for games consoles, the tuner was stripped out of television sets, these were stored in skips as scrap at Atari, and the remaining parts of the monitor were kept and modified for use in computer games. As a result of computer games, people became more used to the idea that video content could be generated locally, in contrast, to the early days of the TV, when people asked how the TV station knew what the knobs on the TV were doing – consumers had a very strong sense of presence of the TV station actually being in their home.

Early video games were coin-operated and contained simpler electronic logic such as counters as at that time the first microprocessors had not yet been invented. Atari was the largest consumer of N-channel LSI (Large-Scale Integration) electronics in the 1970s and 1980s, driving the demand for the ROM and RAM memory chips that then were produced in high volume at a lower cost per chip, enabling personal computers with memory to be priced lower. Bushnell (1996) also highlights that computer games often drove much development in human–computer interaction such as interactive storylines, collaborative computing, anthropomorphism, graphical user interfaces, three-dimensional graphics and the use of non-keyboard and mouse interfaces. There is a prominent utility model in coin-operated games that drives the use of intuitive, easy-to-learn interfaces that require minimum training and instruction – games machine interfaces without these properties were used less and generated less revenue.

Forster (2005) has identified seven different generations of games consoles based upon the technologies they use. Current, seventh-generation, game consoles include the Nintendo Wii which has deviated from the standard D-pad interface,[4] by using a one-handed wireless wand-type games console interface instead. It contains micro-sensors in the form of accelerometers located inside the controller and an infrared detector to sense its position in 3D space. The development and cheaper fabrication of micro sensors and actuators (Section 6.4) can lead to a much wider range of interfaces to ICT devices and ways to interact with these in physical and virtual environments.

---

[3] Nolan Bushnell, the founder of Atari is often credited as being the founder of the computer games industry.
[4] The third-generation games controller introduced in the mid to late 1980s introduced an 8-direction D-pad, a four-way digital cross, with two or more action buttons and later with a thumbstick to give two-dimensional input. This became the standard games controller taking over from joysticks, paddles, and keypads as the default game controller and remains so over twenty years later.

The main focus of many types of electronic games is the interface between the person, the game and an elaborate scoring system. This scoring system is often tuned so that as the game progresses, it is harder and harder to score points but the points amass larger values. If done just correctly, the effect is a near drug-like high in players who are said to be 'in the zone'. If games fail to sufficiently heighten players' experiences, games remain unplayed. Hence, the main focus of games is on heightening the user experience. Sometimes, a games interface requires a great deal of training and experience before it can be heightened. The introduction of a more natural interface, such as the Wii wand, can make it easier for broader audiences to interact with the system and become immersed in the virtual game environment.

5.2.5 *Localised Remote Control: Video Devices*

Many devices that are body-sized or larger such as earlier mainframe and mini-computers in the 1970s and 1980s, large TVs and AV projectors and cars[5]. These often incorporate greater remote and automated control to reduce the degree of manual interaction that would otherwise need to be situated at the device interface. Typically these devices are numeric keypads rather than keyboards. If users need to enter text, e.g., for a caption to an AV recording, then letters are selected by using cursor keys to navigate around a soft (screen) keyboard. Usually, the remote control requires a line of sight to a display in order to see the feedback and status of the device being controlled.

Some large devices have no remote control partly because the operation of the device requires local manual input which cannot yet be robustly automated. One type of device for which a profusion of remote controls exists is the AV player recorder whether in the form of mobile camera, fixed set-top box player, etc. There are several design and wider engineering issues here. There is a profusion of remote control devices which have overlapping features and in some cases several devices need to be orchestrated with respect to a common feature, e.g., setting the volume for a home entertainment centre. A universal localised remote control may be useful (Section 2.3.2.3). Devices may have limited network connectivity and may not always be connected to an IP network. Device control could be usefully operated remotely, e.g., because someone has forgotten to set a timer recording but this may cause conflicts if the device is being used by another local user who wants to perform a different AV function. If devices were universally connected to a common network, their status and utility could be more easily remotely accessed and configured.

5.3 Hidden UI Via Basic Smart Devices

The WIMPS-style interface which dominates PCs is considered by many computer scientists to be obtrusive in the sense that it requires users to consciously think about how to operate a mouse pointer interface and which keys to press to use the computer. The computer itself is localised and users must go to its location to use it. In contrast, systems which can be situated where our activities are based and which directly make use of natural human sensory input offer a less obtrusive computer interface.

In the following sections, several of these more natural interfaces are considered, beginning with multi-modal interaction. Some of the four commonly used devices, mentioned in Section 5.2, now support more natural inputs and interaction such as touch-sensitivity, gestures to control the display and speech input control.

---

[5] The projector itself can now be quite small but it is the screen itself that users interact with and this is much larger. Current cars often uses a remote control to lock and unlock it.

5.3.1 Multi-Modal Visual Interfaces

The *modality* of interaction refers to a mode of human interaction using one of the human senses and to the type of computer input. The categories of human senses are sight, touch, hearing, smell, and taste. ICT systems have modalities that are equivalent to some of the human senses such as cameras (sight), input devices such as touchscreens, keypads and pointer devices (touch, haptic), microphones (hearing) and the use of various chemical sensors and analysers (smell and taste).

The majority of interactive systems predominantly use a single visual mode of output interaction between a system and a human user but this can overload humans as the world becomes more digitally interactive and as more objects can seek to interact with the user at any one time. The use of multiple sensory channels can alleviate this bottleneck by increasing the bandwidth available. Human interaction is naturally multi-modal in the sense that users can use multi-modes of input and output to an extent concurrently. Systems typically use multiple instances of haptic modes, for input – mouse and keyboard. Human users can receive multiple inputs from other humans and systems, i.e., listening to a voice and looking at someone. Users can also transmit multiple outputs, i.e., gesturing and talking, at the same time.

The way in which human sense inputs and controls motor outputs is multi-modal. There is an additional richness to human senses that are not mimicked adequately using current computer input and output interfaces. The input mode of interaction is affected not only by the content but also by the context, i.e., by the tone of a voice, by the perceived stance and by the physical proximity between sender and receiver, by facial expression, eye contact, and body language and by smell and touch. ICT systems such as video conferencing have the potential to allow one party to see and hear one another and provide some limited support for nonverbal and multi-modal communication but as they are often designed to be a single source recording, it is difficult to conceive how they can capture the equivalent multi-modal interaction typically used by humans.

Jaimes and Sebe (2005) classify modalities into two types: the human senses and other ICT device modalities such as mouse and keyboard. They define several types of multi-modal interfaces. Perceptual interfaces seek to leverage sensing (input) and rendering (output) technologies in order to provide interactions not feasible with standard interfaces and common I/O devices such as the keyboard, the mouse and the monitor. *Attentive interfaces* or iHCI are context-aware interfaces that rely on a person's attention as the primary input. The goal of attentive interfaces is to use gathered information to estimate the best way to communicate with the user. Wearable interfaces include a combination of ICT devices modalities that are worn by the user, typically a video camera and a microphone (Section 5.4.1). Their focus is on multi-modal interaction which includes visual interaction. Visual modal systems are divided according to how humans can interact with the system: large body movements, gestures, and eye-gaze. Visual interaction can be classified into either command interaction and noncommand interfaces in which actions or events are used to indirectly tune the system to the user's needs. Vision-based human motion analysis systems generally have four stages: motion segmentation, object classification, tracking, and interpretation.

Integrating multiple modes is complex because the signals sensed are in different forms and have a different frequency and in humans they are fused at different levels such as data, feature and decision, depending on what is being sensed or actuated. Sharma *et al.* (1998) present the basic theory and models for fusing multi-modes at these different levels. There are two main approaches. Data for each modality can be processed separately and only combined at the end. However, human use of multimodal interaction often employs individual modalities which are complementary and redundant. Therefore, to accomplish a human-like multimodal analysis of multiple input signals acquired by different senses, the signals cannot be considered mutually independent and cannot be combined in a context-free manner at the end of the intended analysis, but need to be processed in a joint feature space, according to a context-dependent model (Jaimes and Sebe, 2005).

5.3.2 Gesture Interfaces

Gestures are expressive, meaningful body motions involving physical movements of the fingers, hands, arms, head, face, or body, with the intent of conveying meaningful information about interacting with the environment. There are three main types of body gestures: hand and arm gestures, head and face gesture and full body movement (Figure 5.2). Gestures can be sensed using: wearable devices such as gloves or body suits; by attaching sensors such as various magnetic field trackers, accelerometers and gyroscopes to the surface of the body; by using cameras and computer vision techniques. A gesture may also be perceived by the environment as a compression technique for information to be transmitted elsewhere and subsequently reconstructed by the receiver, e.g., to support telepresence and telecontrol (Mitra and Acharya, 2007). Gestures can also be classified into contactful gestures, e.g., handshake, touchscreen gesture input or contactless gestures, e.g., waving at a person or camera to attract attention. This type of classification also reflects 2D gestures on a touch-sensitive planar surface as opposed to 3D gestures using a hand-held device incorporating gyroscopic or accelerometer sensors.

Gesture recognition has wide-ranging UbiCom applications such as: developing aids for those with impaired hearing enabling very young children to interact with computers; designing techniques for passive identification such as gait analysis (Section 12.3.4.1); recognising sign language used by deaf people; monitoring people's emotional states or stress levels (as part of affective computing, Section 5.7.4); classifying body behaviour such as lie detection; navigating and/or manipulating virtual environments (telecontrol and telepresence) and to enhance interaction with ICT devices such as games consoles.

The first basic gesture interfaces were the pen-based gesture interfaces based upon light-pens in the mid-1960s, allowing pointers to be moved on the screen, synchronised to the movement of the pen. This can also emulate regular hand-written characters and can be used as a substitute for keyboard-based text input. Once of the first commercial applications of this was for the Apple Newton PDA in 1992. From the mid-2000s, gestures were being used in several types of games consoles, mobile phones, cameras, etc.

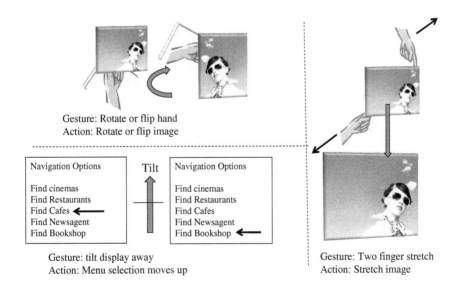

Figure 5.2   Use of rotate, tilt and stretch gestures to control a display

Sony's EyeToy, a peripheral device attached to its PlayStation games console is a motion-sensitive camera which can record a person and then project them into a virtual landscape on a video display. Movements of the person in the physical environment can be copied to allow the virtual display character to enable them to selectively interact with the virtual landscape.

Several current hand-held devices incorporate sensors such as inbuilt micro-gyroscopes to sense the orientation of the devices, to sense if they are being tilted or flipped from horizontal to vertical. Tiling can be used as a gesture to scroll up and down a list, e.g., a list of menu commands. Flipping can be used to rotate a view by 90 degrees from horizontal to vertical.

Specific physical gestures between humans can be used to automatically trigger actions in a virtual system. For example, a handshake within a certain context[6] can signal the agreement of a business deal. Physical gestures can be linked to trigger exchanges and interactions between Body Area Network or BAN (Section 11.6.4) devices and between people using inductive or capacitive electric field effects. This can in turn be used to trigger authentication, digitally signed agreements and order requisitions to be circulated across companies (Figure 5.3). A prototype BAN system which allows users to exchange electronic business cards by shaking hands was first demonstrated by Zimmerman (1996).

There are several design challenges in using gesture interfaces including how the ICT system recognises, classifies, defines the start and end of the gesture and binds it with particular virtual

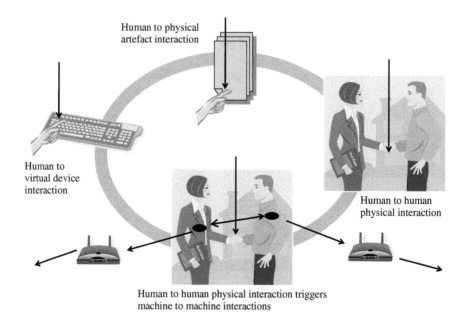

Human to physical artefact interaction

Human to virtual device interaction

Human to human physical interaction

Human to human physical interaction triggers machine to machine interactions

Figure 5.3 Human to virtual device interaction, human to physical device interaction, human to human physical interaction, which can in turn trigger human to virtual device interaction

[6] The context is vital to reduce the inherent nature of one-to-many relations between gestures and possible actions into a one-to-one relation between a gesture and an action. In this example, the context enables a handshake to be used as a greeting or a forced handshake to trigger a specific business transaction.

environment objects. If someone is pointing at several things in the field of view, it may not be clear what is being pointed at. There often exist many-to-one mappings from concepts to gestures and vice versa. Gestures may be static or dynamic. e.g., stop using a horizontal cutting motion of the hand versus holding the palm in a vertical position in front. Gestures can be ambiguous and are often incompletely specified. Similar to speech and handwriting, gestures can vary for the same individual between different sessions of use can vary between individuals and can be culturally specific. Gestures can vary within an individual because spatial information varies due to: where it occurs (spatial information); the path it takes which may contain spatial-temporal variability (pathic information); the sign it makes (symbolic information) and the use of affective information (its emotional quality) (Mitra and Acharya, 2007).

Computation approaches to deal with the ambiguity, uncertainty and variability of gestures include mathematical models based on Hidden Markov Chains (HMC) and tools or approaches based on soft computing such as fuzzy logic (Section 8.6.2). Most approaches first wait until the gesture is completed, then perform the gesture segmentation and then the recognition which leads to a delay between performing the gesture and it being classified. An alternative technique is to seek to recognise each segment of a gesture as it happens by computing a competitive differential observation probability (CDOP), the difference of observation probability between the gesture and the non-gesture.

There are two further characteristics of gestures which will challenge its uptake for ubiquitous computing. Although the gesture can be detected, the gesture may have a one-to-many relationship to the object it applies to, e.g., pointing at objects in the physical world can refer to several objects within the same depth of field of vision or in different depths of field of vision. Gesturing is a relatively slow method of input, particularly for repetitive actions, compared to direct text entry[7] and direct manipulation.

5.3.3 Reflective Versus Active Displays

Ebooks are light weight,[8] thin, long-lasting powered, pocket-sized devices with touch screens enabling pages to be turned by touch. A key difference between computer displays and ebooks is the type of display used. Ebook screens are designed to be more like paper, reflecting rather than transmitting light, requiring no energy to reflect light, to be readable in direct sunlight and equally viewable from any angle.

The current material of choice to realise ebooks is based upon Electrophoretic Displays or EPDs (Dalisa, 1997; Inoue et al., 2002). EPDs are reflective displays using the electrophoretic phenomenon of charged particles suspended in a solvent. Displayed text and images can be electrically written or erased repeatedly (Figure 5.4). EPDs had a serious problem with reliability in early products. The performance of their displays degraded due to the lateral drift of particles caused by gravitation, and the agglomeration of particles caused by direct contact between particles and electrodes. Consequently, EPDs were not used at first in actual devices on a commercial basis. In order to solve this problem, in 1987, Inoue proposed the use of microcapsules to embody the electrophoretic material (Inoue et al., 2002). EPD particles are actually contained in thousands of microcapsules deposited onto a substrate. Microcapsules contain

---

[7] The average computer user can type about 30 words a minute, while the best typists can type four times that number. Keyboard entry is a quicker method to input text and give commands than gestures when keyboard input is deemed to be not too obtrusive.

[8] They are also far lighter and more portable than sets of analogue books. An ebook could easily store the content equivalent of hundreds, if not thousands, of physical books.

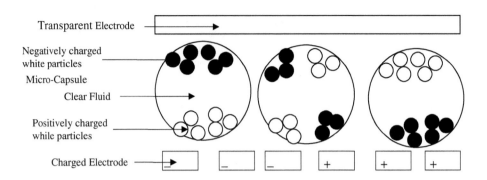

Figure 5.4 Electrophoretic displays are reflective type displays using the electrophoretic phenomenon of charged particles suspended in a solvent

negatively charged black particles and positively charged white particles suspended in a clear fluid. When a positive electric field is applied to the bottom of the substrate, the black particles move to the top of the microcapsule, causing that pixel to appear black and when a negative field is applied, the pixels move to the top and that appears white. Microcapsules are designed to be bi-stable. Once the field has been applied to attract the particles to form black or white areas on the substrate, it can be removed. Unlike active displays, no energy is required to maintain the black and white areas. Products based upon EPD tend to focus on the concept of printing the ink rather than on the 'paper' substrate or display, highlighted by the use of the term Electrophoretic Ink or e-ink (Jacobson *et al.*, 1997).

Readability and legibility continue to be a critical issue for ebooks (Harrison, 2000). Additional requirements are also portability, ability to annotate while preserving context, ability to skim or quickly move through pages and to do cross-referencing across several documents and have instant accessibility with a quick boot-up or start from hibernate mode. These requirements will vary depending on the different types of reading needed. Ihlström *et al.* (2004) discuss future scenarios of use and design of the e-newspapers, i.e. newspapers on epaper. Their findings were that functional requirements for the e-newspaper should consider mobility, interactivity, adjustment for special target groups and personalisation. There were also several issues regarding navigation, pagination, structure and use of overviews discussed during the prototyping, leading to layout suggestions for a one-page display. The design of future electronic news device and services, called epaper is also reported by Inbar *et al.* (2008).

5.3.4 Combining Input and Output User Interfaces

In the UIs discussed so far, input devices are separated from output devices. For example, a pointer input device such as a mouse or tracker ball only outputs a single position (x, y) events at any given time together with one, two, or three button presses. The state of the input is available as a visual cue only. One of the earliest combinations of user input and output devices is the touchscreen where the user can touch the display in order to activate a selection on the screen. Two further examples where the input device and output UI are merged are tangible user interfaces and organic user interfaces. Rekimoto (2008) compares and contrasts tangible versus organic UIs. Tangible UIs tend to be domain-specific whereas organic UIs are more generic and less application-oriented. Tangible UIs are more logical, or manipulation-oriented, whereas organic UIs are more emotional, or communication oriented.

5.3.4.1 Touchscreens

Touchscreen are displays where the position of contact with the screen, generally with pointed physical objects such as pens or with fingers, can be detected. An event can then be generated for an associated visual object at that position and an associated action triggered. To an extent, a touchscreen behaves as a two-dimensional planar smart skin. Wherever it is touched, some virtual object can be activated. A touchscreen can behave as a soft control and display panel that is reprogrammable and which can be customised to suit a range of applications and users.

Touchscreens can operate using a variety of electromagnetic principles such as resistive, capacitive, surface acoustic waves, etc. Resistive touchscreens are composed of two thin metallic electrically conductive and resistive layers separated by a thin space. When some object touches the screen, the layers connect at the contact point and using basic electricity principles such as Kirchoff's Laws, the position can be deduced from the position-dependent current flow.

Touchscreens are easy to learn to use, are durable, require no additional work space and have no moving parts. These characteristics make them ideal for many workplaces and public spaces. Touchscreens were initially developed in the 1970s but came into widespread use more in the 1980s (Pickering, 1986). Touchscreens today are used routinely in many applications and many devices such as point of sales terminals, mobile devices and games consoles, vehicle navigation systems and information, kiosks. Not only can these devices support pointing, they can also support particular movements of single fingers to indicate a next screen command and support multiple finger gestures such as moving two fingers resting on an object away from the object indicating zooming out, magnifying or stretching the object.

5.3.4.2 Tangible Interfaces

A *Tangible User Interface*[9] (TUI) is a user interface that augments the real physical world by connecting digital information to everyday physical objects and environments. A first application was the Tangible Bits project led by Ishii, reported in 1997.[10] Other examples of projects include Sony DataTiles and to an extent ambient wood (Section 2.2.4.4). Tangible interfaces can be realised by attaching micro sensors and actuators (Section 6.4) to physical objects. These can then be used as input devices to allow their manipulation to generate data streams in an output device or virtual view in a related virtual environment, e.g., RFID (Radio Frequency Identifier) tagging a physical object enables the physical object to be moved and it to be tracked in a virtual view (Section 6.2).

Fishkin (2004) provides a taxonomy of TUIs based upon embodiment and metaphors. Embodiment refers to something taking a physical form and metaphor refers to an understanding of actions in the physical domain having equivalent actions in some other virtual conceptual domain. Four types of embodiment are described. Full embodiment is when the output device is the input device: the state of the output device is fully embodied in the input device, e.g., a robot (Section 6.7). Another example is the Claytronics Project (Golstein *et al.*, 2005) in which collections of MEMS components are able to adapt their physical form. Nearby embodiment means the output takes place near the input object, typically, directly proximate to it. The output is tightly coupled to the focus of the input, e.g., a light-pen, computer mouse, games console, steering wheel, wand, etc. Environmental embodiment means the output is 'around' the user, e.g., an ambientROOM device

---

[9] Tangible user interfaces are also referred to as passive real-world props, graspable user interfaces, manipulative user interfaces and embodied user interfaces (Fishkin, 2004).

[10] The use of digital RFID-tagged physical objects which represents a tangible UI to ICT applications have been around much longer and preceded much computer science HCI research into tangible UIs (Section 6.2).

uses audio and light of physical objects to relate to states in virtual views. Distant embodiment is when the output is distant from the input, e.g., a TV remote control, in which visual attention is switched between the input (the control) and the output (the TV screen). Fishkin (2004) groups metaphors into two types: those which relate to the shape of an object, a noun metaphor, and those which relate to the motion of an object, a verb metaphor. Systems may also be neither noun nor verb or can be both noun and verb metaphors. A full metaphor is when the physical system is the virtual system, e.g., a robot or collection of MEMs devices (Goldstein *et al.*, 2005).

5.3.4.3 Organic Interfaces

Rekimoto (2008) uses the terms organic and organic interaction for such interfaces, because they more closely resemble natural human–physical and human–human interaction such as shaking hands and gesturing. Vertegaal and Poupyrev (2008) define three characteristics which characterise organic UIs. First, the display can be the input device. Second, the display can take on any shape. Third, displays can change. According to Ishii (2008), who was the first to propose the concept of tangible UIs, first-generation tangible UI components could not change shape but only the spatial relationships between the components. However, the second-generation Tangible UIs can change shape.

Schwesig (2008) considers how the analogue physical world can interplay with the digital virtual world. One approach is to imbue real physical objects with digital properties, e.g., Tangible UIs and augmented reality. An alternative approach is to simulate physical environments in virtual environments (virtual reality). But physics real or simulated can also be very limiting: because virtual systems are not bound by the laws of the physical world, e.g., hypertext links break when printed on paper. Organic interface design represents a less literal approach which, rather than focusing on physical objects or metaphors, emphasises the analog, continuous, and transitional nature of physical reality and human experience. This perhaps enables organic systems to combine the strengths of both the physical and virtual or digital world: by combining sensitive analog input devices with responsive graphics. User experiences that acknowledge the subtleties of physical interaction can be created.

Organic Light-Emitting Diode (OLED) displays are based on organic polymer molecules that compose emissive and conductive layers of the display structure melded together through a form of printing (Co and Pashenkov, 2008). These are widely seen as a successor to the ubiquitous Liquid Crystal Display (LCD) display. OLED displays have several benefits compared to LCDs. OLED displays do not require backlighting (saving power). They are viewable at oblique angles. OLED displays are constructed out of transparent layers so that red, green, and blue layers can be stacked such that a full-colour (RGB) pixel is a full colour-mixed single pixel with depth rather than a cluster of red, green and blue pixels, thus potentially giving better resolution. These displays can also be produced using flexible substrates which allow any physical object which is covered with these substrates to behave as a smart skin. This has many applications for novel[11] types of display. One example is that displays can be folded or rolled to reduce the size of the physical form during transit and then can be expanded again when operated, thus realising a more portable version of a large screen.

---

[11] In theory, this can support the illusion of invisibility. If a car or a cloak is coated with this material and if many MEMS cameras could be embedded into the material, then they could record the scene from behind an object and then project it onto the object's front view display. To an observer at the front of the display or object, it looks as if they are directly viewing the scene behind the object, the object has disappeared – truly disappearing computing.

It is interesting to note that for the Pileus umbrella which uses the umbrella as a display (Section 2.3.3), OLED could have been used as the fabric of the umbrella and hence as the display. However, because of the physical environment requirements of using the umbrella device in windy conditions (Section 5.5.3), it was felt that OLED fabrics would not be robust enough to be used in severe conditions, might get torn or suffer from circuit breaks and hence OLED was not considered suitable. Instead, a projector system was used to project content onto the surface of the umbrella.

5.3.5 Auditory Interfaces

Auditory interfaces[12] have the benefit of supporting hands-free input and output which can be used as an additional interaction modality when the visual senses are already being consciously used. Auditory interfaces can support natural language text input as a replacement for keyboard text input. There are several main design challenges, interpreting auditory input in the presence of audio noise,[13] access control for voice activation (voice recognition) and natural language processing of voice input, e.g., to recognise commands. There are two basic auditory interfaces: speech-based and non-speech based (Shneiderman and Plaisant, 2004, pp. 374–385).

Auditory interfaces can provide an interface between the user and a computer-based application that permits spoken interaction with applications in a relatively natural manner compared to WIMPS interfaces. In 1969, John Pierce of Bell Labs said that automatic speech recognition would not be a reality for several decades because it required developments in artificial intelligence. However, in the early 1970s, work started on the development on Hidden Markov Models (HMMs) and this led to the release of the first commercial speech recognition products in the 1980s. Today, its accuracy under ideal conditions is about 95% or better. A good overview of speech systems issues is given by McTear (2002).

There are several important applications for voice recognition software. Generally discrete word recognition systems with a limited vocabulary are more accurate and used more than continuous speech recognition systems. Speech-based authority interfaces are used in telephony interactive voice response, or IVR, a computerised system that allows a person, typically a telephone caller, to select an option from a voice menu and to interact with a computer system. In addition, auditory interfaces can be used in voice command user interfaces, dictation and transcription.

Together with haptic interfaces, auditory interfaces are particularly important for visually impaired users. There are a variety of auditory interfaces which are not speech-based. Non-speech auditory cues include earcons, synthetic audio sequences associated with a structural context in a UI such as a menu hierarchy, which can also be used to indicate the state during an interaction. Sounds are used in a variety of machines and devices to give warnings and to highlight abnormal behaviour.

5.3.6 Natural Language Interfaces

Generally, interaction can be more easily processed and understood if it is defined using an expressive language that has a well-defined syntax or grammar and semantics but this requires

---

[12] The focus here is on the use of audio to actively control devices and services rather than on passive audio devices which just receive and transmit sound content.

[13] There is a story about the first show-case demonstration of a well-known speech recognition system in the 1990s being suddenly curtailed when during audience participation, someone shouted out the MS-DOS system hard disk formatting command as input which the system then acted upon.

that users already know the syntax. Assuming users have been educated to read and write, users could use the languages they already use to converse with each other to converse with machine, i.e., use a natural language interface (NLI),[14] thus avoiding users having to learn a specific system interface language. NLI can also be useful for intermittent users who may not be able to remember the system command syntax. Although NLI has these benefits, its use also has several disadvantages[15] (Shneiderman and Plaisant, 2004, pp. 332–341). The semantics can be ambiguous, e.g., consider the question 'Did you see the man in the park with a camera?' which makes it complicated for machines to parse the query. The meaning could be clarified through further dialogues between the user and machine but this reduces the throughput. NLI can be especially verbose and slow when a series of commands needs to entered. Because the computer is far quicker at displaying information, it may be far quicker to do so and let users select from the information available to construct some query or response. Different users may also have a different understanding of the domain in terms of its scope and of the meaning of terms. It is also complex for a system to have to support several natural languages and for the system to find equivalences between these. Natural language commands can be given in three forms: it can be typed, it can be hand-written and it can be spoken. The latter two forms require additional pre-processing to transform handwriting to text and speech to text respectively.

5.4 Hidden UI Via Wearable and Implanted Devices

5.4.1 Posthuman Technology Model

In the *Posthuman* model (Hayles, 1999), technology can be used to extend a person's normal conscious experience and sense of presence, across space and time. There are three types of posthuman technology: accompanied, wearable and implants. Accompanied technology is technology external to the body which accompanies it but is not directly attached to it, e.g., personalised mobile devices, smart cards, smart keys, etc. (Chapter 4). Wearable technology is technology external to and directly attached to humans, e.g., hearing aids and wireless earpieces attached to mobile phones to support hands-free use of phones. Implants are technology internal to the body. The obvious applications for implants are medical, to use various prosthetics and bio-implants to overcome paralysis in limbs and muscles and to help regulate and treat irregular biological phenomena such as heart activity. Future prospects for the Posthuman model are also discussed in Section 13.7.1.

5.4.2 Virtual Reality and Augmented Reality

Most computers currently present visual information in two dimensions, although simple three-dimensional or 3D effects can be created by using shadows, object occlusion and perspective. These are an important element of games consoles which heightens user satisfaction in the main group of users. More realistic 3D effects can be created by mimicking the stereoscopic vision of eyes

---

[14] The Loebner Prize for AI, http://www.loebner.net/Prizef/loebner-prize.html, accessed May 2008, is the first formal instantiation of a Turing Test and is awarded to a computer each year which is judged to be the most human like. The Turing Test was named after Alan Turing who asked the question 'If a computer could think, how could we tell?' Turing's suggestion was, that if the responses from the computer were indistinguishable from that of a human, the computer could be said to be thinking. This field is generally known as natural language processing.

[15] These limitations of HMI are clear to most users because if a search is done in application-specific help or via a Web search engine in response to a query, many results returned seem irrelevant.

where each eye sees a slightly different perspective of the same scene. For example, a 3D headset or goggles that contains two miniature screens, each one showing the same scene from a slightly different perspective. Alternative techniques to simulate 3D are to either use polarised light or to blank out each eye synchronised to the computer frame rate so that each eye sees alternate images. As the head moves, sensors detect the change in angle to view the scene and the changing scene perspective is calculated and presented.

Virtual reality seeks to immerse a physical user in a virtual 3D world whereas augmented reality seeks to make interaction in the physical world more virtual by digitally enabling relevant objects in the real world. Both virtual reality and augmented reality seek to enable humans to interact using a more natural interaction than humans use in the real world such as using voice and gestures, rather than using the keyboard mouse interface of the PC. Virtual reality (VR) uses a computer simulation of a subset of the world and immerses the user in it using head-mounted displays, goggles, gloves, boots and suits (Section 5.4.1). In augmented reality systems, electronic images are projected over the real world so that images of the real and virtual world are combined. To this extent, virtual reality can be considered as a sub-set of augmented reality in which there is no real world but just an artificial reality. One of the first examples of augmented reality was the head-mounted display by Sutherland (1968). Similar systems are in use today in types of military aircraft.

5.4.3 Wearable Computer Interaction

The essence of wearable computing is to embed computers into anything that we normally use to cover or accessorise our bodies. This includes clothes, jewellery, watches, eye wear, teeth wear, ear wear, headwear, footwear, and any other device that we can comfortably attach to our bodies and allow to behave as hidden computers. In a broader sense, devices can also be embedded in the environment that accompany us, in our transport vehicles are extensions of wearable computing. Wearable computers are especially useful when computer access is needed while a user's hands, voice, eyes or attention are actively engaged within a physical environment.

One of the first examples of wearable computers was a concealed card-sized analog computer designed to predict the movement of roulette wheels (Thorp, 1998). An observer used micro-switches hidden in shoes to indicate the speed of a roulette wheel. The computer would indicate an octant to bet on by sending musical tones via a radio to a miniature speaker hidden in a collaborator's ear canal. The system was successfully tested in Las Vegas in the early 1980s, but hardware issues with the speaker wires prevented them from using it beyond their test runs. About the same time Steve Mann also developed experiences with MTOS ICT devices rather than ASOS devices called WearComp and WearCam (Section 2.2.4.5) which are still ongoing. Mann (1997) specified three criteria to define wearable computing.

- *Eudaemonic*[16]*criterion* (in the user's personal space): the ICT device appears to be part of the user as considered by the user and observers of the user.
- *Existential criterion* (iHCI Control by user): ICT devices are controllable by the user. This control need not require conscious thought or effort, but the locus of control must be such that it is within the user's domain.

---

[16] This is named after a group of West-Coast physicists, known as the Eudaemons, who independently developed the first wearable computers in parallel with Mann in the late 1970s. One of their first applications was a shoe-type embedded computer which was used to assist a roulette player predict where the ball would land. Another early shoe device was the shoe phone as used by Maxwell Smart in *Get Smart*, the 1960s spy spoof series.

- *Ephemeral criterion* (responsiveness): interactional and operational delays are non-existent or very small.

  - ○ Operational constancy: It is always active while worn.
  - ○ Interactional constancy: One or more output channels are accessible (e.g. visible) to the user at all times, not just during explicit HCI.

Wearable computers, because they can accompany users everywhere, represent a clear form of UbiCom. A more complete conceptual framework for wearable computing, called Humanistic Intelligence (H.I.) considers the informatic signal flow paths between the individual and the computer (Mann, 1998).

5.4.3.1 Head(s)-Up Display (HUD)

Head(s)-Up Display or HUD, is any type of display that presents data without blocking the user's view (Sutherland, 1968). This technique was pioneered for military aviation and is now being used in commercial aviation and cars. There are two types of HUD. In a fixed HUD, the user looks through a display element attached to the airframe or vehicle chassis, the system projects the image with semi-transparency onto a clear optical element and the user views the world through it (augmented reality). In a head-mounted display, the system precisely monitors a user's direction of gaze and determines the appropriate image to be presented. The user wears a helmet or other headgear which is securely fixed to the user's head so that the display element does not move with respect to the user's eye.

5.4.3.2 Eyetap

EyeTap is a device that is worn in front of the eye that acts as a camera to record the scene available to the eye, and acts as a display to superimpose a computer-generated imagery on the original scene available to the eye (Mann and Fung, 2002). An EyeTap is similar to a HUD but differs in that the scene available to the eye is also available to the computer that projects the HUD. This enables the EyeTap to modify the computer-generated scene in response to the natural scene. The EyeTap uses a beam splitter to send the same scene (with reduced intensity) to both the eye and a camera. The camera then digitises the reflected image of the scene and sends it to a computer. The computer processes the image and then sends it to a projector. The projector sends the image to the other side of the beam splitter so that this computer-generated image is reflected into the eye to be super-imposed on the original scene. One use, for instance, would be a Sports EyeTap that enables the wearer to follow a particular player in a field and have the EyeTap display statistics relevant to that player as a floating box above the player.

5.4.3.3 Virtual Retinal Display (VRD)

Virtual Retinal Display (VRD), also known as a retinal scan display (RSD), draws a raster display (like a television) directly onto the retina of the eye (Johnston and Willey, 1995). The user sees what appears to be a conventional display floating in space in front of them. This is in contrast to past systems that have been made by projecting a defocused image directly in front of the user's eye on a small 'screen', normally in the form of large sunglasses. The user focused their eyes on the background, where the screen appeared to be floating. The disadvantages of these systems were: the limited area covered by the 'screen'; the heavy weight of the small televisions used to project the display, and the fact that the image would appear focused only if the user was focusing at a particular 'depth'. Limited brightness made them useful only in indoor settings.

5.4.3.4 Clothes as Computers

Unlike HUD, EyeTap and VRD that focus on single sensors, clothes as computers use a network of sensors that can be worn and the data from them fused to allow other types of non-visual context-awareness. Van Laerhoven *et al.* (2002) have reported their experiences with a body-distributed sensor system that integrated tens of accelerometers spread over the body into a garment with the majority on the legs and the rest divided over the arms and upper body. The accelerometers for the legs were integrated into a harness to enable testing and capturing of data from multiple users of different figure heights, while the others were attached on regular clothing using Velcro. The experiments indicated that it is feasible to distinguish certain activities of a wearer whose clothing has an embedded distributed sensor network. These activities could also include gestures made by the user, and basic events related to garments, such as putting on a coat or taking off a coat. These can be recognised with a reasonably high precision.

Current sensors require rigid physical substrates to prevent de-lamination, and the mechanical incorporation of bulky switches. This drastically limits the physical form, size and tactile properties of objects using these sensors (Orth *et al.*, 1998). This has led to the creation of fabric-based computers and a product – the Musical Jacket that is being marketed by Levi in Europe. The Musical Jacket incorporates an embroidered fabric keypad out of a sewn metallic conducting fabric BUS and non-conducting cotton and nylon tulle (the insulating layers), a battery pack, a pair of commercial speakers and a miniature MIDI synthesiser. When the fabric keypad is touched, it communicates through the fabric BUS to the MIDI synthesiser, which generates notes. The synthesiser sends audio to the speakers over the fabric BUS as well. Power from the batteries is also distributed over the fabric BUS.

5.4.4 Computer Implants and Brain Computer Interfaces

The opposite of wearing computers outside the body is to have them more directly interfaced with the body. Many people routinely use implants, for example, pace-makers are used to regulate the electrical activity of the heart, artificial limbs can increase mobility and contact lenses inserted into the eye can improve the contrast to track balls in sports even for people with good sight.

Of specific interest is developing devices that can adapt to signals in the human nervous system. By connecting electronic circuitry directly to the human nervous system, physiological signals that represent our thoughts, our emotions, and our feelings can be directly input to computers allowing humans to operate machines by thought power (Warwick, 1999). This represents the ultimate natural interface, thought control instead of motor control of devices. Brain–Computer Interfaces (BCI)[17] or Brain–Machine Interfaces (BMI), in contrast to Human–Machine Interfaces or Human–Computer Interfaces which support indirect interfaces from the human brain via human actuators i.e., haptic interfaces and machine sensors, are direct functional interfaces between brains and machines such as computers and robotic limbs. There are several design choices here: the human–device interface could be situated to have a direct versus indirect connection to the nervous system and the device could be situated at the brain or

---

[17] The term brain–computer interface was first coined by Jacques Vidal in the 1970s.

situated elsewhere in the body. Lebedev and Nicolelis (2006) classify BMIs as whether or not they utilise invasive versus non-invasive neural signals.[18] Invasive techniques where electrodes are implanted cranially can record from single or multiple sites and within these sites can sample signals from small groups of neurons or larger groups of hundreds of neurons. First, invasive, non-brain BMI, then non-invasive brain BMI are discussed and then invasive brain BMI is considered.

Warwick *et al.* (2003) reported that in 2002 an array of 100 electrodes was surgically implanted into the median nerve fibres of his left arm. A number of experiments were carried out that showed he was able to control an electric wheelchair and an intelligent artificial hand and that he was able to create artificial sensation by stimulating individual electrodes within it.

Non-invasive BMI was first reported by Vidal (1973) whose early work laid much of the groundwork to collect and computer process high-quality EEG signals. Features are extracted from the signals and a translation process converts these features into symbols or commands to control electrical devices. Non-invasive brain BMI is more mature and has less safety issues than invasive brain BMI and is considered more suitable for use in daily activities. A typical wearable computer system for non-invasive brain BCI consists of an EEG recording cap, and a Body Area Network (Section 11.6.4) to communicate the EEG signals to a mobile device which also acts as a gateway between the BAN and the Internet.

Navarro (2004) discusses the use of BCI to accurately predict brain activity triggers in a highly changing environment as commonly is the case in performing daily life activities. For example, there are differences between the real-world environment and the experimental environments when identifying and matching brain pattern(s) to trigger specific things. BCI experiments in Virtual Reality (VR) environments have also been done in order to recreate more precisely the situations when triggering a BCI. Such experiments are able to minimise the uncertainty of the brain inputs originated from the 'outside' environment, by being able to recreate an identical wider multi-sensorial experience for the user. This potentially engages the user's perceptions in a broader manner giving similar mental states and brain activity patterns when an identifiable event which triggers the BCI occurs.

Invasive BMI was first demonstrated in experiments by Chapin *et al.* (1999). He showed that rats could use their motor cortex to control the movement of a robot arm to dispense drinking water. Research has been initiated to discover if this approach could help restore the motor abilities in physically disabled humans, via the use of artificial limbs or by bypassing neural network failures in humans by functionally stimulating paralysed muscles.

Lebedev and Nicolelis (2006) envision the neuroprosthetic developments for invasive systems which might emerge in the next ten to twenty years. Invasive brain BMI will use a fully implantable recording system that wirelessly transmits multiple streams of electrical signals, derived from thousands of neurons. BMIs will become capable of decoding spatial and temporal characteristics of movements and intermittent periods of immobility. BMIs will be able to utilise a combination of high-order motor commands to control an artificial actuator with multiple degrees of freedom or directly stimulate multiple peripheral nerves and muscles through implantable stimulators. Much research is needed before any of these types of BMI, non-invasive and invasive, can be used more routinely as a UbiCom system in order to improve the convenience of the accuracy of use, the

---

[18] Non-invasive systems which primarily exploit electroencephalograms (EEGs) to control computer cursors or other devices such as wheelchairs have the advantage of not exposing the patient to the risks of brain surgery, however, they use signalling channels with a very limited capacity rate, typically about 5–25 bits per second per source. Gamma-type EEGs which represent certain cognitive and motor function have the highest frequency range, 25 Hz to 100 HZ.

robustness and sustained use of systems, and allay safety concerns about implants for the host, concerning the psychological and physiological effects of using BMI systems.

5.4.5 Sense-of-Presence and Telepresence

The Posthuman model is related to a discussion of alternative realities which seek to extend the experience of the here and now conscious sense of presence of human beings situated in the physical world, to experiences of being somewhere else, possibly in another time. A feeling of presence in the experience provides feedback to a person about the status of his or her activity. The subject perceives any variation in the feeling of presence and tunes his or her activity accordingly. People can experience alternative realities depending on the type of environment people are situated in and on their perception of the environment. Reality can be technology mediated, chemically mediated and psychologically mediated. A discussion of the experience of presence in mediated environments is given by IJsselsteijn and Riva (2003).

Non-interactive AV media such as films, theatre and music, and interactive AV media such as electronic games and even the sense of smell and touch, possibly relating to previous experiences, can transport the viewer into another realm or state of perception, transporting the sense of presence from a human's immediate physical locality to another remote presence or experience. Hyperreality, for example, characterises the inability of consciousness to distinguish between reality and non-reality such as fantasy, particularly when someone becomes immersed in the experience. A sense of immersion can often be achieved when playing either real or electronic games in which someone is enclosed and embraced by the AV medium and transported to another realm or state of perception. In this kind of immersion, a person is affected by the environment at multiple levels – perceptual, sensory, psychological and emotional.

Telepresence allows a person in one place to feel as if they are present in a remote place, to give the appearance that they are present and have an effect, at a location other than their true location. *Telecontrol* refers to the ability of a person in one place to control objects remotely. These can be linked to sensing body movements locally which can then be used to control such objects remotely coupled to telepresence. For example, it allows humans to manipulate dangerous substances in a remote environment, from a safe environment, as if they were present. Wilson and Shafer (2003) describe the development of a wand-like device which the user points at the device to be controlled, and then makes simple gestures or speech to control the device. The intelligent environment system interprets the user's manipulation of the wand to determine an appropriate action in context. For example, the user may turn on a light in the room by pointing the wand at the light and pressing the button. Alternatively, the user may point the wand at the light and say 'Turn on'. Kim and Kim (2006) have also developed a gesture recognition method to open and close curtains and to turn on and off lights in a smart home environment sensed using three CCD cameras, which are attached at angles of 0, +45, and −45 degrees.

5.5 Human-Centred Design (HCD)

Conventional system design focuses on designing the core of the system to support sets of service actions or functions in order to handle service requests. In contrast, human-centred design, also called user-centred design, focuses on the design of the part of the system human users interact with, the user interface. For many users, the user interface represents the product. Whereas conventional system design mostly involves users only at the start and end of system development life-cycle, human-centred design involves users throughout the whole of the design life-cycle.

5.5.1 Human-Centred Design Life-Cycle

The basic phases of development of user-centred interactive system design are similar to those used in conventional functional system design (Figure 5.5). This development cycle comprises four phases of development: (1) requirement gathering and analysis; (2) modelling and design; (3) implementation; and (4) testing. Interaction design or user-centred design is closest to the proto-typing cycle model. Preece *et al.* (2006), Dix *et al.* (2004) and Shneiderman and Plaisant (2004) offer comprehensive advice and instruction on human-centred design, human–computer interaction and how to design user interfaces.

The ISO 13407[19] human-centred design process for interactive systems specifies four principles of design: (1) the active involvement of users and a clear understanding of user and task requirements; (2) an appropriate allocation of function between users and technology based upon the relative competence of the technology and humans; (3) iteration is inevitable because designers hardly ever get it right the first time; and (4) a multi-disciplinary approach to the design. The human-centred design life-cycle involves four main sets of activities (Figure 5.5):

1. Define the context of use in terms of scenarios, use cases, task models, and the ICT, physical and social environment context of use.
2. The stakeholder and organisational requirements[20] must be specified.

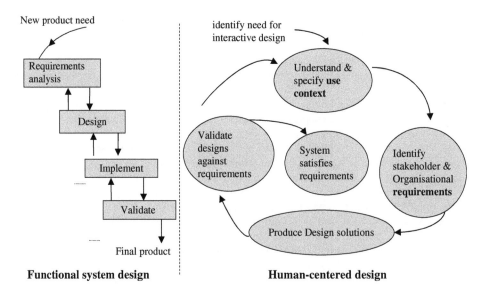

Functional system design   **Human-centered design**

Figure 5.5 Comparison of a conventional functional system design approach with a human-centred design approach

---

[19] ISO 13407:1999 Human-centred design processes for interactive systems, available from http://www.iso.org/iso/iso_catalogue.htm, accessed Jan. 2008.
[20] Requirements are defined as unambiguous, specific and clear statements of what the system should do or how it should perform.

3. Multiple alternative UI designs need to be built. Designers can suggest alternatives to help the user to 'break out of the box', to identify better alternatives rather than sticking with current familiar designs. Alternatives are generated through research and synthesis, creativity and through looking at similar products for inspiration.
4. Designs need to be validated against user requirements.

Harper *et al.* (2007) propose extending the four-phase interactive development life-cycle of study, design, build, evaluate to understand, study, design, build, evaluate. While the understand stage provides a framework to guide design and research the human values which need to be supported by a system, the study stage involves fleshing out the details of how individuals and social groups pursue and achieve those particular aspirations.

5.5.2 *Methods to Acquire User Input and to Build Used Models*

The basic techniques used to acquire user input for the formative evaluation of designs can also be used throughout the human-centred design life-cycle from requirements gathering, to model user behaviour. Selection of appropriate user input techniques will depend on a variety of factors, such as the different sub-types of environment requirements, cost and time constraints, and availability of different types of users.

There are two basic types of techniques to gather user input depending on what type of human environment the user input is acquired in. In field studies or ethnographic studies, feedback from users can be acquired in users' natural settings. In usability testing, users can be tested in controlled settings, often in a usability lab, where cameras and input filters can observe what users do and what they input into a system. There are two basic types of evaluation with respect to whether or not direct interaction or indirection with users is used. These can be combined. Field studies and lab studies can involve direct or indirect user interaction. For example, the use of tagging techniques, highlighted in Section 6.2, can enable the changing spatial-temporal context of subjects including humans and animals to be tracked in the field.

Direct interaction techniques include questionnaires, interviews, inspections and focus groups and cognitive walk through. There are several types of questionnaire and interview. Unstructured interviews are not directed by a script. This are rich but may not be able to be easily replicated. This is a way to get informal feedback from users. This is also called quick and dirty feedback. Structured interviews are tightly scripted, often like a questionnaire, are replicable but may lack richness. Semi-structured interviews are guided by a script but interesting issues can be explored in more depth. These can provide a good balance between richness and replicability. Focus Groups or Group interviews are groups guided by an interviewer who facilitates group discussions on a specified set of topics.

Inspection or heuristic evaluation: experts and experts as users inspect the user interface and are asked to consider and document how a set of usability heuristics apply when carrying out pre-specified user tasks. There are different types of usability heuristics that apply when evaluating different types of interactive products such as online communication, web-sites, desktop computers and mobile devices.

Cognitive walk through: this is an alternative approach to inspection in which designers and expert evaluators as individuals or groups walk through the sequences of a task and document whether or not a user will know how to achieve a task, what actions are available, whether or not users can interpret the system response to an action.

Observing users is the main technique used to gather indirect input about users. The same tools that are used to observe users in controlled environments (in the lab) can be used in uncontrolled environments (in the field) but the way in which they are used differs. In the lab, the details of user behaviour can be recorded using fixed observation equipment, e.g., key click analysis can be performed to track users in lab. In the field, we can focus more on the user context but can collect less detailed information. Note that observations of user tasks make it difficult to tell what the users intend to do. Hence, observations can be supplemented by asking users to perform a cognitive walk through and to think aloud.

Predictive models provide ways of soliciting user feedback about products or designs, without necessarily directly involving users. Quantitative models of user interaction sequences such as clicking keys, moving a mouse pointer, thinking and moving between mouse and keyboard, can be used to calculate task efficiency such as how long it takes to complete a task and to quantitatively compare user interactions. is the most efficient methods. An example of such as a method is the GOMS-KM model (Raskin, 2000). This can seem less expensive than methods which quire user input but its usefulness is limited to systems with routine or predictable tasks.

5.5.3 Defining the Virtual and Physical Environment Use Context

In the PVM scenario, the virtual computing or ICT context of use has an internal and external context (Figure 5.6). The internal use concerns the internal storage capacity for playback and review of recordings, etc. The external ICT context requires the AV recorder to share content with external systems in order to view, archive and retrieve recordings. The physical environment context concerns the physical environment operating conditions for the system).

5.5.4 Defining the Human Environment Use Context and Requirements

5.5.4.1 User Characteristics

The human environment context concerns user characteristics, social context and the usability context. The meaning of user characteristics is illustrated by example. In the PVM scenario, users need to be able to safely hold the camera, to be able to intuitively operate the basic features of the

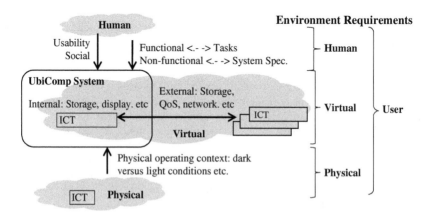

Figure 5.6 Requirements for interactive design considers a wider set of requirements beyond functional and non-functional requirements

camera, etc. A configuration or set of user preferences for the use of a device or service, referred to as a user profile,[21] can be specified. In many current devices, such user profiles are device-specific and localised within the device. Sometimes it is useful to be able to define and share profiles across multiple applications and devices. However, currently, sharable cross-application user profiles are not supported in the majority of consumer devices. Other types of user profile such as personas can be used to characterise types of imagined users. Stereotypes users are similar to personas but are types of user that are derived from collecting, analysing and clustering real user interaction.

In the PVM scenario, ICT devices are used in personal social and public spaces. Humans may also act as part of organisations and hence the organisational context may constrain the individual context of use. The device may need to be operated remotely so that the owner of the AV recorder can also appear as part of the scene. The camera may signal to people in the scene that it is about to take a photograph in order that people can pose for the photograph and it may signal that the recording is complete. The AV recorder could be designed as a hidden recorder in order to capture recordings of people who act naturally rather than pose for recordings. The human context of use may also define how the user is instructed and trained to operate the device and to resolve operating problems.

According to the ISO 9241-11[22] standard for guidance on usability, usability is defined as the extent to which a product can be used by specified users to achieve specified goals with effectiveness, efficiency (time to complete a task) and satisfaction in a specified context of use. Usability is not a single, one-dimensional property of a user interface. Usability is a combination of factors. ISO 940-11 explicitly mentions effectiveness, efficiency and satisfaction. But these can often be expanded further. For example, effectiveness is often expressed in terms of learnability (ease of learning), memorability (ease of recalling how to use a previously learned UI) and error frequency (the number of errors per unit time in operating a UI to achieve a task). User satisfaction can also be expanded into part of a wider set of user experience criteria (Preece *et al.*, 2006). Specific types of UI components also have more specific usability. For example, dialogue systems should have the qualities mentioned in ISO 9241-110[23] such as suitability for task; self-descriptiveness; controllability; conformity to user expectation; error tolerance; suitability for individualisation; and suitability for learning.

5.5.5 Interaction Design

Interaction design focuses on designing systems to be usable by their human users. It focuses on the design of the user interface part of the system. Two main differences with conventional design are, first, that several alternative prototype designs may be produced and, second, that there may be some (formative) evaluation of these prototype designs. The overall design of a UI generally projects some simpler, higher-level view of the system, referred to as the user interface conceptual model.

---

[21] The term user profile is multi-faceted. It can mean user preferences for device or service access or it can represent the configuration of a resource used. User profiles may or not be device or service specific. Devices may support single or multiple user profiles.

[22] ISO 9241-11:1998 Guidance on usability, available from http://www.iso.org/iso/iso_catalogue.htm, accessed Jan. 2008.

[23] ISO 9241-110:2006 Ergonomics of human–system interaction – Part 110: Dialogue principles, available from http://www.iso.org/iso/iso_catalogue.htm, accessed Jan. 2008.

5.5.5.1 Conceptual Models and Mental Models

There are an amazing number of everyday things, Norman (1988) states that there are about twenty to thirty thousand of these. In addition, many of these, such as a camera, are in turn made up of other distinct parts. It seems very challenging for people to learn to operate and understand tens of thousands of such devices of varying degrees of complexity if the interaction with each of them is unique. This complexity of interacting with new machines is reduced because people tend through experience to build up mental models that predict the behaviour of different types of objects and base their interaction of familiar and unfamiliar objects on these. The user's mental model develops when a person starts to learn how to use a system that is unfamiliar to them. Once a user becomes familiar with system operation, the mental model is not used and habits develop that link user actions to sensory input without thinking.

The complexity of interacting with new systems is also reduced if they have parts that provide strong clues on how to operate themselves. These are referred to as *affordances* (Norman, 1988). For example, a camera lens has a rotatable ring to adjust the degree of zoom. The usability of affordances is to an extent based upon experience, is subjective and cultural. Imagine you are in a foreign country and need to access some machine which has a slot as input. It may not be obvious what is to be inserted into the slot e.g., a token proprietary to that machine or some denomination of local currency, unless the slot is labelled unambiguously in a language which you understand. Some affordances which are mechanical or electromechanical may invite trial and error interaction but this may permanently damage the device or be dangerous.

Designers of systems could also aim to design interactive systems to support appropriate conceptual models. A conceptual model is an abstraction of a system or service at a level that is understandable to a user and matches the user's mental model of how he or she thinks the system operates. A common way to define a conceptual model (virtual objects) in a human computer interface is to link the virtual objects or widgets in it to related physical world objects.[24] However, extensions are often added to virtual representations and behaviours of physical objects within a particular application domain to increase their flexibility and functionality. For example, a bin placed on the virtual desktop makes it easier to discard information. However, a bin is not normally placed on a physical desktop in the physical world environment. Also it is difficult to mimic the continuity of the analogue world in the digital world. In the analogue world we do not close all open files if we plan to continue working on them but we must do so in the digital world if we need to power down the system (unless we use the system sleep or hibernate mode).

Usability is reduced when a system projects a conceptual model from which a user cannot build a mental model they understand, for example, when systems fail, low-level error code messages and help may be presented that are only meaningful to the developer of the system but not to the user. Usability is also reduced when users have built up an erroneous mental model of the conceptual model of a system. There is not necessarily any single generic conceptual model that will fit all applications.

5.5.6 Evaluation

There is a strong motivation for evaluating products during development, formative evaluation. It enables problems to be fixed before a product is shipped. It enables developers to focus

---

[24] Note this mapping of the manipulation of virtual objects to the manipulation of physical objects is the opposite of the mapping used in tangible user interfaces that links the manipulation of physical objects to the manipulation of virtual objects.

on real problems for which there is evidence from use, not on problems that they perceive or imagine but which may never arise. The business case for formative evaluation is equally strong, the time to market can be reduced sharply, leading to substantial cost savings. Upon final release, the sales department has a rock-solid design to sell without having to sidetrack into how it will work in a future release. Evaluation of the completed product called summative evaluation can be used to verify and demonstrate that a system complies with regulatory requirements such as safety and physical environment requirements. Evaluations need to be planned and the selection of the different types of user doing the evaluations needs to be carefully considered. The methods for acquiring user input for evaluations can be based upon those used for user requirements gathering. (Section 5.5.2).

5.6 User Models: Acquisition and Representation

User context models can be viewed from two perspectives: users' models of systems and systems' models of users.[25] Users have a (mental) model of the UbiCom system, of how the user understands how the system works, to facilitate their eHCI interaction with it (Section 5.5.5.1). The focus of user modelling in this section is on how UbiCom Systems can use a model of the user (the user context) in order to facilitate the iHCI interaction with it. The focus of user modelling for UbiCom is on how UbiCom systems can acquire models of the user. There are several different design choices for user modelling.

- *Implicit vs. explicit models*: implicit or indirect acquisition of the user model e.g., by observing the user versus explicit or direct user interaction.
- *User instance (individual) modelling versus user (stereo)type modelling*: stereotypes are derived through data mining and clustering and classifying past interaction of many individual ones into groups or through hypothesising about types of users by experts (Han and Kamber, 2006, pp. 285, 382). Individuals can then be associated with their stereotypes based upon matching a few initial interactions against those of groups.
- *Static versus dynamic user models*: static user models are often acquired in a single-shot model. Some user features, once created, tend to be invariant, such as a person's fingerprints, gender, birthplace while other user features vary with respect to time, location and context such as the user mood and experience, etc. Dynamic user models vary across user tasks, time and/or space and may need to be updated from time to time. Machine-learning algorithms (Chapter 8) enable a user model to semi-automatically or automatically adapt to a changing environment context. If user models are complex, they can also be acquired in a multi-shot mode.
- *Generic versus application-specific models*: the latter applies to a specific task or application domain whereas the former can be used across tasks and application domains.
- *Content-based versus collaborative user models*: content-based user models depend upon prior characterisations of content and then matching these to an individual user's preferences for characteristic content. Collaborative user models depend on matching individual user preferences to a stereotype user's preferences and then using the latter to help complete the preferences of the individual.

---

[25] If there is bidirectional modelling of the UbiCom system, in terms of the UbiCom system modelling the user and the user forming a mental model of the system operation, this increases the complexity of the model and can introduce cyclic instabilities.

One major problem in user modelling is the acquisition of knowledge about the user. There are two main ways to do this. Systems can either ask the user explicitly for such interests (explicit feedback) or it can observe a user's system usage (implicit feedback) and then infer and anticipate certain behaviour. Often these can be combined, either different sub-types of implicit or explicit models or explicit models can be combined with implicit models in order to improve the user model.

Hybrid user models may also be used. Stereotype user models may be used in conjunction with explicit user modelling in order to try to reduce individual user values by trying to suggest default values from matching user stereotypes, see Section 5.6.3. Stereotype user models are also often used in conjunction with collaborative user models. An important consideration is to define what to model and how to represent it. Human reasoning is predominantly qualitative or soft, e.g., classifying how much someone likes an item, e.g., using a five-point Likert scale: very much, much, OK, not very much and not at all. Hence, soft computing techniques such as fuzzy logic, probability theory, neural networks and genetic type algorithms seem a good fit for this representation (Karray and De Silva, 2004; Section 8.6).

5.6.1 Indirect User Input and Modelling

Models of users can be formed by gathering and analysing indirect user input and linking these to user goals related to user tasks and activities – context-aware models. Context-aware user models may consider the current user context and interaction or also include the history of previous interaction. This knowledge or model of users' interactions can be used to anticipate, facilitate and simplify interaction (anticipatory computing). It can be used to personalise user input to filter user selections of content and services (personalisation). Models of group behaviour can be used to facilitate individual interaction (recommender systems, Section 9.4.2).

Indirect models of user can be improved by combining several context values such as the location, entity, activity and time, and may generate a more complete understanding of the current situation. These user contexts can also act as indices to other sources of contextual information. For example, knowing the current location and current time, together with the user's calendar, the application will have a pretty good idea of the user's current social situation, such as in a meeting, sitting in a class, waiting at the airport etc.

5.6.2 Direct User Input and Modelling

The main design choice for explicit user modelling is to use single-shot (static versus periodic) feedback. With single-shot feedback, a one-off full interview or questionnaire is undertaken. However, full interviews are often very time-consuming and users may not know how to best fill in a long form and may do it incompletely. The other main disadvantage of a user model is that it can become outdated. Periodic user feedback can utilise piecemeal dialogues. These are less time-consuming and may fit better with session based activities. Further, the system can orientate later dialog sessions by asking more specific questions in a specific context to improve the accuracy of the user model.

Explicitly asking the user would be more precise but it disrupts the user's current task and may be time-consuming and annoying. Moreover, users may be unwilling to fill in forms about preferences and here again the small displays for mobile devices are a restriction. In contrast, indirectly modelling users can be inexact but does not disturb or annoy users. However, there is a potential lack of user control and loss of privacy.

5.6.3 User Stereotypes

The drawbacks of explicitly asking the user can be compensated for by assuming an initial user interest profile from a stereotype, i.e. asking the user only a few questions, such as demographics and other indicators, which allow him or her to be classified. The typical interest profiles for such stereotypes have to be identified in empirical studies. The subsequent user interaction with the system can modify and correct this initial profile.

Stereotypes infer a user model from a small number of facts using a much larger set of facts from a user model (Rich, 1999). A stereotype is a collection of user attribute value types with a confidence in the value as well as a set of reasons why the value is believed. A stereotype is activated by a user's interactions and can in turn predict values with other user attributes. To some extent, collaborative recommender systems make use of a user stereotype. Stereotype design challenges include how to create new stereotypes, how to ascertain the confidence values, how to handle users who may not fit a stereotype and dealing with too many inputs or incomplete inputs.

5.6.4 Modelling Users' Planned Tasks and Goals

Users usually interact purposely with a system in a task-driven way, to achieve a particular goal. There are several ways to analyse and model user tasks. Hierarchical Task Analysis or HTA is a technique **to** decompose a user's goal into a hierarchy of actions. The actions are ordered, starting at the left in Figure 5.7 to follow a sequence to fulfil the goal. Complex actions can be decomposed

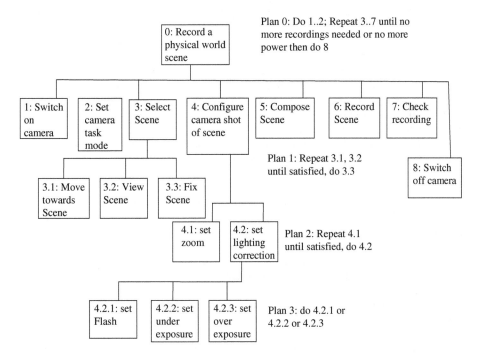

Figure 5.7 A Hierarchical Task Analysis (HTA) model for part of the record physical world scene from the PVM scenario in Section 1.1.1
Note: Parts of the scene have been omitted for simplicity.

into a set of simple actions in order to define actions at a sufficient level of detail so that they can be supported by the system. Plans can be defined to express action choices and repetition.

Consider the PVM scenario; there are several main user tasks that the system must support: recording a physical world scene, annotating a recorded scene, retrieving selected scenes based upon defined search criteria, browsing through recorded scenes. A HTA model for recording a physical world scene task is given in Figure 5.7. This is a partial HTA model for the task. It does not specify the camera focus, face detection, face recognition tasks, etc. This is a simplified type of planning model in which tasks are totally ordered, each task is treated as being independent of other tasks, only single tasks can be executed, not multiple tasks, plans are static, etc.

5.6.5 *Multiple User Tasks and Activity-Based Computing*

Much of what is termed Personal Computing is more suited to office workers who work on single fixed tasks, in a relatively uninterrupted manner, for long periods of time. In contrast, there are other types of human activity, that are prone to be interrupted, nomadic, of short duration, and multiple user activities are likely to be interleaved to achieve multiple user goals. A utility function based upon multiple value attributes can be used to allow multiple goals to be weighted and combined. In the Classroom 2000 project, Abowd and Mynatt (2000) expounded several design principles for everyday computing to support more informal, daily, activities.

- *User activities rarely have a clear beginning or end*: Information from the past is often recycled, e.g., address book, calendar of events, to-do list, etc. Activities are based upon previous experience. Principles such as providing visibility of the current state, freedom in dialogue and overall simplicity in features play a prominent role when designing support for activities.
- *Interruptions are to be expected*: Users should be made aware of the actions that are left uncompleted. In addition, the resumption of an activity may not start at a consistent point, but is related to the state prior to interruption. Resumption may be opportunistic, based on the availability of other people, or on the recent arrival of needed information. Interaction needs to be modelled as a plan that may at some point in time be suspended and be resumed at a later time.
- *Multiple activities operate concurrently*: Activities are continuous and context-shifting among multiple activities needs to be supported. To design for background awareness, interfaces should support multiple levels of 'intrusiveness' in conveying monitoring information that matches the relative urgency and importance of events.
- *Contexts such as time are useful for filtering and adaptation*: However, contexts such as time are rarely represented in computer interfaces. Many current and future personal actions, social interactions and decisions are based upon the outcome of past interactions and events or on the absence of these, e.g., if searches through application help information does not reveal useful information in the past, the user may not attempt to solicit help any time after that.
- *Associative models of information are needed*: Models of information for activities are principally associative since information is often reused on multiple occasions, from multiple perspectives. Associative and context-rich models of organisation support activities by allowing the user to re-acquire and reinforce the validity of information from numerous points of view. These views also support the need for users to resume an activity in different ways.

Bardram *et al.* (2006) characterise activity-based computing as: being application independent, occurring across multiple application tasks, supporting suspend and resume, supporting user roaming, adapting to the resources available, being shared among several collaborators and being context-aware.

In user-centred services, ICT events and service reconfiguration can be expressed at multiple knowledge viewpoints, e.g., using the mental model of different users. The majority of ICT events are either too low level or too detailed for users to understand and handle, or are too high level including some uncertainty about the cause. ICT events are often handled locally within the application code. It cannot be propagated to multiple points, e.g., time is manually set at each device's control panel rather than being automatically set from the network. Rich knowledge portals could be created and maintained that are user-centred, e.g., allowing users to post problem descriptions in user terms rather than selecting them from predefined lists of provider view of problems; these can be used to advise users and to provide self-help to instruct users remotely.

5.6.6 Situation Action Versus Planned Action Models

There are two basic approaches to task design: planned actions that include re-planning, and situated actions (Suchman, 1987, pp. 49–67). In planned actions, user activities are driven by specifying user objectives or goals and plans to achieve that goal (pre-planning), e.g., HTA (Section 5.6.4). Monitoring and awareness of the situation during the plan execution can provide feedback or constraints to possibly modify the plans and make them more resilient. This approach assumes that the user is able to define a goal, that preset plans to achieve the goal can be designed in advance and that contingency plans to adapt to any significant changes can be specified in advance.

An alternative approach is to set a goal but not to predefine a plan to achieve that goal but merely to assess locally at each stage what is the best choice of action that moves closer to the goal, situated actions, e.g., to assess at each step what the optimum next local action is, that brings the user nearer to achieving his or her goal. In addition, feedback on the status of the activity needs to reach the user so that the user can then choose to influence any reconfiguration of the activities, Riva (2005) specifies two main approaches to model situated actions.

The first approach, a situated-cognitive approach builds symbolic models of relations between subjects and the properties of specific environments (affordances and constraints). This does not explain how the choice of possible actions is constrained other than by the situation itself. In addition, the social space influences the activities of the subject. The second approach, an interactional approach such as Wenger's community of practice, says that these develop through joint enterprise as understood and continually negotiated by its members, that these operate though mutual engagement that binds members together into a social entity and that they use a shared repository of communal resources including vocabulary and policies.

A key concern is how a user retains the control of his or her planned activities and perhaps can influence the effect of a detected situation or a changed situation. This can be modelled by a theory of presence that seeks to differentiate between internal and external states and that to experience distal attribution, perceiving an external space outside our boundaries (Riva, 2005). System designs for planned actions are discussed in Section 8.7.

5.7 iHCI Design

5.7.1 iHCI Model Characteristics

A design model of iHCI is characterised by the following properties:

- *Natural (human–computer) interaction can* use a wide variety of physical artefacts, situated throughout the physical world, implicitly linked to virtual computer artefact interaction. Users do not need to be aware that obtrusive computers are present in specific places in the physical environment to access digital services.

- *User models* can be used to anticipate user behaviour based upon:

  o models of past individual user interaction which can be used to anticipate future user behaviour;
  o models of individual user interaction may be grouped into stereotypes of users to anticipate user behaviour based upon the group the individual belongs to.

- *User context-awareness.*

5.7.2 User Context-Awareness

User context-awareness can be exploited to beneficially lessen the degree of explicit HCI needed. The user context-awareness and context adaptation can range from passive to active modes. In a passive mode, the system provides shortlists of tasks and their user constraints which are relevant to the current situation. In the active modes, the system performs the adaptation, e.g., it automates the remainder of a task and lessens users' involvement in the completion of the task.

Users' contexts specify any physical, ICT and human environment context constraints in relation to a user task goal, e.g., to watch a movie. Physical environment context constraints include using particular ambient lighting settings. ICT environment context constraints include watching the movie on a large high-resolution display with a surround sound system. Social environment context constraints determine who to watch the movie with. A user context can include the following properties:

- Users' physical characteristics and capabilities for HCI e.g., how easy they find interacting with a particular type of UI such as a pointing device.
- User presence in a locality or some detected activity within some application context.
- User identity (Section 12.3.4).
- User planned tasks and goals (Section 5.6.4).
- Users' activity situated tasks, which may be spontaneous and unplanned, may be concurrent, may involve composite tasks and may be spread across multiple devices (Sections 5.6.5, 5.6.6).
- User emotional state (Section 5.7.5), e.g., repeatedly, pressing a key may indicate impatience.

5.7.3 More Intuitive and Customised Interaction

MTOS-based devices tend to use a desktop UI metaphor coupled with the use of a filename as an index to organise information. In order to start work on documents, users must remember how they categorised their documents in terms of the name of files and the name of folders and the devices they stored the files in. There are several limitations to this approach. Information does not neatly fall into a category as categories overlap and are fuzzy. It is impossible to generate categories that remain unambiguous over time, and names are an ineffective way to categorise information (Lansdale, 1988). Freeman and Gelernter (2007) propose using virtuality, based upon using unifying visual expressions, as a principle for better personal user interface design, rather than using physical metaphors as the basis for user interaction as the latter can unnecessarily cramp the design.

Freeman and Gelernter identified the following principles for a better personal information system. Storage should be transparent, avoiding the need for filenames and folders. Archiving should be automatic. Reminders should be an integral part of the desktop. Personal data should be automatically available from anywhere and systems should provide useful summaries of documents. In the Lifestreams project, Freeman and Gelernter proposed chronology, i.e., the past, present and future, as an organisational structure for managing information. Freeman and Gelernter also make a distinction between searching and browsing. While there are powerful desktop search engines around, users have to be able to specify what to search for but often they do not do so, hence what is better is supporting a powerful browser engine that instead aids user-directed navigation.

Moran and Zhai (2007) present a wider analysis of surveyed approaches to support more powerful personalised information. This is based upon novel information organisation principles connected with chronology, tasks and multiple user-defined associations for information. It is based upon consideration of a more fluid interplay between individual and social interaction. It is based upon user involvement in multiple rather than single activities. It is based upon high-level activity and goal-directed computing rather than lower-level task-based computing. Moran and Zhai thus propose seven principles to develop the desktop information model into a much powerful model that can be used to underpin user-centred interaction in UbiCom applications:

1. *From Office Container to Personal Information Cloud*: personal information such as preferences is not restricted to specific folders and files that contain the information within specific devices and services but can follow the user around and be used wherever and whenever it is needed, see also user virtual environments (Section 4.2.1). Personal information should be person-aware, persistent, pervasive, secure and able to be referenced and in a standardised format.
2. *From desktop to a diverse set of visual representations*: a variety of visual representations are needed which may be adapted to specific problem domains and different device form factors, complementing the basic conventional desktop metaphor, because work at the desktop is only a part of the range of UbiCom interaction.
3. *From interaction with one device to interaction with many*: to support use of multiple instances and multiple types of devices, resources and services, design patterns are needed that separate the view from the model. Designs are also needed to deal with the main inherent characteristics of multiplicity (Section 9.2).
4. *From mouse and keyboard to greater set of interactions and modalities*: developing more natural interaction with ICT devices is a key part of UbiCom vision, represented by support for the iHCI property in UbiCom systems (Section 5.7).
5. *Functions may move from applications to services*: some desktop applications are clearly too complex for many users to use while, in contrast, Web applications are simpler, this is partly because of the limited standardised interaction support in Web 1. Web 2 technologies enable richer Web interaction. However, this may not necessarily work in volatile service environments (Section 3.3.3.9) and there is still the issue of multiplicity to deal with when binding interaction to services.
6. *From personal to interpersonal to group to social*: much personal interaction is also often part of multiple kinds of interpersonal interaction. Support is needed to make this more seamless by supporting better social networking support (Section 9.4.1) and by enabling more personal information to be shared socially, e.g., a map application where one user can share spatial annotations in a controlled way with others (Liang *et al.*, 2008).
7. *From low-level tasks to higher level activities*: user activities rarely have a clear beginning or end; interruptions are to be expected; multiple activities operate concurrently; contexts such as time are useful for filtering and adaptation: however, contexts such as time are rarely represented in computer interfaces; associative models of information are needed to support and interrelate activities.

5.7.4 Personalisation

Personalisation involves tailoring applications and services specifically to an individual's needs, interests, and preferences. It can also involve adaptation of a consumer product, electronic or written medium, based on personal details or characteristics provided by or on behalf of a user or consumer in the form of a user profile, e.g., a favourites list for viewing AV content. The profile may not necessarily be provided by the user subject but be gathered by observing a user's interactions without their knowledge. There are several prominent uses of personalisation including targeted marketing, product and service customisation, information filtering and personalised customer relationship management (CRM).

Personalisation aims to develop a more complete model of user-context that is more reusable and persists: across different user sessions or instances of the same type of user tasks, e.g., repeat orders; across different user tasks, e.g., user personal details such as home address may be reusable across tasks; across different users, e.g., enabling a user with little experience of a new task or having little knowledge of the context to exploit the experience of other users who are more familiar with a task or have more knowledge. Two key issues are, first, to design a personal preferences model so that it can be distributed and shared across multiple applications and users; second, to consider dynamic task-driven user preference contexts versus using a static user context that acts as a lowest common denominator context that applies for multiple users in multiple tasks.

The perceived benefits of personalisation include efficiency. Service interaction that can identify users and associate previously gathered user information, e.g., it can make one-click service invocation possible. A system that makes it easier for the customer to invoke additional services and to repeat the same service can potentially increase the amount of business. It can better serve customers by anticipating their needs. It can reduce the information overload on the user (Maes, 1994). This is of particular importance in resource-constrained environments and interactions. It can improve the usability of services. It can make the interaction user-centred and more satisfying for the user. It provides more matching choices or ones of interest to the customer. It has a social dimension, treating the customer as a known individual rather than just another anonymous customer.

Personalisation can aid users who have a partial view of the environment, using access devices with limited display and networking capacity of mobile devices such as cell phones or hand-held computers that have variable degrees of blindness or have partial control of the environment. A possible solution for this is the adaptation of services and contents to users' personal interests in addition to adaptation to their current location and type of access terminal. The adaptation of services and contents to personal interests filters the available information. The filtering process is based on a user profile describing the interests, abilities and characteristics of this user. Personalisation tailors content to specific user viewers. A flexible way to do this is to match the characteristics of the content to the preferences of a viewer stored in a user viewer profile.

The possible disadvantages of personalisation include a loss of anonymity, a loss of privacy, loss of user control, disadvantageous discrimination for consumers, reduced choice and more rather than less information load. There is a trade-off between a loss in privacy versus the potential gains from personalisation. When a provider's organisation knows about the individual, a service provider can sell this knowledge on to other providers. A provider could manipulate the individual by making recommendations to the detriment of the customer, e.g., providing tempting recommendations coupled to easier credit enabling impulse buying by customers who may struggle to pay back the amount borrowed. There is a loss of user control because the provider could create a model without the user's knowledge and the provider could create a model without the user's permission, i.e., that is illegal. Finally, there is the potential for sellers to maximise their profit by offering different deals to different types of customer, e.g., a more favourable deal for the same product is offered to new customers than existing customers.

Personalisation tends to filter information and services to the user that are instances of or are similar to the personal preferences. This by definition hinders users from widening their choice and experience of new things – personalisation can keep us in our comfort zone. Personalisation can cause more rather than less of an information load because whenever customers selectively invoke a service, e.g., buy an item, several additional services, which in the worst case scenario may be non-discriminatory, are triggered by providers that are perceived to match their interests. This can be distracting. In addition, because providers also hold personal details of customers, they can push their service remotely and often across several channels. Hence, in practice, multiple levels of filters are needed to filter choices and to filter unsolicited offers.

5.7.5 Affective Computing: Interactions Using Users' Emotional Context

One important human trait used in human–human interaction is the ability to recognise, interpret, process and share human emotions. In 1995, Picard at MIT proposed the idea of affective computing that relates to, arises from, or influences emotions. Affective computing applications included computer-assisted learning, perceptual information retrieval, arts and entertainment, and human health and interaction. Emotional responses make a core contribution to human behaviour. The design challenges for affective computing overlap to some extent the design issues for determining the user context and those for developing more complex human-like intelligence models for use in UbiCom Systems. In addition, models of human intelligence (Russell and Norvig, 2003) are a core model for building artificial intelligence along with rational intelligence models. Picard has reviewed some of the main design challenges for this paradigm and identified six design challenges (Picard, 2003):

- *The range of means and modalities of emotion expression is very broad*: many of these modalities may be inaccessible (e.g., blood chemistry, brain activity, neurotransmitters), and many others cannot be differentiated.
- *People's expression of emotion is so idiosyncratic and variable*: accurately recognising an individual's emotional state from the available data is challenging. However, emotions can be more accurately classified if they are determined over time and if they are correlated to other factors such as time of day.
- *Cognitive models for human emotions are incomplete* (little real progress has been made with cognitive modelling). Existing models of emotion often use highly stylised stereotypes of personality types and emotional responsiveness, which do not correspond to real behaviour in real people. The role of situational factors in emotion expression is poorly understood.
- *The sine qua non of emotion expression is the physical body but computers are not embodied in the same way*. Hence because computers are not embodied, they cannot reliably and believably express emotion. Computers can express emotions without having physical bodies as seen in some film and animation characters. Note also that people with varying degrees of physical disabilities can also express emotions in a range of ways with even very limited modalities.
- *Emotions are ultimately personal and private*: they provide information about the most intimate motivational factors and reactions. Attempts to detect, recognise, and manipulate a user's emotions can invade user privacy. However, humans often attempt to manipulate the emotions of other humans and this is not considered unethical.
- *There is no need to contaminate purely logical computers with emotional reactiveness*. However, several studies have supported vital roles for emotion in many background processes: perception, decision-making, creativity, empathic understanding, memory, as well as in social interaction (Picard, 2003).

5.7.6 Design Heuristics and Patterns

UI design should seek to support a system conceptual model based upon HCI principles which supports good usability and which maps to a clear user's mental model. There are many overlapping ad hoc sets of higher-level HCI design heuristics proposed by a number of different HCI designers to promote good design of HCI interaction. Specific guidance is needed to apply these heuristics to design UIs to comply with these HCI principles. UI design patterns define a set of high-level design heuristics to support HCI usability principles and then map these into lower-level more concrete design patterns (Tidwell, 2005).

A set of lower-level UI design patterns which are oriented to visual information accessed via desktop Web browsers and mobile phones has been identified by Tidwell (2005). Design patterns based upon these but oriented more towards iHCI UbiCom interaction are given in Table 5.1. These can be related to the higher-level design heuristics given in Table 5.2.

Table 5.1 UI design heuristics for UbiCom based upon the high-level heuristics proposed by Tidwell (2006)

Heuristic	Description	Design implications
Safe exploration	Let users explore or browse without getting lost or getting into trouble	Provide the ability to undo and try something different
Satisficing[1]	Searching to find options that are good enough rather than the best option	Make descriptions and labels informative
Changes in midstream	Users can change their mind about what was being done	Ease start, pause and re-entry of processes; Remember previously typed information
Deferred choices	Users may not want to make choices now, may not have enough information, may wants to skip unnecessary questions	Mark, select small set of mandatory inputs; Use good defaults enable users to return to deferred fields
Incremental construction	Many things evolve, they require the experience of doing something several times, they are not got right first time	Make it easy to achieve goals piecemeal; constantly give user feedback
Habituation or unification	Users can make false choices if they get used to a pattern of input which varies across applications and devices	Standardise common gestures or interaction everywhere; support user customisation of interaction
Constituted actions	Lower level actions within a context counts as another high-level action, e.g., use of keyboard short-cuts, multi-tap, gestures, etc	Support definitions of sets of low-level actions to have a higher-level meaning
UI proxy	Users may have to use multiple individual devices and interfaces in order to access a composed service or UI	Support use of proxies to simplify access to multiple individual interfaces
Context-based Memory recall	Users often manipulate objects and based upon context such as when and where they used something, not by what it is named	Store context with objects. Use predictable contexts
Prospective memory Context-aware trigger	People tag objects in their environment to remind themselves to deal with them later	Support prospective tags Support proactive and situated task tips
Situated help	Help access is tailored to different levels. Provider's help may not be understood by users. Help may not relate to user experiences	Support multi-level help. Allow help concepts to be tailored to user. Link help to communities of practice
Instant feedback	Users can unnecessarily interrupt or reconfigure tasks or give redundant input if they cannot observe the effect of their input	Allow users to see immediate effect of actions they take and give positive feedback; Avoid unneeded steps
Context-driven Explanations	It is often not clear why UI options are not available or why actions do not work	Enable users to get explanations within a context about permitted or forbidden actions
Prospective, anticipated actions	Once a task sequence is started or progressed beyond a certain point it may be clear which tasks will follow to achieve a goal	Support proactive automated task suggestions and partial task completion by the system

Table 5.1 (*continued*)

Heuristic	Description	Design implications
Streamlined replay	Facilitate user actions to be repeated by detecting redundancy & streamlining replay	Ease repeating. Detect repeated patterns and advise users
Streamlined input	Input may be limited because lack of UI space, inability of user to give detailed input	See techniques in Sections 5.2.3.1, 5.3
Streamlined output	Output may be limited because lack of UI space, inability to access a specific UI model	See techniques in Sections 5.2.3, 5.3 Show extras on-demand

Note: [1] Satisficing, derived from a combination of satisfying and sufficing, was proposed by social scientist Herbert Simon in the 1950s. When faced with uncertainty about the future and costs in acquiring information in the present, the extent to which agents can make a fully rational decision is limited. Thus agents have only a

Table 5.2 Some examples of lower-level HCI design patterns which are linked to higher-level HCI design heuristics, based upon Tidwell (2005)

Pattern group	Low-level design patterns	High-level heuristic
Organising of the whole:	Wizard: lead user through UI in steps in prescribed order	Help
	Extras On Demand: show important content, hide the rest	Minimalist
	Multi-level Help: use mixed help to support varied user needs	Help
Getting around	Clear Entry Points: present few descriptive entry points	Safe Explore
	Global Navigation: put clear navigation links everywhere	Habituation
	Breadcrumbs: on each page show map to parent and top pages	Safe Explore
	Escape Hatch: add a clear link to known place	Safe Explore
Organising the layout of a part	Visual Framework: design each page using same basic layout	Habituation
	Card Stack: use separate panels, stack so only 1 shows	Minimalist output
	Responsive Disclosure: start with minimal layout, guide user step by step, showing more of the UI as it is completed.	Minimalist
	Responsive Enabling: start with mostly disabled UI, enabling more of it as user progresses	Minimalist
Doing things	Smart Menu Items: adapt menu labels to show what they do when invoked	Minimalist
	Cancelability: user can cancel long tasks without side-effects.	Error handling
	Progress Indicator: show users the degree of progress	Visible status
	Multi-Level Undo: provide a way to easily reverse a series of actions performed by the user	Error handling
	Macros: users can create their own sequences of commands	Streamlined repetition
Getting input from users	Forgiving Format: permit user to enter input in variety of formats, application interprets it intelligently	Error handling
	Input Prompt: pre-fill a text field with a prompt that tells user what to do	Help
	Auto-completion: complete the entry by anticipating what the user will type next	Minimalist

(*continued overleaf*)

Table 5.2 (*continued*)

Pattern group	Low-level design patterns	High-level heuristic
	Dropdown Chooser: extend menu by using drop-down panel with a more complex value-selection UI	Minimalist
	Good Defaults: pre-fill forms with best guesses the user wants	Satisficing
	Linked error messages: and if possible put indicators next to originating controls	Error-handling

The UI patterns listed in Table 5.2 are at a lower level than the UI heuristics or guidelines but are still too high level to be implemented using the API of a particular UI toolkit to create the display elements such as buttons and a text-field and set up event-handlers to support the user interaction. A more understandable way to link the UI objects defined in a UI toolkit to the HCI pattern is to use a 'back-end' software design pattern such as the MVC[26] or Model, View, Controller (Krasner and Pope, 1988) (Figure 5.8). This decouples the view (screen presentation) from the model (application object) via the controller (the reaction of the UI to user interface). The controller can use either a subscribe pattern or notify interaction pattern to link to the view when the model changes.

The design heuristics summarised in Table 5.2 can also be used to partially support designs for implicit interaction. This is termed partial support for iHCI because support for some characteristics of implicit interaction, such as users' emotional context, users' understanding and physical characteristics have not yet been modelled in terms of design patterns. Activities across a multiplicity of devices and applications are supported by the design patterns: safe exploration,

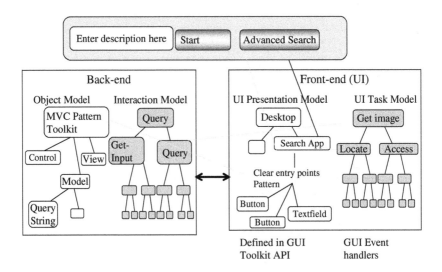

Figure 5.8   Relating the HCI design heuristic

[26] The MVC pattern is supported in many developer frameworks such as Microsoft ASP.NET and WinForms.NET and in Java using the open-source Struts and Spring frameworks.

satisficing, changes in mmidstream, deferred choices, incremental construction, habituation or unification, constituted actions and UI proxy. Context-aware interaction is supported by: context-based memory, context-aware stigmergy, situated help, context-driven explanations, streamlined repetition, prospective actions, streamlined input and streamlined output.

EXERCISES

1. Why is it important to study human–computer interaction for ubiquitous computing?
2. Compare and contrast the user interfaces and user interaction used with four common types of device: personal computer, hand-held mobile devices used for communication, games consoles and remote-controlled AV displays.
3. Discuss designs to overcome the inherent limited input and output capabilities of hand-held mobile communication devices.
4. Compare and contrast the use of explicit human interaction, implicit human interaction and no human interaction for the following human activities: preparing a meal, driving a vehicle from start to finish, washing clothes and washing your pet or yourself.
5. Describe some activity that you think cannot be digitally automated and then undertake a Web search for ICT solutions to support it.
6. Discuss the disadvantages of using more natural interaction in a digital universe. Consider hand-writing with respect to its throughput, accessibility, interoperability, etc. Then weigh up both the pros and cons of primarily supporting natural human–computer interfaces.
7. Using the classification of Fishkin (2004) for tangible user interfaces, classify the following types of UI: computer mouse, any RFID-tagged object such as food items, passports, etc., robots and smart travel cards.
8. Debate whether or not the prevalence of MEMS components could lead to users being able to create their own customised tangible UIs to applications, moving what is a currently a topic of computer science research into mainstream consumer products.
9. Outline scenarios where tangible UIs are a benefit and scenarios where their limitations outweigh their benefits.

References

Abowd, G.D. and Mynatt, E.D. (2000) Charting past, present, and future research in ubiquitous computing. *ACM Transactions on Computer-Human Interaction*, 7(1): 29–58.

Bush, V. (1945) As we may think. *The Atlantic Monthly*, Vol. 176: 101–108. Reprinted and discussed in *ACM Interactions*, 3(2) (1996): 35–67.

Bushnell, N. (1996) Relationships between fun and the computer business. *Communications of the ACM*, 39(8): 31–37.

Chapin, J.K., Moxon, K.A., Markowitzet, R.S., *et al.* (1999) Real-time control of a robot arm using simultaneously recorded neurons in the motor cortex. *Nature Neuroscience*, 2: 664–670.

Co, E. and Pashenkov, N. (2008) Emerging display technologies for organic user interfaces. *Communications of the ACM*, 51(6): 45–47.

Dalisa, A. (1997) Electrophoretic display technology. *IEEE Transactions on Electronic Devices*, 24: 827–834.

Derrett, N. (2004) Heckel's law: conclusions from the user interface design of a music appliance – the bassoon. *Personal and Ubiquitous Computing*, 8: 208–212.

Dix, A., Finlay, J., Abowd, G. *et al.* (2004). *Human-Computer Interaction*, 3rd edn. Englewood Cliffs, NJ: Prentice Hall.

Fishkin, K.P. (2004) A taxonomy for and analysis of tangible interfaces. *Personal and Ubiquitous Computing*, 8(5): 347–358.

Forster, W. (2005) *The Encyclopedia of Game Machines: Consoles, Handheld and Home Computers 1972–2005*. New York: Gameplan.

Freeman, E. and Gelernter, D. (2007) Beyond Lifestreams: the inevitable demise of the desktop metaphor. In V. Kaptelinin and M. Czerwinski (eds) *Beyond the Desktop Metaphor: Designing Integrated Digital Work Environments. Cambridge*, MA: MIT Press, pp. 19–48.

Goldstein, S.C., Campbell, J.D. and Mowry, T.C. (2005) Programmable matter. *Computer*, 38(6): 99–101.

Han, J. and Kamber M. (2006) *Data Mining: Concepts and Techniques*, 2nd edn. New York: Morgan Kaufmann Publishers.

Harper, R., Rodden, T., Rogers, Y. and Sellen, A. (eds) (2007) *Being Human: Human-Computer Interaction in the Year 2020*. Technical Report, Microsoft Research Ltd. Available from http://research.microsoft.com/hci2020/downloads/BeingHuman_A4.pdf, retrieved March 2008.

Harrison, B.L. (2000) E-books and the future of reading. *IEEE Computer Graphics and Applications*, 20(3): 32–39.

Hayles, N.K. (1999) *How We Became Posthuman: Virtual Bodies in Cybernetics, Literature, and Informatics*. Chicago: University of Chicago Press.

Hofstede, G. (1997) *Cultures and Organisations: Software of the Mind*. New York: McGraw-Hill.

Ihlström, C., Åkesson, M. and Nordqvist, S. (2004) From print to Web to e-paper – the challenge of designing the e-newspaper. In *Proceedings of ICCC 8th International Conference on Electronic Publishing (ELPUB 2004)*, Brasilia, pp. 249–260.

Inbar, O., Ben-Asher, N., Porat, T., *et al.* (2008) All the news that's fit to e-ink. Paper presented at Conference on Human Factors in Computing Systems, CHI '08, session on Research Landscapes, pp. 3621–3626.

Inoue, S., Kawai, H., Kanbe, S., *et al.* (2002) High-resolution microencapsulated electrophoretic display (EPD) driven by poly-si TFTs with four-level grayscale. *IEEE Transactions on Electron Devices*, 49(9): 1532–1539.

IJsselsteijn, W.A. and Riva, G. (2003) Being there: the experience of presence in mediated environments. *Emerging Communication*, 5: 3–16.

Jacobson, J., Comiskey, B., Turner, C., *et al.* (1997) The last book. *Systems Journal*, 36(3): 457–463.

Jaimes, A. and Sebe, N. (2005) Multimodal human computer interaction: a survey. In Proceedings of IEEE International Workshop on Human Computer Interaction in Conjunction with ICCV 2005. In: *Lecture Notes in Computer Science*, 3766: 1–15.

Johnston, R.S. and Willey, S. (1995) Development of a commercial virtual retinal display. In W. Stephens and L.A. Haworth (eds) *Proceedings of Helmet- and Head-Mounted Displays and Symbology Design*, pp. 2–13.

Jones, M. and Marsden, G. (2006) *Mobile Interaction Design*, Chichester: John Wiley & Sons, Ltd.

Karray F. and De Silva C. (2004) *Soft Computing and Intelligent Systems: Design: Theory, Tools and Applications*. London: Pearson Books.

Kim, D. and Kim, D. (2006) An intelligent smart home control using body gestures. In *Proceedings of 2006 International Conference on Hybrid Information Technology (ICHIT'06)*, 2: 439–446.

Krasner, G.E. and Pope, S.T. (1988) A cookbook for using the model-view controller user interface paradigm in Smalltalk-80. *Journal of Object-Oriented Programming*, 1(3): 26–49.

Kung, S.Y., Mak, M.W. and Lin S.H. (2004) *Biometric Authentication: A Machine Learning Approach*. Englewood Cliffs, NJ: Prentice Hall.

Lansdale, M.W. (1988) The psychology of personal information management. *Applied Ergonomics*, 19(1): 55–66.

Lebedev, M.A. and Nicolelis, M.A.L. (2006) Brain-machine interfaces: past, present and future. *Trends in Neurosciences*, 29: 536–546.

Lenoir, T. (2002) *Makeover: Writing the Body into the Posthuman Technoscape. Part One: Embracing the Posthuman: Configurations*. Baltimore, MD: Johns Hopkins University Press and the Society for Literature and Science, 10: 203–220.

Liang, Z., Poslad, S., Meng, D. (2008) Adaptive sharable personalized spatial-aware map services for mobile users. Paper presented at GI-Days 2008 Conference.

Maes P. (1994) Agents that reduce work and information overload. *Communications of the ACM*, 37(7): 30–40.

Mann, S. (1997) An historical account of the 'WearComp' and 'WearCam' inventions developed for applications in 'personal imaging'. In *1st International Symposium Wearable Computers*, pp. 66–73.

Mann, S. (1998) Humanistic intelligence: WearComp as a new framework for intelligent signal processing. In *Proceedings of the IEEE 86(11)*, pp. 2123–2151.

Mann, S. and Fung J. (2002) EyeTap devices for augmented, deliberately diminished, or otherwise altered visual perception of rigid planar patches of real-world scenes. *Presence: Teleoperators and Virtual Environments Archive*, 11(2): 158–175.

Marcus, A. and Gould, E.W. (2000) Crosscurrents: cultural dimensions and global Web user-interface design. *ACM Interactions*, 7(4): 32–46.

McTear, M. (2002) Spoken dialogue technology: enabling the conversational user interface. *ACM Computing Surveys*, 34(1): 90–169.

Mitra, S. and Acharya, T. (2007) Gesture recognition: a survey. *IEEE Transactions on Systems, Man, and Cybernetics, Part C: Applications and Reviews*, 37(3): 311–324.

Moran, T.P. and Zhai, S. (2007) Beyond the desktop metaphor in seven dimensions. In V. Kaptelinin and M. Czerwinski (eds) *Beyond the Desktop Metaphor: Designing Integrated Digital Work Environments.* Cambridge, MA: MIT Press, pp. 335–354.

Myers, B., Hollan, J. Cruz, I. *et al.* (1996) A brief history of human computer interaction technology. *ACM Computing Surveys*, 28(4): 794–809.

Navarro, K.F. (2004) Wearable, wireless brain computer interfaces in augmented reality environments. In *Proceedings of International Conference on Information Technology: Coding and Computing, ITCC 2004*, 2: 643–647.

Norman, D.A. (1988) *The Psychology of Everyday Things.* New York: Basic Books.

Orth, M., Post, R., and Cooper, E. (1998) Fabric computing interfaces. In *Proceedings of Conference on Human Factors in Computing Systems, CHI 98*, Los Angeles, pp. 331–332.

Picard, R.W. (2003) Affective computing: challenges. *International Journal of Human-Computer Studies*, 59: 55–64.

Pickering, J.A. (1986) Touch-sensitive screens: the technologies and their applications. *International Journal of Man-Machine Studies*, 25: 249–269.

Pignotti E., Edwards, P. and Grimnes, G.A. (2004) Context-aware personalised service delivery. In *European Conference on Artificial Intelligence, ECAI 2004*, pp. 1077–1078.

Preece, J., Rogers, Y. and Sharp, H. (2006) *Interactive Design: Beyond Human-computer Interaction*, 2nd edn. Chichester: John Wiley & Sons, Ltd.

Pressman R.S. (1997) *Software Engineering: A Practitioner's Approach*, 4th edn. Maidenhead: McGraw-Hill.

Raskin, R. (2000) *The Human Interface.* Reading, MA: Addison-Wesley.

Rekimoto, J. (2008) Organic interaction technologies: from stone to skin. *Communications of the ACM*, 51(6): 38–44.

Rich, E. (1999) Users are individuals: individualizing user models. *International Journal of Human-Computer Studies*, 51: 323–338.

Riva, G. (2005) The psychology of Ambient Intelligence: activity, situation and presence. In G. Riva, F. Vatalaro, F. Davide and M. Alcañiz (eds) *Ambient Intelligence.* IOS Press. Available from http://www.ambientintelligence.org, accessed December 2005.

Russell, S. and Norvig, P. (2003) *Artificial Intelligence: A Modern Approach*, 2nd edn. Englewood Cliffs, NJ: Prentice Hall.

Schwab, I. and Pohl, W. (1999) Learning user profiles from positive examples. In *Proceedings of International Conference on Machine Learning and Applications*, Chania, Greece, pp. 15–20.

Schwesig, C. (2008) What makes an interface feel organic? *Communications of the ACM*, 51(6): 67–69.

Sharma, R., Pavlovic, V.I. and Huang, T.S. (1998) Toward multimodal human-computer interface. *Proceedings of the IEEE*, 86(5): 853–869.

Shneiderman, B. (1983) Direct manipulation: a step beyond programming languages. *IEEE Computer*, 16(8): 57–69.

Shneiderman, B. and Plaisant, C. (2004) *Designing the User Interface: Strategies for Effective Human-Computer Interaction*, 4th edn. Reading, MA: Pearson Addison-Wesley.

Suchman L.A. (1987) *Plans and Situated Actions: The Problem of Human Machine Communication.* Cambridge: Cambridge University Press.

Sutherland, I. (1968) A head-mounted three-dimensional display. In *Proceedings of Fall Joint Computer Conference*, pp. 757–764.

Thorp, E.O. (1998) The invention of the first wearable computer. In *2nd International Symposium Wearable Computers: Digest of Papers, IEEE Computer Society*, pp. 4–8.

Tidwell, J. (2005) *Designing Interfaces: Patterns for Effective Interaction Design*. New York: O'Reilly.

Van Laerhoven, K., Schmidt A. and Gellersen, H.-W. (2002) Multi-sensor context-aware clothing. In *Proceedings of 6th International Symposium on Wearable Computers* (ISWC 2002), Seattle, IEEE Press, October.

Vertegaal, R. and Poupyrev, I. (2008) Organic user interfaces. *Communications of the ACM*, 51(6): 26–30.

Vidal, J.J. (1973) *Annual Review of Biophysics and Bioengineering*, 2: 157–180.

Warwick, K. (1999) Cybernetic organisms: our future? *Proceedings of the IEEE*, 87(2): 387–389.

Warwick, K., Gasson, M., Hutt, B.D., *et al.* (2003) The application of implant technology for cybernetic systems, *Archives of Neurology*, 60(10): 1369–1373.

Wilson, A. and Shafer, S. (2003) XWand: UI for intelligent spaces. In *Proceedings of SIGCHI Conference on Human Factors in Computing Systems*, pp. 545–552.

Yee, K-P. (2003) Peephole displays: pen interaction on spatially aware handheld computers. In *Proceedings of SIGCHI Conference on Human Factors in Computing Systems*, Ft. Lauderdale, FL, pp. 1–8.

Zimmerman, T.G. (1996) Personal area networks: near-field intrabody communication. *IBM Systems. Journal*, 5(3–4): 609–617.

6

Tagging, Sensing and Controlling

6.1 Introduction

As electronic components become smaller, faster, and cheaper to fabricate, more low-power and more low maintenance, they can be more easily deployed on a massive and pervasive scale.[1] Ongoing work on *Micro-Electro Mechanical Systems* (MEMS)[2] will enable sensing and actuation down to a scale of a nanometre. The possibilities for miniaturisation extend into all aspects of life, and the potential for embedding computing and communications technology quite literally everywhere is becoming a reality. IT will eventually become an invisible component of almost everything in everyone's surroundings, extending the Internet via embedded networks of MEMS deep into the physical environment, making greater use of the expanded IPv6 address space. MEMS form a powerful enabler for the vision of smart ubiquitous computing environments. Although many mechanisms can be made usefully smaller in order to embed them and make them blend into the physical world, some macro mechanisms are still needed, often to support human activities, e.g., displays, human transport vehicles and many household appliances.

Embedded systems are a second IT trend to enhance natural or artificial physical objects with digital, networked embedded devices to provide improved manual, semi-automatic and automatic sensing and control, e.g., in motor vehicles.

A third trend is the increasing annotation of the physical world, often driven by a need to enrich human and other natural interactions in these environments, to be better informed about these

---

[1] There are vast societal issues here (Chapter 13). If surveillance cameras become smaller, rather than using them to deter crime (crimes are less likely if people know they are being observed), cameras can be used for serendipitous monitoring. This could be used for business ethnographic studies (Section 5.5.2), e.g., to observe what people look at in public spaces, such as shopping stores. It can also be used more actively than passively, to signal events to change the appearance of the physical environment to distract people if businesses perceive them to observe things that have too low a sales potential.

[2] Note although MEMS primarily refers to micro-sized mechanisms, it is also often taken to include millimetre and nanometre-sized devices in practice.

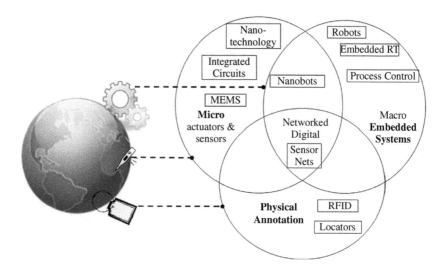

Figure 6.1   Enabling ubiquitous computing via micro, macro embedded and annotation of physical objects in the world

interactions at both a local and global level. For example, we may want to locate where we have left things such as our keys or phones, or we may want to learn where an insect we have never seen before, originates from. We often need to discover where things are in the physical world, to discover their characteristics and to the track them, to virtualise views of the physical world and its physical objects. The design of virtual annotation models may also learn from nature in this respect. Social insects such as ants lay chemical trails to support pheromone-based interaction with their environment and hence indirectly with each other (Hölldobler and Wilson, 1990).

6.1.1   Chapter Overview

In this chapter, three complementary enabling trends for pervasive computing are described (Figure 6.1). First, tagging of objects in the physical world, using a mixture of sensors and tags, enables these features to form part of a discourse in a virtual computer environment. Second, mechanisms can be reduced to millimetre, micrometre and nanometre sizes to be more easily blended into the physical world. Design issues in creating and operating much smaller devices are considered from an ICT engineering point of view. Third, digitally enabling and networking macro mechanisms and devices are accompanied by embedded process control computer systems including robots to automate interaction with humans and to allow systems to operate autonomously.

6.2   Tagging the Physical World

Physical tags refer to digital tags, which are networked electronic devices with an identity, e.g., a RFID (Radio Frequency Identifier) tag. When these are attached to or linked to physical objects, they provide a way to audit physical spaces and processes.

   An important motivation for the use of physical world tags is to support context-based querying and tracking of physical world objects. Tagging is also an enabler for augmented reality (AR) which deals with mapping physical world objects into computer artefact objects in order to more conveniently access them and to manipulate them in the virtual environment. AR environments can be discussed from a HCI perspective (Section 5.4.2) or from an information viewpoint perspective (Feiner *et al.*, 1993). Here the focus is on the latter.

Several techniques can be used to identify objects in the physical world and to capture virtual views of the world. Digital identifier tags can be added to objects perhaps when artificial objects are created or even when natural objects are hatched. A virtual view of the world can be captured in terms of an image, video or audio recording, that is located in space and time, and then the features of interest in these views can be identified and extracted. Specific physical phenomena in the world can be sensed and captured, e.g., a change in temperature over space and or time, and then linked to virtual views and to associated data. Increasingly, mobile capture devices such as phones and cameras incorporate location sensors to add location data to the timing annotation for recordings.

6.2.1 Life-Cycle for Tagging Physical Objects

Virtual tagging involves several separate sub-processes for data storage and creation, data processing and data presentation:

- *capturing* a physical view or recording of physical objects or some object feature such as:
  - moving or placing a *reader*[3] in range of a tag;
  - moving tags in range of readers;
  - capturing a physical view of an object and its surroundings. This may also involve pre-processing to clean the view, to abstract out an object's main features of interest in a view, e.g., to isolate a voice in a noisy recording.
- *Identifying* physical objects:
  - detecting a pre-assigned object IDs, e.g., RFID, and looking up which object has that ID;
  - assigning an ID to a physical view of the object, e.g., image file ID, and then identifying the object within that view perhaps by relative position, e.g., the object is the top left rectangle in this image
- *Anchoring or relating* objects:
  - defining the attributes and relationship of objects with respect to a physical view, e.g., object marked on photograph of part of world;
  - defining the attributes and relationship of objects with respect to a virtual view, e.g., object marked in some abstract view such a spatial route.
- *Organisation or structuring*: objects to inter-relate with different objects within the same view, e.g., a map, and between different views.
- *Presentation*: superimposing graphic annotation on physical world views such as maps or by different degrees of detaching annotations in different forms from the physical views.
- *Management*: managing the annotation processes and data including, creating editing, removing, recycling, storing, querying and access control to the annotation data.

6.2.2 Tags: Types and Characteristics

There are many ways we can characterise and classify tagging (Figure 6.2). Mackay (1998) describe three basic approaches to augment physical world objects for use in virtual (computer) environments: (1) augment the user by generating a digitised view of part of the physical world while the user is onsite in that part, e.g., use AR; (2) augment the physical object by directly attaching to, or

---

[3] A *tag reader* is a device that can receive and interpret the RF signal transmission from an electronic tag. If the tags are passive, the reader may also have to supply the energy for the tag to transmit its information.

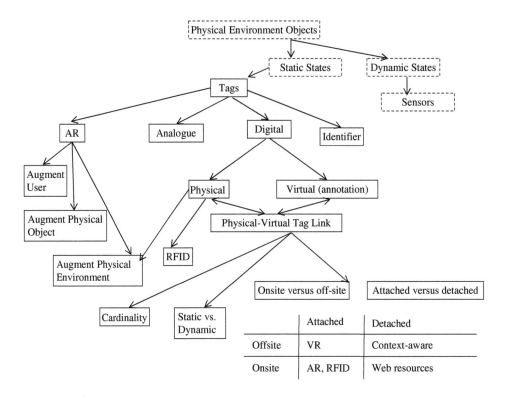

Figure 6.2   Taxonomy for types and characteristics of tags

embedding a physical (digital) tag into, the physical object, e.g., use RFIDs; and (3) augment the environment surrounding the user and object, e.g., digitise and present a view of the physical environment in a virtual environment while the user is collocated in that environment. Mackay did not differentiate whether or not the virtual tag and physical tag should both be together onsite.

Hansen's (2006) analysis focuses on the presentation of the annotation. This also focuses on whether or not the user of the annotation is onsite (co-located or local) with the physical object versus offsite (not co-located or remote). The virtual tag may be (linked to a physical tag which is) attached (or anchored) directly to the object it refers to. In contrast, the annotation may be detached (or not augmented) or not collocated with the object it refers to. Augmented reality is considered as an example of users being onsite using an attached virtual tag. However, the use of augmented reality can be broader: the focus is not on whether or not the user is situated in close proximity to a physical object they are interacting with, but whether or not the activities of users are situated in the physical world rather than in a virtual world. Virtual reality is taken as an example of user situated offsite, where annotation is attached to particular representation of physical objects. Note, with VR, the annotation is not to physical objects but to a virtual view of them. Virtual views may also be imaginary not real views of a physical environment and may have no links to any physical object. A Web-based information system is an example of a type of tagging where virtual tags can be offsite and not anchored to any specific physical tag. A context-aware application is where the user is onsite but being presented with some annotation that is linked to the physical object but may be accessed remotely via some mobile wireless device (Section 6.2.5).

Hansen's onsite versus attached classification of tags is complementary to Mackay's analysis. For example, Hansen's classification cannot differentiate the different ways for augmenting the physical world. In Hansen's analysis augmenting the user (e.g., AR use) and augmenting the object (e.g., RFID use) are both classed as onsite and anchored. Further issues concern the cardinality of the relations between physical objects and their tags and annotations of how physical tags can act as identifiers and tag management issues.

Physical tags can be used to enable physical things to be identified, described and represented in a virtual environment as virtual tags. Virtual tags are metadata or descriptions or annotations about the physical object, which can then be manipulated in the virtual environment. These annotations are linked to the physical tag identifier. The annotation may be stored onsite in the physical tag in the physical world or stored offsite in part of the virtual environment, e.g., in a database. Tags as physical identifiers could be linked to annotations and virtual tags which can also be identifiers, e.g., W3C XML Universal Resource Identifiers (URI), enabling physical tags to be managed as virtual tags. Physical tags could linked to multiple virtual tags.

A physical tag can be linked to multiple virtual tags. However, not all virtual tags may be associated with a physical tag. Physical tags tend to be attached to physical objects with a static state. If the state of physical objects changes, then it is better to also use sensors to determine the dynamic state of the physical object. The tags themselves need to be detected and identified. This is performed with the aid of a tag reader device.

6.2.3 Physical and Virtual Tag Management

Physical object annotation for business still faces many management challenges that inhibit their mass consumer use. Many of these issues are also similar for sensors too. Tags can be bound to objects when objects are created versus attached later. Tags can be permanently bound to objects versus being removable. Tags could be removed from objects at the end of some usage life-cycle versus never removed. Tags can be simply disposed of when the object is disposed of (Section 12.4). Tags may need to be read outdoors in noisy, wet, dark or very bright environments. The annotation data needs to be stored, distributed and integrated starting from the captured data; data management must start as soon as the data is captured (by tag readers).

There may be tens of tags and readers per cubic reader. Readers may interfere with each other. Many redundant annotations of similar items may be captured, many times over. One way to handle redundant tag records is to use the application to filter events, to correlate events to business events, e.g., the package is here means that the package has been delivered to the local depot. Applications and businesses need to define the level of aggregation, reporting, analysis. These may need to be dynamically reconfigured. Interoperability may also be required between different annotation systems. These design challenges are the focus of ongoing research and development. Organising, structuring and management of (annotation) metadata are considered in more detail in Section 12.2.9.

A further dimension includes whether or not the physical object or the human owner of the physical object, if appropriate, is aware of the tag or not, if the owner has or has not sanctioned the use of the tag, i.e., this leads to privacy issues associated with smart environments (Section 12.3.4).

6.2.4 RFID Tags

RFID (Radio Frequency Identifier) tags can be attached to objects to enable identification of objects in the world over a wireless link. Unlike earlier barcode technology, it does not require a line of sight and manual orientation to read the identifying tag (Want, 2006). RFID tags are increasingly being used to tag high value goods. However, in 2009, bar codes are still more common than RFID tags for use with many low cost retail items. Nath *et al.* (2006) consider a range of applications of static readers to automatically tag, interrogate and track rental cars, luggage, mail packages and hospital patients. RFID tags can also be used in mobile readers embedded in phones to support quicker no-touch access to pay for local goods and to access local resources such as hotel rooms. The primary information that can be stored and retrieved in an RFID tag is the identifier data field[4] itself. RFID tags may also contain limited additional information such as its batch manufacturing information including its manufacture date. RFID chips often operate in a promiscuous mode: they reply to a generic scan rather than wait for a reader to provide an activation code or some other form of authentication. Many early major adopters of RFID tags were retail businesses that used them in logistics operations such as Wal-Mart and Tesco. However, they are also used by the military starting with the Second World War in the early 1940s when they were used in the Identification Friend or Foe (IFF) systems by British military aircraft.

RFID tags may be classified into whether or not they have their own energy supply (active RFID tags) or whether or not they are passive RFID tags, using the energy supply of the reader. Active tags are more expensive and require more maintenance but tend to have a longer range compared to passive tags. A typical RFID system consists of two main components: the tag itself and a reader that scans the tag for its ID and contains additional computation, data storage and communication facilities (Figure 6.3).

Figure 6.3 RFID tag application: (a) transponders in cars cause toll barriers to automatically lift as cars approach; (b) tags on pallet of goods tell distributers where goods are located; and (c) tags on clothes in retail outlets can signal alarms if they are removed without permission

---

[4] RFID tags used in the supply chain are encoded with an *Electronic Product Code*, or *EPC*, which is a globally unique identifier for the object being tagged This is typically a 96-bit field which sets aside some bits for a manager ID, some bits for type of objects ID (24 bits, >16 million types of object) leaving 36 bits which can represent over 68 billion different instances of objects.

6.2.4.1 Active RFID Tags

Active RFID tags are used on large assets, such as cargo containers, rail cars and large reusable containers, which need to be tracked over distances such as a distribution yard. These usually operate at 455 MHz, 2.45 GHz, or 5.8 GHz frequencies and they typically have a read range of 20–100 m, costing from 10 to 50 US dollars (2005), depending on the amount of memory, the battery life required, whether the tag includes an on-board temperature sensor or other sensors and the ruggedness required.

There are two types of active tags: transponders and beacons. Active transponders are woken up when they receive a signal from a reader. An important application of active transponders is in toll payment collection and checkpoint control. When a car with an active transponder approaches a tollbooth, a reader at the booth sends out a signal that wakes up the transponder on the car windshield. The transponder then broadcasts its unique ID to the reader. Transponders conserve battery life by having the tag broadcast its signal only when it is within range of a reader.

Beacons are used in a location-based systems (Section 7.4) where the precise location of an asset needs to be tracked within a region such as a distribution yard or along a transport route. Longer-range location-based systems could utilise GPS or mobile phone trilateration (Section 7.4). In a location-based system, a beacon emits a signal with its unique identifier at pre-set intervals (it could be every three seconds or once a day, depending on how important it is to know the location of an asset at a particular moment in time). The beacon's signal is picked up by at least three reader antennas positioned around the perimeter of the area where assets are being tracked. More complex active tags could also incorporate sensors, e.g., a tag on a logistics item could sense if it has been dropped or if its surrounding temperature has become too hot or too low.

6.2.4.2 Passive RFID Tags

Passive RFID tags contain no power source and no active transmitter, their power to transmit their information, typically between 10 μW and 1 mW, comes from the reader. They are cheaper than active tags, currently costing about 20–40 US cents. They are lower maintenance and much shorter (read access) range than active tags, typically from a few cm to 10 m. A passive RFID transponder consists of a microchip attached to an antenna. Transponders can be packaged in many different ways, e.g., sandwiched between an adhesive layer and a paper label to create a printable RFID label; embedded in a plastic card, a key fob and in special packaging that can resist heat, cold or harsh cleaning chemicals.

Passive tags can operate at low frequencies (124 kHz, 125 kHz or 135 kHz), at a high frequency (13.56 MHz) and at ultra-high frequencies (UHF: 860 MHz to 960 MHz). Low-frequency tags are ideal for applications where the tag needs to be read through certain soft materials and water at a close range. As the frequency of radio waves increases, radio signals start to behave more like light. They cannot penetrate materials as well and tend to bounce off many objects. Waves in the UHF band are also absorbed by water. The big challenge facing companies using UHF systems is being able to read RFID tags on cases in the centre of a pallet, or on materials made of metal or under water.

There are two different approaches to transfer power from the reader to passive tags: near field and far field (Want, 2006). The major advantage of far-field tags, as the name implies, is that they can signal information over greater distances compared to near field. Near-field passive RFID interaction is based upon electromagnetic induction. An RFID reader passes a large alternating current through its electromagnetic coil (antenna), resulting in an alternating magnetic field in its locality. If a tag that incorporates a smaller coil is placed in this field, an alternating voltage will appear across the tag. This voltage can then be rectified and coupled to a capacitor which can then accumulate sufficient charge to power the tag chip. Similarly, the tag reader can then use that energy to vary the magnetic field through its antenna to send a signal containing the tag ID to the reader antenna. Far-field passive RFID interaction is based upon capturing EM waves propagating from a dipole antenna attached to the reader. A smaller dipole antenna in the tag receives this energy as an alternating voltage difference and again can use this to charge itself with energy. However, near-field magnetic induction cannot reverse the process to transmit a signal from the tag to the reader as the field reduces inversely with respect to the cube of the distance between them so they use back-scattering instead (Want, 2006).

6.2.5 Personalised and Social Tags

The main examples of physical world annotation considered so far, RFID tags, are targeted at business and organisational users. As long ago as 1945, before the advent of the digital computer age, the idea of personal annotation and its use as part of electronic personal information were postulated by Bush (1945) in his Memex system as 'a device in which an individual stores all his books, records, and communications, and which is mechanized so that it may be consulted with exceeding speed and flexibility. It is an enlarged intimate supplement to his memory.' What he did not explain was how the information was collected and used for annotation.

There are several current initiatives to annotate personal views of the physical world, e.g., Gemmell *et al.* (2006) discuss a project called MyLifeBits to record all the personal experiences of an individual. Other initiatives such as Semacode (2005) propose a scheme to define labels that can be automatically processed from captured images and linked to a Web-based spatial information encyclopaedia. At the most fundamental level, a semacode encodes URLs as part of 100 character string encoded into 2D barcodes. To create semacodes, a URL is entered and software converts this into a semacode image that can be printed and attached to physical objects (Figure 6.4). Some management may be needed to control malicious removal, movement and attachment of tags. Mobile devices can incorporate a semacode scanner, consisting of a camera to photograph the code, plus software to read those codes, parse them and load the associated resource onto a user device.

There are several important differences in use between business use of annotation and personal or social use. Physical artefact annotation is often driven by business goals, i.e., to reduce costs by detecting and recuperating misplaced assets. Annotation for business often uniquely identifies objects, tagging specific types of artefact when they are manufactured, using very simple alphanumeric codes to represent the artefact. Annotation for personal use is less specific, deterministic and

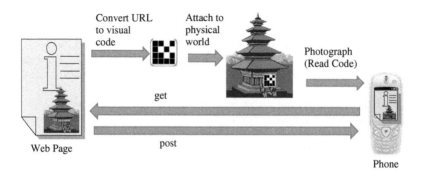

Figure 6.4 The processes of augmented reality tagging

more subjective, multi-modal (using multiple sensory channels) and represented using multimedia. Subjective annotations are used in multiple contexts, multiple applications and multiple activities by users. An important challenge here is the so-called semantic gap between the low-level object features that can be automatically extracted from a record of physical objects in their environment and their high-level meaning with respect to a context of use.

6.2.6 Micro Versus Macro Tags

Most physical world tags such as RFID tags are macro-sized tags to fit macro-sized objects. RFID tags have the potential to be reduced in size based upon MEMS techniques to hundredths of a millimetre which are much cheaper to mass produce but which are invisible to the unaided human eye. The size of a wireless antennae transceiver for the micro tag, which can be designed to be external to the tag, is dependent on the wavelength of the wireless signal transceiver and is typically of the order of about 5 cm for a 2.45 GHz signal. Using MEMs and nano technology markers, we have the potential to mark up the physical world and observe and affect the world in unprecedented ways at very fine, levels of granularity, at the micro and at the nano or molecular level.

6.3 Sensors and Sensor Networks

Sensors are a type of transducer that converts some physical phenomenon such as heat, light, sound into electrical signals. Sensors often act as an enabler, as inputs to a system behaviour so that it can more favourably adapt, often embedded as part of a control loop in pre-programmed systems that perform specialised rather than general purpose functions, e.g., a temperature sensor may be hard-wired into a heating system. Sensors can be used to do the following: instrument and monitor environments; track assets through time and space with respect to some workflow or process; detect changes in the environment defined to be of significance that humans are unable or are put at risk to perceive; control a system with respect to the environment within a defined range of changes; adapt services to improve their utility.

Sensors like RFID tags are networked. The basic architecture is similar: sensors can act as data generators, intermediate or services nodes receive, post-process and store data, possibly remotely. However, whereas tags just generate a fixed electrical signal, sensor data may change because it is the output from a transducer that converts varying physical phenomena into varying signals. The data processing is also likely to be more complex for sensors than tags, see below. Sensors may range in scale from nano sensors to macro sensors such as a windsock used to indicate the wind direction.

6.3.1 Overview of Sensor Net Components and Processes

The main components of a typical sensor network system, given in Figure 6.5, are sensors connected in a network that is serviced by a sensor access node.[5] A slightly different but compatible view of a sensor network is to view sensors as being of three types of node: (1) common nodes

---

[5] The concepts of a sensor node and sensor net can be a bit ambiguous. A sensor can act as a node in a network of sensors (Akyildiz et al., 2002) versus a special sensor network server that receives data from multiple sensors which is often also referred to as a sensor (access) node (Zhao and Guibas, 2005). There are sensor nets connected and served by a single server node and sensor nets consisting of multiple server or access nodes and sensor networks, i.e., networks of networks.

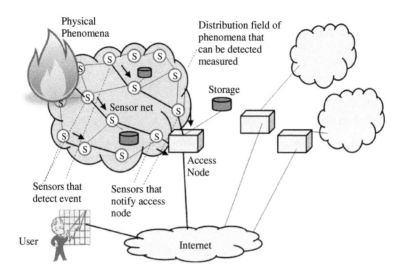

Figure 6.5 A sensor network used to detect increases in heat and report these to a user

mainly responsible for collecting sensor data; (2) sink nodes that are responsible for receiving, storing, processing, aggregating data from common nodes; and (3) gateway nodes that connect sink nodes to external entities. Common nodes are equivalent to sensors, and access nodes combine the functionality of sink and gateway nodes. In , some sensors in the network can act as sink nodes within the network in addition to the access node. Section 6.3.5 considers the advantages of using a more distributed data storage, querying and processing approach.

Sensors are transducers that convert some physical phenomenon such as heat, light, or sound into an electrical signal. Energy is a central concern and communication rather than processing is the primary consumer of scarce energy resources. Sensors often have a low-power, short-range wireless interface that enables them to communicate with other sensors within their range and with data receivers or readers, also called sensor nodes. In Figure 6.5, sensors could collaborate so that only a single sensor source and sensors along a single-path forward the data, in order to conserve energy.

Sensor access nodes multiplex data from multiple sensors, often also supporting a local controller with a microprocessor and signal processor. Sensor nodes may also support local data logging and storage. Sensors range in size from micro to macro sized. Sensor nodes range in size from shoe-box size to, pencil-case size to match-box size. The sensor access node acts as 'Base station' that will route queries to other appropriate sensors nodes in a sensor network. In the illustrative example given in Figure 6.5, three sensors are in range of an event, two sensors are damaged by the event, and two sensors are in range of the access node. As the input event sensors are not in range of the access node, they must route their data through other sensors to get to the access node in order for the events to be accessible over the Internet.

Sensor nets could contain large numbers of sensors and nodes. Sensor nets can be heterogeneous in terms of the types of sensor nodes and types of sensor; this makes interoperability a key concern. Managing all of these constraints and creating a system that functions properly for the application domain while remaining understandable and manageable by human operators and users who may be casual passers-by, is a big challenge. Estrin *et al.* (2002) characterise the design issues of sensors in terms of spatial and temporal scale, variability, degree of autonomy, functionality and complexity. This characterisation has been extended and the property of sensor autonomy has been merged into functional complexity. Sensors can be characterised in terms of:

- *Characteristics of the phenomena being sensed*: The type of physical phenomenon sensed includes location, temperature, etc. The user context defines the threshold, range, history, e.g., when a resource is being accessed more than a set number of times or when a threshold value is breached. The spatial-temporal distribution in the environment may be defined. If a single sample is measured, it may be insufficient because some phenomena varies in space and time, i.e., are moving. Hence, this concerns the sampling interval, the extent of overall system coverage, and the relative number of sensor nodes to input stimuli.
- *Variability*: of the type of environment being sensed; application tasks; the spatial distribution, i.e., objects of interest in the environment which may move.
- *Sensor physical characteristics, including power, mobility, and size.* Passive sensors transfer power from a reader whereas active sensors require their own power source. Sensors may be anchored in a fixed location or may be mobile. Sensors may be anchored in a part of the environment and that moves with it such as an animal or sensors may be untethered, i.e., airborne or waterborne. Sensors may vary in size from nanometers and up.
- *Functional complexity*: Some sensors have no autonomy and have simple functionality. Sensors merely convert some physical phenomena into data that is simply reported to human users. Sensors can have more autonomy and can be pre-configured to automatically detect pre-defined events. Sensors can be reconfigurable, self-configurable and self-optimising. Multiple sensors may collaborate in situ, e.g., often more than one acoustic sensor is used in various acoustic systems, one to detect background noise and one placed to detect a signal, combining data from both can help to improve the signal to noise ratio. Sensors can also be deployed as part of an embedded control system.

The challenges of designing and deploying sensors and solutions are summarised in Table 6.1.

The main functions of sensor networks can be layered in a protocol stack according to the physical network characteristics, data network characteristics, data processing and sensor choreography (Figure 6.6). Each of these is discussed in turn. Other conceptual protocol layered stacks could also be used instead to model sensor operation, e.g., physical, data link, network, transport, layer and application horizontal layers together with some generic power management, mobility management and task management vertical layers that cross-link each of the horizontal layers (Akyildiz *et al.*, 2002). However, the protocol stack in Figure 6.6 seeks to emphasise the need to flatten the horizontal protocol stack to support fewer errors in data transmission and to consolidate management control for sensor data collection and processing. Note this does not restrict each of these components, e.g., data processing, from being distributed.

6.3.2 Sensor Electronics

A block diagram of a circuit for a sensor is given in Figure 6.7. It split into four main functions: sensing, processing, transceiving and power related. The signal from the sensor is filtered and amplified, converted into a digital signal by the analogue to digital converter (ADC), some simple digital signal processing (DSP) is performed at the sensor before the signal is modulated for transmission. This particular sensor design also supports input configuration for the DSP. The MEMS design of the sensor is able to decrease the size and power consumption of the sensor by aggregating multiple separate electronic components into a single chip. Sensors often need to be able to operate unattended, long-lived, low-duty cycle systems.

Table 6.1 Challenges in designing and deploying sensors and some corresponding solutions

Challenges of a sensor net system	Design solutions
Sensor energy is a scarce resource for data transmission	Use a sensor net that deploys, low-power, short range transmissions
	Network sensors into mesh networks and use multi-hop transmissions
	Filter data in-situ and transmit only filtered data
Limited memory and computation power in sensors	Harvest renewable energy from the environment and store
Dynamic and non-deterministic spatial-temporal distribution of events. May not be able to pre-determine how to optimally deploy individual sensors	Use a sensor net to increase the sensor density around estimated signal source positions when deterministic;
	Design sensor distributions to be reconfigurable, self-organising, to be mobile
	Support variable sampling and support bursty data collection
Sensor failure is common due to a lack of power, physical damage, active (jamming) or passive environmental interference of the transceiver, access node or non-optimal positioning	Use dense networks of low power sensors with redundant paths to route data through the network
	Use of counter measures and frequency shifts
	Locators and trackers are needed to locate (moving) sensors and can be used to position them
Multi-hop sensor networks may have a dynamic topology. No global knowledge about structure of network may be known	Use specialised routing protocols to work over dynamic mesh topologies
Sensors can be too costly to update once deployed	Design sensors and sensor access nodes to be low maintenance. Support sensor redundancy
Sensors can generate huge quantities of data	Use in-situ data processing both in the sensor and the sensor access node

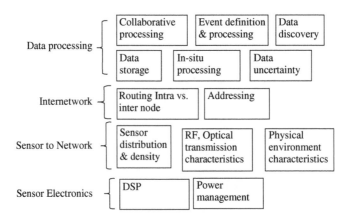

Figure 6.6 The main functional characteristics for sensor net deployment

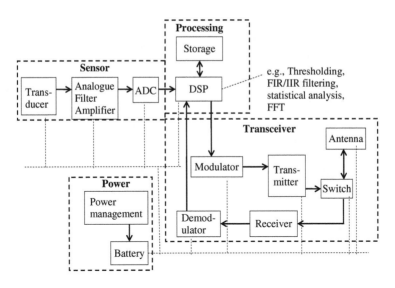

Figure 6.7   Block diagram for a sensor electronics circuit

6.3.3 Physical Network: Environment, Density and Transmission

Sensors may be deployed in three phases: (1) a pre-deployment phase where group dispersal from multiple release and scatter points or individual placement occurs; (2) a deployment phase in which sensors may move after deployment because of signal phenomena changes, energy optimisation or task changes; and (3) a redeployment phase where additional sensors are used.

There are several reasons motivating the use of multiple low cost, short-range, low-power sensors rather than using a few long-range, high-power and high-cost sensors. First, dense networks of sensors can improve the signal-to-noise ratio (SNR) by reducing the average distance of sensor to signal sources of interest. Each sensor has a finite sensing range determined by the baseline (floor) noise level. A dense sensor field improves the odds of detecting a signal source within a range. Once a signal source is inside the sensing range, further increasing the sensor density decreases the average distance from sensor to signal source, thus improving the SNR. As an example, consider acoustic sensing in a two-dimensional plane (Zhao and Guibas, 2004, pp. 6–9). The acoustic power received, Pr, at distance d from a power source, Ps, varies inversely to the square of the distance between them (Equation 6.1):

$$Pr \propto Ps/d^2 \tag{6.1}$$

The SNR of the power received signal to the noise level signal power level, Pn is expressed in a logarithm decibel scale:

$$SNR = 10 \log (Pr/Pn) = 10 \log Ps - 10 \log Pn - 20 \log d \tag{6.2}$$

Increasing the senor density by a factor of k reduces the average distance to target inversely by the square root of r. Thus SNR advantage of denser network is given in Equation 6.3.

$$SNRdif = SNR(d/\sqrt{k}) - SNR(d) = 20 \log(d/(d/\sqrt{k}) = 10 \log k \tag{6.3}$$

Thus, an increase in the sensor density by a factor of k improves the SNR at a sensor by 10 log dB for acoustic type signals. Sensors need to be distributed in such a way that they can maximise the coverage areas and be local enough to detect strong enough signals from the physical phenomenon of interest. They must be distributed so that there is some overlap between adjacent sensors network coverage so that they find a data transmission path to a sensor node. The distribution and coverage of the sensors should be arranged to match the distribution and coverage of the physical phenomena of interest in order to optimise detection.

However, using a uniform sensor distribution may not always be optimal, for example, because signal attenuation and obstacles are non-uniform. Increasing the sensor density may have no effect when the density of obstacles is similar to the density of sensors. It may not always be possible to pre-determine an optimal sensor deployment distribution. Hence, there is a degree of sensing uncertainty (Kaiser *et al.*, 2005). Random deployment could also be used, e.g., in inaccessible terrains or disaster relief operations. In addition, sensors that are mobile may be able to self-organise themselves into an optimum configuration.

Second, the energy efficiency for communication can be increased through the use of a multi-hop topology for the sensor network (Zhao and Guibas, 2004). In an N hop network, overall distance for transmission is Nd where d is the (average) one-hop distance. The minimum receiving power is Pr and the power at the transmission node is Ps; a is the RF attenuation coefficient which is typically in the range 2–5 because of multipath and other interference effects:

$$Pr \propto Ps/d^a \tag{6.4}$$

$$Ps \propto d^a\,Pr \tag{6.5}$$

Therefore the power advantage of an N-hop transmission versus a single hop transmission over the same distance Nr, Pdif, is given in Equation 6.6:

$$Pdif = Ps(Nd)/(N \cdot Ps(Nd)) = (Nd)^a\,Pr/(N \cdot d^a\,Pr) = N^{(a-1)} \tag{6.6}$$

However, the power increase must be balanced against the disadvantages of using more relay nodes: the increased power use by all the components, the increased cost in using more sensors and the increased latency in forwarding messages over multi-hops.

In addition, power management can also be supported by optimising routing or processing management to consider powering down transceivers in redundant sensors and sensor routes. Relevant information can be aggregated during multi-hop data exchange and if sensors support data exchange via multi-paths, the system has some resilience against individual sensor node failures. Akyildiz *et al.* (2002) consider in more detail network access protocols including signal modulation and demodulation.

6.3.4 Data Network: Addressing and Routing

Nodes in any kind of distributed system may need to be uniquely addressable. In most distributed systems, the address of nodes makes use of the topological location of a node in the network. The (virtual) network topology may also be different from the physical topology. In fixed IP type networks, the addressing scheme involves IP assignment and hierarchical host name lookup, multicast packet data interleaving and routing with service registration and lookup. Node discovery uses an attribute based addressing scheme that is independent of the network topology. Typically, network nodes are addressed in terms of the resource characteristics at that node defined by a set of attributes, e.g., those at the 'southmost' edge of the region covered. Service-oriented networks typically use multiple levels of indirection to address nodes.

A key design issue here is to how make the node attribute to physical node address resolution efficient to support the low power requirements of massively distributed real-time, sensor networks. An efficient attribute-based node addressing scheme has been proposed by Heidemann *et al.* (2001), among others, called directed diffusion. This supports in-network processing to leverage CPU-communications trade-offs for sensor networks, reducing the number of indirections and operates directly over low-level (hop-by-bop) communication protocols.

The basic idea of directed diffusion, i.e., data centric routing, is to name data (not nodes) with externally relevant attributes such as data type, time, location of node and SNR. This can support in-network aggregation and processing of data sources, publishing data by sensor sources and subscription to data by data client nodes. A node may play multiple roles, e.g., aggregating, combining and processing incoming sensor node data and becoming in itself a source of new data. A node may act as a client for triggering event data and then as a server that only publishes data when a combination of conditions arises.

Other routing protocols to directed data fusion include the following ones. *SMEC* which creates a subgraph of the sensor network that contains the minimum energy path. *Flooding*[6] sends data to all neighbour nodes regardless if they receive it before or not. *Gossiping* sends data to one randomly selected neighbour. *SPIN* sends data to sensor nodes only if they are interested. *SAR* creates multiple trees where the root of each tree is one hop neighbour from the access node. *LEACH* forms clusters to minimise energy dissipation. Routing algorithms for sensor networks can also be classified according to type of network structure such as flat or hierarchical, or classified according to interaction protocol such as multipath, query-based or negotiation-based (Al-Karaki and Kamal, 2004).

6.3.4.1 Sensor Networks Versus Ad Hoc Networks

There are several differences between sensor networks and ad hoc networks (Akyildiz *et al.*, 2002). In contrast to ad hoc networks, sensor networks are denser and contain several orders of magnitude more (sensor) nodes that may not be uniquely addressable at an application level because of the large number of instances of nodes of the same type. The topology of sensor nets can be very dynamic. The topology may need to adapt to locally detected sensors, to unknown signal characteristics, to the (self-) reconfiguration of in situ nomadic sensors and to dynamic routing, in order to optimise a low SNR and power usage (Section 6.3.2). Sensor nodes often use a broadcast communication paradigm, whereas ad hoc networks often use point-to-point communications. Typically sensors use a mesh topology network (Section 11.7.8.5), that is dynamic, mobile, and unreliable, that assumes no universal routing protocols and no central registry of sensor locations and routes. Each sensor acts a router and as a data source and depending on design also as a data sink and gateway.

6.3.5 Data Processing: Distributed Data Storage and Data Queries

Sensors as data sources are naturally distributed and hence the computation becomes distributed as a consequence of performing it locally in order to reduce it. Processing involves collection of events from massively distributed sensors. It would be simplest to allow each sensor to act in isolation, to keep all data events but this requires a large amount of storage. A traditional centralised data storage approach could be used with many sensors generating data transactions that cause updates in a data warehouse. Data events extracted from sensors could be stored in a RDBMS server. Query processing takes place at data server nodes only. In addition, multiple data from RDBMS could be aggregated into a data warehouse system and SQL could be used to query sensor data identified using an attribute-based addressing and routing[7] scheme.

---

[6] *Flooding* means broadcasting. However, in the case of the low resource (bandwidth and processing) nodes, broadcasting can easily overwhelm the capabilities of nodes. Note also that broadcasting is used in other dynamic networks such as P2P networks and ad hoc networks in order to locate resources, peers, services, users, etc., because fixed middleware services may not exist.

[7] Conventional addressing and routing, *address-centric routing*, involve passing of data from a source to its destination without interpreting the content of the data. In contrast, *attribute-based addressing* and routing evaluates the contents of the transmitted data at each hop in the network in order to determine how to route the data.

However, many events may contain little information of value because readings are constant. Further, multiple completing events may be generated to characterise a signal. Another alternative to storing each event is to filter data events and only transmit the filtered events. This requires data filtering to occur at each sensor. In addition, data processing such as data aggregation can be more expediently performed in the network, at or near the data source sensor nodes (Bonnet *et al.*, 2000). This is exploited to reduce communication in the data sink-type nodes. It depends on the design whether or not each common sensor node behaves as a sink node. Sensor database systems need to support distributed query processing over sensor network and to consider how the sensor database design can represent sensor data, represent sensor queries, process query fragments on sensor nodes and distribute query fragments. In theory, distributed database techniques could be used here.

To support energy-efficient and scalable data transmission, sensor nodes can be autonomously clustered using attribute-based sensor addressing. Data aggregation processes could also be recursively applied to form a hierarchy of clusters, e.g., the SINA (Sensor Information Networking Architecture) system of Shen *et al.* (2001). In SINA, the Sensor Query and Tasking Language, SQTL which can be aligned to the Structured Query Language (SQL) acts as an API between sensor applications and the SINA middleware supporting three types of events: (1) events generated when a message is received by a sensor node; (2) events triggered periodically by a timer; and (3) events caused by the expiration of a timer. A simple user query is as follows, 'SELECT avg(getTemperature()) AS avgTemperature'. Another data storage system for sensor nets is Cougar (Bonnet *et al.*, 2000).

Application tasks may require node collaboration because an isolated event may not be significant or may be too easily recognised as a false positive or false negative. Node collaboration may also be used to filter out duplicate events and to switch off active sensors that produce or forward redundant or non-significant information. Communication of nonsignificant data at the application level also wastes the scarce power and bandwidth resources available. Thus sensor data access models could use utility functions that weigh up the gain of one or more sensor nodes creating and transferring data versus constraints against the cost of power and bandwidth use in doing so. This is the idea behind information-based sensor tasking (Zhao and Guibas, 2004).

6.4 Micro Actuation and Sensing: MEMS

MEMS (Micro-electro mechanical systems) are micron- to millimetre-scale electronic devices fabricated as discrete devices or in large arrays (Berlin and Gabriel, 1997). MEMS perform two basic types of functions, acting as sensors or actuators. Both actuators and sensors act as transducers converting one signal into another. Of specific interest are transducers that covert some environmental phenomena such as temperature, humidity, pressure, etc., into a digital electrical signal and vice versa. MEMS actuators convert an electrical signal into physical phenomena to move or control mechanisms such as motors, pneumatic actuators, hydraulic pistons and relays. MEMS sensors work in reverse to actuators, they convert some environmental phenomena such as temperature, humidity and pressure into an electrical signal. These relatively small components have high resonant frequencies leading to higher operating frequencies.

Collections of millions of cooperating sensing, actuation and locomotion mechanisms can be viewed as a form of programmable matter because these can self-assemble in arbitrary three-dimensional shapes (Goldstein *et al.*, 2005) (Figure 6.8), forming the basis of much more fluid and flexible computers and human–computer interfaces, for example, flexible devices and tangible computer user interfaces (Section 5.3.4). A long-term goal of the use of such collections is to be able to achieve a synthetic reality which, unlike virtual reality and augmented reality, allows the physical realisation of all computer-generated objects. This has the benefit that users will be able to experience synthetic reality without any sensory augmentation, such as head-mounted displays and be able to physically interact with any object in the system in a natural way (Goldstein *et al.*, 2005).

Figure 6.8 Some examples of MEMS devices, size of the order of 10 to 100 microns (left to right): mite approaching the gear chain, polysilicon mirror, triple-piston microsteam engine. Reproduced by permission of © Sandia National Laboratories, SUMMIT(TM) Technologies, www.mems.Sandia.gov

6.4.1 Fabrication

MEMS design differs from that of the equivalent macro devices which comprise mechanical and discrete electronic component design because these micro components are based upon silicon-based Integrated Circuit (IC), also called chip, design. Analogue devices may also be replaced by IC versions, e.g., whereas a traditional thermometer is based upon a liquid, such as mercury, expanding along a tube referenced to a calibrated scale, an electronic thermometer can be built out of a thermocouple and IC amplifier.

ICs consist of several layers of p-type and n-type doped silicon which have been added to a substrate. An optical microfabrication approach, photolithography, is then used to fabricate the circuit. This first covers a layer with a photoresistant chemical. Then the circuit pattern to be fabricated is drawn onto a photomask. The photolithography systems shines the UV light through the photomask, projecting a shadow onto a layer that then reacts with the photoresistant chemical and hardens, allowing the selective removal of parts of the substrate to be chemical etched away. Engineers thus design a new circuit by designing the pattern of interconnections among millions of relatively simple and identical components. It is the diversity and complexity of the interconnections between these that produce the accompanying diversity of electronic components including memory chips and CPUs. The miniaturisation of IC-based MEMS processing has two important advantages over macro electromechanical devices and systems: batch fabrication and power reduction. Multiplicity makes it possible to fabricate ten thousand or a million MEMS components as easily and quickly as one, such economies of scale are critical for reducing unit costs.[8] Second, IC performance is enhanced when components are closer together because they can be switched quicker and use lower power.

The second part of MEMS to supplement the microelectronics part is the micromachines part. Interestingly, these micromachines can be fabricated just like ICs. MEMS-type ICs can be fabricated in different ways using: bulk micro-machining (etching into the substrate); surface micro-machining (building up layers above the substrate and etching); and by machining LIGA[9] deep structures.

6.4.2 Micro-Actuators

The mechanisms involved in micro-actuation while conceptually similar to the equivalent macro mechanisms may function fundamentally differently. They are engineered in a fundamentally

---

[8] Gershenfeld (1999) has even proposed the idea of MEMS fabrication not in specialised plants but by self-contained desktop printers distributed among product developers. Instead of printers depositing ink from ink cartridges, these printers deposit materials from cartridges to form MEMS devices. He calls this Printed EMS or PEMS.
[9] LIGA is the German acronym for X-ray lithography, electrode position, and moulding.

different way using integrated circuit design and nanotechnology. MEMS actuator applications include:

- *Micro-mirror* array-based projectors (micro-projectors) can be used to generate large screen display content from smaller devices. This has applications in navigations systems (Heads-Up Displays or HUDs), as a positioning aid in medical diagnosis and treatment and in manufacturing to produce reference points for drilling. A micro-sensor system may also be needed to complement the use of micro-actuators, e.g., to detect and compensate for noise motion such as camera shake, ensuring a steady picture, even when moving.
- *Inkjet printers heads*: MEMS can be used to control ink deposits onto paper. Fuller *et al.* (2002) also propose that MEMS devices can be printed onto and distributed with paper using an ink-jet printer.
- *Optical switches*: optical cross-connect switches (OXC) are devices used by telecommunications carriers to switch high-speed optical signals in a fibre optic network. OXC commonly have electronic cores and as data rates increase may become a bottleneck in the communication, stimulating the development of an all-optical MEMS switch (Yeow *et al.*, 2001).
- *Micro-fluid pumps*: The essential components include a fluid actuator, a fluidic control device, and micro plumbing, e.g., for use in delivering medicine. One of the major challenges here is to choose materials that can be used to fabricate integrated circuits that are biocompatible, e.g., Parylene (Meng and Tai, 2003).
- *Miniature RF transceivers*: can replace passive low-Q, where the Q-factor indicates the rate of energy dissipation relative to the oscillation frequency, components in communication devices such as vibrating resonators, switches, capacitors, and inductors and put them on a single high-Q MEMS RF transceiver chip. This enables a greater miniaturisation of communicators, see Mansour *et al.* (2003) and Rebeiz (2003).
- *Miniature storage devices*: can support gigabytes of non-volatile data storage in a single IC chip, low power, and low data latency, e.g., worst-case rotational latency 5–11ms, sub-millisecond average access time. Yu *et al.* (2007) consider how to optimise RDB storage on MEMS storage devices.

6.4.3 Micro-Sensors

Kahn *et al.* (2000) regard size reduction as paramount to make sensor nodes as inexpensive and easy-to-deploy as possible, e.g., to incorporate the requisite sensing, communication, and computing hardware, along with a power supply, in a volume no more than a cubic millimetre, while still achieving a suitable performance in terms of sensor functionality and communications capability. These millimetre-scale nodes are called 'Smart Dust'. Smart Dust can be small enough to remain suspended in air, circulated by air currents, sensing and communicating for hours or even days. Smart dust motes contain micro sensors, an optical receiver, passive and active optical transmitters, signal-processing and control circuitry, and a thick film battery power source.

A critical part of the design is very efficient power management in terms of the power storage and power consumption for both sensing, processing and data transmission. Stored energy is about 1J, that is targeted to be consumed at 10 μW throughout the day. Power-optimised CPUs typically consume 1 nJ per 32-bit instruction and some RF data transmission is relatively power-hungry, e.g., Bluetooth radio-frequency (RF) communication chips will use about 100 nJ per bit transmitted, hence lower power data transmission is needed. RF also presents another challenge in that there is very limited space for antennas, thereby requiring extremely short-wavelength, i.e., high frequency data transmission.

Some common MEMS sensor applications are as follows: *accelerometers* can be used to control the safety airbag release in almost all cars today. Accelerometers detect the rapid negative acceleration of the vehicle to determine when a collision is occurring and the severity of the collision. *Angular rate sensors* and *gyroscopes* can be used to measure the rotational velocity or angular rate of an object. Compared to classic gyroscopes based on optical or (macro)-mechanical principles, they do not need a fixed point for referencing, are very cheap and can withstand harsher environment. The principle of operation is based on the *Coriolis effect*. When a micro-electro mechanical system (MEMS) resonator is driven at about 10kHz, due to the angular rate, the Coriolis force excites a second oscillation perpendicular to the first one. This oscillation is proportional to the angular rate and measured using capacitive methods. Applications include image stabilisation and orientation in devices such as cameras and mobile phones and in navigation systems used in game consoles.

6.4.4 Smart Surfaces, Skin, Paint, Matter and Dust

MEMS can be permanently attached to some fixed substrate forming smart surfaces or be more free-standing, forming smart structures that can reorganise. An example of a smart surface is a paint that is able to sense vibrations because it is loaded with a fine powder of a piezoelectric material called lead zirconate titanate (PZT). When PZT crystals are stretched or squeezed, they produce an electrical signal that is proportional to the force (Berlin and Gabriel, 1997). MEMS could be mixed with a range of bulk materials, such as paints, gels, and spread on surfaces or embedded into surfaces or scattered into and carried as part of other media such as air and water. For example, coating bridges and buildings with smart paint could sense and report traffic, wind loads and monitor structural integrity. A smart-paint coating on a wall could sense vibrations, monitor the premises for intruders, and cancel noise (Abelson *et al.*, 2000). Smart surfaces can also be woven out of organic polymers that have light-emitting and conductive properties. Organic computing can be used to form smart skin and smart clothes (Section 5.3.4.3). Similar to sensor nets, MEMS can also be networked in MEMS nets. The Smart Dust project led by Kris Pister produced prototypes of many novel types of low-powered networked MEMS sensors (Section 2.2.3.2).

In the Claytronics Project,[10] Goldstein *et al.* (2005) have proposed using masses of thousands to millions of sensor, actuator and locomotion MEMS devices that can behave as malleable programmable matter and can recreate artefacts for a wide range of physical shapes and objects. A long-term goal of such MEMS ensembles is to enable these to be self-assembled in any arbitrary 3D shape, to achieve a synthetic reality. *Synthetic reality*, unlike virtual reality or augmented reality, allows users to experience synthetic reality without any sensory augmentation, such as head-mounted displays and so to be able to physically interact with any object in the system in a natural way. The programmable matter idea introduced in the Claytronics Project uses re-assembly and the use of moving electronics around for communication to produce new forms for matter. Perhaps the ultimate programmable matter is to base it upon nanotechnology, to be able to engineer matter on the molecular level, moving molecules around not just electrons.

Others refer to such ensembles of computational particles, dispersed irregularly on a surface or throughout a volume where individual particles have no *a priori* knowledge of their positions or orientations, as *amorphous computing* and *spray computing* (Zambonelli *et al.*, 2005). These particles are possibly faulty, may contain sensors and effect actions, and in some applications might be mobile and referred to as amorphous computing particles (Abelson *et al.*, 2000).

---

[10] The Claytronics Project is a joint project of researchers at Carnegie Mellon University and Intel Research. Website: http://www.cs.cmu.edu/~claytronics/, accessed Feb. 2008.

New design and fabrication models are needed to engineer such systems (Section 6.4.1). Novel techniques are needed to manage groups of MEMS devices perhaps by incorporating behaviours based upon self-organising interaction mechanisms (Section 10.5.1). However, there are still safety issues that concern these management techniques which limit their effectiveness to manage scattered collections of MEMs in practice.

6.4.5 Downsizing to Nanotechnology and Quantum Devices

Gordon Moore (1965), Intel's co-founder, made a prediction, now popularly known as Moore's Law, which states that the number of transistors on an IC chip doubles about every two years.[11] This of course does not mean that the software processing capability will also increase in this same way. Graham (1989) identifies two reasons why software performance may not increase in this proportion: it is not just a transistor density increase, the computation architecture may also need to change to take advantage of this and this can take 5–10 years. In addition, the communications capability does not necessarily increase in proportion to the computation increase. With respect to the hardware, typically, this type of increase in IC chip transistor density is possible because of breakthroughs in photolithography that occur every six to seven years, and that each supports three size reductions in ICs. The shorter the wavelength of the light used, the thinner you can make the mask and hence the smaller you can make the parts of transistor and wiring. Recently, however, problems with using the latest breakthrough technique, called extreme ultra-violet photolithography, to generate shorter wavelength light of 13.5 nm, in between the end of the Sun UV spectrum and the start of the X-rays, may lead to delays in increasing the transistor density (Santo, 2007). In addition, decreasing transistor size and increasing transistor density are far more complex than only reducing the optical wavelength to draw the circuit mask. Bohr *et al.* (2007) have stated that the thin layer of insulation that electrically insulates the transistor gate is down to a width of a few atoms and is facing problems as the insulation is breaking down. Hence, new materials need to be modelled and designed at the molecular level and this is a key aspect of the field of nanotechnology.

Nanotechnology[12] can be defined as the manipulation, precision placement, measurement, modelling, and manufacture to create systems with less than 100 nm (Poole *et al.*, 2003). Nanotechnology is not just smaller MEMS. It is also based upon a broader range of materials and mechanisms and sizes down to the molecular level, e.g., carbon-based nanotubes have much better conductivity than silicon-based semiconductors at the nanometer level (Banerjee and Srivastava, 2006). In contrast, MEMS focuses on semiconductor-based single IC chip technology and micro-machining. Whereas MEMS tends to use a top-down approach to device design, nanotechnology seeks also to use a bottom-up design.

The drive to switch transistors faster and to be low-powered has been to make them smaller. However, when electronic components approach the nanometer size, odd things begin to happen as electrons begin to reveal their quantum nature, e.g., electrons have the potential of crossing a transistor even if it is switched off. This raised an early concern about the feasibility of nanotechnology arising from quantum uncertainty about whether or not it would make these systems impossibly unreliable. A second severe limitation is thermal noise which causes local molecules' movement because of heat. This phenomenon limits what can be done mechanically at the

---

[11] There is some controversy over whether Moore's First Law applies every two years or every eighteen months. There is also Moore's Second Law which says that the cost of doubling circuit density increases in line with Moore's First Law.

[12] Nanotechnology overlaps with the terms *nanocomputers*, *molecular nanotechnology* and *molecular manufacturing*.

molecular scale (Peterson, 2000). In addition, advances must also be made in positioning and in the control of structures at this level (Devasia, *et al.*, 2007).

Nanotechnology at first proposed to use a bottom-up approach to design, to be able to assemble custom-made molecular structures for specific applications, for example, to create stiff molecular materials to reduce the effects of thermal-induced molecular movement. A major challenge for this design process is the complexity and novelty in understanding and being able to model materials at this level. More research is needed to understand how combinations of materials, in particular compounds, give materials at the molecular level certain physical and functional properties. Chang (2003) gives an overview of some devices of nanotechnology that have been built. Nanotechnology requires two main types of engineering support: molecular positional assembly, manipulating and positioning individual atoms, and massive parallelism because otherwise it would take one robotic arm forever to build a kilogram-sized object one molecular at a time, a huge numbers of robotic arms working together in parallel is needed.

6.5 Embedded Systems and Real-Time Systems

Embedded systems are used mainly online for task enactment in the physical world, in contrast to general purpose computers which are often used offline for information access and sharing. Thus embedded computer systems differ from general purpose (MTOS) systems in three main ways. Embedded systems focus more on single task enactment. Safety-criticality may be important because actions affect the physical world. Third, tasks often need to be scheduled with respect to real-time constraints. An embedded system is a component in a larger system that performs a single dedicated task. This can use a far simpler and cheaper operating system and hardware because there is only one process. This simplifies memory management and process control and omits inter-process communication, which typically need to be supported in an MTOS. An embedded system may or may not be visible as a computer to a user of that system. It may or may not have a visible control interface for human users. Embedded device(s) may be local and fixed, e.g., a printer or AV record or playback unit, or may be mobile and distributed in aircraft, ship and in Internet appliances.

An embedded system is programmable and contains one or more dedicated task computers, microprocessors, or microcontrollers. A microprocessor is an integrated circuit which forms the central processing unit for a computer or embedded controller, but requires additional support circuitry to function. A microcontroller is a microprocessor plus additional types of processor that supports other devices and is integrated into a single package. Other types of device support devices may include serial (COM), ports, parallel ports, USB ports, Ethernet ports, A/D and D/A, interval timers, watchdog timers, event counter/timers, real-time clocks (RTC), Digital Signal Processing (DSP) and numeric coprocessors. Even a general purpose computer may itself contain additional dedicated computers to control battery charging and discharging and for AV recording and playback. Embedded system may contain programmable logic elements such as FPGA (Field Programmable Gate Arrays), or application-specific integrated circuit (ASIC) which are in contrast to non-programmable processing chips such as CPUs.

Embedded single process systems are not generally linked to other externals systems without exposing and re-designing their control interfaces – it is hard to interlink these to become part of a bigger system. Embedded computer systems differ from a MTOS computer in several ways. They use a rich variety of microprocessors, hundreds of types that are dedicated to a specific task or tasks, e.g., peripherals, networking, etc. Almost every embedded design (hardware and software) is unique. Each embedded computing devices may be designed for its own rigidly defined operational bounds, e.g., heating system. Designs are often engineered for the highest possible performance at the lowest cost. Performance may not be an important consideration. They often operate under moderate to severe real-time constraints. These characteristics suggest that interoperability and internetworking of embedded system devices may be more challenging than that for MTOS computer devices.

Software failure can have life-threatening consequences if systems fail. Hence some types of embedded system are designed to be fault-tolerant. Tolerance for bugs may be a factor a thousand times better in embedded systems than in desktop computers. Embedded systems may often have constraints concerning power consumption. They are often designed to operate in a wider range of physical environmental conditions than a personal computer, e.g., in damp, hot and cold, and dark conditions. They often use fewer system resources than desktop or laptop systems. There may be no hard disk or removal media. All codes might need to be stored in ROM. Embedded systems may or may not have some common Operating System (OS) services available, e.g., there is no standard C language function printf() for debugging when there is no terminal. Embedded systems hence may require specialist development tools, e.g., may use on-chip debugging resources instead of printf(). Embedded computer systems are not always easy to program because they may not separate the hardware from the control software which is fundamental to making the network more programmable. The hardware and software are highly vertically integrated and interdependent, such as ASICs, microcontrollers.

There are many examples of the use of embedded system applications. Transport vehicles are not only an important application area for the deployment of new sensor technology and for robots used in manufacturing and for unmanned (robot) self-steering vehicles, they are also a good application area for embedded systems. Modern cars network multiple embedded systems for antilock brake systems (ABS), cruise control, climate control, wing mirrors, locomotion and drive sensor data monitoring, etc. (Leen and Heffernan, 2000). For example, in the ABS sub-system, sensors measure the speed at which the wheels are turning. If the wheel speed decreases rapidly, the ABS detects a blocking danger and immediately eases the hydraulic pressure to the brakes and then raises it until it is just under the blocking threshold. This easing off and raising of the pressure can be repeated several times per second.

The different embedded sub-systems need to be interconnected into a holistic control and monitoring system. In the mid-1980s, Bosch developed the Controller Area Network (CAN), one of the first and still one of the most widely used vehicular networks with more than 100 million CAN nodes sold yearly. A typical vehicle can contain two or three separate CANs operating at different transmission rates. A low-speed CAN running at less than 125 Kbps usually manages a car's comfort electronics for seat and window movement controls. These control applications are not real-time critical and use an energy-saving sleep mode. A higher-speed CAN runs more real-time-critical functions such as engine management, antilock brakes, and cruise control.

Some embedded systems are networked but are hard-linked to specific networked services only. For example, cars and vacuum cleaner can be designed to alert the manufacturer's service centre when it malfunctions or needs a service. Cash-point machines, Electronic Point of Sale or EPOS terminals, ticket booking machines, communication, information and entertainment devices and GPS are other examples of networked embedded devices. System support for (robotic) organisations and coordination of multiple autonomous process controllers is discussed further in Chapter 9.

6.5.1 Application-Specific Operating Systems (ASOS)

Embedded systems require an application-specific operating system (ASOS) that is customisable and reconfigurable to meet the requirements of specific applications in order to provide lower cost and higher performance by eliminating general-purpose MTOS features (Section 3.4.3) that are unnecessary for specific applications and through better tailoring those features that are included (Friedrich *et al.* 2001). Software or hardware processing techniques such as the use of data compression can extend the data storage capability for a given memory capacity, reducing the cost of the memory and the energy needed to access the data stored in memory.

A key design issue is whether or not the configuration of the MTOS is fixed at design time or can be changed during the operational lifetime to support new requirements. For example, an embedded environmental control system for a smart home uses a microcontroller executing an ASOS which does not require many features of the file system, inter-process communication or control, networking and security facilities provided by a MTOS. However, the addition of a burglar alarm system to the smart home may require networking support to interlink different security monitoring and alarm components and the addition of a file system for system logging. Some devices running ASOS may need to be extensible. Friedrich *et al.* (2001) review a number of component-based OSs to support this.

6.5.2 Real-Time Operating Systems for Embedded Systems

Real-time embedded systems applications are a subset of embedded system applications which perform safety-critical tasks with respect to time-constraints because if these are violated, the system may become unsafe, such as the operation of cars, ships, airplanes, medical instrumentation monitoring and control, multimedia streaming, factory automation, financial transaction processing and video games machines. Real-Time Operating Systems or RTOS can be considered to be a resource-constrained system where the primary resources, such as data transfer, are constrained in time. This in turn constrains the number of CPU cycles but other resources may also be constrained such as memory. A RTOS reacts to external events that interrupt it.

RTOS design focuses on scheduling efforts so that processes can meet real-time constraints and optimise memory allocation, process context switching time and interrupt latency (Section 3.4, Li *et al.*, 1997). As multiple interrupt events may occur, an RTOS must have a mechanism for priority scheduling of interrupts. An RTOS may also use additional process control to lock specific processes in memory to prevent the process swapping overhead. There are a range of real-time design concerns to support critical response time of a task, the time to detect an event and trigger a corresponding action to handle it, critical data transfer rates, optimising both response times and data transfer rates and optimising these when there are simultaneous tasks. The two key factors that affect the response time are process context-switching (to switch between different processes) and interrupt latency (the time lag before the context switch is possible).

Timeliness is the single most important aspect of a real-time system. A real-time system (RTS) is one where the timing of a result is just as important as the result itself. A correct answer produced too late is just as bad as an incorrect answer or no answer at all. An RTS is one in which the correctness of the computations not only depends upon the logical correctness of the computation but also upon the time in which the result is produced. If the timing constraints are not met, system failure is said to have occurred.

Timing constraints can vary between different real-time systems. Therefore, an RTS can fall into one of three categories: soft, hard or firm. In a *soft RTS*, the timing requirements are defined by using an average response time. A single computation arriving late is not significant to the operation of the system, though many late arrivals might be. An example of this is airline reservation systems. If a single computation is late, the system's response time may lag. However, the only consequence would be a frustrated potential passenger.

In a *hard RTS*, the timing requirements are vital. A response that's late is incorrect and system failure will result. Activities must be completed by a specified deadline, always because if a deadline is missed, the task fails. Deadlines can be a specific time, a time interval, or the arrival of an event. This demands that the system has the ability to predict how long computations will take in advance. An example of a hard RTS is a pacemaker. If the system takes longer than expected to initiate treatment, biological damage could result.

In some types of (firm) RTS, timing requirements are a combination of both hard and soft ones. Typically the computation will have a shorter soft requirement and a longer hard requirement. An example of a firm RTS is a patient ventilator that assists patients with breathing problems to breathe for them. The system must ventilate a patient so many times within a given time period. A few seconds delay in the initiation of the patient's breath is allowed, but no more than this.

6.6 Control Systems (for Physical World Tasks)

The simplest type of control is activated only when defined thresholds are crossed,[13] e.g., a thermostat switches the heating on when the temperature falls below the lower threshold or switches the cooling on when it rises above the upper threshold. However, this frequently leads to oscillation of the temperature between the lower and upper thresholds.

Feedback control uses continuous monitoring of some output using sensors and reacts to their changes in order to regulate the output. There are two basic kinds of feedback. Negative feedback seeks to reduce some change in a system output or state whereas positive feedback acts to amplify a system state or output. There are several basic designs for feedback control. Negative feedback can use a derivative of the output to combine with the input to regulate the output. In a simple proportional control system, a signal is negatively fed back to a system in proportion to the degree the system output diverges from the reference value. This leads to a much smoother regulation of a heater about its reference temperature and in a voltage controlled electric motor so that speed can be more smoothly regulated.

6.6.1 Programmable Controllers

Programmable controllers have been developed to support much more configurable and flexible control, e.g., microcontrollers. The hardware architecture of microcontrollers is much simpler than that of the more general purpose processor mother-boards in PCs. Microcontrollers do not need to have an address bus or a data bus, because they integrate all the RAM and non-volatile memory on the same chip as the CPU. The CPU chip can be simpler, 4-bit processors as opposed to 64-bit processors. I/O control can be simpler as there may not be any video screen output or keyboard input. However, micro-controllers can range in complexity, for example, they can include digital signal processors (DSPs). Originally, microcontrollers were only programmed in assembly language and then later in C code. Programs are often developed in an emulator on a (PC) development platform that is then downloaded to the target device for execution, maintenance, debugging and validation.

More recent microcontrollers can be integrated with on-chip debug circuitry accessed by an in-circuit emulator that enables a programmer to debug the software of an embedded system. The in-circuit emulator typically has a JTAG interface.[14] This is a special four- or five-pin interface added to a chip, designed so that multiple chips can have their JTAG interfaces pipelined together. Hence, a test probe only needs to connect to a single JTAG interface to have access to all chips on a circuit board. Although it is most often used to access test-dedicated logic embedded in integrated circuits, it can also be used as an FPGA (Field Programmable Gate Array) programming interface (de la Torre *et al.*, 2002).

---

[13] *Threshold control* systems are similar to event-condition-action or *ECA systems*.

[14] The JTAG (Joint Test Action Group) refers to the IEEE 1149.1 standard entitled *Standard Test Access Port and Boundary-Scan Architecture*.

6.6.2 Simple PID-Type Controllers

There are certain circumstances in which the simple proportional or P-type controller output is not regulated correctly, e.g., when there is a lot of delay in the plant between changing the input and seeing a change in response of the output. In this case a parameter can be regulated below its reference value. To solve this problem, either integral or differential control or both can be added to the control. A *PID controller* is so called because it combines Proportional, Integral and Derivative type control (see Figure 6.9). A proportional controller is just the error signal multiplied by a constant and fed out to a hardware drive. An integral controller deals with the past behaviour of the control. It is almost always used in conjunction with proportional control. It sums all the preceding inputs and is used to add longer-term precision to the control loop by making use of the past history. A derivative controller is used to predict the plant behaviour, then this might be used to stabilise the control via a feedback loop. Such control algorithms, P, PI, PD or PID, are often simple enough to be hard-coded into controllers along with some adjustment controls, e.g., the gain of the proportional element.

6.6.3 More Complex Controllers

The PID controllers described in the previous section can be used for coarse control of robots' arms, e.g., to support palletising, and to support coarse-controlled locomotion. The control design algorithms are based on the assumption that some basic knowledge exists of the plant to be controlled was available and that some approximating models can be used to form a starting point for the design of the controllers and can then be tuned using feedback to cope with the uncertainties of the plant parameters.

However, the algorithms are not well suited to tasks that need finer control, dynamic control, dealing with uncertainties in the control and the need to support adaptive control. There are several main sources of uncertainty in control that can occur:

- *environment dynamics* is a stochastic form of physical objects that naturally varies;
- *random action effects*: some actions can have different effects because of variations in physical characteristics of things when they are formed, complex effect propagation;
- *inaccurate models*: some characteristics of the physical environment, e.g., the weather, are inherently non-deterministic;
- *approximate computation*: e.g., non-linear behaviour is often approximated by a linear solution in a part of its range;
- *sensor limitations*: improper sensor placement, partial views of world, and poor signal reception, see Table 6.1.

Two of the main techniques for controlling uncertain systems are adaptive control and robust control. Adaptive control uses online identification in which either the plant parameters of interest are identified using error predictions (indirect adaptive control), or the parameters are monitored and are adjusted by tracking errors (direct adaptive control) in an attempt to learn the uncertain parameters of the system. Adaptive control is a nonlinear feedback technique which performs identification and control simultaneously in the same feedback loop. This can be used to deal with systems with large uncertainties (Xie and Guo, 2000).

The adaptive approach is applicable to a wide range of parameter variations, but is sensitive to unstructured uncertainties. A robust controller is designed to make the system insensitive to all uncertainties and the final controller has a fixed structure. A robust controller is suitable for dealing with small uncertainties (Yu and Lloyd, 1997).

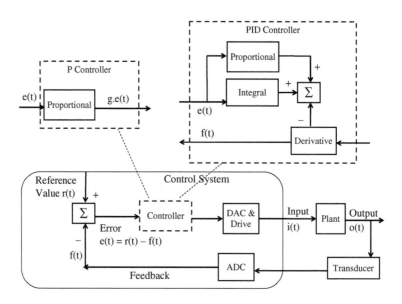

Figure 6.9 Two simple control systems: a proportional type controller (top) and a PID-type controller (bottom)

6.7 Robots

An important type of application which combines control and sensors is robots. An associated term is cybernetic which is a Greek term meaning self-steering such as a governor or pilot. In the early 1960s, robots[15] started to be used to automate industrial tasks particularly in manufacturing, freeing human workers from repetitive, risky and harmful tasks and from tasks that require skill, strength, or dexterity beyond the capability of humans, e.g., in terms of positioning and movement. The main components of robots are:

- *End effectors or actuators*: an end part of a robot that acts on the world in some way such as grabbing, drilling or stacking and that is attached to some moveable arm, gantry or chassis.
- *Locomotion in the form of a moveable arm or chassis*. For example, a robot arm consists of a chain of rigid bodies connected at joints to position its end-effector, whereas a moveable chassis can be propelled by wheels or tracks.
- *Drive*: powered by air, water pressure, or electricity. The power is used to drive some controlled motion defined by the number of degrees of freedom of movement. The major electronic component that produces motion is the electric motor. There are different types such as AC motors and DC motors such as servomotors and stepper motors.
- *Controller*: governs the movement of the end effectors based upon inputs from sensors, motion planning and control theory, such as to drill or punch holes, to palletise (stack and organise

---

[15] In fact, the word 'robot' is derived from the Czech word for forced labour or serf and was first coined about 40 years before the first industrial use of robots, by the Czech playwright Karel Capek in his play *Rossum's Universal Robots* that opened in Prague in January 1921.

factory products onto a transportable frame) physical objects, and to control or open apertures to control the flow of a liquid or gas. Controllers are programmable so that robots can be configured to handle a range of tasks. The control may be quite simple: it may reflect some fixed pattern in a static environment or it may need to be variable because of uncertainty, see Section 6.6.3.

- *Sensors*: provide input information about the state of the physical world, i.e. to determine position and orientation and to track the motion of some end-effector and input these into the controller in order to adapt the behaviour of the effector and the state of the world. Often robots use multisensory perception of the state of the controlled process and its environment (Brooks and Stein, 1994).

Robots differ in focus from embedded systems in that the programmable control for the latter is fixed and task-specific. However, sometimes it is not so clear to define exactly what is a robot and what is not a robot.[16] It seems many everyday devices such as washing machines, elevators, various individual car parts and optical and magnetic spinning multimedia storage devices contain the key components of robots but they are embedded systems as they can be programmed to perform one task. Commercial floor cleaning robots such as the Rumba are also embedded systems by this definition. However, since their locomotion, including collision avoidance, is under autonomous control, such devices are also often referred to as (mobile) robots. More variable robots include automated floor cleaners, etc. Garcia *et al.* (2007) have classified physical world applications of robots along three main dimensions:

- *a robot manipulator or robot arm*: is a linked chain of rigid bodies connected by joints that lead to an end effector, e.g., manufacturing assembly robots, medical robots that assist doctors by performing precision tasks, rehabilitation robots such as certain types of artificial limbs, etc.
- *mobile robots*: carry out tasks in different places using locomotion, e.g., unmanned airborne, waterborne and land-based reconnaissance robots used for geographic map generation, carpet cleaning robots, surveillance and security robots, etc.
- *biologically inspired robots*: take inspiration from biology for their manipulation and mobility, e.g., humanoid robots that can walk, robots that can act as pets, etc.

6.7.1 Robot Manipulators

A manipulator consists of a linked chain of rigid bodies that are linked in an open kinematic chain at joints. A rigid body can have up to six Degrees of Freedom (DOF) of movement. This comprises three translational movements such as moving up and down (heaving), moving left and right (swaying) and moving forward and backward (surging). It also comprises three rotational degrees of freedom: tilting up and down (pitching); turning left and right (yawing) and tilting from side to side (rolling). Joints can be designed to restrict some of the freedom of movement.

A manipulator uses motion planning to calculate the trajectory of the robot between the current or start position and the goal position. A control algorithm, e.g., based upon PID or adaptive control, is used to actually control the trajectory. Kinematic calibration is used to compare the theoretical estimated position from motion planning against the actual sensed position. This can then be used to tune the motion planning for more accurate future use. The contact force at the

---

[16] Joseph F. Engelberger (born 1925) is often credited with being the 'Father of Robotics'. Along with George Devol, Engelberger developed the first industrial robot in the United States, the *Unimate*, in the 1950s which was used by General Motors in 1962. He is once purported to have said when asked to define a robot, 'I can't define a robot, but I know one when I see one' (Carlisle, 2000).

end-effector of the manipulator may need to be regulated so that the effector is fit for purpose, e.g., so that a robot arm does not grip the object it is holding too tightly so that it is crushed or grip it too loosely so that the objects slips or drops from the arm. In addition, manipulators need to cope with variations in components and objects that they are manipulating. There are two solutions here, either to use adaptive AI techniques or to allow human operators to be in the loop to either use remote control or send remote commands to a local manipulator.

6.7.2 Mobile Robots

Mobile robots use various kinds of locomotion systems to move around based upon propellers or screws in aerial and aquatic environments, and based upon capillary tracks, wheels and legs in terrestrial environments. The simplest types of mobile robots to control are ones that follow a predetermined trajectory in a controlled, static, environment. In dynamic non-deterministic environments, control is more complicated as robots need to able to navigate to avoid obstacles. In addition, the number of DOF is much less compared to a robot manipulator. In the simplest case, terrestrial vehicles can have one DOF, forward and back. A simple way is to navigate obstacles is to use collision detection and then either to reverse or to choose a random direction but this can make it inefficient to reach a goal. A more complex approach is to anticipate and avoid collisions and to replan paths to reach goal destinations. One of the most well-known and highly successful uses of mobile robots was the Mars Explorer Robots that started development in 2000, were launched in 2003 and landed on Mars in 2004 to fulfil their mission (Erickson, 2006). Since 2002, one of the most successful consumer-oriented task-based mobile robots has been iRobot's Roomba vacuum cleaning robot which includes odometry determination to estimate their position travelled relative to a starting location and bump sensors and IR sensors to detect IR tags which are used to confine the Roomba's movement within space (Tribelhorn and Dodds, 2007).

Localisation is used to determine a robot's position in relation to its physical environment. Localisation can be local or global. Global localisation is discussed more in Section 7.4. Local localisation concerns a robot which corrects its position in relation to its initial or to other current reference locations.

6.7.3 Biologically Inspired Robots

The basic components of a robot described in Section 6.5.1 can be compared with human components. Humans use end-effectors in the form of hands, feet and other parts of their body. A muscle system is used for locomotion to move the whole body or parts of it. A brain system is a controller that processes sensory information and tells the muscles what to do. A drive system based on stored energy derived from periodic food intake is used to power the human body. Five basic types of sensors are used on the human body surface to see, hear, touch, taste and smell in order to gather information about its environment. In general, mechanical design today falls far short of the performance achievable by biological systems. An 80 Kg person uses about 100 W of power and carrying a 45 Kg bag of cement uses about 350 W of power. In contrast, a 'spot-welding robot' which can carry a 45 Kg payload weighs 450 Kg and consumes 5–10 KW. (Electrical) 'Horsepower' (hp) can be defined as 750 W and in comparison humans can produce about 0.1 to 0.3 hp for periods of several hours or longer. Machines can outperform humans but humans have quite an efficient weight to power ratio (Carlisle, 2000).

Biologically inspired robots are more complex and combine legged locomotion capabilities with robot manipulators. There are two main focuses to these robots: legged locomotion and Human-Robot Interaction (Garcia et al., 2007). The use of legs enables legged robots, compared to mobile robots, to more effectively travel over irregular terrain, stairs, loose and sandy terrain and over ditches. Biped robots often have more DOF than either the mobile robot or robot manipulator, making them more complex to control. For example, in the lower body alone, it may have has six DOF on each leg, three DOF on the hip, one DOF on the knee and two DOF on the ankle. A particular design challenge for biped robots is stability, to maintain overall balance, of a larger body mass compared to the leg mass and in the use of leg and body movements, gaits, in which individual legs are not continually in contact with the ground at any one time. Biped robots often aim to control this balance by controlling the Zero Moment Point (ZMP) so that the dynamic reaction force at the contact of the foot with the ground does not produce any moment or any rotational force.

A second main focus is social human robot interaction (Fong *et al.*, 2003). Robots can assist humans by sensing situations of interest to particular human groups such as intruder detection when the locus of attention of a human cannot be present and assisting the less physically capable. Robots can also fulfil a social role: in terms of engaging humans via interaction at the level of emotions, (affective computing), e.g., artificial pets; social guided learning by imitation or by tutelage (individual and safe tuition) and through interaction using more human-oriented interfaces such as speech recognition.

6.7.4 Nanobots

Nanobots can be manufactured in the same way as MEMS, see Section 6.4.1 or at a molecular level. Although the microscopic world is governed by the same physical laws as the macroscopic world, the relative importance of the physical laws changes in how it affects the mechanics and electronics at this scale, and this must be taken into account (Abbott *et al.*, 2007). The balance between volume ($\sim$length^3) and surface ($\sim$length^2) changes as an object is scaled because these depend on the length differently. Electrostatic forces are widely considered to scale well to the microscale, magnetic effects seem to scale well or not depending on scale, and fluid mechanics becomes simpler in that the effect of turbulence is less important, e.g., the flow pattern does not change appreciably, whether the flow is slow or fast, and the flow is reversible (Abbott *et al.*, 2007). Nature in terms of micro-organisms can be harnessed in order to provide a host body for nanobots to move about., e.g., MC-1 magnetotactic bacteria (MTB), bacteria contain internal compasses that result in them swimming persistently in one direction along a magnetic field (Martel, 2006). Shrinking device size to these nano dimensions leads to many interesting challenges such as nanoscale imaging for viewing these devices, manipulating nano-objects with nanotools, measuring mass in femto-Gram ranges, sensing forces at pico-Newton scales, and inducing GHz motion (Dong and Neleson, 2007).

6.7.5 Developing UbiCom Robot Applications

Most industrial types of robots, although programmable, are expensive and are too specialised to perform just a few tasks and developed for use in static and clean environments. They have accurate actuators and sensors in order to determine their position, to perform precise movements and to actuate robot effectors in very carefully controlled physical environments. Most low-cost consumer-type robots such as robot pets, although able to sense the physical environment and to sense simple interactions with humans and adapt to them, tend not to be readily programmable by consumers.

A number of robots toolkits are available which enable skilled developers to develop their own applications. In order to illustrate some of the issues in deploying programmable robots, an overview is given of the use of a programmable robot to automate changing a part of the physical world on behalf of a human owner. Because resources and costs are limited, a low-cost robot

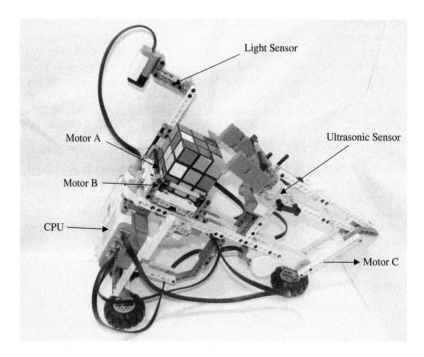

Figure 6.10   Using the Lego Mindstorm NXt robot to solve Rubik's Cube

platform was chosen, the Lego Mindstorm NXT Robot[17] Toolkit (Ferrari and Ferrari, 2007). This consists of a dual processor board, allowing one coprocessor to be dedicated to a task such as monitoring an optical tachometer to determine the rotation of a servo motor while the main processor performs other tasks. A basic kit has three servo motors and has four sensors, a simple touch sensor, a light sensor to measure light intensity, a sound sensor and an ultrasonic sensor to determine distance.

The task chosen was to get the robot to manipulate a Rubik's Cube (Figure 6.10) (a mechanical puzzle invented by Erno Rubik in 1974) into its end state where all the sides show just one colour (the solved state). The goal could be to delegate this whole task to the robot or to design the robot to interact with humans, to advise humans and demonstrate to humans how to perform particular changes to the cube. Both of these goals involve several subtasks which must be designed. First, the mechanics of the robot must be designed, so that it can hold the cube in a way that the robot arm can rotate the whole cube or just one side of the cube and is able to move the optical sensor over the cube in order to determine the colour of the sides of the cube.

Second, a design for how and when the robot senses the state of the world, in this case, the cube, is needed. In order to reduce the amount of sensing and movement of sensors across the whole face, it can be decided just to sense the state of the cube, once, at the start. The sensors could also be used to indicate that the cube is correctly positioned in reach of the robot. A robot could sense its internal

---

[17] Home Page http://mindstorms.lego.com/, accessed Nov. 2007. There are several third party programming languages and environments such as Java for Lego Mindstorms http://lejos.sourceforge.net/, accessed Nov. 2007.

state in order to later estimate if it has enough energy to complete the task. A representation is also needed for the state of the cube that a cube-solving algorithm can understand.

Third, a planning algorithm is needed to perform a series of rotations to the cube and its sides of the cube to end up with the cube in the solved state. There are about 43 million, million, million permutations for the positions of the cube to search through for a brute-force search (Section 8.7.1) in order to construct a plan. The use of informed searches such as a corners first cube algorithm can, however, drastically reduce searches. For example, if there are two opposite faces with different colours, there will be no corner or edge that contains both these colours. The plan derived must define cube rotation actions that the cube can practically execute and this may need to take into account the fixed energy reserve that the robot has.

Finally, an overall architecture that integrates these different sub-tasks is needed such as the vertical one pass hybrid architecture (Section 8.3.7). This uses a lower reactive layer to take sensor inputs about the world and a higher search and planning deliberation layer to derive the plan to solve the cube. Then the overall software to implement these tasks needs to be programmed, downloaded into the robot processor to be executed when it is triggered.

There are several practical issues which arise when physical robots carry out tasks. The optical sensor sensitivity is limited and it may be insufficient to differentiate the light intensity between some different colours such as orange and red. If the cube is not positioned correctly in the robot arm or cradle, the sensors will sense gaps and erroneously take these as the state of part of the cube. Cubes have variable amounts of friction to rotate different sides. In some cases, the robot may only half-rotate a cube and stop if it detects too much friction. Also there is some elasticity in the robot arm because of the Lego bricks which means that parts of the robot can fall off as the robot arm moves. The mechanics of the robot need to be maintained. The grip of the robot arm can slip when the robot holds the cube so that a base side can be rotated. Hence, much effort is needed at a low level to tell robots to carry out specific tasks. Tasks must be carefully designed to fit the robot's capabilities. In the real physical world, the robot must be able to handle incorrectly applied physical forces and incorrectly sensed physical world states.

There does not yet exist, flexible general purpose UbiCom robots, which can act as autonomous assistants or servants for mass human use.

EXERCISES

1. Does location determination involve tagging or a sensing? Explain.
2. Annotation or tagging of physical objects can be classified along two dimensions: if the user of the annotation is co-located (on-site) with the physical object versus not co-located; if the anchoring of the annotation is attached directly to the object it refers to versus the annotation is detached. Explain the implications of this and give some examples of use.
3. Compare and contrast the Semacode scheme to local interaction in the physical world with other schemes such as different kinds of RFID tags and the Cooltown Project (Section 2.2.2.4).
4. The IPv6 address is huge. Is it large enough to uniquely identify each living thing on earth? Propose schemes by which we could attempt to tag each living thing on earth. What would the pros and cons be in doing so?
5. Discuss how we can create and organise an address space to identify all useful artificial objects in the physical world. Compare and contrast the EPC identification scheme to identify things with the IPv6 addressing scheme. Could EPC be used in the Internet instead of IPv6 to also represent virtual as well as physical objects?
6. Compare and contrast tags versus sensors.

EXERCISES (continued)

7. Discuss the design of the address space and routing algorithms used in sensor networks.
8. Discuss why the MEMS (micro) design of a mechanical device is so different from the equivalent macro design.
9. Embedded systems generally do not need a full general purpose operating system to function. Why not?
10. Define the main elements of a robot. Define the properties which characterise a robot. Define three major types of robots. How does a robot differ from an embedded system, and a feedback control system? (More exercises are available on the book's website.)

References

Abbott, J.J., Nagy, Z., Beyeler, F., *et al.* (2007) Robotics in the small. Part I: Microrobotics. *IEEE Robotics & Automation*, 14(2): 92–103.

Abelson, H., Allen, D., Coore, D. *et al.* (2000) Amorphous computing. *Communications of the ACM*, 43(5): 74–82.

Akyildiz, I.F., Su, W., Sankarasubramaniam, Y. *et al.* (2002) A survey on sensor networks. *IEEE Communications Magazine*, 40(8): 102–114.

Al-Karaki, J.N. and Kamal, A.E. (2004) Routing techniques in wireless sensor networks: a survey. *IEEE Wireless Communications*, 11(6): 6–28.

Banerjee, K. and Srivastava, N. (2006) Are carbon nanotubes the future of VLSI interconnections? *Proceedings 43rd Annual Conference on Design Automation*, San Francisco, USA, pp. 809–814.

Berlin, A.A. and Gabriel, K.J. (1997) Distributed MEMS: new challenges for computation. *IEEE Computational Science & Engineering*, 4(1): 12–16.

Bohr, M.T., Chau, R.S., Ghani, T. *et al.* (2007) The High-K solution. *IEEE Spectrum*, 44(10): 23–29.

Bonnet, P., Gehrke, J. and Seshadri, P. (2000) Querying the physical world. *IEEE Personal Communications*, 7(5): 10–15.

Brooks, A. and Stein, L.A. (1994) Building brains for bodies. *Autonomous Robots*, 1(1): 7–25.

Bush, V. (1945) As we may think. *The Atlantic Monthly*, Vol. 176: 101–108. Reprinted and discussed in *ACM Interactions*, 3(2), Mar. 1996: 35–67.

Buttazzo G. (2006) Research trends in real-time computing for embedded systems. *ACM SIGBED Review* 3(3): 1–10.

Carlisle, R. (2000) Robot mechanisms. In Proceedings of the 2000 IEEE Intrenational Conf.erence on *Robotics and Automation, ICRA 2000*, San Francisco, USA, pp. 701–708.

Chang, C-Y. (2003) The highlights in the nano world. *Proceedings of the IEEE*, 91(11): 1756–1764.

De laTorre, E., Garcia, M., Riesgo, T. *et al.* (2002) Nonintrusive debugging using the JTAG interface of FPGA-based prototypes. In *Proceedings 2002 IEEE International Symposium on Industrial Electronics*, ISIE 2002, 2: 666–671.

Devasia, S., Eleftheriou, E., Moheimani, (2007) S.O.R. A survey of control issues in nanopositioning. *IEEE Transactions on Control Systems Technology*, 15(5): 802–823.

Dong, L. and Nelson, B.J. (2007) Tutorial - Robotics in the small Part II: Nanorobotics. *IEEE Robotics & Automation*, 14(3): 111–121.

Erickson, J.K. (2006) Living the dream – an overview of the Mars exploration project. *IEEE Robotics and Automation*, 13(2): 12–18.

Estrin, D., Culler, D., Pister, K. (2002) Connecting the physical world with pervasive networks. *IEEE Pervasive Computing*, 1(1): 59–69.

Feiner, S., Macintyre, B., Seligmann, D. (1993) Knowledge-based augmented reality. *Communications of the ACM*, 36(7): 53–62.

Ferrari, M. and Ferrari, G. (2007) *Building Robots with LEGO Mindstorms NXT*. ISBN-13 978-1-59749-152-5.

Fong, T., Nourbakhsh, I., Dautenhahn, K. (2003) A survey of socially interactive robots. *Robotics and Autonomous Systems*, 42: 143–166.

Friedrich, L.F., Stankovic, J., Humphrey, M., *et al.* (2001) A survey of configurable, component-based operating systems for embedded applications. *IEEE Micro*, 21(3): 54–68.

Frost, G.P. (2003) Sizing up smart dust. *Computing in Science & Engineering*, 5(6): 6–9.

Fuller, S.B., Wilhelm, E.J. and Jacobson, J.M. (2002) Ink-jet printed nanoparticle microelectromechanical systems. *Journal of Microelectromechanical Systems*, 11(1): 54–60.

Garcia, E., Jimenez, M.A., De Santos, P.G. *et al.* (2007) The evolution of robotics research. *IEEE Robotics & Automation Magazine*, 14(1): 90–103.

Gemmell, J., Bell, G. and Lueder, R. (2006) MyLifeBits: a personal database for everything. *Communications of the ACM*, 49(1): 88–95.

Gershenfeld, N. (1999) The personal fabricator. In *When Things Start to Think*. London: Henry Holt & Co, pp. 63–75.

Goldstein, S.C., Campbell, J.D. and Mowry, T.C. (2005) Programmable matter. *Computer*, 38(6): 99–101.

Graham, J.H. (1989) Special computer architectures for robotics: tutorial and survey. *IEEE Transactions on Robotics and Automation*, 5(5): 543–554.

Hansen, F.A. (2006) Ubiquitous annotation systems: technologies and challenges. In *Proceedings of the 17th Conference on Hypertext and Hypermedia*, Odense, Denmark, pp. 121–132.

Heidemann, J., Silva, F., Intanagonwiwat, C., *et al.* (2001) Building efficient wireless sensor networks with low-level naming. In *Proceedings of 18th ACM Symposium on Operating Systems Principles*, Banff, Canada, pp. 146–159.

Hölldobler, B. and Wilson, E.O. (1990) *The Ants*. Cambridge, MA: Harvard University Press.

Kahn, J. M., Katz, R.H. and Pister, K.S.J. (2000) Emerging challenges: mobile networking for Smart Dust. *Journal of Communications and Networks*, 2(3): 188–196.

Kaiser, W.J., Pottie, G.J., Srivastava, M. *et al.* (2004) Networked Infomechanical Systems (NIMS) for Ambient Intelligence. In W. Weber, J.M. Rabaey and E. Aarts (eds) Ambient Intelligence. Berlin: Springer Verlag, pp. 83–114.

Koester, D.A., Markus, K.W. and Walters, M.D. (1996) MEMS: small machines for the microelectronics age. *Computer*, 29(1): 93–94.

Krueger, M.W. (1993) Environmental technology: making the real world virtual. *Communications of the ACM*, 36(7): 36–37.

Kumar, R., Farkas, K.I., Jouppi, N.P. *et al.* (2003) Single-ISA heterogeneous multi-core architectures: the potential for processor power reduction. In *Proceedings of the 36th Annual IEEE/ACM International Symposium on Microarchitecture*, Washington, DC, USA, pp. 81–92.

Leen, G. and Heffernan, D. (2000) Expanding automotive electronic systems. *Computer*, 35(1): 88–93.

Li, Y., Potkonjak, M. and Wolf, W. (1997) Real-time operating systems for embedded computing. In *Proceedings of 1997 International Conference on Computer Design* (ICCD'97), pp. 388–392.

Mackay, W.E. (1998) Augmented reality: linking real and virtual worlds: a new paradigm for interacting with computers. In *Proceedings of the Working Conference on Advanced Visual Interfaces* (AVI'98), pp. 13–21.

Mansour, R.R., Bakri-Kassem, M., Daneshmand, M., *et al.* (2003) RF MEMS devices. In *Proceedings. International Conference on MEMS, NANO and Smart Systems*, pp. 103–107.

Meng, E. and Tai Y-C. (2003) Polymer MEMS for micro fluid delivery systems. Paper presented at ACS Polymer MEMS Symposia, New York, USA, pp. 552–553.

Martel, S. (2006) Towards MRI-controlled ferromagnetic and MC-1 magnetotactic bacterial carriers for targeted therapies in arteriolocapillar networks stimulated by tumoral angiogenesis. In *Proceedings IEEE International Conference of Engineering Medicine Biology Society*, pp. 3399–3402.

Moore, G.E. (1965) Cramming more components onto integrated circuits. *Electronics*, 38(8): 114–117.

Nath, B., Reynolds, F. and Want, R. (2006) RFID technology and applications, *IEEE Pervasive Computing*, 5(1), 22–24.

Norman, D.A. (1988) *The Psychology of Everyday Things*. New York: Basic Books.

Petersen, K. (2005) A new age for MEMS: solid-state sensors, actuators and microsystems. In Digest of Technical Papers. The 13th International Conference on Transducers '05, 1: 1–4.

Peterson, C. (2000) Taking technology to the molecular level. *Computer*, 33(1): 46–53.

Poole, C.P., Jones, F.J. and Owens, F.J. (2003) *Introduction to Nanotechnology*. Chichester: John Wiley & Sons, Ltd.

Rebeiz, G.M. (2003) *RF Mems: Theory, Design, and Technology*. Chichester: John Wiley & Sons, Ltd.

Santo, B. (2007) Plans for next-gen chips imperiled. *IEEE Spectrum*, 44(8): 8–11.

Shen, C-C., Srisathapornphat, C. and Jaikaeo, C. (2001) Sensor information network architecture and applications. *IEEE Personal Communications*, 8(4): 52–59.

Tanenbaum, A.S. (2001) *Modern Operating Systems*, 2nd edn, Englewood Cliffs, NJ: Prentice-Hall.

Tribelhorn, B. and Dodds, Z. (2007) Evaluating the Roomba: a low-cost, ubiquitous platform for robotics research and education. Paper presented at IEEE International Conference on Robotics and Automation, pp. 1393–1399.

Wallich, P. (2007) Deeply superficial. *IEEE Spectrum*, 44(8): 56–57.

Want, R. (2006) An introduction to RFID technology. *IEEE Pervasive Computing*, 5(1): 25–33.

Xie, L-L. and Guo, L. (2000) How much uncertainty can be dealt with by feedback? *IEEE Transactions on Automatic Control*, 45(12): 2203–2217.

Yeow, T-W., Law, K.L.E. and Goldenberg, A. (2001). MEMS optical switches. *IEEE Communications*, 39(11): 158–163.

Yu, H. and Lloyd, S. (1997) Variable structure adaptive control of robot manipulators. *IEE Processes and Control Theory and Applications*, 144(2): 167–176.

Yu, H., Agrawal, D., and El Abbadi, A. (2007) MEMS based storage architecture for relational databases. *The VLDB Journal*, 16(2): 251–268.

Zambonelli, F., Gleizes, M-P., Mamei, M. and Tolksdorf, R. (2005) Spray computers: explorations in self-organization. *Pervasive and Mobile Computing*, 1(1): 1–20.

Zhao, F. and Guibas, L. (2004) Wireless Sensor Networks: An Information Processing Approach. New York: Morgan Kaufmann.

7

Context-Aware Systems

7.1 Introduction

Context-aware systems are systems that are aware of their situation (or *context*[1]) in their physical, virtual (ICT) and user environment, and can adapt the system to this in some way, benefiting from knowledge of that situation. For example, in the personal memories scenario, the camera can detect the distance of the camera to the subject of the photo and automatically adapt the focus of a camera lens when recording the image.

The term context-aware was first used in 1994 by Schilit and Theimer to refer to a system that can provide context-relevant information and services to users and applications.[2] Schilit *et al.* (1995) also defined a context-aware system as a system that adapts itself to the context. There are many other similar definitions. For example, Dey (2000) defines a system to be context-aware if it uses context to provide relevant information and, or services to the user, where relevancy depends on users' task. Ryan *et al.* (2008) define a context-aware system as having the ability to detect and sense, interpret and respond to aspects of a user's local environment and to the computing devices themselves.

There are a range of definitions for context. Some are more concrete and others are more abstract. Dey and Abowd (2000) define context as 'any information that can be used to characterise the *situation* of an entity that is considered relevant to the interaction between a user and an application'. An example of a more concrete definition of context is to define it as a member from the set of context types, such as location, identities of nearby people, objects and changes to those objects (Schilit and Theimer, 1994). There are many possible useful sets of contexts – these often depend on applications' use of the context. Environment monitoring uses multiple types of distributed sensors to determine an environment context such as air pollution, temperature,

---

[1] Context is often used loosely as a synonym for situation. The concept of a context is modelled more particularly here. A context is defined here in terms of associations between the states of three types of system environments defined in Chapter 1, physical, human (user) and virtual (ICT), usually in relation to a goal context.

[2] This preceded noteworthy applications that used the concept of context-awareness but did not specifically name this use as such, e.g., the location-aware application by Want in 1989 to support call routing to mobile users (Section 2.2.1.2).

Ubiquitous Computing: Smart Devices, Environments and Interactions   Stefan Poslad
© 2009 John Wiley & Sons, Ltd

humidity. Time, location and person can be combined into a diary context for a personal scheduler application.

7.1.1 Chapter Overview

In this chapter, a range of context-aware applications are discussed. Context-aware applications can be classified in two main ways (Section 7.1.2): in terms of the type of environment context they are aware of, e.g., location, time, person, ICT system, etc., and in terms of how applications adapt themselves to the context. Models and architectures for context-awareness are discussed in Section 7.2. Next several major types of context-aware applications are considered in order to demonstrate the utility of context-awareness in practice: mobile user-awareness (Section 7.3), spatial or location awareness (Section 7.4), temporal-awareness (Section 7.5) and ICT system awareness (Section 7.6).

Other chapters also cover context-awareness as follows. Context-awareness is a type of application that uses sensor-based systems (Section 6.3) and can be used to control and regulate the physical environment (Section 6.6). Designs for context-aware systems can be based upon intelligent systems (Section 8.3). Awareness of the external context of the environment is often usefully combined with internal system awareness or self-awareness (Section 10.3).

7.1.2 Context-Aware Applications

There are several surveys of systems and applications by Chen and Kotz (2000) and by Baldauf et al. (2007). Loke (2006) gives a good introduction to many types of context-aware systems including: mobile services; context-aware artefacts – everyday objects and hand-held devices that can be made context-aware and context-aware virtual environments such as Gelernter's abstraction of a *Mirror world* (Gelernter, 1992). Some of the most widely deployed applications such as *mobile context-aware* systems use multiple types of context-awareness at multiple levels (Section 7.3). Brown et al. (2000) classify context-aware applications into six types: proactive triggering, streamlining interaction, memory for past events, reminders for future contexts, optimising patterns of behaviour, and sharing experiences.

Here, context-aware applications are discussed with respect to the type of context that is used and what it is applied to. One of the dominant applications for context-awareness since the *Active Badge* system of Want et al (Section 2.2.1.2) has been that of location-awareness for mobile users. Examples of such projects include the Cyberguide project (Long et al., 1996), the GUIDE project (Davies et al., 1999), SmartSight and the CRUMPET project (Section 7.3.4). Cyberguide is a system to provide tourist information to visitors to the campus at the Georgia Institute of Technology.[3] Cyberguide users carry an Apple handheld computer that connected to a central PC via infrared to beacons around the centre. The University of Lancaster created GUIDE to provide tourist information on a hand-held Fujitsu. Carnegie Mellon University has a tourist assistant system called SmartSight (Yang et al., 1999). This system uses a wearable computer to answer spoken questions about local landmarks, provide translations and aid navigation.

There are many other human activities and automatic services that orientate themselves to physical phenomena. Here are a few examples. Transport systems can automatically brake if

---

[3] Abowd's group at the Georgia Institute of Technology has been particularly active in leading research into context awareness, researching context-aware applications since the mid to late 1990s. Its projects include Classroom 2000 (Section 2.2.2.1), the *Aware Home* which started in 1998, which is still ongoing, and Cyberguide.

they detect obstacles ahead (collision detection). Weather-aware applications include outdoor sports activities, such as sailing, agriculture and commerce. Food supermarkets use weather predictions to help decide the quantity of certain foods to stock because certain foods such as salad ingredients, soft fruits and grilled meat are consumed more in sunny weather.

Spatial awareness is included as a core type of physical environment awareness because objects are inherently located in space (and time). Spatial awareness enables spatial information such as maps to relate to the current location and destination location of their users. Temporal-aware applications schedule their services with respect to time. Personalised systems allow services to be tailored to users, to their personal preferences, e.g., users can get restaurant recommendations based upon their food preferences.

A second way to classify context-aware applications is into sub-types of passive versus active context-awareness.[4] In a *passive context-aware system*, the context is presented to users to make them aware of the current context or to a change of context. The system is not active in terms of adapting any usage or application to the context. In an *active context-aware system*, the adaptation to the context is performed by the UbiCom system, not the human users. Two main types of passive context-aware application are simple context presentation of context information and context-based tagging. Three main types of active context-aware application are context-based information and task filtering, context-based task activation and context-based task control and regulation.

Simple *context presentation* of post-processed information includes observing the value of some varying physical phenomena, e.g., the temperature over some time and locations. A manual decision can be made to trigger an action when the value passes through or remains above some threshold. Some contexts are not always readily machine-readable, e.g., the level of an analogue thermometer, need some post-processing to be useful. The absolute temperature needs to be equated to a graduated ruler of levels of temperature.

Context tagging or annotation of physical environment objects refers to recording views of physical objects, e.g., visual images. Context-based tagging of physical, human and virtual world entities can enhance later retrieval via processing the tags rather than the virtual views themselves. Tagging can also be used to track objects, including people and goods in real time.

The design of active context-aware systems is considered in detailed below (Section 7.2.1). *Context- filtering systems* filter information based upon the context, e.g., to automatically adapt content to fit a mobile user's limited display device and location. Content adaptation may also be useful for less physically able people who require information services to be adapted to users with limited sight and limited motor control. Context-aware systems are particularly useful because they can lessen the cognitive load to access information, devices and to operate physical devices. This type of information filtering is one of the essential elements of the automated personal assistant software agent envisaged by Maes (1994). This is because users do not need to manually sift all possible data to find the data relating to their situation, the system can prioritise and select only the data that relates to their context, e.g., only presenting the maps that contain the current location and destination location.

Context-based task activation focuses on user centred task activation provides policies for when sub-tasks or a task configuration is activated, e.g., a road haulage firm could configure black boxes in their trucks to instruct drivers to take a break after a certain period of time on the road. Another example is flood sensors in a river basin may be set to trigger a flood defence initiative when a certain water level is reached. Real-time sensing of physical contexts such as location and

---

[4] There is also a hybrid context adaptive system where the full active adaptation by the system may need manual operator assistance. The human user guides or corrects the automatic adaptation.

temperature can be used to control the movement of a robot arm during manufacturing and to regulate the temperature in a heating system. Context ware systems act as a type of adaptive distributed event-driven system in which context events are sensed, filtered and user tasks adapt to the context. For example, a logistics type spatial-awareness application allows a new route to a destination to be calculated and used when it detects that someone or something has deviated from their pre-determined route. Context-awareness is an element of types of control system where a context is sensed in order to control a context within defined operational limits.

7.2 Modelling Context-Aware Systems

Models of context-aware systems need to define what a range of contexts describe and how contexts are created, composed and used for adaptation. Context-aware system models need to define how to represent contexts in a computation form and how to support an operational life-cycle in using context-aware systems.

Baldauf *et al.* (2007) have characterised context-aware frameworks with respect to: distributed system architecture, sensing design, context data representation (referring to the context data classification by Strang and Linnhoff-Popien, 2004), type of context processing which is in turn is dependent on the context data representation, context resource discovery, historical context data, and security and privacy. The frameworks discussed here have most of these characteristic elements except some do not have explicit context resource discovery, historical context data and security and privacy support. They also differ in terms of the detailed design and implementation of these characteristics.

7.2.1 Types of Context

Several ways to classify contexts have been proposed as follows. Prekop and Burnett (2003) and Gustavsen (2002) refer to external and internal types of context which are similar to these physical and user contexts[5] respectively. Hofer *et al.* (2003) refer to physical and logical contexts where the logical context is similar to the user context. Dey and Abowd (2001) have proposed places such as rooms and buildings, people, either individuals or groups, and things such as physical objects and components as types of context. Schilit *et al.* (1995) classify context into three categories where you are (location context including which physical environment resources are located with the user), who you are with (social context), and what (ICT) resources are nearby. Chen and Kotz (2000) and Henricksen *et al.* (2002) make a distinction between passive and active context-aware systems. A static context describes aspects of a pervasive system that are invariant, such as a person's date of birth. A dynamic context refers to a user or an environment context. These contexts can be highly variable over space and time, e.g., temperature. Morse *et al.* (2000) have modelled context in terms of six main dimensions pertaining to what, who, where, when, how it is accessed and why it is useful.

In Table 7.1, the types of context are classified here with respect to the type of environment: physical, human and ICT. Table 7.1 also relates this classification of contexts to the one of Morse *et al.* (2000). The human *user context* defines the user's tasks to achieve goals and defines the social and the physical or environment context.

---

[5] It is important to differentiate several different kinds of user context awareness such as awareness of identified versus unknown individuals, of user stereotypes, of personal preferences and of user goals (see Table 7.1).

Table 7.1 A classification of the main types of context by type of UbiCom system environment and according to that of Morse *et al.* (2000)

Characteristic	Description
(Physical) Environment Context	
What	Type of physical environment or physical phenomena context-awareness such as awareness of temperature, light intensity, chemical or biological concentration etc
Where	*Spatial awareness* or *location awareness*: where an awareness of context can be exploited This can be at the current location or in terms of one location in relation to one or more other locations, e.g., the current location in relation to a start or destination location or to a route
When	*Temporal awareness*: when context-awareness is useful – now, later and during some activity This can be defined in terms of an absolute time or in terms of a relative time to some other event or condition
ICT Environment System Context	
How	*ICT awareness*: awareness of how any context is created and adapted over an ICT infrastructure, e.g., a context or context-aware application can be accessed over a wireless link and via a mobile terminal
User Environment Context	
Who	*User context-awareness:* who might benefit from an awareness of someone's context. The user context is divided into personal and social user contexts *Personal Preferences*[1]: e.g., a referee at a sports activity may prefer to blow the whistle for minor versus major sports offences. Personal user contexts may in turn be subdivided into: *personal identity context*, e.g., a particular person who is a referee; *personal stereotype context*, e.g., the pattern of actions at a sports event indicates the individual is a type of referee *User Activity or Task Context*: describes a user's current situation, e.g., person is standing versus running *Social (User) Context*: describes how the actions of someone may affect others, e.g., the whistle blown at a sports match or by a referee, policeman or spectator has different effects on others depending on their societal roles
Goal Context for Applications or Users	
Why (Task)	*User or application goal*: why a context is useful, the higher-level application or user purpose the context is used for, e.g., a location serves to show someone or something in relation to their destination *Context Adaptation*: how the current context can transition to the goal context

Note: [1]Personalisation is user-awareness of a person's preferences (Section 5.7.3).

Each individual context is itself defined by a *meta-context*, information that describes the context. For example, the location context must also define which type of location coordinate system and units are used. In addition, the relation of the current context to other contexts may be important, e.g., the relation of someone's home to the nearest medical centre. Note that intrinsically the focus is on varying contexts. The major invariant context is the user and his or her characteristics such as their identity and perhaps their contact details. The transition from the current to the destination or goal context follows a fixed or dynamic mapping. This represents a form of pre-planned context adaptation (Section 7.2.3).

Context often implies a situation which affects a system which is external to the system, the conditions of the environment surrounding the system (*external context*) which is described by Morse *et al.* (2000). Equally, pertinent are the internal system conditions and the conditions of use of which is a system is aware (*internal context*). More generally we can refer to a context as something which relates to, or affects, the use of the system, i.e., either through internal or through

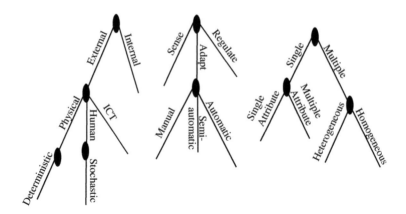

Figure 7.1   Multidimensional multi-level support for a UbiCom property, e.g., context-awareness

external interaction The multidimensional features of types of contexts which can be modelled are summarised in Figure 7.1.

7.2.2 Context Creation and Context Composition

New contexts can be created in real time using sensors situated in the physical environment, e.g., temperature sensors. Often lower-level *raw contexts* output from sensors need some post-processing into higher-level contexts that are relevant to users and applications. Sensed values may need to be scaled or transformed into different value ranges or domains, e.g., an electrical signal from a temperature sensor gets mapped to a temperature value on a Celsius or Fahrenheit temperature scale. Sensed values may also need to be related to other entities to be useful, e.g., absolute position coordinates may be mapped to a positional context such as a particular building, zip or postcode or other land area identifier. Rather than use the location coordinate itself as a low-level context, it is often more useful to access some abstraction of the context, e.g., 'this photograph was taken at the home of these people'.

Some contexts such as location, entity, activity and time, act as sources of contextual information from which other contexts can be derived – *derived contexts or context reuse*. Combining several individual context values may generate a more accurate understanding of the current situation than taking into account any individual context.[6] This is particularly useful in iHCI system applications as it can avoid users having to explicitly identify themselves or to explicitly identify their tasks and goals. For example, knowing the current location and current time, together with the user's calendar, can enable an application to infer a user context, such as having a meeting, sitting in a class, waiting at the airport, etc.[7] Approaches to determine individual user contexts include:

- *Combining several simpler contexts* (context composition). Systems may need to handle mediation and interoperability between multiple contexts. The ordering in which contexts are combined may be important. The determination of and combination of contexts may not scale because uncertainty and complexity increase.

---

[6] Combining contexts may also lead to conflicts (Section 7.2.7).

[7] There are several challenges with this simple calendar context approach, e.g., users may have recorded multiple appointments and it may not be recorded which appointment has priority.

○ *Combining homogeneous contexts*: e.g., from multiple independent sensors because of variations in individual measurements.
○ *Combining heterogeneous contexts*: this can be used to determine a composite or high-level context, e.g., user, context.
○ *Deriving high-level context from lower-level ones*: For example, from knowing someone's weight and approximate location, a user's identity could be derived.
○ *Deriving a lower-level context from a higher-level one*: a position can be determined from annotated positions such as a street name and a building name and information supplied by a passer-by.

- *Consulting a user profile* or user preferences, e.g., a user's calendar to find out what the user is supposed to do at a certain time. A user profile or calendar may not contain all the relevant information to relate to activities. Users may also choose to deviate from their calendar or profile in a spontaneous manner. Contexts may not be fully deterministic or fully observable.
- *Asking users* (Section 5.6) to define their user preferences: closed queries (choose from a set which best describes the context) versus open queries can be used.
- *Observing users* (Section 5.6), e.g., use machine vision based upon camera technology and image processing to identify features such as faces, fingerprints, etc. and then link these to the user context.

Context-aware systems may need to support the maintenance of the context annotation including creation, modification and deletion. A challenge here would be to harmonise or standardise the annotation so that they can be consistently used by all users. Security, in particular access control, could be useful in certain applications to protect privacy or to limit access. For example, users may wish to share their location with friends but not with the general public.

Context-aware systems may need to be able to interlink heterogeneous contexts. Contexts may be of the same type but be represented differently and be defined using different meta-contexts. For example, there are over 100 different location coordinate systems in use; the same type of context may be annotated differently. The same context may have different semantics depending on the application that uses it. For example, a person who stops in front of a building may be waiting for somebody or something or may be simply interested in gazing outdoors. A composite-context may be needed to help interpret the meaning and use of an individual context, e.g., in this case to ascertain not just the location of someone but their gaze. Stopping and gazing at the building may signal an interest in architecture whereas stopping and gazing elsewhere may signal that someone is waiting for something else.

7.2.3 Context-Aware Adaptation

The current context often seems to be the main focus in a context-aware system but it is really the relation of the current context to a goal context that is the essence of context-awareness and adaptation. Out of all possible contexts, that can be determined or are available, only those current contexts that affect the goal context of the specific application are important. The execution of a plan of actions moves the current context forward to the goal context. There are often constraints on the goal context and on how much the actual current context can deviate from the planned current context. Although the value of the current context changes and different parts of the same overall plan are executed depending on the context, the overall planned behaviour of the system often remains constant, it is not adaptive (Figure 7.2).

There are several issues which make context adaptation more complex in practice. First, sometimes the plan of actions to move towards the goal state is conditional and may fail in various ways: the plan may be incorrect, the actions may fail, the system state may have changed, etc. Hence, *conditional plan* execution is needed to capture deviations in the plan which can be detected by

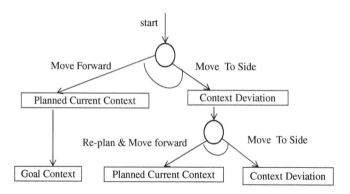

Figure 7.2 A conditional planning model of context-awareness based upon pre-planned actions that move the system towards a goal context

sensing the actual context and comparing it against the planned context. Planning is discussed in more detail in Section 8.7. If a context deviation occurs, the planned behaviour of the system needs to be changed in order to still meet the goal context, this is called *context-aware adaptation*. Second, context-awareness may require multiple independent or interdependent goal contexts to be supported, e.g., collecting revenue by maximising the number of goods transported and being on time.

In some applications, e.g., the personal memories scenario, there are multiple user goal contexts: first, to capture and record people and, second, to annotate the recording for enhanced later retrieval of the recording using the contexts of where, when and who is in the recording. In both cases, inherent in this application is the ability to associate a recording with the combined current location, time and identified people in the recording context for the recording. The location and time part of the current context can be generated automatically by clock and location determination sensors respectively. If a goal of the context-aware personal memories application is also to record good quality images of people (goal context), a combined facial recognition and auto-focus on face system is needed to transform images of people in the current photo shoot (current context) into one where the camera auto-focuses on the recognised faces of people in the image.

Context adaptation from the current to the goal context may be constrained. Context constraints in the personal memories scenario are that people must face the camera and be in the near field for the auto focus and detect people context adaptation to work. Sometimes the context constraints apply to the goal context, e.g., just focus on the people in the AV recording, and sometimes the context constraints apply to the adaptation, e.g., just use main roads to move goods or people from the start or current location to the destination location. Sometimes there may be a *single-shot* or *single-pass* context adaptation, e.g., calculate a preset route from the start location to the destination location and sometimes *multi-shot* context adaptation may be used such as in an automatic collision avoidance system for a vehicle.

The physical world context may itself describe other associations of interest other than the relation of the current context to the goal contexts. The relative location, e.g., where someone lives or where some sports activity took place, may be of more interest than the absolute location. Sometimes the absolute time is recorded. At other times the relative time is more of interest, e.g., how old someone was when the recording was made. These relations in space and time can be specified at the time of the recording or could be generated at a later date. In an active context-aware system, the context adaptation is done by the system but in a passive context-aware system, the system presents the context to human users and they must manually perform any context adaptation.

Context-aware users and applications may require adaptation towards a composite goal context such as a combination location, person, time and ICT access system. Design issues concern how to compose independent and dependent individual contexts, e.g., does a scheduled bus that is late try to minimise its lateness by avoiding picking up more passengers on route even although that would generate more revenue for the bus company. A work-flow for context adaptation is sometimes needed to serialise the adaptation of several individual context adaptations in turn. The order in which the individual adaptations are performed may be important, e.g., determine location and adapt to location, then determine user preferences and adapt to user preferences or vice versa, as this can greatly impact the performance, i.e., the size of the search space and the complexity of the context mediation.

An example of the issues in combining multiple independent local contexts is the use of a location-aware meeting service that may propose multiple joint meeting places based on the GPS current locations of the participants and their route to the destination meeting place. The determination of a proposed joint context for meeting can be complex because different multiple weightings of multiple parameters may need to be used to reach an agreement, for example, if travellers have different modes of transport, different current locations, different routes, and different preferences for the type of meeting place (Meng and Poslad, 2008).

7.2.4 Environment Modelling

The simplest type of context-aware system operates in an environment that it can fully observe and determine. It usually operates in an episodic environment that depends upon the current state of the environment in relation to the goal state but not on the past state of the environment. More sophisticated context-aware systems may also store each or some of the current contexts which will then become the *historical context* in order to present, explain, and reason about, how current or past goal contexts were actually achieved (Section 8.2.2).

7.2.5 Context Representation

Strang and Linnhoff-Popien (2004) have surveyed different types of representation for context models and identify six different types of representation. These have been discussed and graded with respect to how well they support a distributed context model and composition; partial validation (with respect to a formal schema for the context model); richness and quality of information (the sense of sensor heterogeneity and time-variance of the environment context); incompleteness and ambiguity of context gathered; level of formality (to precisely describe and share the meaning of context) and applicability to existing environments (particular to Web service oriented frameworks). This is summarised in Table 7.2.

Strang and Linnhoff-Popien conclude that out of the six context representations, an ontology is the most promising representation for context models. However, in practice, this analysis is oriented more to a Web-like service-oriented infrastructure rather than to a very distributed, heterogeneous, scalable, low resource embedded systems infrastructure that need to be designed to handle ontologies. Some ontology representations seem to be closely coupled to one specific type of logic reasoning that may not easily be applied to support temporal, uncertainty and spatial reasoning.

There are several practical issues in using an ontology-based context model for mobile users. Thin-client devices and other low-resource computer nodes that acquire contexts still lack the processing power to handle the parsing of ontology instances. In addition, some context-aware applications such as location-awareness require the ability to handle large complex spatial data structures. The level of granularity at which semantic relationships are exposed and manipulated needs careful thought. It can quickly become computation intractable to reason about either the detailed semantics of large data structures or about high volumes of context data measurements on-line. It may be better for scalable context-aware systems for much of the data to be handled at a syntactical rather than a semantic level, using simpler event-condition-action handling at the syntactic level.

Table 7.2   Different types of context representation according to Strang and Linnhoff-Popien (2004)

Model	Type of structure and how retrieved	Comments
Key-value	Simple, flat, data structure for modeling contextual information Get by: linear search	Pros: Easy to manage and parse in embedded systems Cons: Uses exact matches, lacks expressive structuring, lacks efficient context retrieval algorithms, has weak formalism, handling incompleteness. May need multi-values
Markup scheme	Hierarchical data structure, e.g., XML, consisting of user defined markup tags with attributes that can be arbitrarily nested Get by: Markup Query Language	Pros: distributed model, uses underlying resource identifier and namespace model; XML Web services are becoming pervasive; handling heterogeneity, handling incompleteness Cons: expressive structuring and weak formalism
Graphical	Graph data structures and richer data types, e.g., UML (Unified Modelling Language), ORM (Object-Role Modeling) Get by: transformation algorithms	Pros: more expressive than key-value, and hierarchies Cons: support for distributed context model, handling incompleteness, lack of formalism for on-line automated access
Object oriented (o-o)	Context processing is encapsulated, hidden to other components. Access is through specified interfaces only Get by: algorithms	Pros: distributed o-o is mature, some partial validation but often not very formal. Reuse can be supported through inheritance and composition Cons: handling incompleteness
Logic based	A logic defines the conditions in which a concluding expression or fact may be derived) from a set of other expressions or facts Get by: reasoning	Pros: Strong formalism, expressive structuring Cons: handling uncertainty, time varying instances, heterogeneity, often difficult to partially validate – its either all or nothing, simple structuring, handling incompleteness
Strong ontology	A combined expressive conceptual model with a logic Get by: reasoning	Pros: expressive structuring, handling heterogeneity, partial validation Cons: handling uncertainty, scalability in searching large data volumes, use in low resource embedded environments

7.2.6 A Basic Architecture

The characteristic elements identified by Baldauf *et al.* (2007) are also present in the high-level architecture model of a general context-aware system given in the basic context-aware model[8] (Figure 7.3). This consists of four main components: current context capture, goal

---

[8] In order to simplify the model, it does not support self-awareness and focuses on episodic environments in which the system only needs to be aware of the current context.

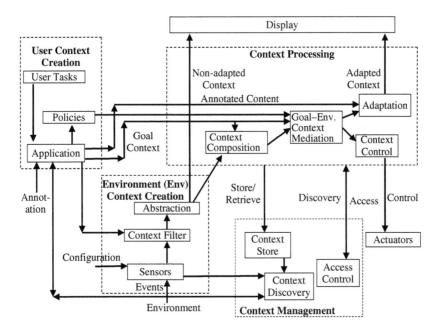

Figure 7.3 A general architecture for context-aware systems

context creation, adaptation of the current context to the goal context and context management. The core design features of this architectural model are as follows. It is independent of, and can be reified into, a variety of architectural styles (Section 3.2). It articulates in detail the distribution and interaction of four main groups of context-based functions that form a life-cycle process for operating a context-aware system. It is also independent of how these different functions can be partitioned and distributed. It separates the concerns of the use of context from the context itself.

The context-aware operational life-cycle consists of: (1) the physical, human and ICT environment context; (2) data is created; (3) the user goal context needs are created; (4) the environment contexts are used to adapt some user or application goal context; and (5) the context data need to be managed. This life-cycle is discussed in more detail below.

1. Context Determination

   Acquisition of the environment context: sensors convert the (current) state of the environment such as time, location and physical phenomena such as temperature, light, sound, etc. into data for computation. This may entail many sub-processes such as sensor calibration and sensor configuration, e.g., to select how frequently measurements are made.

   a. Acquisition of user contexts such as human user identity, stereotypes and personal preferences: these can be acquired directly or indirectly from user interaction.

   b. Encapsulation and abstraction (post-processing of raw contexts): of a context into a service: encapsulation to enable the context to be accessed via a published interface; abstraction of context data through interpreters, e.g., to harmonise heterogeneous context values into a common representation; convert between different representations and structures for the same type of context.

   c. Filtering: of the environment context with respect to user context. Filters only consider events within a certain range that adhere to defined context constraints or policies.

2. User Context Acquisition

   a. Acquisition of the goal user context: this can be derived from users' application tasks.
   b. Policy or constraints acquisition: created from users' tasks to determine how user contexts are mediated by environment contexts.
   c. Encapsulation and abstraction of user goal contexts.

3. Context Processing

   a. Context-composition: an application may govern the use of multiple environment contexts that may be acquired from multiple context sources. One type of context may be used to affect another type of context based upon a user context policy.
   b. Mediation: multiple environment contexts are linked and interrelated.
   c. Adaptation: passive or active or control:

      i.   Passive (or Presentation): in a passive context-aware system, the environment context is used as a constraint to select or query information from an application or user context and simply to present it to the user. The user manually controls an application or adapts their actions to the context.
      ii. Active (or Automatic): The application or user context adaptation automatically adapts to the environment context. This may be of push or pull type.
      iii. Control: the user context may be used to control the environment.

4. Context Management

   a. Discovery: directory services enable context sources, stores and users to be registered and discovered.
   b. Storage: of context data into some data resource, Storage may include history-based organization of the stored context and support for fast query-based retrieval.
   c. Sharing of environment and goal contexts so that they can be distributed and accessed.
   d. Access control: protects the privacy of any context information that can be linked to human owners.

In this architecture, the mediation or adaptation of the application or user context to the environment context is performed in a separate sub-system (middleware) to the application itself so that elements of the mediation can be reused across applications. In contrast, a context-aware system can also be designed so that the adaptation is part of the application, because the application has specific requirements for the adaptation, rather than being factored out into middleware.

Different processes and algorithms may need to be used for context-awareness. Environment context determination may require information from multiple sensors, e.g., for positioning, to be aggregated. Determining a user context may require aggregation and composition of multiple contexts from multiple sources in possibly multiple application domains at multiple levels of abstraction. Uncertainty reasoning, e.g., there is only a probability that it may rain and that arrangements at an outdoor event need to be changed, and conflict resolution, e.g., the diary indirectly says that someone should be present in a room but a door entry authentication system has indicated that no one has entered the room. This may be needed when multiple heterogeneous contexts are combined.

Adapting a user context to an environment context may require search algorithms (Section 8.7.1) to search a very large context data space to find particular matches. Context matching may also need to use complex (semantic) metadata models (Section 8.4) and mediation processes (Section 9.2.2) that are able to undertake matches in heterogeneous context spaces.

More specification details and examples of these processes and algorithms are given with respect to the sub-types of context-awareness: mobility awareness, spatial-aware (anywhere), temporal-awareness (anytime), human user-awareness and ICT awareness.

7.2.7 Challenges in Context-Awareness

Some of the key challenges in modelling context have been identified by Henricksen *et al.* (2002), Strang and Linnhoff-Popien (2004), Table 7.3 extends this set. Of the challenges in Table 7.3, challenges 2 and 6 are highlighted. The combination of automatic context adaptation with incorrect, incomplete or imprecise context determination can lead to many false positives and false negatives for user contexts, producing unsatisfactory user experiences, with users preferring a manual system without automatic context adaptation. Some possible solutions are given in Table 7.3 – some of these will be quite challenging to deploy. For example, maybe the system could give feedback as to why a particular user context is expected to hold but this dialogue may distract users from their tasks. Further, the inner working of algorithms may need to be exposed but these may require specialist knowledge to understand. In addition, the sheer volume of context information that may be collected, i.e., challenge 5, may be costly to administer and may contain much data that is of very little significance.

Table 7.3 The main challenges in modelling contexts

Challenge	Causes	Possible solutions
1. User Contexts may be incorrectly, incompletely, imprecisely determined or predicted, *ambiguous*	Implicit observations of user contexts may be incorrect, incomplete or imprecise	Use context composition to improve context accuracy
	Users may provide faulty information when explicitly asked	Use iterative user interaction processes; give more explanation
	User contexts modelling from too little input data, over too small a time period	More accurate user contexts require time to tune. Use machine learning, and
	Context users may define their needs imprecisely, e.g., find X that is *near to* Y	simulation to improve the determination and prediction of user context over time
2. Environment Contexts may be incorrectly, incompletely, imprecisely defined, determined or predicted	Delays can occur in exchanging dynamic context information	Use context composition to improve context accuracy
	Disconnections or failures can mean that the path between the context producer and the consumer is cut, i.e., part, or all, of the context is unknown	Handle context uncertainty Partial contexts may need to reason about, to predict the part that isn't readily accessible Further context reasoning may be needed to define a context
3. Contexts may exhibit a range of spatial-temporal characteristics: local effects in time and space	Some types of context are invariant whereas others types are highly temporal variable Context generators may be mobile or may vary across regions (spatial variance)	Define how often to acquire each context based upon its variability Define interpolation and aggregation to combine values Handle context unpredictability

(continued overleaf)

Table 7.3 (*continued*)

Challenge	Causes	Possible solutions
4. Contexts may have alternative representations	Degree of post-processing of sensor data	Support multiple representations of context and be able to mediate between them
	Multiple representations exist for same context in different forms, at different levels of abstraction and perspective depending on usage	
5. Contexts may be distributed and partitioned, composed of multiple parts that are highly interrelated	Context relationships maybe evident or less obvious. They may be related by derivation rules that make a context dependent on other information (context)	Use a rich conceptualisation to represent context
	Composites contexts may need to be partially validated as all parts cannot be always be accessed	Partial contexts may need to reason about to predict the part that isn't readily accessible
6. context-awareness may generate huge volumes of data	Large complex environments may be studied, many sensors may be used	Filter context information before storing
	Focus is on archiving context data rather than on applying the context	Use appropriate search and archiving techniques for storage
	Many contexts may be available but the users are not clear which ones are useful to apply	Use data mining techniques to analyse data
		Use active associations to reveal contexts, else keep them hidden
	Many remote events may displace fewer local events	Design locality awareness to decrease with distance
7. Context sources and local processes may need to be embedded in low resource infrastructures	Context sources such as sensors are highly distributed, mass produced and embedded in cheap low cost, low resource embedded systems	Consider the trade-off in the expressivity of context messaging against the resources needed to handle such messages
8. Context use can reduce the privacy of humans	Contexts are often naturally linked to humans to be of use	Information Security is needed to protect context information
9. Awareness of context signals and shifts can overload users or distract users	The autonomy and awareness of the status many individual system components and the urge to self-maintain and upgrade	Ensure context shifts if automated occur safely and do not disrupt users
	If context-shifts are automatic, user control is reduced, use can be disrupted	

Challenge 6 highlights the glut of data and metadata which can be generated by context-aware systems. One useful technique seems to be to support the ability to set *conditional activated contexts* or *future context-awareness* but to keep these hidden from users and applications, using context-aware middleware services, until the conditions that activate them become true. An example of this was given by Mann in Section 2.2.4.5. In which a particular user shopping list context only becomes activated when a user is detected to be in the proximity or entering a particular shop. The conditional activated context concept also allows contexts to be set by one process and consumed by another process, and the setting versus activation of the context to be temporally (delayed) and spatially separated.

7.3 Mobility Awareness

7.3.1 Call Routing for Mobile Users

One of the earliest examples of context-aware application is the *Active Badge Location System* of Want *et al.* (1992), begun in 1989. This application was intended to be an aid for a telephone receptionist before mobile phone networks became widespread so that employees could be contacted when they were away from their desk or home location. Once a person was located, phone calls could be forwarded to a desk-phone closest to where the person was located. The system uses an awareness of the location of users to route calls through to their nearest fixed line phone Sensors were mounted in the offices, common areas and major corridors to detect the signals from active badges that were worn; of course, some private areas where people were free from being monitored were also defined. The system provided a table of names against a dynamically updating field containing the nearest telephone extension and a description of that location – it is an early example of digital call routing to mobile users.[9] Want *et al.* (1992) did not specify the more general concepts of context and context-aware because they focused on location determination. Today, two decades later, many mobile applications now routinely use location awareness to provide additional types of location-dependent services.

7.3.2 Mobile Phone Location Determination

Basic mobile phone location determination involves determining which mobile phone transmitter and its area of operation (its cell), a phone is nearest to. Mobile phone users tend to be registered in a Home Location Register (HLR) database by a Mobile Switching Centre (MSC) that is maintained by a mobile network operator. A Visitor Location Register (VLR) database in a MSC is also maintained for a cluster of mobile phone cells within a location area (Figure 7.4). When users pass between areas, a cell notifies its VLR that the user is entering or leaving its location area. When a call is made by user B to user A, the call first queries the VLR of user B to see if it knows the location of user A. If it is not there, a call is made to the user A's HLR as each VLR that A visits will notify A's HLR. A's HLR will then notify B's VLR which VLR A is in. A type of query called paging can be used to locate the particular cell user A is in (Pashtan, 2005).

Emergency call statistics show the increasing predominance of calls by wireless phone users. However, unlike a fixed location phone, mobile phone users can sometimes provide inaccurate positioning information. In the late 1990s, a U.S.A. government mandate U.S E911, initiated programs to set a minimum location accuracy for mobile phone users in emergency situations. Similarly in the early 2000s an equivalent mandate in the European Union, EU E112, was initiated (Pashtan, 2005). While mobile phone networks need to routinely determine the phone location within a cell[10] in order to route calls (Figure 7.4), enhancements, to mobile phones (hardware and software) or to the network or to both, are needed to support more accurate location determination within a cell. GSM mobile phone network operators can use

---

[9] This early mobile context-aware application is essentially a type of local digital mobile or cell phone call routing. Although global mobile phone systems became available in the mid-1980s, these were analogue or 1G systems. Digital or 2G mobile phones only started to became available in the early 1990s (Chapter 11).

[10] Cells may be about 150 metres or less in size in urban areas but about 30 kilometres in rural areas. TDOA, GPS and hybrid techniques can improve this accuracy to 30–50 m but GPS does not work indoors.

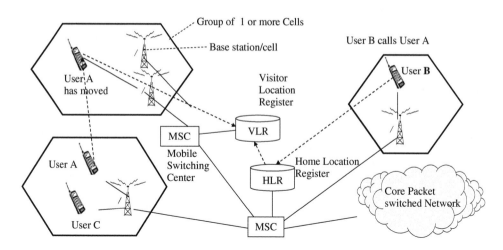

Figure 7.4   Location determination in mobile networks

three different improved position determination methods based upon enhanced observed time difference of arrival (TDOA) between the handset and multiple base-stations, GPS (support in handsets, for outdoor use) and hybrid techniques such as A-GPS (assisted GPS) which combines GPS with network information.

7.3.3 Mobile User-Awareness as an Example of Composite Context-Awareness

Mobility context-awareness is an example of composite context adaptation. First, spatial awareness is used to adapt activities with respect to their locality. Spatial awareness may not require any remote data service access other than position determination as the position could just be used to select static data that is held locally, e.g., maps used in navigation systems, see Section 7.4. In modern digital phones, cameras and other recording devices, location determination that allows users to automatically annotate not just when but also where photos were recorded will becomes common. Information retrieval from remote sources can be personalised to users' preferences. ICT context-awareness (Section 7.5) is useful for mobile users as it adapts remotely accessed content so that it fits better the characteristics of mobile access devices and better fits the bandwidth available in the local wireless access loop. Content adaptation can also occur with respect to personal preferences.

7.3.4 Tourism Services for Mobile Users

The CRUMPET, Creation of User-friendly Mobile services Personalized for Tourism, system (Poslad et al., 2001) is an example of a composite context adaptation application. In this system, tourism information services such as maps, routes and sight recommendations can be adapted to a spatial context that pertains to the current location, the personal context of a service, the network context and the terminal context, see Figure 7.5.

The CRUMPET System Context-aware Architecture is based upon a Multi-Agent type distributed system design. A typical user's interaction (use context) for a particular service triggers the composite

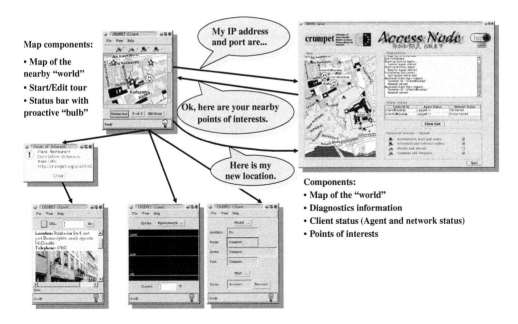

Figure 7.5   A composite (location, person, terminal and network) context-aware application

context (of network, terminal, location and user) adaptation (Figure 7.5). In the CRUMPET system, a, particular ordering of the context-aware adaptation is used as follows. A user's access terminal profile of memory and display capabilities is exchanged with the system during the start of a user session. Localisation is, for example, used twice (Poslad *et al.*, 2001). First, the current position of a user can be used to define a user's request and to further filter the relevant information. Unless the relevant location is specified explicitly, the user gets information relevant for his or her current spatial context. Second, a user's movements within a region can indicate their interests. If, for instance, a user visits a number of old churches, then he or she is probably interested in churches and perhaps also other historic buildings in this town, like an old city hall. Users generate a lot of potential events of interest as they move. This can be exploited for user modelling and to detect and anticipate relevant user interests. Hence, the combined location and personal model context can be used to produce a map of things of interest at a location. This is an example of environment context composition in which one type of context (location) may be used to determine another type of context (personal preferences) based upon a user context policy. Finally, the network profile based upon monitoring the performance of the local mobile terminal to access node, the content, e.g., a personalised, location-aware map is adapted to the terminal and network profile respectively.

7.4 Spatial Awareness

According to many, location or spatial awareness[11] is one of the main drivers for mobile services (Marcussen, 2001). This enables services to be remotely accessed anywhere, i.e., to enhance the

---

[11] Spatial-ware is taken to be more a general and inclusive term than location-aware. Whereas spatial-aware is aware of a specific 3D space and possibly its structure, location-aware focuses on an 1D point in that space.

Table 7.4 Some types of SAS application with illustrative examples

Type of SAS application	Examples
Navigation	Where is the nearest petrol station? How do I get to X?
Context change	Long traffic queue ahead, go next right on the A1 to avoid this
Query location context	What is the speed limit on this road here? When does this building open?
Personal Emergency	Medical: I'm having a heart attack! Roadside: My car has broken down!
Enterprise Asset Tracking	Why does it always take twice as long to deliver to that customer? Why is our delivery van deviating from the standard route?
Public Asset Tracking	Transport: Display the estimated train or bus arrival times based upon their current location determination
Personal Asset Tracking	Tell me if my child strays beyond the neighbourhood. Where has Smudge the cat gone this time? My car has been stolen, where is it?
Location/time based offers	Free calls on your mobile phone, while you are in location X; special offers on food, etc.
Location and time synchronisation	Users can share & synchronise their context (location, activity, mood) publicly and privately with buddies; arranging an ad hoc meeting nearby

experience of mobility. Spatial Aware Systems (SAS) are commonly referred to as Location-Aware Systems (LAS) but the term spatial-aware is a more accurate term as this type of system is often not just aware of locations but needs to be aware of spatial features that can range from simple point location features, through to two-dimensional polygon-type areas, to more complex irregular three-dimensional spaces that vary with time and their spatial relations.

A passive type of SAS presents only a relative or absolute location context to the user. A more active SAS automatically adapts the application output to the location context, e.g., a 'SatNav' application that uses a satellite-based global positioning system can show updated current positions and routes on maps. A more active system can seek to use feedback to control a local environment, e.g., to vary the human population density in an area with time, e.g., by making available attractive commercial deals specific to an area. There are many examples of SAS applications, see Table 7.4.

A typical Spatial-Aware System consists of spatial (environment determination) context determination, user context determination, spatial-based adaptation of the user context and management of the spatial context data stored in a GIS or Geographical Information System.

7.4.1 Spatial Context Creation

7.4.1.1 Spatial Acquisition

There are a variety of methods are used to capture structured or vector spatial data in the GIS. A digitiser can produce vector data as an operator traces points, lines, and polygon boundaries on a map. A scanner produces a raster map that could be further processed, e.g., using edge detection to produce vector data. Survey data can be entered directly from digital data collection systems; survey instruments or other remote sensors such as satellite remote sensing provide another important source of spatial data. Satellites may use different types of sensors to passively measure the reflectance from parts of the electromagnetic spectrum or radio waves that are sent out from an active sensor such as radar. Remote sensing collects raster data that can be further processed to identify objects and classes of interest, such as land cover. Analogue aerial photographs can be digitised and features extracted. Stereo pairs of digital photographs allow data to be captured in two and three dimensions, with elevations measured directly from a stereo pair using principles of photogrammetry. As high quality digital cameras become cheaper, these can be used to capture spatial objects directly. Location determination systems can also be used to determine location coordinate. But these need to be used in combination with annotations, also called geo-attributes or metadata, linked to location coordinates.

7.4.1.2 Location Acquisition

There are several basic location acquisition methods based upon triangulation, lateration, scene analysis and proximity (Hightower and Borriello, 2001). Triangulation is one of the most universal positioning techniques. This uses the geometric properties of triangles to calculate locations of objects based upon distance (lateration) and angles or bearing (angulation), see Figure 7.6. Lateration determines the location of point O with respect to three reference points A, B and C. Measuring R_A determines the point to be on circle A, measuring R_B determines the points to be where one of the two points where circle A and circle B intersects. Measuring a third point determines it to be location O. An angulation algorithm can be used to determine the coordinates of position of location O by determining the line of sight angles at two points A and B and from knowing the distance between A and B.

Lateration and angulation are often usefully combined in order to know not only where an object is but from which orientation it is being viewed. Lateration uses two main techniques to determine distance in practice based upon attenuation and Time of Flight (TOF), also referred to as Time of Arrival (TOA). Attenuation, also referred to as Received Signal Strength (RSS), estimates the

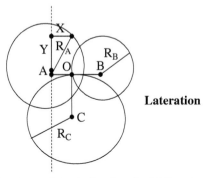

Lateration

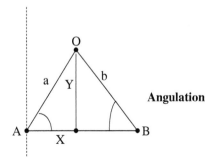

Angulation

3 Equations to determine location of point O w.r.t. known locations A,B, and C on a 2D plane

$R_A{}^2 = X^2 + Y^2$
$R_B{}^2 = (X - (AO + OB))^2 + Y^2$
$R_C{}^2 = (X - AO)^2 + (Y - OC)^2$
Use substitution to get X and Y

$X = (R_A{}^2 - R_B{}^2 + (AO + OB)^2)/2\,(AO + OB)$
$Y = (R_A{}^2 - R_C{}^2 + AO^2 + OC^2)/2OC) - AOX/OC$

If distance AB, angles at A and B are known then X and Y can be determined using basic trigonometry

Sin A = Y/a
Sin B = Y/b
Y = a * Sin A = b * Sin B
Cos A = X/a
X = a * Cos A = AB – b * Cos B

Figure 7.6 Using lateration to determine the location of point O with respect to three reference points A, B and C (left). Using angulation to determine the location of O with respect to two angles for the *line-of-sight* from two points A and B and knowing the distance between A and B

Radio Frequency signal strength at the receiver knowing the attenuation of the signal as some function of distance and signal transmission strength, e.g., $1/d^2$ for the simple free space case where d is the distance between the sender and receiver. However, distance and timing measurements are complicated in practice by variable attenuation, due to moisture in the air, etc, multi-path effects, reflections, spot interference and by not knowing the time of transmission accurately. An example of a RSS technique is the use of attenuation measurements of WLAN (Saha *et al.*, 2003) that can be performed in mobile devices with WLAN access. They report an accuracy of 1–7 metres depending on the signal processing used and the number of and distribution of WLAN access nodes or transmitters. In spaces such as indoors and certain outdoor areas where there are many obstacles that can cause reflection, refraction and multiple path effects, the attenuation is non-uniform, causing it to be a less accurate method than TOA. However, some TOA systems such as GPS may not always be accessible because they require an unimpeded path between transmitters and receivers, see below.

An example of a TOA system is a satellite-based *Global Positioning System* or *GPS*. TOA measures the time the signal is sent versus the time it is received. The distance d between the sender and receiver can be estimated if the signal propagation is known (distance d = time * signal propagation speed). This assumes accurate clock synchronization and that the sender knows the time of transmission but this may not be practical. Time Difference Of Arrival (TDOA) measurement at two or more receivers or sent from two or more senders can be used to provide improved estimates of the difference in distances between the senders and receivers. Trilateration uses absolute measurements of time-of-arrival from three or more sites. It is a method of determining the relative positions of objects using the geometry of triangles in a similar fashion to triangulation. Unlike triangulation, which uses angle measurements (together with at least one known distance) to calculate the subject's location, trilateration uses the known locations of two or more reference points, and the measured distance between the subject and each reference point. To accurately and uniquely determine the relative location of a point on a 2D plane using trilateration alone, generally at least three reference points are needed. Signals can be measured with respect to multiple transmitters to correct for any variable attenuation.

An example of a triangulation system is the *VHF Omni-directional Ranging* (*VOR*) system used for navigation by commercial aircraft. This uses fixed ground transmitter stations that transmit two different signals: an omni-directional station identifier signal and a rotating transmitter signal that is swept in 360 degree arcs so that the signal is in phase due North and out of phase due South. Using phase shift measurements from two or more VOR stations, aircrafts can determine their position. VOR is supplemented by GPS for aircraft navigation.

Scene analysis refers to the use of abstractions or markers of a terrain scene that can be cross-correlated with previously recognised and stored markers to infer the location, e.g., a vehicle can abstract features in a view of the terrain such as a church spire, canal or railway line to orientate itself. Someone with a mobile phone and camera can take a picture and transmit it to a remote service for recognition. The advantage of scene analysis is that no active signal transmission network is required which requires power and which can reduce privacy, rather it relies on passive observations.

Proximity analysis makes use of short-range transceivers at fixed positions to determine when the appropriate networked objects such as RF IDentifier (RFID) Tags come within range. Proximity LAS by definition in the simple case determines a relative rather than an absolute position. The use of RFID systems for proximity-based location of RFID tagged assets such as vehicles, people and goods is now fairly common (Section 6.2.4).

Selecting a location determination technique depends on several factors such as location accuracy, range, availability, coverage, cost and limitations (Hightower and Borriello 2001). Hazas *et al.* (2004) has also classified location acquisition devices with respect to accuracy of determination and the degree of deployment grouped into research labs, customised deployment and consumer deployment. GPS, Bluetooth, infrared and WiFi are the most widely deployed devices but have relatively low accuracy of 10-100 Meters. Devices with a short range of RF transmission such as Bluetooth and infrared have an accuracy of 1–100 m and are often used for proximity location determination. A major issue with TOA and TDOA systems is that although they have a global range, they require a line of sight with three to four transmitters depending on accuracy, probably restricting use to outdoor use where there are no high-sided obstructions. Hybrid systems or assisted systems seek to combine strengths and minimise weaknesses of several location determination systems. For example, a terrestrial cellular network could be combined with GPS in an assisted Global Positioning System (A-GPS) and support both an improved, accuracy of several metres or better in an open air environments with an accuracy of 20m indoors (Djuknic and Richton, 2001) or it can be combined with WiFi trilateration techniques. Van Greunen and Rabaey (2005) give a comparison of location acquisition algorithms based upon scalability with respect to number of network nodes and communication energy.

7.4.2 Location and Other Spatial Abstractions

A location coordinate in itself is often not so useful, it is too low level: it is the location in relation to the location context that is useful and gives it its meaning. The location coordinate and its associated geographical area within a bounding box or region and any geo-assets of interest in that area, e.g., petrol station, restaurant, are needed. In forward-tracking, the relation of the current coordination to the end coordination, the future route is of use, e.g., how far away the destination is and how to get to it. In backward tracking, the relation of the current location coordination to the start coordination and to past routes is of interest. It is also challenging to automate application-specific types of context, e.g., a building can function as a restaurant, sports hall, conference room, etc.

An abstraction or service, such as a Geospatial Information System (GIS) service, is needed to answer spatial queries such as 'Is there a type of service X within 1 km of here?'. GIS services represent real-world objects such as roads, land use and elevation and associate these with digitised spatial data. Real-world spatial objects can be discrete objects such as buildings or continuous fields such as rainfall and elevation. GIS data consists of the geometrical object, e.g., point, line, polygon, etc; the geo-attributes or metadata such as ID, address in terms of a ZIP or postcode, postal addresses, mobile phone cell location, x, y, z, etc) and types of feature, and associated attributes such as size and colour.

7.4.3 User Context Creation and Context-Aware Adaptation

7.4.3.1 Cartography: Adapting Spatial Viewpoints to Different User Contexts

A major SAS application is cartography (Clarke, 2001), the use of maps to show different views of spatial relationship between regions and points. Maps can show a selection of restaurants within a region or show a route between a location and a destination. There are different types of maps for pedestrian use, road and transport use and for showing boundaries and land use (cadastral maps), e.g., to indicate potential sites to site new commercial premises. Maps are a reduction and abstraction of the geographical physical space that we inhabit that describe locations, regions, and their attributes. Cartography or map applications require the use of specialised data management systems to store, manage and search large spatial structures in GIS Systems (Section 7.4.4) and to process and link locations and regions to the attributes of interest (Section 7.4.3.2).

In contrast to raster or image maps that are represented as only one layer, vector maps can be stored as multiple layers, from simple to more complex spatial views of features such as points, line segments representing buildings, roads and rivers, to regions that contain collections of these. Maps can be dynamically created to show only the layers and objects of interest to simplify the spatial view for users. Software such as Geotools, an open source GIS toolkit (Garnet, 2007)) provides APIs to add, remove and manipulate layers to form a map that are retrieved from a GIS, e.g., using the Open Geospatial Consortium (OCG) Geographical Markup Language (GML). Maps may need to adapt to fit the resources of the display such as a mobile device and to support the download of the map over a wireless connection (on-demand) versus wired connection (pre-cache).

7.4.3.2 Geocoding: Mapping Location Contexts to User Contexts

Algorithms such as those based upon Geocoding (Ratcliffe, 2001) can be used to associate a usage context with a spatial context. Geocoding maps spatial locations (x, y coordinates) to and from street and postal addresses. Individual address locations that are points, rather than segments, can then be derived by interpolating, or estimated, by examining address ranges along a road segment. Individual addresses can then be associated with other metadata such as the type of businesses operating at a particular address to support queries about an instance of a business. Other algorithms are used to help with address matching when the spelling of addresses differ. Address information that a particular entity or organisation has data on, such as the post office, may not entirely match the query terms. There could be variations in street name spelling, community name, etc. Consequently, the user generally has the ability to make matching criteria more stringent, or to relax certain parameters. Semantic approaches could be used here to make the matching more automated.

7.4.4 Spatial Context Queries and Management: GIS

Different sub-system designs are needed for the different main functions of a spatial aware system: to determine the spatial context; to search the spatial context space and to match a spatial context to a user context; to store the spatial context and to exchange spatial context information. The determination of a higher-level query relationship, such as what is the next part of this route, or where are the petrol stations in this region, requires different kinds of algorithms than those used to determine location, mainly because the search space can be much larger and have many more spatial independencies that need to be evaluated. Spatial data structures may be very large complex structures, requiring sophisticated data search and data organization methods in order to optimise the associated computation and reduce the time it takes to fulfil search requests. Spatial contexts are stored and retrieved from a GIS.

Algorithms to support spatial data processing typically associate unbounded spatial structures such as line segments that represent roads, such as segments a–g in Figure 7.7, and irregular bounded polygon spatial structures that represent areas or buildings, such as Objects C and D (Figure 7.7) with bounding boxes that are arranged as hierarchies into various tree data structures such as quad trees (adjacent non-overlapping bounding boxes) and R Trees (overlapping bounding boxes) (Samet, 1990). A spatial query, e.g., determining the route from Object C to Object D, involves querying the R–tree to determine which bounded boxes the spatial objects of interest are in and then identifying the intermediate bounded boxes that contain the route segments, e.g., c to i in Figure 7.7.

The Open Geospatial Consortium[12] has proposed several conceptual ways for extending an SQL RDBMS to support the storage and queries of spatial data including defining operators such as

---

[12] OCG 1997, see http://www.opengeosapial.org/, accessed Dec. 2008.

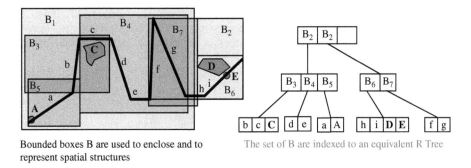

Bounded boxes B are used to enclose and to represent spatial structures

The set of B are indexed to an equivalent R Tree

Figure 7.7   Storing and indexing spatial structures in an R-tree to support efficient spatial queries

length, the number of points in a path, distance between spatial objects and is-right-of. Many relational databases include spatial extensions.

Data after entry usually requires editing, e.g., vector data must be made 'topologically correct' before it can be used for some advanced analysis; errors such as undershoots and overshoots must also be removed. In addition, there exist many ways to specify spatial point and more structures, e.g., over 100 different coordinate systems exist for positions. One of reasons for different coordinate systems is that there are different corrections for 2D projections of the Earth not being a perfect sphere in different regions, e.g., North American Datum 1983 (NAD83) works well in North America, but not in Europe. It is likely that measured location co-ordinates (e.g., from GPS system) and geospatial object (e.g., building or point in road) coordinates in GIS will vary across regions and will vary between a map representation. Newly acquired location data may require conversion before use.

7.5 Temporal Awareness: Coordinating and Scheduling

Time is often recorded as a core attribute in context-aware systems, it is inherent in human work and leisure activities. Freeman and Gelernter (2007) proposed in the Lifestreams project that time-oriented streams ought to be used for managing personal information rather than the static file name folders that are used in most information systems today. Many activities require spatial-temporal awareness. They need to be synchronised to a particular space and time, e.g., going to school, university, work, shopping, theatre, etc. Gardner (1983) regards the complex spatial-temporal awareness and problem-solving by an individual when using his or her physical body: to perform a complex surgical procedure, to execute a series of dance steps, or to catch or hit a flying ball, as an important form of human intelligence called bodily-kinaesthetic intelligence. Aspects of temporal awareness concerning temporal context creation, distribution, abstraction and adaptation of user contexts to the temporal context are considered in more detail in the following sections.

7.5.1 Clock Synchronization: Temporal Context Creation

Clocks and embedded timer devices need to be synchronised and resynchronised so that different users, services and processes can share a common time context. Timers may be configured manually because they are not network enabled and because there is no global absolute time. Time may be set manually using a proprietary local set-clock device interface by someone who often reads time from another non-authoritative clock source – this introduces human error. Periodically, device times need to be reconfigured when the timing circuits loose power or because these do not accurately keep the same time over a medium or long time interval and drift because of temperature fluctuations around the clock chip.

Clocks and timers that are network enabled can be automatically synchronised to external authorative time sources such as those based on Universal Time Coordinated (UTC) services. The *Network Time Protocol* or *NTP*[13] (Mills, 2003) defines both an architecture and a set of protocols for time synchronization that keep time accurate to tens of nanoseconds. NTP uses the *Intersection Algorithm* to enable NTP clients to select from multiple time sources. Multiple time sources are needed in case one fails. The intersection algorithm is essentially a type of agreement algorithm for estimating accurate time from a number of noisy time sources and then finding their intersection. NTP is used in conjunction with an organization hierarchy of primary NTP servers connected to atomic clocks. Secondary servers are synchronised to primary servers. NTP clients can configure their time to make a step change to or to make a gradual change to the accepted UTC time if their time differs. The latter method avoids major disruption to an ongoing time-sensitive process.

Van Greunen and Rabaey (2005) note that traditional time synchronization algorithms such as that of Mills (2003) achieve maximum accuracy without regard to the computation and communication costs and hence the energy expended by the algorithms. They discuss some synchronization algorithms based upon lightweight tree synchronization and algorithms used in sensor nets that trade off synchronization accuracy against utilising less energy and communication.

7.5.2 Temporal Models and Abstractions

Time can be modelled as a linear sequence that points from past to present across multiple parallel events and activities, the arrow of time, and then branches to different future times. It is also known that time varies for different users depending on their relative velocity difference to each other as defined in the Theory of Relativity. This effect is not significant for users moving at speeds significantly less than the speed of light (Hawkins, 1988) and hence is discounted here. Time may be modelled as an instant, as a period that represents the set of instances between two defined instants and as an interval or duration, a length of time that has no associated start or end instance. Instant events or activities with a very short period are defined as an instant whereas activities are often defined in terms of a start instant and an interval.

Temporal models are, in practice, complicated by the coexistence of multiple times. Multiple times exist and vary across different geographical regions to enable time to be in harmony with natural daylight (astronomic time) in different geographical regions. Time-zones corresponding to different geographical regions, oriented in an easterly direction, are standardised as offset times from a reference time such as UT1 or GMT (Greenwich Mean Time). This simplifies time-keeping for mobile users who travel and users who trade across time-zones. This is in comparison to earlier rail travel in the nineteenth century that could involve people travelling on trains operated by different train companies that kept different times because their end terminuses or head offices were situated in different time zones – quite confusing for the traveller. Different syntaxes for date and time-stamps exist, e.g., a year-month-day-hour-minute versus day-month-year-hour-minute with or without a time-zone offset format, etc. could be used. Hence translators are needed to convert between different date and time formats.

---

[13] NTP represents the longest-running, continuously operating, distributed application in the Internet that started in the early 1980s.

7.5.3 *Temporal Context Management and Adaptation to User Contexts*

Given a set of tasks to perform (the user context), a set of resources to use and a set of time constraints (the temporal context), the objective of task scheduling is to allocate times and resources to user tasks. Task scheduling is simplest when tasks are totally ordered, they start at predefined anchor times, they take predefined time-intervals to complete, they are non pre-emptive such that once started they are not interrupted and resumed, and they use reusable resources with no resource constraints or resource sharing, Simple scheduling can involve deriving a personalised schedule that is a subset of another schedule known a priori, such as a university course schedule, broadcast multi-channel TV entertainment schedule and travel timetable. A simple scheduling process starts with an initial request to identifying tasks and their time constraints which are known a priori, then allocate them to resources and time-slots then executing the schedule. When sub-sets of tasks are to be selected from a schedule known *a priori*, time can be used as a single-dimensional index to store and, together with preferences, to form a composite constraint to select tasks (Figure 7.8). Allocation of tasks to human and other resources can be abstracted to form a graph of task nodes. Paths can be constructing between nodes to represent the schedule. A partial ordering of tasks occurs around those that have specified anchor times and hard time constraints. Other approaches include considering tasks as a history of cause and effect.

The scheduling of tasks is harder when there is variability and uncertainty concerning the time constraints, when tasks can be pre-empted and where there are task resource and task coordination constraints, when resources are consumable and when the semantics of a temporal event vary.

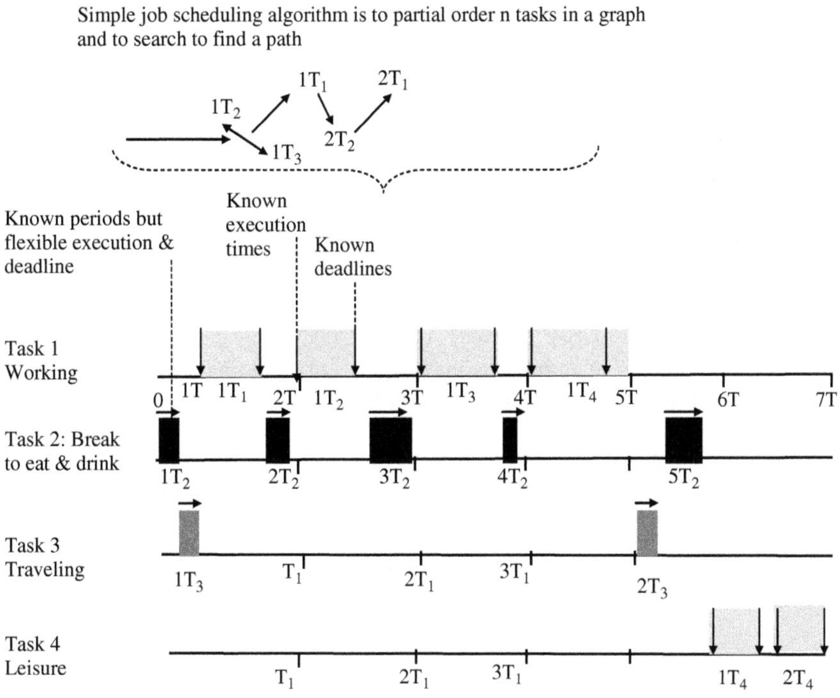

Figure 7.8 Simple task scheduling for non pre-emptive tasks with execution times, deadlines and periods known *a priori* without resource restrictions

For example, commerce may distinguish different transaction times according to the time an item is ordered, the time an item is paid for and the time it was delivered. Generally, a time-stamp is stored as an attribute of any environment or user context, i.e., the value of this environment variable is valid at this time-stamp.

7.6 ICT System Awareness

To support the vision of ubiquitous services delivery over heterogeneous networks and access via heterogeneous computers and terminals, service delivery is improved if it takes in account the characteristics of the network and the receiving user's terminal. Otherwise, content may be delivered that does not display well or even at all, or that takes a very long time to be delivered. Service adaptation should normally be transparent to the majority of users because users may not define their system characteristics correctly. For example, communication services for mobile users involve the use of automatically configured data routes. However, for particular services, management policies may restrict automatic adaptation. For example, banking and secure messaging services may need to be more restricted over relatively unsecure networks.

7.6.1 Context-Aware Presentation and Interaction at the UI

A User Interface or UI facilitates presenting and entering information by human users, supporting information queries to specific tasks, including instructions for entering information, as well as presenting the response. Universal content access entails content access via a proliferation of interactive devices with diverse capabilities. These may range in size, weight and mobility. They may have different display capabilities such as screen resolution, size and colour depth. They may use different forms of input, including different types of keyboard, pointer devices, speech and gesture input (Section 5.2).

7.6.1.1 Acquiring the UI Context

The UI context can be defined in a UI device profile. There are several different specifications for representing the UI profile. W3C has defined CC/PP the Composite Capabilities / Preferences Profile: Structure and Vocabularies 2.0 (W3C CC/PP, 2006). This defines a client profile data format, and a framework for incorporating application and operating environment-specific features including the terminal hardware, the terminal software and the terminal Web browser. CC/PP does not define how the profile is transferred, nor does it specify what CC/PP attributes must be generated or recognised or what context mark-up language is used. CC/PP is designed for use as part of a wider application framework. CC/PP is represented in RDF/S and an XML serialization. An example CC/PP device profile is shown in Figure 7.8. Strang and Linnhoff-Popien (2004) discuss the expressivity of CC/PP Version 1 and some attempts to enhance its expressivity. Several standards groups are actively developing standards related to Web service access on mobile devices.[14]

---

[14] See, for example, the W3C Mobile Web Initiative, http://www.w3.org/Mobile/, accessed 2008, which includes device profile specifications. It is not clear how backward compatible these specifications are with CC/PP.

7.6.1.2 Content Adaptation

A process framework is needed to adapt content to UI profiles. Most of the focus for presentation adaptation has focussed on adapting content designed for decimetre sized screens and for small centimetre sized displays. Adaptation of content designed for small screens to be displayed on large screens, e.g., projectors, is simpler and could involve interpolation and smoothing to calculate additional intermediate pixel values to smooth out jagged edges when display pixels groups get magnified. There are two dimensions to content adaptation: transformation of the created content representation to a different one used in the access device, adaptation of the interaction, adaptation to use a particular device display convention and adaptation of the content itself. Each of these is considered in turn.

- *Transformation of the created content representation to that used in the access device*, e.g., a map-defined GML is transformed to an image for display. Content that is adapted can range from being passive, to being active because it contains little scripts of code in an application specific program that generates output. Much research and development has taken place to develop Web-based content including content-based script languages that can span wireless mobile devices and wired PC type devices. Pashtan (2005) discusses the variety of content languages representations used for mobile devices derived from three bases: Hand-held Device Markup Language (HDML), HyperText Markup Language (HTML) and Compact HTML (cHTML). Different mobile devices will support different Web Browsers that implement support for particular Web content languages. The continued evolution and variety of pervasive devices being developed probably require multiple content representations.
- *Adaptation of the interaction*: Efforts to standardise content languages have largely focused on image and text layouts. W3C has created a Multimodal Interaction Activity whose mission is 'to allow users to dynamically select the most appropriate mode of interaction for their current needs, including any disabilities, whilst enabling developers to provide an effective user interface for whichever modes the user selects. Depending upon the device, users will be able to provide input via speech, handwriting, and keystrokes, with output presented via displays, pre-recorded and synthetic speech, audio, and tactile mechanisms such as mobile phone vibrators and Braille strips' (W3C MMI, 2002).
- *Adaptation of the content*: e.g., to display a large map on a small screen. This often involves more than simple scaling because vital detail may be lost when content is reduced. A common approach when detail needs to be retained in content that is too large to display on a small screen is to split the content into multiple screens and to support techniques to navigate between these such as stacking windows or scrolling windows. Text display could reduce full text to only displaying the title or a summary on a mobile device.
- *Adaptation to a particular presentation style*: to adhere to the layout and the presentation of a page for a given device profile, e.g. put the menu on the right for desktop landscape screen and on the top for a portrait screen. The navigation style for a particular device profile may also need to be adhered to. The presentation style may be in part determined by the presentation representation.

Adaptation to different heterogeneous terminals should preferably automatically adapt content to the terminal capabilities. There are two main approaches to this:

- *Lowest Common Denominator (LCD) approach*: content is created that can be used on a few categories of devices that cover a large number of devices. Each device in the category supports a lowest common denominator profile for that category. For example, J2ME in 2005 currently defines two categories (configurations) of pervasive devices. A Connected Device Configuration (CDC) supports constantly connected network devices and a Connected, Limited Device

Configuration (CLDC) supports personal, intermittently connected mobile devices. CDC targets devices above 2MB of both RAM and ROM. CLDC Mobile Information Device Profile (MIDP) targets personal devices, with a screen size of 96×54 pixels, a display depth of 1 bit, one- or two-handed keyboard, a touch-screen input device, 128 KB non-volatile memory and 8 KB non-volatile memory for application-persistent data. The LCD approach sacrifices richness of some devices for the limitations of others. This can lead to non-intuitive use for some complex interactions in order to keep interactions simple.

* *Transcoding of content to adapt it to specific types of access devices*: this transforms content from one form to another via clearly defined mapping functions. Here the terminal input and output capabilities must be distributed to services that require access to them. Then the content may either be statically or dynamically adapted to the terminal capabilities. Note if the terminal capabilities are incorrectly identified, e.g., by the human user, non-intuitive, unusable interactions and unusable content presentation can occur. Transcoding can be static (content is prepared for particular target devices in advance) or dynamic (content is dynamically prepared for particular target devices).

The ease with which content can be transcoded depends upon how annotated the structure of the content is. Annotation can be used as inputs for automated rules to convert content. HTML content contains no standard annotation of the structure of the content whereas database content does. A typical approach is to convert content including active content into an intermediate canonical format, e.g., Java Bytecode, .Net Common Language Runtime (CL) code or to the W3C platform and language-neutral Document Object Model (DOM) used to define the structure of an XML document. Three different sub-processes are then involved to display it: *Manipulation* which involves filtering, ordering and sorting the parts of a document that will be displayed; *Transformation* of elements and document structure with respect to a particular user and display context into another format, e.g., using the XSLT (Extensible Stylesheet Language Transformation); *Rendering* of elements that are suitable for a particular display, e.g., using XSL the Extensible Style Sheet Language. Pashtan (2005) describes some good examples of the use of XSLT and XSL for transforming content for mobile devices.

In an ideal world, content should be agnostic to the delivery context. An author creates content once and publishes it in many ways across a range of devices including desktop devices and other mobile devices. It is currently still very difficult, if not impossible, to efficiently design text and image-based content and interaction that can be viewed on both desktop and mobile devices, let alone across the greater range of pervasive devices that support multimodal interaction.

7.6.2 Network-Aware Service Adaptation

User mobility requires some services to know the address of the user's terminal as the user roams in the network in order to route data to the user's new location, see Section 7.3.2. Mobile users are often situated in an environment, where there may be multiple data communication networks available. Because of the variety in the different network types and characteristics of the networks, the Quality of Service (QoS) may change dramatically based on the network that the user is currently connected to. Users on the move may come across variations in networking conditions that can sometimes be quite dramatic. One of the strategies used is that the wireless link is monitored and the content can be adapted to a lower fidelity when a lower bandwidth link is detected. For example, the round trip time can be monitored and it can be used as an indicator to optimise the images transmitted to match the bandwidth available. Content can be adapted by using degradation mechanisms to reduce the quality of the content. Context-awareness could also try to opportunistically procure extra resources in order to maintain the quality of the content exchanged (Couderc and Kermarrec, 1999).

A service that is aware of the characteristics of the physical network is called *underlay-network* aware (Chapter 11) and can adapt its transfer of content to the physical network characteristics, e.g., its bandwidth. Raz *et al.* (2006) have analysed the requirements posed by context-aware services on the network layer. The predominant IP-based network layer design in current use supports a simple unreliable packet forwarding mechanism combined with more complex routing protocols and the end-to-end TCP transport protocol. This requires enhancements to support more flexible context-aware QoS delivery. To support this type of adaptation, firstly, some form of network context needs to be defined and acquired, i.e., the local network state needs to be extracted from the various network elements. Second, local network states needs to be composed to produce a global or end-to-end state. Third, these types of service-aware network models need to be supported. This can involve distributing and accessing network state information via open network APIs. Finally, active network adaptation models, in contrast to passive network adaptation models, can be designed to allow networks to be reconfigured by services to better justify their needs. For example, in a programmable network, networks can be reprogrammed by injecting code into them (Section 11.7.8.3).

Although service adaptation to networks can be end-to-end, in the application layer, network configurability is limited. A more flexible type of network configuration is to modify data transfer and routing in the intermediate network nodes between the sender and receiver. An example of local control of a network by applications is to set up filters for certain kinds of data traffic to block disruptive traffic or to change the QoS priorities. Because of the lack of support for application-specific routing in standard network elements at the network layer, this is supported in intermediate nodes at the application layer. Examples of such service-oriented network models include programmable networks, content-based networks and overlay network (Section 11.7.8). Raz *et al.* (2006) describe a system architecture for network context-aware services based upon use of a programmable model. The heart of the system is the service layer that supports context-aware service creation and policy-based service management. The service layer uses an underlying active application network platform to control and access an IP-based network. This system has been applied to several scenarios such as a delayed-write scenario (Section 3.3.3.4) that waits until a high-bandwidth link is available before transferring a large file; prioritised use of a network in a medical emergency, and a virtual conference where the system evaluates the QoS of the network connections of each of the participants before advising participants how best to configure their network connections for participation.

A central design issue for network awareness is how network contexts are accessed and distributed. Ocampo *et al.* (2005) define the concept of context flow in which a flow defined as a network data stream that adheres to a defined network protocol is described using a (network) flow context. The flow context tag information can be pushed together with the flow itself and can be received by any intermediate network elements that can act on it. This has the advantage that flow context can influence short-lived network flows.

Whereas context-awareness can improve information filtering, it can also do the opposite. For example, computer hardware and software are increasingly aware of their version in relation to the latest released version. It is becoming routine for the hardware manufacturer, operating system vendor and every major application vendor on a daily basis to distract user tasks by signalling the updates automatically. If updates are automatic, user control is reduced and systems can automatically reboot without closing down applications properly. What's more these updates can rarely be prevented, they can often only be delayed. The autonomy and awareness of the status of many individual system components and the urge to maintain these can be very distracting for active applications and users. A second major problem is that the context can be incorrectly determined and any resulting adaptation can be incorrect, e.g., early SatNav systems on occasion routed vehicles down dead-end roads. Finally, the expectations of users must be realistic.

Acquisition of non-deterministic environment contexts, e.g., the weather, or non-deterministic user contexts such as what a user's leisure activities will be, will inherently make automatic context-awareness semi-deterministic or non-deterministic.

EXERCISES

1. Compare and contrast a general distribution system, an embedded control system, a sensor-based system, a general context-aware system, a location-aware system and a personalised system.
2. Outline the design of a simple state-based model for context-awareness based upon current context state being driven by pre-planned state transitions to move it to the goal context state.
3. Discuss the privacy and ethical issues of location determination. Is it acceptable for a provider to be able to determine the location and context of a requestor or customer when an incident occurs, when desired, and vice versa?
4. Discuss the design issues in designing systems to be aware of multiple contexts for an application of your choice. In particular, consider how your design deals with: conflicting and overlapping context information; whether or not multiple contexts are adapted in one stage or into a sequence in which the order is in important; how uncertainty is dealt with and how different semantics for contexts are handled.
5. Discuss whether or not it is useful for systems to be aware of their external environment without being self-aware.
6. Compare and contrast different methods of location determination with respect to accuracy, indoor and outdoor use and local versus global location determination.
7. Discuss the motivation for ICT environment awareness for mobile users. Outline designs to support context-aware content adaptation to the characteristics of the access devices and its local network link.

References

Baldauf, M., Dustdar, S. and Rosenberg, F. (2007) A survey on context-aware systems. *International Journal of Ad Hoc and Ubiquitous Computing*, 2(4): 263–277.

Brown, P.J., Burleson, W., Lamming, M., *et al.* (2000) Context-awareness: some compelling applications. In CHI2000 Workshop on The What, Who, Where, When, Why and How of Context-Awareness.

Chen, G. and Kotz, D. (2000) A Survey of Context-Aware Mobile Computing Research. Technical Report TR2000-381. Available from http://citeseer.ist.psu.edu/chen00survey.html. Accessed November 2006.

Clarke, K.C. (2001) *Getting Started with Geographic Information Systems*, 4th edn. Upper Saddle River, NJ: Prentice Hall.

Couderc, P. and Kermarrec, A.-M. (1999) Improving level of service for mobile users using context-awareness. In *Proceedings of 18th IEEE Symposium on Reliable Distributed Systems*, pp. 24–33.

Davies, N., Cheverst, K., Mitchell, K. *et al.* (1999) Caches in the air: disseminating information in the guide system. In *Proceedings of 2nd IEEE Workshop Mobile Computing Systems and Applications*, (WMCSA 99), pp. 11–19.

Dey, A.K. (2000) Providing Architectural Support for Building Context-Aware Applications. Ph.D. thesis. Department of Computer Science, Georgia Institute of Technology, Atlanta, November 2000. Available online from http://www.cs.cmu.edu/~anind/context.html, Accessed June 2007.

Dey, A.K. and Abowd, G.D. (2000) Towards a better understanding of context and context-awareness. In *Proceedings of the Workshop on the What, Who, Where, When and How of Context-Awareness.*

Dey, A.K. and Abowd, G.D. (2001) A conceptual framework and a toolkit for supporting rapid prototyping of context-aware applications. *Human-Computer Interactions (HCI) Journal,* 16(2-4): 97–166.

Djuknic, G.M. and Richton, R.E. (2001) Geolocation and assisted GPS. *Computer,* 34(2): 123–125.

Freeman, E. and Gelernter, D. (2007) Beyond Lifestreams: the inevitable demise of the desktop metaphor. In V. Kaptelinin and M. Czerwinski (eds) *Beyond the Desktop Metaphor: Designing Integrated Digital Work Environments.* Cambridge, MA: MIT Press, pp. 19–48.

Gardner, H. ([1983] 2003) *Frames of Mind: The Theory of Multiple Intelligences.* New York: Basic Books.

Garnet, J. (2007) Geotools: the open source Java GIS toolkitUser Guide. Retrieved June 2007, from the Geotools Website: http://www.geotools.org/.

Gelernter, D. (1992) *Mirror Worlds, or the Day Software Puts the Universe in a Shoebox... How It Will Happen and What It Will Mean.* Oxford: Oxford University Press.

Gustavsen, R.M. (2002) Condor – an application framework for mobility-based context-aware applications. In *Proceedings of Workshop on Concepts and Models for Ubiquitous Computing,* Göteborg, Sweden.

Gutta, S., Kurapati, K. and Schaffer, D. (2004) From stereotypes to personal profiles via viewer feedback. In W. Verhaegh, E. Aarts, and J. Korst (eds) *Algorithms in Ambient Intelligence.* Dordrecht: Kluwer Academic Publishers.

Hawkins, S.W. (1988) The arrow of time. In *A Brief History of Time.* New York: Transworld Publishers Ltd, pp. 143–154.

Hazas, M., Scott, J. and Krumm, J. (2004) Location-aware computing comes of age. *IEEE Computer* 37(2): 9597.

Henricksen, K., Indulska, J. and Rakotonirainy, A. (2002) Modeling context information in pervasive computing systems. In F. Mattern and M. Naghshineh (eds) *Proceedings of Pervasive 2002,* Springer-Verlag, LNCS 2414: pp. 67–180.

Herring, J.R. (ed.) (2006) OpenGIS Implementation Specification for Geographic information - Simple feature access - Part 2: SQL option. Document No. OGC 06-104r3, Version: 1.2.0. Available from http://www.opengeospatial.org/standards/sfs.

Hightower, J. and Borriello, G. (2001) Location systems for ubiquitous computing. *Computer,* 34(8): 57–66. Extended paper: A Survey and Taxonomy of Location Systems for Ubiquitous Computing, available on line from http://citeseer.ist.psu.edu/hightower01survey.html, Accessed June 2007.

Hofer, T., Schwinger, W., Pichler, M., *et al.* (2003) Context-awareness on mobile devices – the hydrogen approach. In *Proceedings of 36th Annual Hawaii International Conference on System Sciences,* pp. 292–302.

Kasanoff, N. (2001) *Making It Personal.* New York: Persus Publishing.

Loke, S. (2006) *Context-aware Pervasive Systems.* Auber Publications.

Long, S., Kooper, R., Abowd G.D., *et al.* (1996) Rapid prototyping of mobile context-aware applications: the Cyberguide case study. In *Proceedings of 2nd International Conference on Mobile Computing and Networking* (Mobi-Com 96), pp. 97–107.

Maes, P. (1994) Agents that reduce work and information overload. *Communications of the ACM,* 37(7): 30–40.

Marcussen, C.H (2001) Mobile data and m-commerce in Europe – a mobile network operators' revenue perspective, 1999–2003. August 2001. Available online from http://www.crt.dk/UK/Staff/chm/P_CHM.htm, accessed May 2007.

Meng, D. and Poslad, S. (2008) A reflective context-aware system for spatial routing applications. Paper presented at 6th International Workshop on Middleware for Pervasive and Ad-Hoc Computing, Leuven, Belgium, December 2008.

Mills, D.L. (2003) A brief history of NTP time: memoirs of an Internet timekeeper. ACM SIGCOMM *Computer Communication Review,* 33(2): 9–21.

Morse, D.R., Dey A.K. and Armstrong, S. (2000) The What, Who, Where, When, and How of Context-Awareness. Workshop. Abstract. In *Proceedings of the 2000 Conference on Human Factors in Computing Systems* (CHI 2000), The Hague, The Netherlands, April 1–6, p. 371.

Ocampo, R., Cheng, L., Lai, Z. and Galis A. (2005) ContextWare Support for Network and Service Composition and Self-adaptation. MATA 2005: 84–95.

Pashtan, A. (2005) User mobility and location management. In *Mobile Web Services.* Cambridge: Cambridge University Press, pp. 54–80.

Poslad, S., Laamanen, H., Malaka, R., *et al.* (2001) CRUMPET: Creation of User-friendly Mobile services Personalized for Tourism. In *Proceedings of 3G2001 Mobile Communication Technologies*, London, 2001, pp. 28–32.

Prekop, P. and Burnett, M. (2003) Activities, context and ubiquitous computing. *Computer Communications*, 26(11): 1168–1176.

Ratnasamy, S., Francis P., Handley, M., *et al.* (2001) A scalable content-addressable network. In *Proceedings of Conference on Applications, Technologies, Architectures, and Protocols for Computer Communications*, San Diego, California, pp. 161–172.

Ratcliffe, J.H. (2001) On the accuracy of TIGER-type geocoded address data in relation to cadastral and census areal units. *International Journal of Geographic Information Sciences*, 15(5): 473–485.

Raz, D., Juhola, A., Serrat-Fernandez, J., *et al.* (2006) *Fast and Efficient Context-Aware Services*. New York: John Wiley & Sons, Ltd.

Ryan N., Pascoe J. and Morse D. (1998) Enhanced reality fieldwork: the context-aware archaeological assistant. In V. Gaffney, M. van Leusen, and S. Exxon (eds) *Computer Applications in Archaeology*. British Archaeological Reports, Oxford: Tempus Reparatum.

Saha, S., Chaudhuri, K. Sanghi, D., *et al.* (2003) Location Determination of a Mobile Device Using IEEE 802.11b Access Point Signals. *IEEE Wireless Communications and Networking*, 3: 1987–1992.

Samet, H. (1990) *The Design and Analysis of Spatial Data Structures*, Reading, MA, Addison-Wesley.

Schilit B., Adams, N. and Want, R. (1995) Context-aware computing applications. In *Proceeding of 1st International Workshop on Mobile Computing Systems and Applications*, pp. 85–90.

Schilit, B. and Theimer, M. (1994) Disseminating active map information to mobile hosts. *IEEE Networks*, 8(5): 22–32.

Strang, T. and Linnhoff-Popien, C. (2004) A context modeling survey. In 1st International Workshop on Advanced Context Modelling, Reasoning and Management. UbiCom 2004, pp. 34–41.

Van Greunen, J. and Rabaey, J. (2005) Locationing and timing synchronisation services in ambient intelligence networks. In W. Weber, J.M. Rabaey, and E. Aarts (eds) *Ambient Intelligence*. Berlin: Springer Verlag, pp. 173–197.

W3C CC/PP (2006) Composite Capabilities / Preferences Profile (CC/PP) 2.0 Home Page. Available from http://www.w3.org/Mobile/CCPP. Accessed July 2007.

W3C MMA (2002) Multimodal Interaction Activity, Home Page. Available from http://www.w3.org/2002/mmi/. Accessed July 2007.

Want, R. (2006) An introduction to RFID technology. *IEEE Pervasive Computing*, 5(1): 25–33.

Want, R., Hopper, A., Falcao, V., *et al.* (1992) The Active Badge Location System. *ACM Transactions on Information Systems*, 10(1): 91–102.

Watson, R.T. (2006) Spatial and temporal data management. In R.T. Watson, *Data Management: Databases and Organizations*, 5th edn. Chichester: John Wiley & Sons, Ltd, pp. 415–430.

Yang, J., Yang, W., Denecke, M. *et al.* (1999) Smart Sight: a tourist assistant system. In *Proceedings of 3rd International Symposium on Wearable Computers*, pp. 73–78.

8

Intelligent Systems (IS)

With Patricia Charlton

8.1 Introduction

Intelligent systems (IS) are systems which use artificial intelligence (AI), also referred to as machine intelligence, computational intelligence and include (intelligent) agent-based systems, software agents and robots. Many 'clever' and highly flexible algorithms require iterations of refinement, sometime including manual input, to 'mature' into algorithms that can be reused. There is also intelligence is in the development process for constructing such an algorithm. The output of an intelligence system can be highly complex and flexible but is often viewed as just computation rather than computation intelligence. For example, Optical Character Recognition (OCR) used in scanners was once considered part of the computer vision AI but now is considered merely a part of document scanning. This has led Rodney Brookes of the MIT AI Lab to coin the term, 'the AI effect', for algorithms that mature and disappear from being regarded as AI.

In modelling systems that exhibit intelligent behaviour, researchers often draw upon studying the way humans problem solve using analogies that match the reasoning and knowledge processing that have taken place. This process allows us to think about how problem solving occurs, using the brain as a concept to test out theories. It is clear that machines do not problem solve in the same way humans do. In fact, machines can be designed to be very good at problem solving in ways that humans are not and vice versa.

There is also debate about whether or not the biological components, e.g., the twenty billion or so neurons that constitute the brain, are individually intelligent or intelligent in clusters, and whether or not other animals such as insects, e.g., ants or bees, exhibit intelligent behaviours or whether or not even simpler single- and multi-cellular organisms can be used as building blocks for intelligent systems. Surprisingly simple behaviour can lead to the emergence of far more complex collective behaviour that may be termed intelligent.[1] Such systems will not have an explicit and rich notion of knowledge. Systems built upon this type of

Ubiquitous Computing: Smart Devices, Environments and Interactions   Stefan Poslad
© 2009 John Wiley & Sons, Ltd

behaviour are referred to as artificial life and are often included as a topic in many artificial science textbooks, e.g., Coppin (2004, pp. 363–383).

8.1.1 Chapter Overview

This chapter is the first of three chapters that comprise the smart interaction section of the book. Smart interaction is split into three main chapters: individual intelligent systems of different types; intelligent systems consisting of multiple interaction intelligent entities: Chapter 9); autonomous interaction and artificial life (Chapter 10) This chapter (Section 8) focuses on different types of individual intelligence as proposed by Russell and Norvig (2003, pp. 44–54). This chapter continues with a discussion of basic concepts. The next main section (Section 8.3) focuses on architectural designs. Then three main representations and models of IS are considered: Semantic KB IS Models (Section 8.4), Classical Logic IS Models (Section 8.5) and soft computing models (Section 8.6). Then some of the main generic IS model operations are discussed.

8.2 Basic Concepts

8.2.1 Types of Intelligent Systems

Because of the diversity of different types of AI and different properties of AI, no single general definition for AI is given here. Instead, some dimensions for classifying specific types of AI and then some typical design models for intelligent systems will be discussed. There are several dimensions along which AI can be classified (Table 8.1). Each of these is considered in more detail below.

Table 8.1 Dimensions along which intelligent systems can be classified

Dimensions for classifying types of intelligent System
Strong or weak intelligence
Physical (embodied) hardware, e.g., robots or virtual software, e.g., software agents
Fundamental properties such as autonomous, social, reactive, proactive etc
Thinking (cognitive) or acting (behaviour)
Human or rational
Complex organisms (explicit, high-level, knowledge-based action selection) or simple cellular organisms (implicit low-level action selection)
Type of design architecture: reactive, model-based, goal-based, utility-based etc
Learning or non-learning
Certainty or uncertainty
The environments in which intelligent systems operate: observable, deterministic, sequential etc.
Individual intelligent entities or as multiple, collective, intelligent entities

---

[1] Honey bees can communicate directions to a field of flowers up to a kilometre away using a 'dance language' (Riley *et al.*, 2005). This was first proposed by Karl von Frisch who was later awarded the Nobel Prize in 1973. Ants are able to regulate nest temperature within limits of 1°C, can co-operate in carrying large items and can find the shortest routes from their nest to a food source.

There is a philosophical difference between the notion of Weak AI where machines can act or simulate intelligence versus the notion of Strong AI that machines actually think. An alternate view of strong versus weak intelligence is to define a set of sub-properties for intelligence and group these into strong or weak. Wooldridge and Jennings (1995), for example, define two types of intelligent system called a weak agent and strong agent. A weak agent is defined as supporting system properties such as autonomous,[2] social, reactive and proactive. A strong agent is defined as being mobile, veracious or truthful, benevolent (an agent will always try to do what is asked of it) and rational. A distinction can be made between an intelligent system which is embodied physically, e.g., a robot which acts locally, versus an intelligent agent which exists in a virtual computing environment, e.g., a software agent which can act globally, being free to (remotely) act and roam within the whole of the virtual computing space, i.e., the Internet.

Several other researchers have classified the properties of IS as follows. Nwana (1996) identifies three core properties of agents: autonomy, learning and cooperation, and then seven types of agents such as collaborative, interface, mobile, information/Internet, reactive, hybrid and smart in terms of their support for their three core properties. Franklin and Graesser (1996) give the core properties of agents as reactive, autonomous, pro-active, temporally continuous, communicative, adaptive, mobile, flexible (using unscripted behaviour) and human character. Other commonly used classifications are mobile agents versus static intelligent agents, reactive versus deliberative and individual agents versus multiple agent systems. Each of these classifications for intelligent entities overlaps to some extent but there are important differences.

In part to handle the difficulty with the strong versus weak AI type of classification and the inherent difficulty in distinguishing between simulating thinking versus actual thinking, Russell and Norvig (2003) distinguish types of intelligence[3] along two dimensions, first, in terms of systems acting (behaviour) versus systems thinking. Second, types of intelligence are distinguished in terms of whether or not systems that act or think like humans (embodied, mentalistic, emotional, moralistic, etc.) called *human intelligence*, versus systems that think and do the right or ideal thing called, *rational intelligence*. Rationality can be defined more exactly in terms of achieving some defined performance measure given what it knows about its environment and about the effects of its own actions and about the current and past states of the system and the environment. There are several different designs to support rational behaviour, for example, based upon reactive behaviours (Section 8.3), internal system models (Section 8.1.1), goal-based and utility-based functions (Section 8.6) and learning, as proposed by Russell and Norvig (2003, pp. 32–54). Acting also often implies thinking in order to do the right thing.

8.2.2 Types of Environment for Intelligent Systems

It is a challenge for any system to act in open system environments. Russell and Norvig (2003) have categorised open system environments[4] along several dimensions (see Table 8.2). The simplest types of system environments are those that are fully observable, episodic and static. Static here means from the perspective of the system that the environment does not change to affect the system during its action selection and execution. Environments naturally change state

---

[2] The link between autonomy and intelligence is explored in more depth in Section 10.2.1.1.
[3] This Human versus Rational AI classification also overlaps with that of *Strong AI* versus *Weak AI*. Strong AI is AI that matches or exceeds human intelligence. Weak AI refers to the use of software to mimic some specific sub-types of human intelligence.
[4] The classification, although specifically referring to environments for intelligence systems, can also be applied more generally. The term 'world' is often used in place of environment.

Table 8.2 Environment models for UbiCom systems based upon the classification of environments for intelligent systems by Russell and Norvig (2003)

Environment type	Description of environment	Antonym environment
Fully observable, accessible	UbiCom's sensors give access to the complete state of the environment at each point in time	*Partially observable, accessible*
Deterministic	The next state of the environment is completely determined by the current state and the action executed by the UbiCom system. If the environment is deterministic except for the actions of other UbiCom Systems, then the environment is *strategic*	*Stochastic, non-deterministic*
Episodic	UbiCom System's experiences are divided into atomic 'episodes'. Each episode consists of the system perceiving and then performing a single action. The choice of action in each episode depends only on the episode itself	*Sequential*
Static	The environment is unchanged while an UbiCom system selects and execute its actions i.e., to adapt to its environment	*Dynamic*
Discrete	A limited number of distinct, clearly defined states and actions characterise the environment	*Continuous*
Passive	The environment of the UbiCom system can be dynamic, stochastic, etc but the environment itself is not active, in the sense of modelling the system that is acting in it	*Active*

with respect to time, perhaps because of the operation of active systems acting on it. More complex designs for intelligent systems are needed to think and act in environments that (1) are uncertain and non-deterministic; (2) are partially rather than completely viewed or sensed by the system; (3) are sequential, leveraging the event history to predict future actions rather than just relying current actions; (4) are dynamic rather than static; and (5) consist of other intelligent components rather being surrounded by dumb passive components. Most open system environment are often stochastic because no single system controls the environment. The actions of many other active systems may cause actions and changes to the environment which affect any single system. Of course the environment itself can be stochastic, e.g., the weather 'system' in the physical environment.

8.2.3 Use of Intelligence in Ubiquitous Computing

The use of the term intelligence varies across different fields of computing. Ambient Intelligence (AmI) is used to refer to smart environments but this is quite an abstract model rather than a type of intelligent system specification (Lindwer *et al.*, 2003). Updates proposed by Ramos *et al.* (2008) to specify a more explicit AI model for AmI add an intelligence layer but there are no specific architectures. Human (intelligent) computer interfaces can be designed to adapt to human users (Chapter 5). AI can be used to drive programmable controllers and robots (Chapter 6). AI can be the basic design for context-ware systems: systems which sense and act in environment and adapt to the environment (Chapter 7). AI is associated with programmable networks, called *intelligent networks*, putting intelligence perhaps in the form of logic into network-switching components versus in the end application computers. Agent-type designs can be used for network manager processes, e.g., SNMP agents. Here, it is proposed that specific types of IS architecture models of Russell and Norvig (2003) are used as a basis for specific types of IS design for UbiCom.

In order to explain the usefulness of UbiCom designs based upon IS, some applications and scenarios described earlier in Section 1.1.2 are re-examined in terms of how they could benefit from the use of AI. For the personal memories scenario (Section 1.1.2.1), an IS represents the AV system. Its design uses computer vision to detect eyes and mouth, to determine whether or not someone is smiling, to recognise if recorded faces match faces already classified. Metadata (data which describes other data) to annotate to the image, e.g., using the time, location, the identity of any people recorded, etc., and any visual features extracted from the image, may be represented as explicit knowledge. A system to support this scenario could be based upon a reactive IS design.

For the adaptive bus scheduling scenario (Section 1.1.2.2), an IS representing the bus service can use utility functions to weight the importance of different independent factors that need to be combined to support a goal, e.g., pick up more passengers to generate more revenue versus pick up fewer passengers to minimise the degree of lateness for a bus running late. The prediction of a bus's arrival at scheduled bus-stops is based upon a model of how a bus's environment, such as its road conditions, can dynamically affect the rate of progress of the bus along the route. A system to support this scenario could be based upon a model-based agent and utility-based IS design.

For the foodstuff management scenario (Section 1.1.2.3), the IS represents the foodstuff management system. It requires the use of knowledge mediation to handle the different conceptualisations used on the foodstuff labels versus a recipe instruction book, e.g., the recipe may refer to a weight of zucchini in pounds whereas the item label may refer to courgettes weighed in kilogrammes – these terms are similar. Adaptive sensing and control techniques are needed by a robot in order to plan and execute its operation to pack and unpack food in the food stores. Events need to be monitored and evaluated against pre-set rules. A system to support this scenario could be based upon a combined reactive and goal-based IS design.

For the utility regulation scenario (Section 1.1.2.4), the IS system senses the user context and adapts the system usage based upon the user context, a model of the user's anticipated usage and user management policies. A system to support this scenario could be based upon a reactive and goal-based IS design.

8.3 IS Architectures

An IS system is based upon a conventional distributed system architecture with three enhancements. First, it supports richer models of interaction between IS and their environments such as reactive, proactive and utility-based interaction models. Second, it uses processing models which support a model of advanced system cognition and behaviours such as searching, reasoning, planning and learning. Third, it uses a richer communication model with other intelligent systems and with intelligent environments (Chapter 9).

8.3.1 What a Model Knows Versus How it is Used

There are several ways in which IS models can be classified:

by the type of model architecture, i.e., by what the model represents or what the model knows, the *epistemological level* or structural level;

by how the model is used to solve some problem, the *heuristic level*, e.g., as a logic language that supports reasoning and heuristics to confirm hypotheses and can derive new knowledge;

by what types of environment, a system is situated in, and can operate in (Section 8.2.2).

McCarthy and Hayes (1987) first separated model representations at the epistemological level and then at the heuristic level. A representation is called epistemologically adequate for a person or machine if it can be used practically to express the facts that one actually has about the aspect of the world. A representation is called heuristically adequate if the reasoning processes actually gone through to solve a problem are expressible in that representation. Typically there is a trade-off for the representation between the representational power at the epistemological and the inference of efficiency at the heuristic level.

Often when discussing IS design, many AI practitioners presume that a specific logic-based representation for the environment model of IS is used. This tends to focus on a heuristic rather than on an epistemological representation. The logical model is often referred to as *reasoning-based* or *deliberative type IS*. Semantic knowledge-based (KB) system models focus more on the epistemological level in order to make them reusable across many different applications. In practice, in order to use a general epistemological level KB, additional knowledge relationships and processes must be modelled to relate the general KB model to the specific application domain. If the KB is an ontology, this then refers to specifying application-based ontology commitments.

A learning-based IS can be used to acquire any type of model used in any type of IS, the structural representation for the knowledge will depend upon the type of IS. For example, a learning-based IS can be used for environment context acquisition. This often uses a model representation that can handle uncertainty and heuristics for knowledge acquisition, hence this focuses more on KR at the heuristic level.

Often it may not be clear what the model represents, e.g., the model could represent the internal system design, the external environment design or both. In this text, two separate dimensions of model or knowledge have been separated in terms of what the model represents and how the model is represented in computational form.

8.3.1.1 Types of Architecture Model

Several major types of IS architecture model have been proposed by Russell and Norvig (2003) depending on how the system's actions are driven and what type of environments systems operate (see Table 8.3). In reactive systems, systems' actions depend only on the observed current

Table 8.3 Designs of intelligent systems related to the types of environment they are suited in

Type of IS model	What an IS's actions depend upon (what is modelled in the system)	Types of environments the IS design is suited to
Reflex-based	Current Environment context	Fully-observable, stochastic, episodic, static, physical
Env. model-based, Situated action	Current and past environment context	Partially-observable, deterministic, sequential, dynamic
Goal-based, Proactive	IS's plans of actions to achieve a goal	Partially-observable, deterministic, sequential, dynamic human
Utility-based	IS's weighting of different goals and plans	Partially-observable, Semi-deterministic, sequential, dynamic
Learning	Performance	Partially-observable, Non-deterministic, sequential, dynamic
Multi-IS		Partially-observable, Non-deterministic, sequential, dynamic

environment context. Environment model-based[5] systems are a design for IS which allow systems to take into account past contexts and models of how the environment works in order to anticipate how to act more suitably in changing environments. The actions of a goal-based IS depend upon the model a system has of how its processes and actions lead to some outcome, enabling goal-based IS to be more proactive. For utility-based IS systems, actions depend upon how action selection and goals are weighted. In learning-based IS designs, systems' actions can adapt to improve internal performance metrics over time. With multiple IS designs, each individual system's actions can depend on active interaction with other autonomous IS systems as well as on any internal system model.

In practice, individual designs are often combined into hybrid systems, e.g., environment-model reactive system designs and goal-driven, environment-model reactive systems. Multiple interacting system designs may use multiple hybrid individual IS designs.

8.3.1.2 Unilateral Versus Bilateral System Environment Models

Generally, when ubiquitous system applications are designed, a unilateral model of the environment is used, e.g., a context-aware application can model its environment but not vice versa. The environment is not active. However, as we move to smarter environments, they can be designed to contain a model of the application systems which are situated in them or pass through them (Figure 8.1). A system that models an active environment, which in turn has a model of the systems which use it, can lead to much more complex and cyclic interdependencies between multiple system and environment models. Designers of systems may have to take into the account the degree of intelligence of environments, e.g., humans, other UbiCom systems, etc. and how well the environment can model and understand the systems that inhabit it.

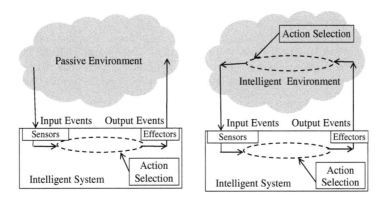

Figure 8.1   Unilateral active system model (left) versus bilateral active system and active environment models

---

[5] Russell and Norvig (2003) refer to one type of agent-based (intelligent) system as a model-based system. The term model-based is regarded as being too general here and is qualified by the term environment as the model refers more specifically to the model a system has about the behaviour of its environment.

8.3.1.3 Model Representations

There are several major concrete representations for IS models, and a variety of model operations which can be performed which depend on these representations. First, IS models can be represented as *process-driven system* models in which prescribed sequences of actions that constitute a process are specified at the design time. Such processing algorithms for complex systems can be difficult to design correctly and to maintain. They may also require the use of heavy computational resources. Computational process models have been dealt extensively in Chapter 3. Second, IS models can be represented as data to support a range of syntactic and semantic conceptualisations, i.e., *knowledge-based models* (Section 8.4). Third, IS models can be represented using various kinds of classical logic (Section 8.4.2.5). Fourth, IS models can be represented using various kinds of soft computing (Section 8.6). These representations for model-based IS are discussed in subsequent subsections.

8.3.1.4 How System Models are Acquired and Adapt

So far, nothing has been said about how an IS acquires a model of its own internal system operation and of its environment. There are two main ways to design how the model is acquired. First, system models can be created by a human designer and built into the system at design time. The limitation here is the designer may not be situated in, or have much experience of, the environment in which the system is used or that the use may vary between users' environments. Hence a design may have a fixed mode of operation, may be incomplete or could lack the ability to be configured for operational use. In order to handle such requirements, systems can be designed to be open to be maintained and upgraded periodically via remote access links from some service repository (Section 12.3.2).

Second, systems can be designed to acquire their models of self-operation and their models of environments themselves – systems that learn. These systems are also referred as *machine-learning* systems. There are several specific designs for machine learning (Section 8.3.6). Processes for hybrid model acquisition by systems can combine manual design and maintenance with automated learning.

Often the property of adaptation for an IS is taken to be synonymous with the property of learning but adaptation or flexibility is a more general concept, learning is one specific type of design to support adaptation. A goal-based design which supports multiple plans may also enable the system to adapt in the sense that if the environment causes one plan to fail, it may switch to another plan. Another type of adaptation is one based upon reflective-type system design (Section 10.3).

8.3.2 Reactive IS Models

In a *reactive system*, intelligent behaviour arises out of the system's interaction with its environment rather than as a result of complex internal knowledge representation or reasoning about events. Action selection[6] is driven purely by the current state of an environment. Rules or conditions are used to filter input events and to determine which system actions are triggered[7] in response to particular events. A reactive type IS is strongly situated in its environment and is highly responsive

---

[6] A reactive system is also referred to as a reflex system, or as a perception-based system or as supporting reactive atomicity.

[7] Action selection is at the heart of the intelligent system. An action selector is also referred to as a *controller*.

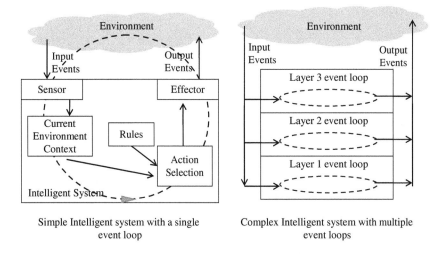

Simple Intelligent system with a single
event loop

Complex Intelligent system with multiple
event loops

Figure 8.2 Reactive type intelligent system

to changes in the environment. Reactive systems tend to be designed as event-based systems (Section 3.3.3.6) which loop through the basic operations of sensing events, filtering events and then triggering system actions (Figure 8.2).

Pre-set actions may be directly triggered from sensor inputs without any conditions, e.g., in the adaptive transport scheduling scenario, if the vehicle arrives at a designated passenger pick-up point, it automatically stops. Alternatively events can be filtered by conditions in order to trigger actions, e.g., a vehicle will stop at a designated stop only if passengers inside the vehicle have requested leaving the vehicle or if one or more passengers are waiting at the stop and request the vehicle to stop.

A reactive system is a natural design for simple sensor control systems (Chapter 6) and for simple context-aware systems (Chapter 7). In a simple reactive system, where there is only one event loop, only one event can trigger an action at any one time. More complex designs need to consider how to respond to multiple simultaneous events and how to handle multiple actions which, when triggered, may conflict.

There are three possible designs to deal with handling multiple concurrent, possibly heterogeneous events: (1) discarding events that cannot be handled in time; (2) supporting event persistence; and (3) supporting event handling concurrency. The first option is the least desirable as it drops events in a non-deterministic manner. However, events may need to be dropped if unforeseen events arrive that cannot be handled. Second, designs can support event persistence such as the use of event buffers or heaps if decisions about action selection in response to some events can be deferred (Section 3.3.3.6). Third, designs can support concurrency by incorporating multiple types of event loop and by layering event loops and giving different event loops priority to trigger actions (Figure 8.2). For example, in the adaptive bus scheduling scenario, the sense 'passengers at bus stop' event triggers the 'pick up passengers' action as a priority over the 'leave passengers at bus stop' because of 'the bus is running late' action.

The purely reactive type of intelligent system works best when it is situated in types of environment that are fully observable, static and episodic. In practice, many systems are designed not to be purely reactive but to combine reactive behaviour with other types of behaviour such as model-based behaviour, goal-based and utility-based behaviour. These are types of hybrid reactive systems. Hybrid IS designs are discussed further in Section 8.3.7.

8.3.3 Environment Model-based IS

An effective way to handle partial observability of the environment is for a system to use some internal model of the environment perhaps by keeping track of past events and analysing them for patterns and using knowledge about the effects of internals system actions (Russell and Norvig, 2003). This *environment model* is also referred to as a *world model*. It is a type of knowledge-based system because the system has knowledge about the world and its actions. The environment model type of intelligent system[8] can model sequential or historical behaviour in the environment. It supports *predictive atomicity*, making decisions based upon current events and based upon a model of past external events which can be used to predict future external events (Figure 8.3).

The environment model IS uses a sequential environment model. This type of IS's actions depend upon the current environment state, past environment states and on knowing the effect of system actions. This type of system design is similar to a situated action type of system design: actions can be unplanned and depend strongly on the context (Suchman, 1987, pp. 49–67). It is also possible to anticipate multiple future environment states, which may never be realised, leading to a theory of multiple possible future environments or world states.

For example, in the adaptive transport scheduling scenario, if someone appears to be running towards the pick-up point but is not there yet, an environment model-based IS will anticipate that if the vehicle slows down or waits a short while, an extra passenger will arrive at the pick-up point to be transported. In the pure (conditional) reactive system, the vehicle will not stop when there are no passengers at the pick-up point (providing no passengers on the vehicle wish to leave).

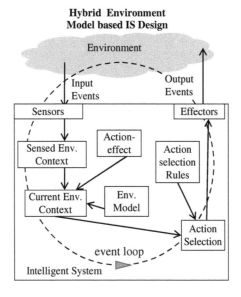

Figure 8.3 Environment model-based IS according to Russell and Norvig (2003)

---

[8] Russell and Norvig (2003) refer to this design of an IS as a *model-based reflex agent*, referring to an IS as an agent and emphasising that the reflex agent design is extended by adding an environment model into the system.

Systems that build such a model of the environment enable their services and applications to optimise and adapt their behaviour to account for behaviours in the environment which are not accessible but which are predetermined (as defined in the environment model). A restriction of the pure environment model type of IS is that it does not include a model of the internal behaviour, e.g., processes of actions by the system.

8.3.4 Goal-based IS

A *goal-based* IS, also referred to as planning-based IS, defines an internal plan or sequence of actions to achieve a future system goal (Figure 8.4). Unlike the environment model-based IS, the action selection for the goal-based IS depends on which is the next best system action to take the system towards a future goal state. In comparison to the environment model-based IS, a goal-based IS tends to dissociate the control of the actions from the environment situation or context of the action, in contrast to a reactive environment model-based system when events trigger system actions as external events. In goal-based systems, internal events, e.g., a scheduled system task that is delayed, can also trigger system actions. The system can also trigger actions when external events are sensed – active external event sensing.

The main benefit of this type of design for users is that users can delegate tasks to this type of IS at a much higher level of abstraction, focusing on what needs to be achieved rather than on the lower-level details of how this is to be achieved. In practice, the distinction between a goal as a state and as a high-level action is similar, e.g., the goal state of a normally powered-down system versus the high-level action of normally powering down a system. A key design issue is how a user

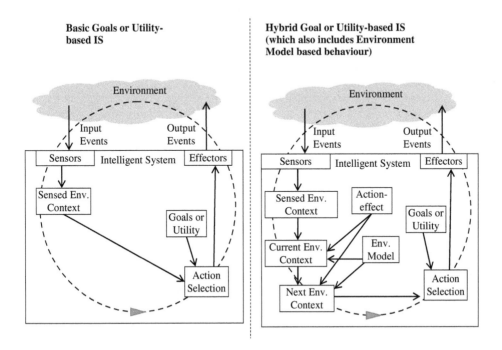

Figure 8.4 Two types of goal-based or utility-based IS design – basic versus hybrid, according to Russell and Norvig (2003)

of a goal-based system knows which goals a system can achieve. A goal-based IS may publicise which goals it can achieve and its plan of action to achieve those goals.

A *hybrid goal-based IS* combines the goal-based action selection of the basic goal IS design with an environment or world model. Such an IS design selects actions not just based upon the (external) environment but also based upon which chain of actions will enable the system to attain its goal state (Figure 8.4). For example, in the adaptive transport scheduling scenario, a goal for the service provider could be for the transport vehicle to arrive at each designated pick-up point on time. A vehicle may therefore choose to ignore a passenger moving towards but not yet present at a passenger pick-up point because if the vehicle waited, it might not be able to meet its commitments to pick up passengers further down the route on time. Finding chains of actions that enables an IS to reach its goal state involves searching and planning (Section 8.1).

Some goal-based design may support multiple plans. This enables the system to adapt in the sense that when the environment causes one plan to achieve a goal to fail, the system may simply switch to another plan. Goal-based design can also involve cooperation and goal sharing with other ISs (Chapter 9).

8.3.5 Utility-based IS

A *utility* refers to a quantifiable measure of the performance or worth or usefulness of a specific goal among a set of possible goals. It can also refer to a specific chain of actions among a set of possible chains of actions. A utility function is used to map a (goal) state or a chain of states to a value which represents its performance or worth. A utility-based IS design is useful when several conflicting goals exist or when multiple goals are possible but only one of them is practical or achievable. For example, for the adaptive transport scheduling service, two conflicting goals[9] to recover from disruptions to a designated schedule are to maximise the revenue by maximising the pick-up load versus maintaining a quality of service by minimising the deviation from a designated schedule. For example, the greater the load that is picked up, the later an already late vehicle becomes. A utility function here could weight revenue generation and maintaining punctuality equally, 50–50%, or it could bias revenue generation higher than maintaining punctuality to be 75–25%.

A utility-based IS design appears similar to a goal-based IS design. A goal-based IS sets a specific goal and selects the action that leads to that goal (goal-based action selection) regardless of its utility. A utility-based IS compares the utilities of different possible outcomes, and selects the action outcome with the highest utility (utility-based action bias). Sometimes it may require some experience or training phases in order to set and tune the utility function values, hence, goal-based action selection may be preferred over utility-based action selection in a new environment.

8.3.6 Learning-based IS

Learning refers to a system improving its performance with experience, with respect to some task (Mitchell, 1997). A system is said to learn from experience E with respect to some class of actions A and performance measure P, if its performance at the set of actions A, as measured by P, improves with experience E. An example from the adaptive goods scheduling vehicle: A='a logistics vehicle

---

[9] There is a modelling choice about representing something as a multi-valued single goal or to consider the same thing as multiple single-valued goals.

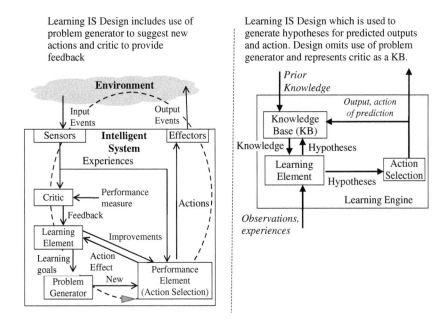

Figure 8.5 Two different learning IS designs, the left according to Russell and Norvig (2003) and the right which focuses more on the use of a KB and on learning to generate hypotheses or heuristic functions

picks up goods on route', E='travelling the route', P='deviation of actual time from predicted time'. The improvement is the measure P reducing to zero.

A basic architecture for a learning-based system, also referred to as *machine-learning*, taken from Russell and Norvig (2003) is given in Figure 8.5. They use a theatrical metaphor to illustrate how a basic learning system works. A learning system comprises four main components: (1) a performance element; (2) a critic or feedback generator; (3) a learning element; and (4) a problem generator. The *performance element* behaves as the (external) action selection element in the other IS architecture models based upon experience, E, of the combined input from the sensors, learning element and the problem generator.

A critic or feedback generator provides feedback of how well the agent is doing, based on a fixed performance standard or performance measure, e.g., in the transport system scenario, the critic provides information about the deviation of the expected arrival time at predetermined route points versus the actual time or uses a questionnaire to ask passengers about their experience of the journey, including punctuality and capacity. The learning element gets feedback from the critic and modifies the performance element, e.g., if customer feedback indicates the transport service was too crowded, the transport service could operate larger capacity or more frequent vehicles. The problem generator provides the performance element with additional input suggestions on new actions to take, otherwise the system's behaviour never changes, e.g., trying different routes to pick-up points to improve punctuality.

8.3.6.1 Machine Learning Design

The design of the learning element depends on: what (which model) is learned, the type of feedback and the model or knowledge representation. Learning may need model representations that can handle uncertainty which is often central to the technique of learning. There are generally three main types of learning or feedback which can be used: (1) supervised learning; (2) unsupervised learning; and (3) reinforcement learning.

Supervised learning, SL, also called *programming by example*, involves learning a heuristic function or hypothesis which transforms inputs into outputs, given examples of inputs and their associated outputs. Then, predictions of the output, based upon the input can be made. If the actual output differs from the expected output, the heuristic function is adapted. The process is repeated until the IS makes accurate and repeatable predictions of output. Specific types of SL include inductive learning (see Figure 8.5), back propagation neural networks and decision trees. In inductive learning, the aim is to learn a good heuristic function which maps the input to the output. With decision trees, the aim is to learn how to classify outputs or actions such as how to classify a user activity based upon observations of time and which devices are active.

Unsupervised learning (UL), involves learning to differentiate input patterns without any outputs (classifications) being defined. An unsupervised learning system can learn to differentiate patterns but not which is the desired output or classification. Examples of UL include probability analysis, data clustering techniques (Han and Kamber, 2006) and Kohonen map neural networks. An example is to mine context data for patterns and then to classify those patterns and therefore contexts. A simple data-clustering algorithm involves finding the frequent sets of things (things could represent the types of AV content people frequently watch, e.g., current affairs, comedy, drama, etc.) with a minimum support[10] count and then from the frequent sets to identify sets of things which also support a minimum confidence[11] count. *Reinforcement learning*, also called *reward-based learning*, involves the evaluation of action based upon rewards for doing an action, e.g., Q-learning.

Learning techniques can be used to acquire any knowledge for any of the models in any of the system architectures. Design issues for learning include the use of background knowledge to boost the system's knowledge base, consideration of the number of ways of the model that is learned, the representation of the learned information. Otherwise, learning systems that acquire all their knowledge from scratch may take too long to acquire enough knowledge to become useful. IS can be designed to contain so-called *background knowledge* or *inbuilt knowledge*, acquired or transferred from other IS systems or humans rather than learnt from scratch. IS can then add knowledge through their experience, adding so-called *foreground knowledge* (Figure 8.5). As the dimensionality of the data increases, there is an exponential increase in the amount of data needed to make good predictions.

8.3.7 Hybrid IS

Hybrid IS models seek to combine the benefits of the individual IS models. There are two basic designs: horizontal concurrent layers versus vertical (sequential) layers (Figure 8.6). Layers consist of single or multiple IS components with a clearly defined interface for input and output.

An example of a multiple concurrent event handling design was given earlier (Section 8.3.2). This uses a horizontal homogeneous layered model to allow multiple events to be handled in parallel.

---

[10] Support is the percentage of all samples where, say, two specific things occur, e.g., a person's viewing log shows they watched comedy and drama. As a function it is Support $(A, B) = P(A \cup B)$, where P is the probability.

[11] Confidence is the percentage of times when a person did one thing they also did a (related) thing, e.g., the percentage of people who watched comedy also watched drama. As a function, it is Confidence $(A, B) = P(B|A)$.

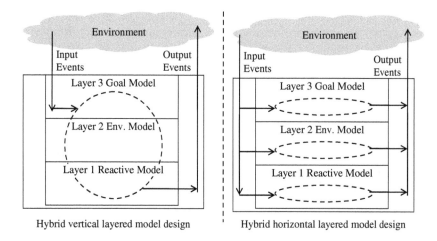

Hybrid vertical layered model design   Hybrid horizontal layered model design

Figure 8.6   Two different designs for a hybrid IS based upon horizontal and vertical layering

This type of model just needs to be generalised to allow heterogeneous models to be layered. For example, a hybrid IS can handle reactive events in a lower reactive layer and handle events which require use of the environment model and reasoning in a higher layer.

A design challenge with this type of model is that multiple output action events can occur for the same input event because each layer independently outputs its own action. This requires some mediator to coordinate and control output actions (Wooldridge, 2001, pp 89–104). A utility function or cost heuristic design for a mediator could be used, allowing the action with the highest utility to be generated when several actions could be triggered.

Heterogeneous models for a hybrid IS can also be vertically rather than horizontally layered. In the designs proposed in the preceding sections (Sections 8.1.1, 8.3.4), different clearly defined IS models are chained together. For example, in the environment model-based IS, environment events are first processed by the environment model before being processed by the reactive (ECA) model (Figure 8.7). In the hybrid goal-based designs, control first flows through an environmental model component or layer before being passed to a goal layer. This chaining of models is a type of single-pass vertical model design in which control flows through each layer in order to generate the action in the last layer. Other types of vertical model design could use multiple passes or flows for control,

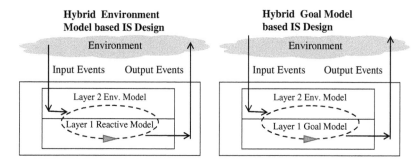

Figure 8.7   Simplified layered views for a hybrid environment model-based IS design and for a hybrid goal-based IS design

information and action generation. Note this type of chained hybrid model does not allow concurrent event processing tasks to occur and can form a processing bottleneck. If any component fails, the whole chain could fail without design support to prevent this. Wooldridge (2001, pp. 89–104) discusses some concrete examples of horizontal layered IS architectures, e.g., Innes Ferguson's TouringMachines and vertical layered IS architectures, e.g., Jörg Müller's InteRRaP.

8.3.8 Knowledge-based (KB) IS

Knowledge-based (KB) models cover a range of IS models in terms of representation, operations and what the model is used for (the type of model). Commonly used types of KB system architectures include production systems or *rule systems*, blackboard systems and semantic KBs such as ontology-based systems (Section 8.4). Currently, because of the prominent use of ontology-based models and their support for rich semantic conceptualisation, the terms knowledge-based, semantic and ontology-based models are often used somewhat synonymously. Noy and McGuinness (2001) summarise the following benefits of KB models: (1) to share a common understanding of the structure of information among humans and machines; (2) to enable the reuse of domain knowledge; (3) to make domain assumptions explicit; (4) to separate domain knowledge from the operational knowledge; and (5) to analyse domain knowledge.

8.3.8.1 Production or Rule-based KB System

In a *production system*, knowledge is represented as a set of rules or productions stored in the KB. A *rule engine* based upon logic reasoning is called a *rule inferencing engine*. A rule engine determines how rules constructed for an IF-fact (also called the condition part or antecedent part) and for the THEN-fact (also called the consequent part or action part) are used. A rule-based KB model can be combined with a reactive-type IS. When new (environment) events are generated, they are represented as new facts (things that are true). The KB system uses the fact in the IF portion of the rule and matches this with current facts contained in the working memory part of the KB (Figure 8.8). When a match is confirmed, the action rule gets activated and its THEN statements are added to the working memory. The new facts added to the working memory can also cause other rules to fire.

The searching of the knowledge base may involve forward or backward searches (Sections 8.7.1, 8.1.1). Rules can be added to the KB manually or automatically through machine

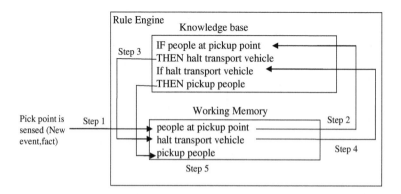

Figure 8.8   A rule type knowledge-based IS

learning. One other complication is that new facts and searches may match several rules which may conflict. Methods for rule conflict resolution include: taking the first match; using more specific matches rather than more general matches; using a higher priority rule over a lower priority rule; using the most recent match or using logic inferencing. Many rule-based engines have been developed, some of which enable rule-based systems to be incorporated as part of more general distributed systems rather than as part of more specialised IS, e.g., *JESS*, the *Java Expert System Shell*, devised by Friedman-Hill (2003).

8.3.8.2 Blackboard KB System

The basic principle of a blackboard KB system is that it functions as a shared knowledge data repository between multiple possibly distributed processes (Section 3.3.3.7). Knowledge sources can be independent and heterogeneous and unlike a reactive IS design, the repository enables selected events to persist in storage rather than be deleted from the event queue once it is consumed. A blackboard KB can take part in multiple asynchronous event condition action rules. A blackboard KB can be combined with a rule-based KB to enable multiple rule-based KBs to interact with the blackboard.

One of the limitations of a rule-type knowledge base is that the same fact can be stored in many rules. This makes facts, and their consistency, difficult to maintain. Further, relationships between the main concepts can only be expressed using rules. Rules may not be very good at expressing rich relationships and conceptualisations of facts. If the main requirement is to maintain and manage descriptive semantic relationships, then it makes more sense to structure concepts not as facts in rules but into rich conceptualisations based upon frame-based KBs and ontology-based KB (Section 8.4).

8.3.9 IS Models Applied to UbiCom Systems

In the introduction, in the section on common myths for UbiCom (Section 1.5.2) it was stated that it is not necessary for UbiCom systems to be AI based. Others such as Thompson and Azvine (2004) argue the contrary, that AI is essential to enable UbiCom. The justification why AI is useful but not essential to enable UbiCom is because, first, it depends upon the nature of the application. Second, it depends upon the specific model of AI being used and, third, there are some alternative computation models to AI that do something similar, e.g., a system can use a plain event-driven system to sense its environment and support context-awareness and a reflective system design implemented using a procedural language can enable a system to adapt itself to its environment.

Specific IS architectures can be used to build systems which support other UbiCom system properties such as context-awareness, autonomy and iHCI as follows. A reactive type IS design is a good design for a minimum context-aware system in which a system's actions depend only on current environment context or on crisp well-defined changes to the context that match some condition. In a more advanced context-aware system design, the actions can also be designed to depend upon the environment model (Section 8.1.1) including past context changes. In practice, context-aware system designs also need support to compose multiple context events, to map sensor events to applications and to manage contexts including storing and querying them (Chapter 7). A system which is (environment) context-aware may also need to be aware of itself, of its own internal system actions in order to accurately control and act on its environment (Section 10.3).

Some environment context changes, e.g., to support iHCI and user context changes, are, however, less well-defined, less crisp. For example, in the personal memories scenario, consider a system which personalises the recording of audio-video content, it may not be able to clearly determine the user context but only to provide an estimate of it. In this case an environment model IS, which represents the model using soft computing techniques, or uses hard computation which aggregates varying values of the environment context into a best estimate, e.g., collaborative filtering systems, seems like a suitable design.

Simple control systems could be based upon the reactive and utility type IS designs. IS designs focus on action selection from a set of possible actions whereas embedded control systems typically have a very limited set of actions. The focus of control systems is on a fine level of granularity of control and the use of specific feedback control mechanisms, which are not explicitly represented in any of the IS designs given. Hence, standard embedded environment control systems tend to use specific designs (Section 6.6). More complex controllers could use uncertainty knowledge models. Complex spatial control systems, i.e. robots, need to incorporate spatial sensors and spatial determination and estimation algorithms in order to move in prescribed ways, coupled with collision detection (reactive IS behaviour) and a spatial memory in order to detect obstacles and to remember how to avoid them in the future (environment model behaviour).

Autonomous systems are similar to IS in the sense that these can be goal-directed and policy-constrained. The self-management properties of autonomous systems, i.e., self-configuring, self-optimising, self-healing and self-protecting, can be designed in terms of unsupervised learning and can incorporate plan-switching and replanning to achieve a goal when internal system and external environment changes would otherwise cause the system to fail to achieve its goals. However, generic goal-based IS models may lack an explicit representation of how to control their environment, i.e., lack explicit control algorithms.

Architectures for UbiCom systems are sometimes needed which can operate in multiple environments. For example, in the adaptive transport scheduling scenario, the transport system needs to be aware of the human environment in terms of the needs of people being transported, the different times they need to get off and the number of people waiting at designated points to be picked up. The transport system needs to be aware of the physical environment of any physical route impediments in order to predict the arrival time at designated points on route. The transport

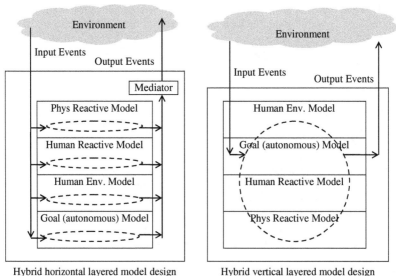

Figure 8.9   Hybrid IS designs to support UbiCom

system may be driven by goals and utility functions to reach all the pick-up points on time and to maximise the goods it picks up. Two different hybrid designs which differ along two different dimensions are given in Figure 8.9. First, different IS models can be executed in a chain (vertical layered model) or in parallel (horizontal layered model).

Second, for the chained model, the order in which the models are executed may greatly affect the computation efficiency, a particular concern for ICT resource constrained systems. In Figure 8.9, the chained design is based upon environment events first processed in a human environment model, then in a goal-based model, then in a human reactive model and finally in a physical world reactive model. For example, in the adaptive transport scheduling system, the IS system first takes into account the world model of the time of day in order to schedule a larger capacity vehicle during peak hours. Next, it takes into account the goals of transport system by maximising revenue by picking up as many fare-carrying passengers as possible. Next, it notes that although passengers are waiting at a pick-up point, its capacity is already full so it does not stop to pick up additional passengers. Lastly, the transport system notices that the road conditions are wet and together with containing a full load, it needs to brake earlier and more gradually in order to bring the vehicle to stop when needed.

8.4 Semantic KB IS

There exist a range of representations that can model knowledge in terms of concepts and their relationships. Often the term *ontology*, expressed informally[12] as a collection of descriptions of the world that helps users define the meaning of their actions on the world, is used synonymously with the term knowledge representation.

8.4.1 Knowledge Representation

According to Davis *et al.* (1993) a *knowledge representation* (KR) can best be understood in terms of five distinct roles it plays. First, a KR behaves as a surrogate, a substitute for the thing itself, used to enable an entity to determine consequences, i.e., by reasoning about the world rather than taking action in it. Second, a KR is a set of ontological commitments, in the sense that one concept generally refers to and is understood through its relationships with other concepts through use. Third, a KR is often used for intelligent reasoning. Fourth, a KR enables efficient machine-readable[13] and machine-understandable[14] computation about knowledge. Fifth, a KR provides a language in order for humans to express things about the world.

There exists a range of ontology models and representations depending on how concepts and their relationships are defined and organised. At one end there are lightweight ontologies such as dictionaries that have simple conceptualisations, having parts such as values of terms that may not be machine-readable and machine-relatable to other terms.

---

[12] Ontologies have many more formal definitions, i.e., as a formal, explicit specifications of a shared conceptualisation (Nicola, 1998; Gómez-Pérez, 1999).

[13] Machine-readable simply refers to data that can be read by a computer and where reading implies the system can understand the structure of the data, for example, in order to extract matching concepts to a query concept. Machine-readable may not be human-readable and vice versa.

[14] Machine-understandable refers to data where a computer can interpret the meaning of the data with respect to (normally a very limited) conceptual framework such as a categorisation of concepts. Machine-understandable may not be human understandable and vice versa.

Currently, the most widely used lightweight KRs seem to be those based upon the W3C Web XML[15] model as a basic node labelled graph representation and extensions to this developed as part of the Semantic Web (SW). XML is a language designed to exchange extensible application and domain-specific hierarchical data structures. This is used by many Web Services designed around a Service Oriented Computing architecture (Section 3.2.4) These data structures are certainly machine readable but XML is a difficult data format on which to build automated machine-understandable processing and to support interoperability between autonomous heterogeneous Web services.

At the other end of the scale are heavierweight ontologies that support more descriptive conceptualisations and more expressive constraints on terms and their interrelationships including logical constraints (Corcho *et al.*, 2003). Regardless of the properties of the specific ontology, heavierweight ontologies generally include the following elements: taxonomic relations between classes (also called categories), data-type properties, descriptions of attributes of elements of classes and object properties, descriptions of relations between elements of classes, and, to a lesser degree, instances of classes and instances of properties. Although heavyweight semantic KRs and logical KRs may have a common representation that supports both semantic queries and logical reasoning, there is the issue of which type of logic should be combined with the semantic representation. If the use of the KR requires temporal reasoning, uncertainty reasoning or crisp logic reasoning, different logic representations will be needed and in some cases multiple logics may be needed to reason about the knowledge. Hence, this is why, in this text, the common non-logical conceptualisation and relationships discussed in this section are separated from a use of the knowledge to support different kinds of logical reasoning (Sections 8.4.2, 8.6).

In terms of computation representations, KR can be displayed in graph notation, also referred to as *Semantic Net* notation. There are many variations of these graphs to represent the different lightweight to heavyweight KRs. An example of a lightweight and a heavyweight one are given in Figure 8.10. Graph notation can then be mapped back to an exact declarative form for computation use such as reasoning about concepts in the knowledge model. Different communities within computer science have also developed similar overlapping models. For example, early versions of the type of KR called frames, developed by the AI community, are quite similar to the object-oriented model developed by the software engineering community.[16]

According to Berners-Lee *et al.* (2001) and Shadbolt *et al.* (2006), the SW started in the 1990s as a vision for the evolution of the Web from making it machine-readable (enabling machines to read and parse the syntax to extract concepts), to making it more machine-understandable (to enable machines to act on the meaning of the concepts) and to support richer service interoperability between heterogeneous service processors, called agents. SW defines a suite of ontology language models based upon RDF[17] as an XML extension which supports edge labelled graph representation, on RDFS[18] and on OWL[19] as an extension to RDFS, which acts as a heavyweight KR and

---

[15] The eXstensible Markup Language (XML), See http://www.w3.org/TR/REC-xml/, accessed Jan. 2007. XML defines an unnamed hierarchy of concepts and properties.

[16] Frames have slots which can represent other frames in the hierarchy or actions. When an event enters a frame, it is matched against relevant slots which may trigger slot updates and further frames. Software objects in an object-oriented model are similar and can be triggered by events as part of an ECA model.

[17] The Resource Description Framework (RDF), See http://www.w3.org/TR/REC-rdf-syntax/, accessed Jan. 2007. RDF adds support for named associations between concepts to XML.

[18] RDF Schema, See http://www.w3.org/TR/rdf-schema/, accessed Jan. 2007. RDFS adds support for category relationships to RDF.

[19] The OWL Web Ontology Language, http://www.w3.org/TR/owl-ref/, accessed Jan. 2007. OWL adds support to RDFS for range and domain constraints, existence and cardinality constraints, transitive, inverse and symmetrical properties and for logic.

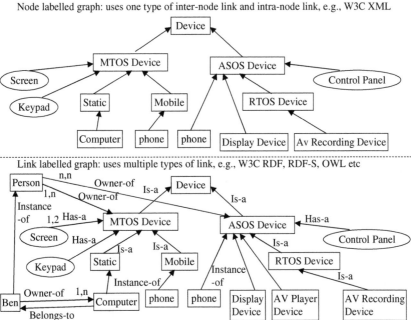

Figure 8.10   Two different graphical KRs for the device domain: a weaker, less expressive, node labelled graph representation and a stronger edge labelled graph representation

includes support for logic reasoning. There is a conceptual model referred to as the Semantic Web layer model presented by Berners-Lee in 2003 that describes how these different models are related (Horrocks *et al.*, 2005). The same type of concepts will have a different structure or syntax in these different types of XML-based ontology models. They will also differ in terms of how explicit they make richer semantic relationships and constraints.

8.4.2 Design Issues

There is still an ongoing debate about what should be specified in a KR model. For example, Horrocks *et al.* (2005) consider how the Semantic Web KR model should support rules and how a closed world versus open world assumption should be modelled.

8.4.2.1 Open World Versus Closed World Semantics

RDBMS-type KB models tend to assume *closed-world semantics*, i.e., if data is not present, it is false (negation as failure). In contrast, semantic-type KBs tend to use *open-world semantics* which regard absence as unknown (negation as unknown). Sometime semantic-type KBs may be mapped onto an RDBMS for more efficient and manageable storage. Providing queries for the semantic type KB are not implemented as relational queries, this difference in open-world versus closed-world semantics should not be a problem. Another point is the semantics of the data is often held in the application code that creates the data storage structures and is not held with the data. Application semantics are often not explicit and not accessible from the data.

8.4.2.2 Knowledge Life-cycle and Knowledge Management

Knowledge-based Management (KM) involves a life-cycle of knowledge creation, deployment and maintenance.

8.4.2.3 Creating Knowledge

In order to create a useful and accurate application domain-based knowledge-based system, a combination of an understanding of the problem and a collection of heuristic problem-solving rules, which experience has shown to be effective in the domain, is needed. The earliest type of knowledge-based system tended to use the knowledge of one or more human domain experts for the source of a system's problem solving strategies and is often referred to as an *expert system*. For example, DENDRAL, developed in the late 1960s (Buchanan *et al.*, 1969) was designed to infer the structure of organic molecules from their chemical formulas and from mass spectrographic information about the chemical bonds present in the molecules.

Today, there are many interactive knowledge creation tools available that enable less expert and specialised developers and users to create knowledge models and then to export the knowledge model out of the tool in a form that can be imported, interpreted or parsed and then invoked via an API by computer applications, for examples, see Protégé[20] and Jena[21] respectively.

The creation of an ontology is a part of a knowledge representation process that is machine-understandable but is also human-understandable based on a common understanding of how people represent, understand and acquire knowledge rather than on how machines do these. For example, all objects that have either a microprocessor, microcontroller or central processor unit, have memory, have an input interface, have an output interface, have a network interface, and are collectively described as 'devices'. The concept of a device is unique in our ontology and has only one definition. A concept represents an idea of something that could be a real-world object or an abstract object such as human behaviour, or feelings, etc.

In general, all such concepts do not have any absolute definition; they are defined in terms of other concepts (ontological commitments from the primary concept to secondary concepts). In the above example, the concept of 'input interface', 'output interface', 'memory' and 'microprocessor', 'microcontroller' or 'central processor unit' have all been used to define the concept 'device' (see Figure 8.10). So concepts are understood through their relationship to other concepts, which have already been understood and remembered. The process of creating a KR for a domain consists of the following steps:

- Defining a concept taxonomy to specify categories in terms of generalisation and specialisation.
- Defining a set of relations used between concepts and between concepts and properties of concepts. This set of relations can itself be organised into a hierarchy.
- Defining constraints on the value in a relation and constraints on values of properties of concepts, e.g., positive integer, etc.
- Defining axioms[22] on relations and concepts (for a logic-based KR).

---

[20] Protégé, an open source ontology editor and knowledge base framework. Home page, http://protege.stanford.edu, accessed Jan. 2007.

[21] Jena, an open-source Semantic Web Framework for Java. Home page http://jena.sourceforge.net/, accessed Jan. 2007.

[22] Where *axioms* are taken as facts, propositions, and sentences (concepts connected by defined operators) which are always true, such as the sentence 'is-owner' is the inverse of 'belongs-to'. However, there is also the issue of whether or not this is true in selected domains or in all domains, or in one possible world or in all possible worlds.

There are many different variations of this process, see, for example, Noy and McGuinness (2001). There are many modelling choices to be made in ontology design as in any kind of design. Some choices are to do with categorisation choices whether or not to model something as a sub-category of a category, as an instance of a category or as a composition, e.g., the 'owner' of a device could be modelled as an instance of a 'user' of a device or a sub-category of user. There are choices of how to model relationships which have different cardinality such as one-to-one, one-to-many or many-to-many. There are choices to be made about the direction of relationships and whether or not two directed relationships are opposites, e.g., 'is_owner_of' relation is the inverse of the 'belongs_to' relation. There are choices about which constraints can be used in relationships between properties and between categories. The ease and explicitness with which complex relationships can be modelled also depend upon the type of ontology models such as lightweight versus heavyweight.

Often a knowledge model which is created requires a process or refinement in order to improve it and validate it, perhaps by using the KB model to solve example problems, letting the stakeholders criticise its behaviour, and making any required changes or modifications to the program's knowledge. This process is repeated until the program has achieved the desired level of performance.

8.4.2.4 Knowledge Deployment and Maintaining Knowledge

The simplest knowledge model to deploy and maintain is for a single world or domain model which has gone through a process of refinement involving all the stakeholders. This knowledge model is understood and agreed by them before being deployed and this then remains fixed during deployment. However, in practice, knowledge deployment is often complicated by the existence of autonomous groups of stakeholders developing overlapping but differing ontology models. In addition, ontological commitments can often only be derived through deployment, which can change the ontology model being committed. This is because an understanding of the domain and applications of a domain varies over time, and with respect to different stakeholders, leading to changes in the domain and changes in conceptualisation. Thus multiple possible semantic models often exist within a domain and need to be related.

The use and interoperation between heterogeneous ontology models during the modelling process of the conceptual world can be roughly classified into two kinds: merging or integration and alignment. Noy and Musen (2000) define view merging as the creation of a single coherent ontology that includes the information from all the sources, and alignment as a process in which the sources must be made consistent and coherent with one another but kept separately. This may entail maintaining local ontology wrappers for each data source leading to a multi-lateral ontology model. The merging approach often leads to the creation of a global knowledge model where individual local ontologies can be mapped onto each other. The alignment approach avoids the process of creating a global knowledge model, instead it maps specific semantic content between the local ontologies, directly (Poslad and Zuo, 2008).

Ontologies within a domain are heterogeneous not just because independent knowledge domains can be modelled differently but also because they face different ontological commitments in different application domains – the *pragmatics* or particular context of use of concepts will vary. Different human users may employ different subsets of concepts with different semantics based upon their understanding. It may be useful if the knowledge base adapts its query responses to users' contexts (Poslad and Zuo, 2008).

8.4.2.5 Design Issues for UbiCom Use

Metadata, data which describes data and which often summarises data, is manually derived or automatically derived, e.g., via summarisation, from data than can be used to annotate data and to enhance information searches. This is because searches can be more efficiently performed by testing a subset of metadata to match data queries rather than testing the whole of the data. If the metadata also uses a representation with a well-defined semantics, then semantic searches on the metadata can be performed in addition to syntactic searches.

The computation resources in some devices, particularly, mobile, ASOS and embedded devices are limited by design. It may not be possible to handle well, semantic information and commands on the device. The length of time computation takes is one resource that crucially affects the semantics of its outputs because contexts are more likely to change in dynamic environments and when resource constrained systems are situated in dynamic environments (Kaelbling, 1991). Care is needed so that the intended semantics of the output of a computation are valid, given the time it takes to perform the computation, and fit the computational resources to use the semantics. Designs for IS operations need to be selected that can fit the ICT resources of devices, e.g., low memory versus longer computation tradeoffs need to be carefully examined.

Devices embedded in a local environment often have only a partial view rather than a global view of their environment. This may be because systems do not have sufficient resources such as time and the ability to acquire and access the necessary global knowledge. Systems today often create information in a far more decentralised manner than they did in the past. They are not isolated systems but bring together large communities often in ad-hoc situations. This social (networking) dimension, such as is found in social networks, creates environments for partial and dynamic information. AI approaches can assist in dealing with incomplete and uncertain information and help in the process of decision making either autonomously or through human intervention.

8.5 Classical Logic IS

Many researchers regard *classical logic* based upon first-order predicate true or false logic[23] as being at the heart of any IS which needs to support reasoning about the system. These types of IS are also commonly referred to as (logic) *reasoning systems, deliberative IS*, and as *symbolic AI* because these systems involve the manipulation of symbols in the form of logic formulae, although in general symbols could also refer to any mathematical formulae including algebraic formulae.

8.5.1 Propositional and Predicate Logic

Propositional logic is the simplest kind of logic model of the world. Here knowledge is represented in the form of propositions, statement or relations which are either true or false. Multiple propositions can be combined using logic operators or connectives on *literals* which represent the things in the KB that are operated on. The standard logic connectives are: *and* (also called *conjunction*, also referred to by A $\wedge$ B, A $\cap$ B), *or* (also called *disjunction*, A $\vee$ B, A $\cup$ B), *not* (also called *negation,* ¬A), *equals* and *implies* (as in A implies B, A $\Rightarrow$ B expressing if A then B rules). These connectives can be used to form sentences that are well-defined formulae adhering to a defined structure or syntax. The meaning or semantics of these operators can be described in Truth Tables.

---

[23] True or false logic is also referred to as *crisp logic* and *binary logic* whose states can be represented as one or zero. It is also the same logic which the underlying digital electronic hardware that ICT systems use.

In *Predicate logic*,[24] *predicates* are defined to support more expressive sentences than propositions which allow a property to be related to some object or a property related to some value. The sentence 'Device A is in hibernate mode', expresses a predicate which may be thought of as a kind of function, also referred to as propositional functions, which applies to individuals (who would not usually themselves be propositions) and yields a proposition, e.g., mode (Device A, Hibernate).

The main difference between propositional and predicate logic is that in propositional logic all relationships are true or false whereas in predicate logic, predicate functions are neither true nor false, although their evaluation may be so. As well as expressing individual relationships, more general relationships can be expressed using two additional logical operators more specifically referred to as quantifiers. The universal quantifier (for all) expresses the notion that all properties have a certain value while the existential quantifier (there exists) expresses the notion that at least one property has a certain value.

8.5.2 Reasoning

Reasoning, also referred to as *inferencing*, involves logical operations on logical sentences or statements within a (logical) model, bounded by what is being modelled, e.g., the world, an application domain, etc., in order to draw conclusions and to derive other sentences, e.g., A entails B, A $\models$ B. Inferencing is used to search for entailments. Sometimes multiple possible worlds or models will be possible, because the IS has an incomplete model of the world, because the IS is not sure which model will become reality in the future, and perhaps because different viewers and applications may also have their own model of the world (Section 8.4.2.4).

Model checking is used to check that entailments of sentences are valid in all possible worlds or models. Valid sentences are called tautologies. Sometimes it is just necessary to check if a sentence is true in some specific model, i.e., it is *satisfied* in that model rather than being able to say it is true in all models, i.e., it is *valid* in all models. Model checking can involve changing logical restructuring, changing the syntax of logical sentence while keeping the semantics the same, in order to make checking the logical equivalence of two sentences easier. Various laws of logic are used to aid model checking, e.g., such as commutative laws which say $A \wedge B = B \wedge A$, associative law which says $(A \wedge B) \wedge C = A \wedge (B \wedge C)$, etc. Model checking is used every time new knowledge is added. If only entailed logic sentences ever get added, then the model is said to be *monotonic*.

There are several standard inference rules in propositional logic. For example, the Modus Ponens rule which says that if sentences $A \Rightarrow B$ and A is given, then B can be inferred. An inference algorithm used for knowledge discovery may use one or more inference rules. If the inference algorithm only finds entailed sentences, then it is called *sound* or truth preserving. Inferencing based upon resolution seeks to show that sentences can be entailed by proving the negated sentence is not satisfiable in that KB by proving the existence of logical contradictions. Resolution-type inferencing can result in lengthy computation. If logic sentences in the KB are restricted to horn clauses which is a disjunction of literals where at most one is positive, e.g., $A \vee \neg B \vee \neg C$ is a horn clause whereas $A \vee B \vee \neg C$ is not, inferencing using backward and forward chaining can be used which is linear as the size of the KB increases, i.e., much less computation is needed than with resolution (Russell and Norvig, 2003, pp. 194–239).

---

[24] The most common form of Predicate logic is called First-Order Predicate Logic or FOPL. Predicate logic is also referred to as First-Order Logic, FOL or FOPL or First-Order Predicate Calculus or FOPC.

Predicate logic in KB can be converted into propositional logic by using rules for instantiating universal and existential quantifier statements in order that propositional logic inferencing can be used, however, the propositional inference approach is often quite inefficient for predicate inferencing because we need to infer each instance of a sentence containing a universal quantifier. The efficiency of propositional inferencing can be improved, first, by lifting first-order inferencing rules such as Modus Ponens so that they can apply more generally. Second, substitutions can be found and used to make logical sentences look semantically equivalent in a process called *unification*.

Specific FOL programming languages, the most popular being PROLOG, PROgramming in LOGic, (Bratko, 2000) tend to support horn-clause type FOL and tend to intertwine the logic with control or ordering for operating on the logic. PROLOG enables applications and tools to be developed which can search stored logic sentences in a KB using backward and forward chaining search algorithms. It can also support planning and machine learning.

Automated reasoners, also called *theorem provers* have also been developed. These, unlike logic programming languages, tend to use full FOL and tend to separate the control of the logic from the logic representation itself more. A wide variety of problems can be attacked by representing the problem description and relevant background information as logical axioms and treating problem instances as theorems to be proved. This insight is the basis of work in automatic theorem proving and mathematical reasoning systems. Many important problems such as the design and verification of logic circuits, verification of the correctness of computer programs, and control of complex systems seem to respond to such an approach.

8.5.3 Design Issues

Of all of the types of logic, first order logic has the most consistent set of axioms, enabling the integrity of the logic model to be more scalable,[25] for use by many systems, compared to, say, soft computing and relational databases. Soft Computing (Section 8.6) relates the truth of something to the amount of evidence and precision about the basic concepts. This can vary according to the context of use including time. Relational DataBase (RDB) type KRs (Section 12.2.9.5) use relational algebra to enable knowledge organised as set or table-based data to be queried to produce subsets or to be joined to form supersets. However, RDBs operate on a closed world assumption and make assumptions about whether something is true based upon its existence in the database. If it is not in the database, then it is regarded as false. Hence, different databases within a domain can disagree about what is fundamentally true or false in a domain because they contain different sets of data for the same domain.

Hence many KR systems are often based on classical logic. However, the main limitations of classical logic must also be noted such as the difficulty in expressing exceptions, imprecision, uncertainty and the degree of computation that is sometimes needed to establish truth of complex logical structures. In practice, logical inconsistencies can occur in a distributed KB unless each new entry is checked against the whole KB and unless any new logic sentences which cause the model to become non-monotonic are prevented. There are also many different sub-types and extensions to classical logic. Sometimes reasoning may require the use of multiple logics, e.g., time and uncertainty. It is far from clear how the completeness and soundness of multiple types of logic model can be checked. Logic-based IS may not operate efficiently in time-constrained environments. In logic-based KB IS, percepts are likely to be symbolic. But for

---

[25] Scalable in this sense means that logic inferences and axioms return consistent results such as truth values across a wide body of possibly heterogeneous knowledge sources.

many environments, it is not obvious how the mapping from the environment to the symbolic percept might be realised (e.g., images). It is not clear whether any single logic will suffice or how hybrid logics can be used. Representing the properties of dynamic, real-world environments (e.g., temporal information) is challenging.

8.6 Soft Computing IS Models

Many decisions which involve interaction with humans and the physical world are soft rather than being expressed as either true or false. These are more qualitative and may involve some imprecision and uncertainty. Such systems can be designed using soft computing techniques, e.g., probability theory to work on problems with states which are uncertain and fuzzy logic to work on problems with states which are imprecise.

8.6.1 Probabilistic Networks

There are several notions which are needed to model the likelihood of indeterminate events happening or to model the degree of belief in a proposition or predicate and to reason about them. There is a *prior or unconditional probability* for a proposition or event in the absence of any information. If information is known, then *conditional or posterior probabilities* must be used expressed as the probability of A given that we know the probability of B, expressed as $P(A|B) = P(A \land B)/P(B)$. A conditional probability is expressed in terms of two unconditional probabilities.

The probabilities for individual values of a proposition are defined in a probability distribution either as a set or vector of discrete functions or as a probability density function. Multiple probability distributions for different propositions may be combined into joint probability distributions, e.g., $P(A \land B)$, the joint probability of weather conditions and traffic conditions. A full joint probability refers to the whole set of different event probabilities which are joined.

Using the product law $P(A \land B)=P(A|B)\ P(B)$ and commutativity so that $P(A \land B) = P(B|A)\ P(A)$ enables $P(B|A) = (P(A|B)\ P(B))/P(A)$ to be derived which is called *Bayes's rule* or law or theorem. This is the basis of probabilistic inferencing. This expresses a conditional probability in terms of another conditional probability and two unconditional probabilities.

A *Bayesian Network* (BN), also called a *belief network* or *probabilistic network*, can be used to represent any full joint probability distribution. A Bayesian network can be used for inferencing in a context-aware UbiCom scenario in which there are both non-deterministic preconditions and non-deterministic outcomes. For example, in the adaptive vehicle scheduling scenario, both passengers and buses can indeterminately arrive at pick-up points and humans and vehicles can indeterminately wait (Figure 8.11).

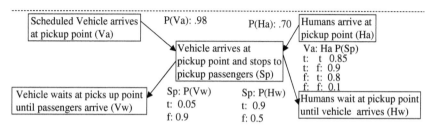

Figure 8.11 A Bayesian network which models vehicles and passengers indeterminately arriving and waiting at pick-up points

8.6.2 *Fuzzy Logic*

Fuzzy logic is useful to represent a model where the outcome of a proposition is deterministic but is somewhat approximate or imprecise such as the vehicle is travelling very slowly, or slowly, or at a moderate speed, or fast or very fast. This kind of imprecision can also be used in fuzzy rules. For example, in the adaptive transport scheduling scenario, a fuzzy logic rule could be if the bus is travelling *slowly* away from the pick-up point and a passenger is moving *quickly* towards the pickup point, then slow down the vehicle to stop *near* the pick-up point. Here, the terms slowly, quickly, and near act as fuzzy descriptors. In contrast, a crisp logic rule, for the same situation in the scenario could not differentiate between the bus moving slowly away from the pick-up point and being able to stop near the pick-up point versus a bus moving fast away from the pick-up point and having to stop far away from the pick-up point when it realises there are additional passengers who could be accommodated.

8.7 IS System Operations

Generic IS model operations are discussed here, include searching and planning. Reasoning as an IS operation is intertwined with the use of specific logic as representation for the IS model so covered both in the Crisp Logic and Soft Computing sections. Learning is also a generic IS operation but requires specific architectural support for feedback, hence is covered in the Learning IS section.

8.7.1 *Searching*

Searching is a problem-solving technique that systematically explores a space of problem states, i.e., successive and alternative stages in the problem-solving process in order to select a goal state or a chain or path through intermediate states to achieve a goal state. This space of alternative solutions can then search to find an answer (Newell and Simon, 1976).

Much of the early research for state space search was undertaken using common board games such as checkers, chess, and the 15-puzzle. In addition to their inherent intellectual appeal, board games have certain properties that made them ideal subjects for this early work. Most games are played using a well-defined set of rules. This makes it easy to generate the search space and frees the researcher from many of the ambiguities and complexities inherent in less structured problems. The board configurations used in playing these games are easily represented on a computer, requiring none of the complex formalisms needed to capture the semantic subtleties of more complex problem domains.

Examples of searching through problem states include: searching the alternative board configurations in a game in order to win the game (goal state); searching through a set of annotated images of faces to see if an unknown face matches an already annotated one (in the personal memories scenario); searching through road junctions in order to find a route between the current location and a destination location (in the adaptive transport scheduling scenario); searching for a route through obstacles in order to guide a robot from one location to a destination location. Hence the problem is expressed as a start state, a goal state, a goal test function to test if each state is the goal state and a utility function that maps a path between two states to some performance or cost metrics which can be aggregated to calculate the cost to reach goal state.

The basic kind of search algorithm is to use an *uninformed search*, also referred to as a *brute-force search* or *blind search* in which no hints are available about how to reduce the search space, e.g., *breadth-first search*, and *depth-first search*. Often the problem search space can be

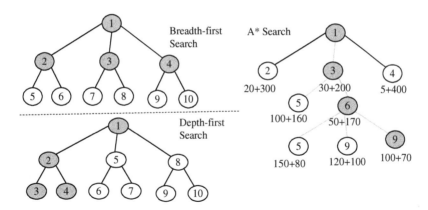

Figure 8.12 Two types of uninformed or brute force search, a breadth first search versus a depth first search and one type of informed search, the A* search

represented graphically (Figure 8.12). Searching involves traversing the graph bread-first or depth-first and testing each node to check if it is the goal state. Uninformed problem space searches tend to operate in the forward direction (progression) from start state to end goal state (called *forward-chaining*).

In some cases, the search algorithm must also handle the case when the IS system cannot identify its start state, e.g., a transport vehicle is lost and requires a route to a destination. The mode of the search space may not be uniform, i.e., different nodes may have different numbers of branches. Also problem spaces may not be single valued but may be multiple-valued.

Some problems such as games can generate extremely large search spaces, requiring large amounts of computation, for uninformed search techniques, e.g., depth-first can fail infinite depth spaces and where loops exist, time-complexity (number of nodes accessed is high), space complexity (maximum number of nodes in memory is linear) and optimality (it may not always find a least-cost solution). For this reason, variations of these informed searches such as depth limited search (depth-first search with depth limit) or an iterative deepening search can be deployed (Russell and Norvig, 2003, pp. 59–93).

A general solution to reduce the computation of an uninformed search is to use informed search techniques which use problem-specific information to limit the problem space. A core component of an informed search is a heuristic function which in its most general form is defined as a function which depends upon the current node in a problem space, e.g., a cost function rich returns a value to reach that current node from a previous node. If nodes represent physical locations, the cost function could be related to the distance between nodes, the time taken to travel between nodes or the energy cost in travelling between nodes. The heuristic function which maps each node to a value depends on information about the problem. A variation of the cost function is to assign a first cost to reach the current node from the previous link and to assign a second cost to go from the current node to the goal node. This cost heuristic is used by the *A*search* or A-star, one of the most well-known types of informed search or heuristic search (Figure 8.12).

A core application for searching in general is information retrieval. Here the concern is in reducing the cost in terms of the number of goal tests for each node. Typically, this is achieved through testing indexes or metadata in place of use of the actual data in search operations and through the use of efficient index structures such as sparse indexes, B-trees, hashing, linked-lists and R-trees to support informed searches (Watson, 2006, pp. 326–341).

8.7.2 Classical (Deterministic) Planning

Much of the early research in planning began as an effort to design robots that could perform their tasks with some degree of flexibility and responsiveness in the physical world. Planning assumes a robot is capable of performing certain atomic actions. It attempts to find a sequence of those actions that will accomplish some higher-level task, such as moving across an obstacle-filled room. Planning research now extends well beyond the domains of robotics, to include the coordination of any complex set of tasks and goals. Modern planners can be used for embodied software robots or agents as well as for complex adaptive control in machines such as particle beam accelerators.

Planning involves searching for a plan and then executing the plan. Searching for a plan uses the following: a planning model representation; backward chaining (from the goal state to the current state) to determine chains of actions which lead to the goal state, forward chaining to reach the goal state from the current state; informed search techniques based upon heuristics (Section 8.7.1) because otherwise uninformed searches would be too inefficient; problem decomposition of a complex problem into simpler more easily solvable problems (e.g., divide-and-conquer).

The planning model represents states, goals, actions which transition states towards goals, chains of actions between non-adjacent states and heuristic cost functions. It also allows the choice of multiple chains of actions to be constrained using some heuristics e.g., a transport system that must pick up multiple passengers in multiple locations may be constrained by the path between locations to minimise fuel consumption or the time taken or some combination of both of these. This can be modelled as a graph where the nodes represent states and the links between nodes represent actions. In Figure 8.12, the links which represent the actions are not labelled to identify the actions but this information could easily be added to form an edge-labelled as well as a node-labelled graph. Actions can be represented in terms of *pre-conditions* which, if true, enable actions to be triggered, *post-conditions* or *effects*, which define what should now should be true if an action is successfully executed.[26]

As mentioned earlier, an important technique to make complex planning problems more solvable is to decompose the problem to reduce the problem space to search. Search techniques often cannot take into account problem decomposition into sub-problems, able to calculate forward or backward chains for sub-problems and then to combine these to form a solution for the whole problem, resulting in linear time-planning rather than exponential time-planning algorithms. A well-known problem decomposition method is based upon hierarchical task analysis or HTA in which a goal is solved by executing a set of high-level actions which then get refined into lower-level actions. In subsequent steps these can then be refined further until only primitive actions remain which cannot be refined further (Figure 8.13). Plans represent decompositions of a high-level action into a set of lower-level actions. Plans are stored in a plan-library from which they are selected to form a path to goals.

One limitation of forward and backward state searches is that they are dependent on the ordering of each current state evaluated in relation to a path to the goal state They also cannot take advantage of actions that can be planned to occur in any order to achieve a goal, providing they do not interfere or interact with each other – *partial-order-planning* or *POP* in Figure 8.13. In some cases, a chain of actions which can occur in any order, can achieve the goal. POP can reduce the size

---

[26] Event-driven models may also be used to define events and action preconditions but they tend not to define action post-conditions or to define an explicit action representation.

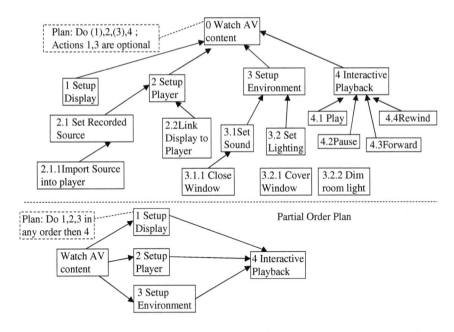

Figure 8.13 Hierarchical Task Plan and Partial Order Plan for watch AV content goal

of the search to find a plan of actions and gives some added flexibility to the plan, e.g., when specific actions are temporarily delayed.

For the goal-based IS design, planning is used to enable a chain of actions to be selected that will achieve the goal. Hence, any action executed is part of the plan. For some types of system interaction, the environment events may trigger actions for which there is no current plan or goal, a situated action. The situated action would trigger a goal-based IS design to form a plan for it.

8.7.3 Non-Deterministic Planning

Classical planning considers planning in static, observable and deterministic environments using systems that can act correctly and whose plans are executed completely. However, some environments may be nondeterministic and partially observable and here classical planning will fail because it cannot determine a course of action. There are two types of planning methods, sensorless or *conformant planning* and *conditional planning* which can work with bounded indeterminacy-type environments, in which actions have unpredictable effects but in which the set of all possible preconditions and effects is fixed. *Contingency planning* or conditional planning, for example, involves constructing plans with different branches representing the different conditions. During plan execution, the agent senses the state and then chooses the appropriate contingency branch. Contingency planning requires the ability to sense actions during executing and the use of a bounded number of contingencies. An example of conditional planning is used in context processing (Section 7.2.4). There are also two types of planning methods, monitoring and replanning and continuous planning, which can work in unbounded indeterminacy and in which the set of preconditions and effects is either unknown or very large (Russell and Norvig, 2003, pp. 417–461).

EXERCISES

1. Compare and contrast the types of architectural models for ISs given in Section 8.3, i.e., reactive, environment model, goals, utilities and learning with respect to how actions are selected, how the model is represented and what types of environments, e.g., deterministic, episodic, etc., these models are suited to.
2. Consider the design of a so-called vertical layered model. Is this a good design if reasoning layers are activated before reactive layers or vice versa? Compare and contrast different orderings for the IS models within a hybrid IS model for different application scenarios.
3. Discuss what kind of expressivity ranging from a lightweight to heavyweight ontology is needed in knowledge modelling used by reactive, environment model-based and goal-based IS designs.
4. Describe the need to make a KR machine readable and machine-understandable.
5. Give the design of a system consisting of a set of limited resource devices that supports machine-readable and machine-understandable data.
6. What are the differences, if any, in how an automatic reasoner operates versus how a logical programming language is used?
7. What is first-order logic? Explain what the difference is between this and propositional logic.
8. Explain why first-order logic (FOL) is so predominant in KB systems.
9. What is soft computing? What is the difference between imprecision and uncertainty? Discuss some specific types of model to model imprecision and to model uncertainty.
10. Show how uncertainty can be handled in planning.

References

Berners-Lee, T., Hendler, J. and Lassila, O. (2001) The Semantic Web. *Scientific American*, 284(5): 34–43.

Bratko, I. (2000) *Prolog Programming for Artificial Intelligence*, 3rd edn. Harlow: Pearson Education.

Buchanan, B.G., Sutherland, G.L. and Feigenbaum, E.A. (1969) Heuristic DENDRAL: a program for generating explanatory hypotheses. In B. Meltzer and D. Michie (eds) *Machine Intelligence*, vol. 4. Edinburgh: Edinburgh University Press, pp. 209–254.

Coppin, B. (2004) *Artificial Intelligence Illuminated*. Sudbury, MA: Jones and Bartlett Publishers.

Corcho, O., Fernández-López, M. and Gómez-Pérez, A. (2003) Methodologies, tools and languages for building ontologies: where is their meeting point? *Data & Knowledge Engineering*, 46(1): 41–64.

Davis, R., Shrobe, H. and Szolovits, P. (1993) What is a knowledge representation? *AI Magazine* 14(1): 17–23.

Franklin, S. and Graesser, A. (1996) Is it an agent, or just a program?: A taxonomy for autonomous agents. In Proceedings of Workshop on Intelligent Agents III, Agent Theories, Architectures, and Languages. *Lecture Notes in Computer Science*, 1193: 21–35.

Frege, G. (1879) *Conceptual Notation: A Formula Language of Pure Thought Modelled upon the Formula Language of Arithmetic*. Halle: L. Nebert, 1879. Trans. S. Bauer-Mengelberg, in Jean Van Heijenoort (ed.) (1967) *From Frege to Gödel: A Source Book in Mathematical Logic, 1879–1931*. Cambridge, MA: Harvard University Press.

Friedman-Hill, E. (2003) *Jess in Action: Java Rule-based Systems*. Upper Saddle River, NJ: Pearson.

Gómez-Pérez, A. (1999) Ontological engineering: a state of the art. *Expert Update*.

Han, J. and Kamber M. (2006) *Data Mining: Concepts and Techniques*, 2nd edn. San Francisco: Morgan Kaufmann Publishers.

Horrocks, I., Parsia, B., Patel-Schneider, P., *et al.* (2005) Semantic Web architecture: Stack or two towers? In *Principles and Practice of Semantic Web Reasoning* (PPSWR 2005), pp. 37–41.

Jasper, R. and Uschold, M. (1999) A framework for understanding and classifying ontology applications. In Proceedings of IJCAI-99 Workshop on Ontologies and Problem-Solving Methods: Lessons Learned and Future Trends, Stockholm, Sweden, August 1999. http://sunsite.informatik.rwth-aachen.de/Publications/CEUR-WS/Vol-18/11-uschold.pdf

Kaelbling, L.P. (1991) A situated-automata approach to the design of embedded agents. *ACM SIGART Bulletin*, 2(4): 85–88.

Karray F. and De Silva C. (2004) *Soft Computing and Intelligent Systems: Design: Theory, Tools and Applications*. Upper Saddle River, NJ: Pearson Books.

Lindwer, M., Marculescu, D., Basten, T., *et al.* (2003) Ambient Intelligence visions and achievements: linking abstract ideas to real-world concepts. In *Design, Automation and Test in Europe Conference and Exhibition*, pp. 10–15.

McCarthy, J. and Hayes, P.J. (1987) *Some Philosophical Problems from the Standpoint of Artificial Intelligence: Readings in Nonmonotonic Reasoning*, San Francisco: Morgan Kaufmann, pp. 26–45.

Mitchell, T.M. (1997) *Machine Learning*. New York: McGraw-Hill.

Newell, A. and Simon, H.A. (1976) Computer science as empirical enquiry. *Communications of the ACM*, 19: 113–126.

Nicola, G. (1997) Understanding, building and using ontologies. *International Journal of Human and Computer Studies*, 46(2–3): 293–310.

Nicola, G. (1998) Formal ontology in information systems. In Proceedings of the International Conference on Formal Ontology in Information Systems, FOIS'98, Amsterdam, IOS Press, pp. 3–15.

Noy, N.F. and Klein, M. (2003) Ontology evolution: not the same as schema evolution knowledge. *Information Systems*, 6(4): 428–440.

Noy, N.F. and McGuinness, D.L. (2001) Ontology Development 101: A Guide to Creating Your First Ontology. Stanford Knowledge Systems Laboratory Technical Report KSL-01-05. Available from http://protege.stanford.edu/publications/ontology_development/ontology101-noy-mcguinness.html, accessed Jan. 2007.

Noy, N. and Musen, M. (2000) PROMPT: Algorithm and tool for automated ontology merging and alignment. In Proceedings of 17th National Conf. on Artificial Intelligence (AAAI'00), pp. 450–455.

Nwana, H.S. (1996) Software agents: an overview. *Knowledge Engineering Review*, 11(3): 205–244.

Poslad, S. and Zuo, L. (2008) An adaptive semantic framework to support multiple user viewpoints over multiple databases. In M. Wallace, M. Angelides and P. Mylonas (eds) *Advances in Semantic Media Adaptation and Personalization*, Series: Studies in Computational Intelligence, Vol. 93. Berlin: Springer Verlag, pp. 261–284.

Ramos, C., Augusto, J.C. and Daniel, S. (2008) Ambient Intelligence – the next step for artificial intelligence. *IEEE Intelligent Systems*, 23(2): 15–18.

Riley, J.R., Greggers, U., Smith, A.D. *et al.* (2005) The flight paths of honeybees recruited by the waggle dance. *Nature*, 435 (12 May): 205–207.

Russell, S. and Norvig, P. (2003) *Artificial Intelligence: A Modern Approach*, 2nd edn. Upper Saddle River, NJ: Prentice Hall.

Shadbolt, N., Berners-Lee, T. and Hall, W. (2006) The Semantic Web revisited. *IEEE Intelligent Systems*, 21(3): 96–101.

Suchman L.A. (1987) *Plans and Situated Actions: The Problem of Human Machine Communication*. Cambridge: Cambridge University Press.

Tarski, A. (1944) The semantical concept of truth and the foundations of semantics. *Philosophy and Phenomenological Research*, 4: 341–375. Available from http://www.ditext.com/tarski/tarski.html.

Thompson, S.G. and Azvine, B. (2004) No pervasive computing without intelligent systems. *BT Technology Journal*, (22)3: 39–49.

Watson, R.T. (2006) *Data Management: Databases and Organizations*, 5th edn. Chichester: John Wiley & Sons, Ltd.

Wielinga, B.J., Schreiber, A. Th. and Breuker, J.A. (1992) KADS: a modelling approach to knowledge engineering. *Knowledge Acquisition Journal*, 4(1): 5–53. Special issue 'The KADS approach to knowledge engineering'. Reprinted in: Buchanan, B. and Wilkins, D. (eds) (1992) *Readings in Knowledge Acquisition and Learning*, San Mateo, CA: Morgan Kaufmann, pp. 92–116.

Wooldridge, M. (2001) *An Introduction to MultiAgent Systems*. Chichester: John Wiley & Sons, Ltd.

Wooldridge, M. and Jennings, N. (1995) Intelligent agents: theory and practice. *Knowledge Engineering Review*, 10(2): 115–152.

9

Intelligent System Interaction

With Patricia Charlton

9.1 Introduction

As more UbiCom devices become better and universally connected, we finally have the building
blocks to enable smart interaction. Smart interaction refers to a richer interaction beyond using
basic universal network communication protocols, e.g., TCP/IP. For example, smart interaction
may make use of coordination to reduce the need for explicit communication when different
autonomous systems interoperate and may make use of semantics to not only share information
and tasks but also to share intentions, goals, plans, knowledge and experiences. In general, the
ability to communicate using a rich language in a meaningful way is often regarded as a clear sign of
intelligence.

9.1.1 Chapter Overview

Next (Section 9.2) interaction multiplicity and attempts to compare and contrast CCI and HHI are
described. Following this, designs for intelligent interacting systems are discussed based upon two
different strategies, making system interaction more intelligent, and making individual intelligent
systems interact, i.e., use of MAS (Section 9.3.3). This is followed by a discussion of specifying
interaction (protocols) using network protocols, semantic exchange and speech acts. Types of
multi-agent system design which support the latter type of interaction are discussed. This chapter
concludes with a brief overview of some intelligent system interaction applications (Section 9.4).

9.2 Interaction Multiplicity

At its simplest level, interaction can be viewed as a *multiplicity* or *cardinality* of communication
relations between independent senders and independent receivers. Interaction multiplicity can lead
to resource contention by multiple users. Multiple transmissions cause conflicts and interference
and can lead to the overuse of operating resources such as energy due to redundant communication.
Interaction multiplicity also increases the complexity of the interaction, and reduces the

Ubiquitous Computing: Smart Devices, Environments and Interactions   Stefan Poslad
© 2009 John Wiley & Sons, Ltd

determinism of and control of the interaction. Hence interaction needs some kind of control mechanisms to prevent, or detect and correct these problems. The mechanisms used for control will depend on whether the interacting entities are dumb, smart yet cooperative or smart yet competitive, i.e., are self-interested. There are several important types of interaction multiplicity (Figure 9.1). This section discusses these types of multiplicity and how multiplicity can be modelled and managed in the different components of ICT systems (Table 9.1). A powerful model for designing this range of types of organizational multiplicity is that of Multi-Agent Systems or MAS (Russell and Norvig, 2003; Wooldridge, 2001; Fasli, 2007).

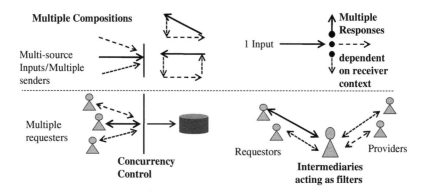

Figure 9.1 Some examples of smart interaction: service composition, concurrency control for shared resources, receiver context dependent responses and active intermediaries acting as filters

Table 9.1 Summary of types of multiplicity and associated designs

Main types of interaction multiplicity	Solving interaction multiplicity in different Viewpoints (V) of systems: Human (H), Physical World (P), Services (S) and Network (N).
Multiple independent senders and receivers	HV and SV: multimodal sensory interaction; Bidders, sellers participate in single vs. multiple markets NV: MIMO, e.g., PSDN, Mesh routing
Use of shared passive intermediaries	AV: RDBMS ACID support NV: IP packet switching NV: MAC, e.g., FDMA, TDMA, CDMA; Token Ring
Use of shared active intermediaries	HV: Personal assistants, brokers etc as agents SV: Proxies, broker models, application caches NV: Routers, wireless base-stations
Shared conversations	HV: Natural language like (e.g., speech act) conversations SV: Service interaction patterns e.g., task delegation, request-response, etc
Organised interactors & interactions	HV, SV: Task delegation, work-flows, social cooperation vs. economic competition
Internally driven (self-organisation)	PV, SV: Self-star systems
Semantic and mentalistic communication	HV, SV: Shared language, meaning, intentions, beliefs, etc.

9.2.1 P2P Interaction Between Multiple Senders and Receivers

Interaction multiplicity (or interaction cardinality) arises when people, tasks and devices distributed in space and time become interlinked and as more of the physical world becomes augmented with digital devices (Figure 9.2). UbiCom can be designed and managed to support basic interaction between multiple independent senders and receivers, these are often referred to as Multiple Input Multiple Output (MIMO) systems.

Communication is complex but can be handled by lower-level network protocols or by higher-level middleware services and hidden from applications. It can involve variable paths, delays, ordering and attenuation or filtering. The same transmission can propagate along variable routes, acting as passive intermediaries between senders and receivers. Transmission paths can be changed in order to adapt to failures and to spatial-temporal congestion (variable paths). Data sent within the same session can take different routes through the underlying network system causing variable delays. Data may be received out of order (variable ordering) or the session may be terminated because it is not clear if it will still arrive at the receiver. A single source may spread a signal, causing different parts of a transmission to scatter and attenuate or to be reflected in variable ways. It may not be possible to determine to what degree, a filter has modified the original source (variable attenuation).

Interaction multiplicity increases in operational and management complexity when multiple senders communicate with multiple receivers which are time synchronised (multiple communications coincide in time) and when they are spatially synchronised (they occupy within the same spatial region). Interaction multiplicity may involve multi-senders (from one-to-one to one-to-many), multicasts (one-to-many) multiple receivers, senders and receivers that are hidden and the use of too few or too many senders or receivers.

9.2.1.1 Unknown Sender and Malicious Senders

Receivers may not always be able to uniquely determine the sender's ID or context. This can cause the following problems. A malicious sender can fake the credentials of someone else (sender masquerade). Malicious senders can send messages which when opened by a message receiver can activate a malicious code. An observed error can be propagated away from the error source and may appear to originate from multiple possible error sources, however, it may be important to determine the source of an error in order to handle it correctly. There are several strategies here. Unknown senders' messages can be held in some quarantine area and tested for side effects.

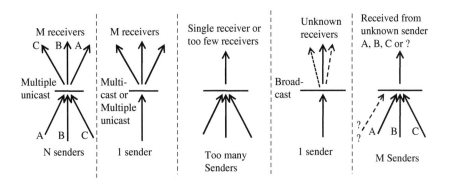

Figure 9.2 Some basic examples of interaction multiplicity

Messages can be tagged by the transmission middleware but must handle the case when senders, e.g., sensors in sensor nets, may not necessarily uniquely or incorrectly identify themselves.

9.2.1.2 Unknown Receivers

Multicast (or broadcast) message transmission can be used by a sender to interact with unknown receivers either as a means to share information or to request a response. Multicasts to unknown receivers are often used for service discovery in dynamic open systems and in particular in P2P interaction. In a directory service or search engine, a requestor often wants to bind to any receiver or the first receiver that can match. There are different designs for selection of providers by a third party directory or search engine. Many communications systems, in addition to humans, are naturally by a multi-modal. These accept multiple modal local inputs and remote inputs over a network. In a Windows desktop, many possible windows associated with different applications can be active, it may not always be clear which window is active and accepts input. When a sender broadcasts or multicasts a request, it cannot predetermine who, if anyone, will respond or fulfil a request. In social models, this is often used as a means to find the first receiver that can assist them, e.g., a call for help.

9.2.1.3 Too Many Messages

Receivers may become over-loaded with events which cannot be processed or stored in a timely manner as more UbiCom systems become connected and available. For example, if a remote location allows remote senders to communicate synchronously anywhere, anytime, a single-tasking receiver, if currently already interacting locally, must decide if it will interrupt local interaction in order to deal with remote interaction. Policy-based management can be used to prioritise incoming events and to handle events sequentially, e.g., messages can be handled or a first-come-first-served policy. Buffers can be used to temporally store received messages that cannot be handled immediately, to allow slow receivers to catch up in less busy periods, to support near real-time synchronous messaging and to support asynchronous messaging. Receivers can also support flow-control, i.e., send signals to senders to ask them to slow down, or else their messages may be discarded. Proxies that act as filters, aggregators or summarisers can be used to off-load event processing from receivers. Social network ratings and recommendations can be used to help filter multiple events and only to accept those with a threshold above a certain level.

9.2.2 Interaction Using Mediators

The use of a *mediator*[1] as a go-between interacting participants has several benefits. It supports greater autonomy between requesters and providers by adding a layer of indirection to enable networked peers to dynamically join and leave interactions while enabling interaction to continue (hot swap-over). It can simplify the number of links that would be needed to link every participant to every other participant, instead each participant links to a intermediary. It simplifies the interaction by acting as proxy to hide the complexity of interaction for one or more of the participants. It enables greater aggregated experience to be exercised to guide interactions when

---

[1] Other general terms that relate to the use of third parties which act as intermediaries between two or more parties include *intermediaries*, *service middleware* and *middle agents*.

intermediaries serve multiple participants. Mediators can be optimised to improve performance e.g., optimised for high throughput in a router or gateway.

Mediators also have disadvantages: performance may fail because another intermediate node is used when transferring data. It uses centralised management for interaction and is a potential bottleneck and a single point of failure. Mediator masquerades can occur. It is easier for a third party to instigate a man in the middle attack to cheat on either or both parties. Because of the perceived failures of relying on mediators, pure P2P systems have been proposed but these have their own problems such as longer latency multi-hop transfers, reliance on broadcasts and more decentralised control. Hence, hybrid designs are often used in practice to combine the advantages of mediated and unmediated interaction, e.g., intermediaries are used for at least part of the interaction, e.g., for node discovery only.

9.2.2.1 Shared Communication Resource Access

Multiple interdependent senders often transmit messages over a shared access or core network channel. For example: wireless local and remote communication; audio, video, data and mobile convergence; multi-modal input human device interfaces. If there is no transmission control or cooperation, multiple transmissions may interfere with each other; e.g., if the same frequency is used independently by multiple transmitters to control multiple devices, such as wireless doorbells and remote control toys – the so-called *hidden node* problem.

A common management technique from a network modelling perspective is to use multiplexing access control to divide the capacity of a communication channel into several logical channels, one for each message signal or data stream. In a reverse process, known as demultiplexing, the original signals can be extracted at the receiver side. Examples of multiplexing type access control are CDMA, FDMA, TDMA. Note much communication multiplicity is often handled at a lower-level network protocol service and hidden from applications, thus applications are unaware of this multiplicity when their data is transmitted. There are two basic schemes to prevent interference: *collision avoidance* schemes (next section) and *collision detection* schemes. Collision detection is normally followed by some scheme to stop the collision progressing, then to repeat the interaction again in such a way to avoid the collision, e.g., the CSMA/CD scheme. Collision detection is also commonly used at the application level in computer games that simulate the physical world to prevent two objects occupying the same space, e.g., to prevent something traversing through a wall. Collision detection is also used in control systems such as robots where a controller seeks to detect and avoid obstacles when carrying out controlled tasks such as palletising in a partially deterministic physical world (Section 6.7).

9.2.2.2 Shared Computation Resource Access

Interaction often entails the use of shared resources when system capacity is planned such that one resource is not able to be dedicated to a single user.[2] This may be because resources are expensive to install and maintain. Instead systems can be designed to handle concurrent resource access as multiple users try to access the same resources simultaneously or to access different resources which are interdependent, e.g., multiple data transactions may try to read and write the same data record in a relational database scheme There are several schemes that can be used to share access such as

---

[2] An example of a design of a system where one resource was dedicated to one user was the circuit switched design used in earlier telecoms systems which used one end-end network connection per user (Section 11.2.1).

multiple access network control schemes mentioned already (Section 9.2.2.1). Information access control schemes can also be used. For example, in order to prevent multiple transactions simultaneously reading and writing the same records in an RDBMS information system (Section 3.2.2.1), concurrency control techniques based upon the first process to access a record, locks access to that record[3] from additional users. A third technique is to use access control schemes based upon circulating one token per resource among users and requiring users to first acquire the token before they are granted access to the corresponding resource, e.g., Token ring network.

Systems may use a MTOS to enable many applications to execute on a single device (Section 3.4.3). This has the benefit of great convenience for the user. The device behaves as a single portal to multiple applications. But these applications can easily conflict in terms of their use of the computer system's resources, e.g., two different applications may wish to stream AV content to the hard-disk drive, but only one application can write to the hard disk at any one time. There are different MTOS process scheduling designs to support concurrent resource access to shared resources based upon fairness, priority and first-come-first-served (Section 3.4.3).

9.2.2.3 Mediating Between Requesters and Providers

In practice, two parties interacting often make use of autonomous *active*[4] *intermediaries* or mediators to facilitate their interaction. The use of intermediaries has several benefits (see also Section 9.2.2). It promotes loose coupling between the initiating (first) party and the responding (second) party. One of the basic problems facing designers of open systems e.g., Internet, is the connection problem (Davis and Smith, 1983) – finding others who may have information or capabilities you need. The use of mediators enhances peer discovery and service discovery (Section 3.3.2). Peers can dynamically share their service preferences or service capabilities, using a generic standardised service representation. Instead of having to request information from each peer, the features of all registered peers can be accessed in one place, a third party at a well-known, static, address. There are two special types of information used in the discovery process: service capabilities and preferences. A *service capability*[5] or service description specifies what types of requests can be serviced by providers. Service preferences define requesters' constraints when selecting a service provider, these constraints may be kept private or public or a mixture of both (Figure 9.3).

There are several different types of mediator and ways of classifying them. Whereas conventional distributed systems often only define one type of mediator, the directory, Decker *et al.* (1997) define a classification[6] for types of mediator by considering which participant knows the capabilities and preferences at the start of service invocation. This scheme produces nine different types of mediator (Figure 9.3). A *blackboard* keeps track of requests: requesters post their problems or requests;

---

[3] The complication here is that multiple transaction application processes may wish to access and then lock several records. Deadlocks can occur because one transaction, Y, can lock some records A and B, but not others such as C and D while another transaction, Z, can lock records C and D, but not A and B. There are several schemes to resolve deadlocks including the use of a transaction manager to monitor, detect and resolve conflicts and the use of *two-phase locking* (Watson, 2006).

[4] Active intermediaries can initiate communication. These are autonomous entities and participate in distributed control whereas passive intermediaries (Section 9.2.2) are typically under centralised control.

[5] Service capabilities are also referred to as service descriptions and as service announcements.

[6] The analysis of mediators here deviates from the classification of Decker *et al.* (1997) in the terminology used to name some of the different types of mediator, how requests are differentiated from preferences and because a wider set of properties is considered. The latter aspect has also been proposed by Wong and Sycara (2000).

Who Knows Capabilities

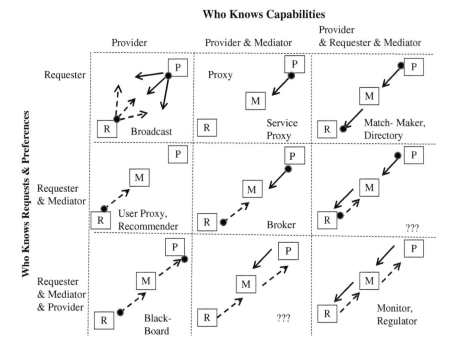

Figure 9.3   Designs for mediators based upon who (requestor, mediator or provider) knows what, i.e., who knows the capabilities (the solid arrows), the service requests and, or preferences (the dashed arrows). Who initiates the flow of capabilities or preferences is indicated by the dot

providers can then query the blackboard agent for requests they are capable of handling. Examples of blackboard systems include newsgroups, bulletin boards and recommender systems.[7] A *broker* is capable of protecting the privacy of both the requester and provider. The broker understands both the preferences and capabilities, and routes both requests and replies appropriately; neither the requester nor the provider need to deal with each other directly in a transaction. A *matchmaker*[8] stores capability advertisements that can then be queried by requesters. Requesters can then choose and contact any provider they wish. A *broadcaster* requires requesters or providers to advertise themselves directly, without requiring use of a mediator, although these could be present and could serve to streamline further interaction.

A *proxy* acts on the client-side, client-proxies (Section 3.2.2.4), on behalf of requesters to simplify access to providers. Proxies may also hide the inner workings of the requester and even anonymise the requester from providers. Client proxies that incorporate AI, e.g., are goal-oriented can interact with users in a much more high-level way, acting as personal assistants (Maes, 1994) enabling humans to delegate goals to them, e.g., configure this room to watch a film, rather than having to specify the detailed actions for them to carry out. Proxies may act on the provider side. A wireless network access node and Web portal are examples of a provider side proxy.

---

[7] A *recommender* system can be regarded as a type of blackboard mediator that anonymises and aggregates service preferences in the form of ratings which can be independently searched by requesters.

[8] A match-maker is sometimes referred to as a directory service or yellow page service.

The final type is a general mediator which sees all messages, service requests and service announcements. This type of mediator may act as a *monitor*, as a *regulator* and as an *arbitrator* to resolve any perceived interaction problems. There are some patterns of interaction marked as '???' in which are left as an exercise to discuss.

Further mediator design issues are as follows. Mediators may be used only when creating a session for interaction, e.g., a requester consults a directory service to locate a provider and then interacts with the provider directly, or can be used throughout the session of interaction, e.g., a broker. Some types of mediator, e.g., brokers and proxies, can afford a degree of anonymity to requestors or providers, hiding their identity of either or both participants. Mediators can be designed to support a range of different representations for capabilities and preferences based upon semantic versus syntactical, different degrees of detail and structure for capabilities and preferences and to store types of an instance of a participant versus the status of a participant. Mediators can be designed to support different types of interaction such as being provider-only (directories) or requester-only (blackboards) or both.

Mediators can be designed to support different types of services such as searching, browsing, single-valued versus multi-valued searches (e.g., type of capability, price, QoS, reputation, etc.) and different degrees of persistence and anonymity for stored data sources. Requesters and providers may expect a certain level of fairness and impartiality by mediators but this depends on the ownership and business model used by the mediator.[9] If mediators are really owned by, or act as a service proxy for, a closed set of providers, requesters need to realise that the service capabilities offered will be limited to a specific set of providers. Even if the mediator claims to be independent and impartial, mediators can be influenced and driven through business incentives, by their business revenue model, by the way information is provided to mediators in order to rank multiple results from a search in a certain way, e.g., providers can adopt names for themselves so that they appear first when multiple provider results are ranked alphabetically.

Mediators can act as a trusted third party. When two parties are relatively unknown or not trusted by each other, mediators can act as a neutral third party they both trust as a regulator for any exchange between them. Mediators should not modify the value of information exchanged, i.e. the fidelity for copying or transfer is unchanged and should not exceed their remit from the party they represent, by committing a party they represent to an excessive liability. Rather than to blindly trust the mediator, participants can use *authorisation delegation* in conjunction with *institutional regulation* when interacting with mediators.

9.2.3 Interaction Using Cooperative Participants

Cooperative interaction enables multiple systems to work together. Cooperation is characterised by two main properties: synchronising activities (*coordination*) and acting together (*cohesion* or *organisational interaction*). Cooperation is easier to manage when: homogeneous designed systems interact; there is centralised control; systems are mandated to serve others (e.g., if a service is

---

[9] The earliest electrical communication systems such as the telegraph and telephone were time synchronized and used human mediators to do the technical job of connecting the caller to the system. The first successful electro-mechanical telephone switch was invented in 1891 by a Kansas City undertaker called Strowger who was purported to be fed up with the local telephone operators diverting his calls from would-be customers to a rival undertaker, and determined to engineer a system to take the operator out of the loop. He devised a switching system that used contact arms rotating on shafts to make contact with any one of ten contacts, representing a digit of a phone number in an arc. Hence, because of the perceived bias of human mediators, an automated telephone exchange system was developed.

Table 9.2 Advantages and disadvantages of cooperative systems

Advantages of cooperative model	Disadvantages of cooperative model
Distributed problem solving: solves it quicker as more parts are processed in parallel. Each part of a system has incomplete information, or capabilities for solving a problem which can then be judiciously combined	Cooperation reduces and competition increases when high concurrency, low capacity resources, non-determinism, prevail
Delegation: Don't need to be able to do everything ourselves. Don't want to do it ourselves, too time-consuming. Instead need to know who to delegate something to that we don't want to do or can't do	Communication efficiency in terms of costs and unreliability may outweigh the extra processing benefits of the distribution
Selection: select best option from a set of candidates, based upon a set of requirements and constraints e.g., work tender or select alternatives	Coordination & management are more complex: parts may be more autonomous & heterogeneous, disruptive group members (insider attack), lack of understanding, ambiguity, conflict
Reliability: there are alternative options	Delegation and session initiation costs too high, e.g., a task may require an overly complex instruction set to achieve it
Social: agents act on behalf, to engage people	Lack of control, privacy

invoked it must complete its execution or fail to complete); systems are designed statically to cooperate; and systems act benevolently and reliably.

Cooperation is harder to manage when: different systems are designed by independent developers; systems are designed to act autonomously; systems support heterogeneous goals; systems need to cooperate dynamically; parties may act in a self-interested manner; systems act malevolently and may non-deterministically malfunction. The advantages and disadvantages of cooperative systems are summarised in Table 9.2.

9.2.3.1 Coordination

Coordination is defined generally as the regulation of diverse elements into an integrated and harmonious operation. The main reasons why the actions of multiple agents need to be coordinated are because dependencies exist between agents' actions and because global constraints exist which a group of agents must satisfy if they are to interact successfully. This requires that an intelligent system reasons about its local actions and the (anticipated) actions of others to ensure that the community acts in a coherent manner (Jennings, 1996).

Without coordination, the benefits of decentralised problem solving vanish and the community may quickly degenerate into a collection of chaotic, loose groups with various kinds of deadlocks prevailing. Explicit synchronisation is needed to handle concurrent sending, receiving and use of share communication resource which would otherwise impede communications, e.g., two peers which start to send to each other at the same time. There are several ways to support explicit synchronisation of senders and receivers such as controlling the communication channel via a third party (Section 9.2.2). Another way to enable explicit synchronisation is through the use of cooperative organisational roles, sending and receiving are then organised via the role. One of the simplest types of organisation defines two roles: the client and server; the server starts receiving requests and sending responses and a client starts off sending request and then waiting for responses. Hence coordination can be considered to be either communication or message-based or processing- (node) based. Coordination may be designed using: synchronisation protocols (message-based); mediators that act as go-betweens for autonomous senders and receivers (message-based); system organisation structures to control interaction via the use of organisational constructs such as roles to constrain interaction (processing-based) and through the use of institutions and norms (processing-based).

In perfectly coordinated systems, systems will not need to explicitly communicate to achieve a common goal because they can synchronise and align their activities, perhaps based upon systems maintaining good models of each other, i.e., *mutual modelling* (Wooldridge, 2001). Perfect coordination is easier to achieve: when systems are designed to act coherently and to be highly cohesive (e.g., systems are designed so that one system has centralised[10] control over other dependent components); when systems use norms and conventions (Section 9.2.3.2). Coordination may also find it useful to maintain a model of itself in relation to their environment.

Perfect coordination is difficult to achieve in practice when systems are open, dynamic, autonomous and intelligent and which are used in a range of system environments which are nondeterministic, episodic, partially observable, etc. However, perfect coordination can be used very effectively for local interactions between local peers in self-organising systems (Section 10.5.1). A main limitation of the mutual modelling approach at the heart of perfect coordination is that the computation needed to define how other dynamic systems act, is complex to define and maintain. Subsequently, there may not be enough system resources, such as time and processing for computation. Hence coordination is used which involves explicit signalling.

Coordination design depends on several factors (Wooldridge, 2001, pp. 189–224). It depends on whether ISs are spatially and or temporally coincident, or not. Temporal and spatial coincident can use local interactions in a physical meeting at a set time. Temporal coincident but spatially separation uses remote interactions within virtual meetings. Temporally separate coordination involves the use of shared spaces and buffers to leave information with guard conditions. Coordination may need to handle inconsistencies and uncertainty which can be handled though: avoidance (using planned interactions), resolution (using various types of agreement and coordination) and fault-tolerance (using alternative plans). Coordination may be driven by no parties (perfect coordination), by one party, by either the requester (pull) or supplier (push), or by multiple parties which may recognise that they could jointly save effort if one and not all perform certain actions.

There are a several designs for coordination of multiple cooperative systems. First, various service composition models can be used (Section 3.3.4). Second, coordination may be based upon interaction protocols with inbuilt coordination mechanisms. Third, models of *joint planning*, also called *Cooperative Distributed Problem Solving* (*CDPS*) as a type of *distributed planning* and can be used (Smith, 1979; Durfee et al., 1989; Durfee, 2001). This process involves decomposing problems into sub-problems, allocating these to different participants, exchanging sub-problem solutions and then creating the overall solution out of the sub-problem solutions.

Finally, *joint intentions* can be used to recognise and establish coordination between agents acting in dynamic environments. Intuitively, joint intentions represent relationships between agents to achieve a set of objectives. The model is seen as a compromise between agents that react, and agents that are deliberative (Jennings, 1993). These two extremes were criticised because reactive systems are regarded as being too inflexible and deliberative systems as too inefficient. The joint intentions belief, desires and intentions (BDI) model considers explicit decisions based up on partial information and partial fulfilment rather than upon more classic models of goals and

---

[10] This is also referred to as *master–slave control* in which the master has unilateral control over the slave components.

actions which tend to focus on complete fulfilment of plans. There is a mapping between goals and desires (an evaluation stance), states and beliefs, and intentions to current goals, to which an agent is committed. The focus on partial information emphasises the principle of allowing an agent to have a commitment to a joint-action but being able to retract that commitment under certain conditions.

9.2.3.2 Coordination Using Norms and Electronic Institutions

Normative behaviour or *norms* expresses behaviour as it should be that can be differentiated from actual behaviour, behaviour as it actually is (Meyer and Wieringa, 1993, described in Boella *et al.*, 2005). If systems are designed to support norms, then it would be clearer if the actual behaviour is legal or illegal, if it is nominal or an anomaly. Institutions and civil societies are designed as normative systems, built upon agreed constraints on individual behaviour (social contracts). These contracts are supported in exchange for quality of service assurances, such as fairness and security, which are backed up by social institutions and norms (Dellarocas and Klein, 2000). In addition to the benefits of an improved quality of service, institutions also offer improved coordination and cooperation because they make available conventions, e.g., rules may exist to avoid conflicts such as people waiting to enter a building or passenger vehicle first allow those to leave before others enter. Institutions reduce the need for explicit communication, provide common control protocols and collective resources.

Communities can be populated by stakeholders who benefit from being members of the community and who participate in regulating behavioural norms within that community. Social institutions set behavioural norms by maintaining and monitoring norms and respond to those who inadvertently transgress from norms to reduce deviations from the norms, e.g., deviations caused by disasters and emergencies, e.g. ambulance system, fire-fighters, coastguards, etc.

There are two main ways to maintain these social norms, either with the incentive of *mutual reward* and benefits that arise from community membership or through fear of the *community retribution* that might result from advertently deviating from norms or by circumnavigating necessary checks or processes.[11] Formal institutes can enforce laws to penalise those who deliberately transgress from the norms, e.g. courts, police, etc., whereas informal institutions may rely on voluntary codes of individuals to comply with norms. For example, deviations from the norms can also be punished in informal institutes in some way such as being named and shamed leading to a fall in reputation, by an increase in a licence fee to operate, by imposing operating restrictions, etc.

Social norms are regulatory sets of rules that regulate interaction and are socially enforced, by social sanctioning. Norms can be expressed in terms of permissions, or prohibitions, obligations and commitments.[12] Normative systems can be designed and modelled in several ways. Normative systems can be designed using *deontic modal logic* founded by Wright in 1951 as the formal study of ought (Boella *et al.*, 2005). They can be based upon agreements of target levels of services (Service Level Agreements). They can be based upon policy management to specify restrictions of what is allowed and what is not. Challenges in designing normative type UbiCom system models are that: they may be multiple dynamic norms in practice; processes are needed for newcomers to discover and learn norms; knowledge about norms may be slow to be acquired if the experience of what the norms are can only be acquired from direct community action and feedback alone; norms may be difficult to define in non-deterministic environments. An additional type of norm does not use institutions in the human sense, instead norms are established through self-organisation (Section 10.5.1).

---

[11] The use of rewards and retribution to guide the way peers learn what the norms are, seems to be similar to the use of reinforcement learning (Russell and Norvig, 2003) but differs because reinforcement learning often only considers positive examples but not negative examples.

[12] A commitment is an intention to complete a goal.

9.2.3.3 Hierarchical and Role-based Organisational Interaction

Some coordinated interaction in organisations is more process-based rather than message-based. The existence of organisations should lead to a reduction in the amount of communication which would otherwise be used in peer-to-peer interaction. There is a *duality* between interaction and organisation in the sense that interactions are constrained by the organisation. Organisations are in part defined and constrained by the type of interactions they support. An *organisation* or *society* is an arrangement of relationships (interactions) between individuals that produces a system with qualities not present at the level of individuals – it is an aggregate or *supra-individual*. Organisations describe both the result of building up the structure and the result of the process of organising something. An organisation has several benefits. It can persist even when faced with a limited number of disruptive individual interactions. Management in terms of coordinating and controlling interaction is eased because the organisation structure constrains the flow of interactions within an organisation.

Physical organisations can be described in terms of the physical structures that arrange the members of the organisation. One of the most common types of physical structures is to use hierarchies or trees. There are different types of structure for hierarchical organisations depending on the type of *hierarchical containment*:[13] composition hierarchies, e.g., DNS, or class hierarchies, object-oriented class hierarchies, etc. The importance of hierarchical containment for management is that it enables members to be easily located in an organisation and can support shorter routes between different members of the hierarchy especially in high breadth, low depth hierarchies that act across flat structures. Other physical structures are linear, graphs, tables and Webs.

Social and service-based organisations are defined in terms of missions and roles and constrains on roles and interlinking or organisation of roles. *Missions* specify the strategic shared goals for the members of an organization, e.g., providing a certain level of quality of service. *Roles*[14] specify the position in the organisation which individuals play. Members' internal actions and external interactions are constrained according to role. A member can play simultaneous roles within an organisation where roles are associated with different activities and different responsibilities. Interactions can be interpreted differently depending on the role.

The types of roles depend on the type of organisation. In P2P imodels, there is a flat organisation of participants which play a peer role to request, forward requests and provide services, e.g., a sensor net node, AV recorder device, etc. In practice, most P2P systems specify additional specialist roles such as: *router role*, e.g., in sensor nets; a *gateway* or *interface role* is used to interoperate between two separate autonomous systems; a *super-peer role* is used to convert a flat P2P structure into an overlay, two-layer hierarchy, structure in order to improve the performance and the use of directory-type mediator to discover services. In (client-) server-based organisation models, there is a flat structure to interlink nodes. Nodes tend to either play a requester or provider role. Additional roles played by each node include access control and monitoring system operations. Special roles are also commonly assigned to nodes to support service discovery and to support system and network management. Social hierarchies are structured as a pyramid with one or more bosses at the top and one or more layers of subordinates below, who are duty bound to respond to, and to obey, the bosses above.

---

[13] A containment hierarchy is a hierarchical set of nested sets. Each member in the hierarchy set represents a member which itself may be another set such that the containing set is the superset, and the member is a subset of it.
[14] A role seems similar to a type of a node in a system. However, roles enable organisational functions to be more dynamically assigned to nodes, to support multiple roles and to switch between them. The use of multiple roles also increases flexibility and fault-tolerance in an organisation. Note that static roles bound to specific nodes are useful because such node can be optimised to support that role, e.g., for data storage.

Organisations themselves can play multiple roles, e.g., they may act more autonomously as peers and they may act in hierarchical organisation. A special type of role called a *boundary spanner*[15] acts as a representative gatekeeper to the organisation, or rather this represents the conduit that connects other external organisations. A boundary spanner performs the *facework* with other organisations, facilitating the establishment of inter-organisational trust (Hexmoor *et al.*, 2006). Members of organisations may give local goals priority over organisational goals so that individuals can refuse or fail to commit to organisational goals even although they have the capability to do so., e.g., an individual device may refuse to participate in certain kinds of interaction because it decides it needs to conserve its energy for a better future opportunity. There are many types of organisation such as: social institutions that seek to regulate behaviour to norms, organisations where its members actions are orchestrated or controlled centrally versus where its members are more autonomous and under distributed control; social organisations where its members cooperate to achieve organisational goals (this section); economic organisations in which self-interested members compete to further their own goals.

Multi-agent systems, distributed AI, provide a flexible way to design a large range of organisations (Gasser, 2001; Wooldridge, 2001). Many UbiCom system applications have been designed and implemented as MASs, for example, to dynamically configure building facilities to support building energy efficiency; for personalised work environments (Yong *et al.*, 2007); for information integration and interoperability (Poslad *et al.*, 2007); information services for mobile users (Poslad *et al.*, 2001) in which agents dynamically adapt information to multiple contexts such as location, person and ICT system.

9.2.4 Interaction with Self-Interested Participants

As an increasing number of diverse smart autonomous configurable and networked devices are introduced into physical spaces, expanding the virtual ICT space more into the physical world, the degree of competitive interaction that occurs will increase. Design models to solve the associated resource conflicts and resource allocation problems will become essential. *Competitive interaction* or *adversarial interaction* is interaction that is driven by *self-interests*. It focuses on autonomous participants or *agents* furthering their own goals, rather than collaborating to help further the goals of others.

An example of competitive interaction for UbiCom systems in the utility regulation scenario (Section 1.1.2.4) is as follows. There are multiple autonomous lighting devices in a smart environment, all seek to switch themselves on but if they all switch on, some are redundant and this wastes energy. Multiple users may seek to configure a shared device or multiple devices that overlap in function in multiple ways, e.g., a high-definition video source inputs its content into a low definition video display or multiple users wish to regulate the heating and lighting levels differently. In some situations, a mix of competitive and collaborative behaviour may be desirable, e.g., peers may collaborate as part of a team but may compete against other teams. There are different types of competitive interaction problems and designs depending on: the number of players, i.e., two player or multiplayer small group or multiplayer large groups; the *interaction protocols* or *interaction mechanisms* used to constrain the self-interested interaction; the *strategies* or long-term plans used to achieve the goals of the participants; the nature of the completion itself in terms of the type of equilibrium that are inherent in games. Self-interested interaction is complicated further when

---

[15] A boundary spanner in a network sense is a type of peer that acts as a gateway but at a social level the boundary spanner has trust relationships inter-organisationally and intra-organisationally. In addition, boundary spanners have more relationships with other peers than the average peer – it behaves as a super-peer.

participants act maliciously, i.e., lie and collude. In contrast to collaborative agents, that make public their goals to their collaborating parties, self-interested agents keep their goals private.

A generic problem for UbiCom is the allocation of limited resources and services to multiple self-interested requestors. Some designs for interaction protocols have already been discussed. Concurrency control can be used to support fair and shared access by essentially serialising access to permit one access per resource at a time (Section 9.2.2.2.). Policy-based management can allow one party to take priority actions based upon some criteria such as a higher priority policy or a more specific policy. Control can be more generally acceded to a third party, e.g., mediator, to determine the outcome of the interaction. Some further designs to manage competitive interaction are considered as follows.

Market-based economic protocols and strategies can be used to allocate resources to individual requesters. Negotiation can be used in marketplaces to agree terms but it can also be more generally used to resolve conflicts. *Agreements* are needed when self-interested peers (and cooperative peers) must synchronise their actions, mutually accepting the individual results of these actions, without recourse to the use of external or third party directed control of participants.[16] Agreements can range from being informal, e.g., a gentleman's agreement, to being legal and binding documents that define the obligations of all parties involved in an interaction. Agreements can be between two participants versus between many millions of participants. Agreements can be used to set common conditions for a deal where participants use common conditions or where participants can independently set their own conditions. *Consensus-based protocols* can be used to reach agreement between multiple participants, but normally for one object at one time.

Another type of protocol and algorithm which can be used to reach agreements is based upon *convergence*. Convergence is regarded as a multi-step processes where two or more entities iteratively reach an agreement. Convergence algorithms and protocols tend to be domain specific. For example, convergence algorithms can be used to reach agreements on router table updates between multiple interconnected routers in networks and on the optimal window size for the amount of unacknowledged data sent on a particular connection before it gets an acknowledgment back from a receiver.

There are several generic models of competitive interaction (and also cooperative interaction). *Task-based models, or task planning models*, which are used to design cooperative interaction (Section 9.2.3.1) can also be used to design competitive interaction. However, whereas plans may be shared in cooperative interaction, they tend to be kept private in competitive interaction in order not to give competitors an advantage. A complementary model to the use of *task-based model* is to use state-based models which focus on determining which state transitions cause a peer's state to move closer to a goal state. *Utility-based models* or *worth-based models* can be modelled to assign some *worth* or some objective quantifiable value to different states of interaction, representing a level of performance or utility value or value-gain for a participant in that state (Russell and Norvig, 2003). An agent may assign a higher utility value or weighting to something which it is closer to its goal state than one which is further away. This helps an agent assess how well it is doing and can help a participant decide when to make an agreement in a competitive interaction.

9.2.4.1 Market-based Interaction and Auctions

Economic interaction concerns the production, distribution and consumption of goods and services. Consumption in economics is used both as a noun in terms of the value of goods and services

---

[16] Multi-party agreements can involve mediators to facilitate agreements, e.g., the use of polling stations, ballot boxes and monitors in voting. Such mediators are there to promote and monitor fairness, to ensure that participants adhere to the voting protocol rules, rather than to control the terms of the agreement.

bought and exchanged for monetary payment, and as a process in terms of the selection, adoption, use, disposal and recycling of goods and services. *Micro-economics*, a sub-field of economics, concerns the behaviour of individuals and organisations (firms) and their interactions within individual markets, given a scarcity of goods or services. Micro-economics behaviour is governed by two main principles: *optimisation*, the participants (human or artificial) try to choose the best deals that they can afford to maximise their payoff, and *equilibrium*, the deals adjust until the amounts of products and services demanded is equal to the amount that is supplied, i.e., a general equilibrium[17] exists (Fasli, 2007). There are several things that can make market interaction more complex. It may sometimes not be clear to consumers what some product is worth and which price gives them an acceptable pay-off. There are often many instances of a type of product supplied by multiple producers which has variable properties or features such as quantity and quality. Consumers may have preferences about the features of product instances. Preferences can be expressed using some utility functions which map each preferred state to some numerical value representing its pay-off. In a market-based economy, each provider and consumer is self-interested and uses their own private strategy to maximise their pay-off.

One of the oldest types but still widely used market-based protocols is the *auction* protocol which is designed to allocate resources such as goods and services to one of the bidders. There are several types of auction protocol depending on the properties for how bids can be made and dealt with; the dissemination policy for defining what information is revealed to whom and when and the clearing policy to decide when sellers and bidders are matched (Wurman *et al.*, 2001). One of the most common types of auction, the English auction, can be classified in terms of the following properties for bids: a single type of goods; single attribute, e.g., price; single sided, e.g., consumers submit bids to the auctioneer or seller; open cry, i.e., bids are public to all; ascending bids are used in which the last remaining highest bid clears the market.

The benefits of managing different types of smart environment and smart device interaction in terms of an auction include a flexible way of matching resource and service provision to demand. Although in theory, auctions often involve only two types of participant, sellers and buyers, in practice, auctions often involve proxies. Applications of auctions and design issues for UbiCom systems are discussed in the negotiation section (Section 9.2.4.2).

9.2.4.2 Negotiation and Agreements

Auctions, which are designed to reach agreements between sellers and consumers in a marketplace, are considered to be a type of more general technique called *negotiation*. The general aims of negotiation is modification of local agent policies which constrain interaction and plans of inter-action, e.g., in the case of negative (harmful) interactions, and identification of situations where new potential interactions are possible and beneficial. Modifications and identification situations trigger the process of negotiation, in the sense that agents communicate in a certain way to reach a common decision. Negotiation is often used for: task and resource allocation; recognition of conflicts; resolution of goal disparities; determination of the organisational structure and hence for organisational coherence.

In general, a negotiation method has three principal components: (1) a public shared interaction protocol that consists of an exchange of a sequence of proposals and counter-proposals until a deal is

---

[17] A general market equilibrium may not exist if monopolies exist, if demand-supply curve is discontinuous, e.g., if small changes in prices leads to big changes in demand or if agent preferences in one market are inter-dependent on other factors, external to the market. *Clearing the market* refers to the aim of the auction to sell off all its goods to bidders.

reached; (2) a deal rule that determines when an agreement or deal is reached after exchanging proposals; (3) the negotiation set that denotes the set of all possible proposals and agreements which can be made including their conditions; strategies that are kept private, e.g., their specific preference for outcomes (Rosenschein and Zlotkin, 1994; Wooldridge, 2001). Design properties for negotiation protocols include being: *pareto optimal*,[18] *stable* and *individually rational* in which there is an incentive for all participants to behave in a certain way, i.e., to maximise social payoffs for all participants and able to support computation and communication efficiency (Sandholm, 1999).

Because of the general nature and complexity of negotiation, there is a range of models, protocols and strategies for negotiation (Krauss, 2001; Jennings *et al.*, 2002). A simple strategy-based bilateral negotiation protocol[19] involves both participants starting by proposing a deal of their choice. If no deal is reached, each participant may either make a small concession or decide to stick to their original proposal. The rounds of making proposals continue until either a mutual agreement is reached between both participants or until both agents refuse to make a concession. Then the negotiation breaks down and a conflict results. Two example strategies for using this simple protocol are to concede often but this may concede too much, with the risk that a participant will not get the best possible deal. If the participant does not concede often enough, a participant risks conflicting with its goal of an assumed pay-off.

Negotiation can be considered to be a distributed search through a space of potential agreements with negotiation proceeding by the participants suggesting specific points (or regions) in the agreement space as potentially acceptable (Jennings *et al.*, 2002). Negotiation can be modelled using game theory[20] because rational participants also need to consider what proposals (moves) their opponents will make in the game and what the private strategy of their opponents could be. Game theory tends to work well when the players' preferences can be quantified in relation to specific outcomes of actions in specific rounds of negotiation. However, humans often find it difficult to do this consistently over time. Humans may wish to update their preferences based upon experience or new information and may find it difficult to assess complex preference and complex outcomes (Russell and Norvig, 2003). In practice, negotiation can be driven by strategies that can be time-limited, and where participants may not act rationally. *Argumentation-based negotiation* allows additional information to be exchanged over and above proposals, e.g., to give more evidence of why something is worth some specific value, to allow parties to exert influence to change their and others' negotiating position (Jennings *et al.*, 2002).

Rosenschein and Zlotkin (1994) define three different problem domain models for negotiation applications depending on whether or not they are task-based, state-based or worth-based. Negotiation can be used in auctions to help allocate resources from providers and to clear the market. Negotiation can be used to in a kind of *reverse auction* in which consumers allocate tasks and providers bid to undertake those tasks. Consumers are contracting out jobs to providers (Davis and Smith, 1983). Both of these types of *task-based domain* negotiation models do not consider any state independencies or state conflicts. One state may be controlled and accessed by one participant that blocks other participants from controlling that state, e.g., an AV device that has agreed to display content from one AV device cannot also simultaneously display content

---

[18] A deal is Pareto optimal if there is no other deal where at least one of the participants is better off and no other participants are worse off.

[19] This protocol is called the Monotonic Concession Protocol (Rosenschein and Zlotkin, 1994).

[20] Game theory is used to develop strategies between competing players who strive to win a game. It is most often applied to competitive interaction that is deterministic, fully observable, involves *rounds* or *turn taking* between two players and uses a *zero-sum* (a zero sum competition, more generally a constant sum competition, is when one participant's loss or gain is exactly balanced by others' gains losses or losses) utility function.

from another device. Negotiation in *state-based domains* must also consider which actions change the state of the world, which actions cause conflict states and how to resolve conflict states. Negotiation in *worth-based or utility-based domains* also considers that certain states may have a greater private utility or worth for participants and that the negotiation strategy or plan by each participant takes this into account.

UbiCom applications of competitive interaction and negotiation models include the following. Data sources could negotiate with data storage resources to determine how much data is stored; this may depend on the storage available, the utility of the data and redundancy of the data. Data sources could negotiate with its communication infrastructure to decide which data from which data sources should be conveyed. For example, sensor nodes which use the *SPIN* (Sensor Protocols for Information via Negotiation) protocol allows them to more efficiently distribute data given a limited energy supply (Kulik *et al.*, 2002).

Key concerns for using multi-round protocols to reach agreement in pervasive computing is, first, support for computation and communication efficiency particularly for use in low ICT resource environments (Park and Yanga, 2008). Second, there is the issue of security and access control, of where the agreement and conflict resolution logic resides, who has control of it and who can monitor it. Without secure access control, specific participants may be able to change their bids and negotiating positions to their advantage. It is a way which can, for example, bypass the negotiation protocol rules about confidentiality of sealed bids.

9.2.4.3 Consensus-based Agreements

Consensus-based interaction can be used to reach an agreement when multiple self-interested participants share a common goal, but may have different plans or constraints to achieve that goal that they may wish to keep private. A consensus is important when different participants or processes interact such that their self-interested goals may conflict, e.g., different participants in a room may wish the lighting or heating to be set at different levels. Consensus may also be useful in situations where there are several alternatives but it is not clear which alternative should be chosen, e.g., a number of devices are networked, all showing different times, it is not known which time should be chosen to set the clock of a new device added to the network. In contrast to negotiation, consensus-based agreements are simpler. They are generally single goal, single object, single value, agreements. A consensus refers both to a state of agreement that is reached by independent participants and to the process to reach an agreement. Some specific protocols for consensus-based interaction include recommender systems (Section 9.4.2) and voting.

Voting involves the following stages: *registration*, compiling a list of participants eligible to vote; *validation*, checking the credentials of those attempting to vote and only allowing those who are eligible and who have not already voted, to cast a vote; *casting* where the eligible voters make their selection; *collection* of the voted ballots; *tallying*, counting the votes. Requirements for consensus protocols can include preserving the privacy of participants, selecting one outcome out of all possible ones versus ranking all possible outcomes and reaching agreements in the presence of faulty interaction.

9.3 Is Interaction Design

As mentioned in the introduction, there are two basic solutions to supporting intelligent interaction, either to design system interaction to become more intelligent and to design individual intelligent systems to interact.

9.3.1 Designing System Interaction to be More Intelligent

Conventional distributed ICT system design often tends to use middleware to make system environment interaction more transparent (hidden) to its users and applications. This simplifies the design of applications but means that application and users remain relatively unaware of how their environment and system infrastructure is changing. Applications cannot benefit from perceiving it and being able to adapt to it. Several system designs give applications more access to the system and environment context. These include simple reactive context-aware system (Chapter 7), IS system designs (Chapter 8) and reflective systems (Chapter 10). Additional benefits to design distributed system interaction to be more intelligent include the following:

- *Mediation and handling heterogeneity*: Semantic KB IS designs can be used to mediate between heterogeneous systems that interoperate.
- *Reflection about communication*: IS can reason about communication failures (Table 9.3). IS can handle interaction in environments which are open, highly dynamic, uncertain, complex or impossible to know *a priori*.
- *Distributed problem solving*: IS solves it quicker as more parts are processed in parallel, e.g., SETI application, etc.
- *Task delegation*: One doesn't need to be able to do everything within a single system. Instead the system needs to know who can do something on our behalf that we can delegate a task to. A specific IS may be designed to do specific tasks, not to do all dependent tasks. It may lack the resources, e.g., power, time, expertise, etc. to do something internally.
- *Flexibility and selection*: select best option in a competition e.g., a work tender, from competing systems. IS systems have built-in mechanisms for flexibly forming, maintaining, and disbanding virtual organizations for systems. Multiple IS provide a variety of stable intermediary forms for flexible system interaction, e.g., brokers, match-makers, message-boards (Section 9.2).

Table 9.3   Causes of interaction errors and some ways to handle these

Interaction failures	How interaction failures can be handled
Network link failure	Use of multiple receivers, communication paths and plans
Receiver down, not ready	Use of push interaction, e.g., subscribe
Wrong message syntax	Use of match-maker to query service description and semantics
Use wrong default values, types	Support queries on metadata, e.g., message parameters
Use of service constraints that cannot be satisfied by provider	Use further interaction such as negotiation, auctions to reach dynamic agreements
Unknown service providers and location	Use match-maker to locate services; Use Previous history of interaction
	Use recommendations from others; Querying a match-maker or directory service in a known location; Broadcast to locate receivers
Client action, e.g., sender cancel	Support cancel by sender
Messages as part of processes not coordinated	Make explicit coordination plans; use predefined interaction protocols
Semantic differences in use of terms at sender and receiver	Use shared explicit semantics for messages

- *Reliability*: means an IS has alternative options if one system fails, it can switch to another system. Systems may be organised as teams (with a leader) versus groups (without a leader).
- *HCI as HHI mediation*: IS acts on behalf of people. If designed to mimic HHI artificially, they can be used to support richer human interaction rather than designing interaction in terms of low-level CCI.

There are also some limitations in designing IS system interaction. ISI can be more complex to design and maintain. The patterns and outcomes of interactions may be less unpredictable. Unplanned or undesirable behaviour can emerge – although this is more of a feature of certain types of AI system particularly adaptive learning systems. As IS systems are proactive, they can refuse requests. An analysis of the pitfalls of developing IS systems has been given by Wooldridge and Jennings (1998).

9.3.2 Designing Interaction Between Individual Intelligent Systems

Part of the motivation for individual intelligent systems to interact with each other is to handle the knowledge bootstrapping problem. A single intelligent entity would need to independently learn everything it needed to know itself (using unsupervised learning). This learning process is much slower than learning from others who are more experienced (using supervised learning and reinforcement learning). It also means that each single intelligent entity would need its own internal, complete, knowledge model of the world and of itself, whereas intelligent entities which can interact could rely on the expertise of others to provide them with new knowledge which they then do not need to acquire.

Interaction design issues concern whether or not a common and extensible message protocol can be designed for use across multiple types of UbiCom system environment interaction such as CCI, HCI, CPI, etc. or for specific use in specific environments. Second, there is the issue of whether or not different types of IS are able to share meaningful information. Third, systems need to share and fix a common understanding of terms or concepts within a domain, and in addition may need to share the some context associated with a message, e.g., the location and the time when events were triggered.

9.3.3 Interaction Protocol Design

Interaction protocols are specifications that define interactions or message patterns. This involves specifying two separate aspects: specifying individual messages versus patterns of multiple messages. Message protocols define the structure and meaning or use for individual messages, e.g., as a request, response, acknowledgement, etc. Messages are defined by a header that is attached to the content, the body, of the message. Typically the message header defines the type of message, the receiver address, the sender address, the message content encoding, the time the message was sent, etc. For example, the HTTP/1.1 protocol specified in IETF RFC 2616 specifies two main types of message: requests from clients to servers and responses from servers to clients.

Messages are often exchanged in practice, as part of an *interaction* (also called a *conversation* or *dialog*) or pattern of messages rather than as individual messages. There are many useful message patterns in practice but two of the common types of message interaction are to support information sharing and task sharing. Information sharing involves get and set type messages, e.g., a query type interaction comprises a *get (query) message* being sent followed by a *set message* containing the results of the query being sent back to the sender, e.g., HTTP request-reply interaction.[21] Set messages can also be sent asynchronously in isolation, e.g., email messages, chat messages or mobile phone text messages. A second type of common message interaction involves task delegation, the simplest example of which is often referred to as a request-reply interaction, defines a request message type to perform a task is sent followed by a receive message containing the result of performing the task.[22]

There are several practical issues to do with the design of interaction protocol. Message interaction may fail, indicated by the status field in the response message, and this needs to be handled. Thus the receiver needs to examine message status codes when failures occur and then to decide how to proceed on that basis. There are a variety of ways to respond to interaction failures (see Table 9.3).

Multiple interaction protocols may need to be predefined and orchestrated, e.g., B requests X from A, then A requests Y from C, followed by C's response to A and A's response to C. This is typically performed in the application and gives applications some flexibility to control the interaction. Interaction protocols are often designed to be service-specific and domain specific – this makes interoperability across services within the same application domain and across different application domains challenging because these will tend to use different interaction protocols. Heterogeneous protocols are may define the same message action, differently.

An additional way to introduce interaction flexibility is through the use of cooperative dialogues. These are interaction protocols which are designed to be extensible at run-time. For example, a non-cooperative query dialogues responds with only 'result' and 'no results' answers. A cooperative dialogue supports alternatives and related information. It does not necessarily expect the communication request to be a direct match. It may anticipate future information exchange based upon past information exchange. Inexact requests can be supported by taking note of the semantics and categorisation of the message content. For example, a request for information about a specific instance of a service could respond with more general information if there is no exact match rather than produce a no information response.

9.3.3.1 Semantic or Knowledge-Sharing Protocols

Semantic or KB protocols are messaging protocols used to share knowledge or semantics. Knowledge and semantics models have been described earlier (Section 8.4). Basic message types for a semantic protocol could be defined as part of the query interaction pattern (Section 9.3.3.). If both the interaction protocol and the content were semantic, then these have the advantage that they could be interlinked, enabling the content to be interpreted within the context of the interaction, e.g., the content that 'the weather is often hot in London' could be linked to a 'set' type message thus conveying the meaning that this message is communicating a new fact.[23]

Often it is only through interacting and through social ontological commitments that some types of knowledge can be defined and used. Searle (1995, pp. 1–29) differentiates knowledge

---

[21] Most HTTP communication is initiated by a user and consists of a request that is applied to a resource on some origin server, followed by a response message. Simple case: user agent (UA) sends a request and reply via single connection or hop (v) to the origin server (O). A more complicated case involves one or more intermediaries being present forming a request-reply chain leading to more hops in between.

[22] A query often involves a computation task to produce the results of the query behind the scenes. However, the primary sender message is a query not a task request.

[23] Of course, if this were being added to a monotonic knowledge base, it would have to contend with the unlikely event of a contradictory fact such as 'the weather is seldom hot in London'.

which pertain to brute facts, e.g., this equipment is hot, and institutional facts which require human institutional interaction for their existence, e.g., this equipment is too hot, it is operating outside its safety conditions. Often semantic content represents conceptual or structural relationships with possible constraints representing the brute facts. However, semantic content can be used to represent and exchange a much wider range of semantics than this. It can be used to represent and exchange metadata, rules, models and processes, plans and goals, and experiences in order to represent the institutional knowledge (Figure 9.4).

In practice, network protocols often tend to be specified in terms of human-readable semantics but not in terms of a machine-readable semantics.[24] One of the reasons for this at one stage was perhaps the lack of a standard semantic representation. Although standard representations such as OWL do now exist (Section 8.4.1), a couple of usage challenges remain. First, the logic that OWL supports is not a temporal logic so explicitly expressing temporal relationships such as 'this message follows' cannot be done. Second, there are performance limitations and several computational costs in handling such rich protocol messages, such as the amount of CPU cycles and memory

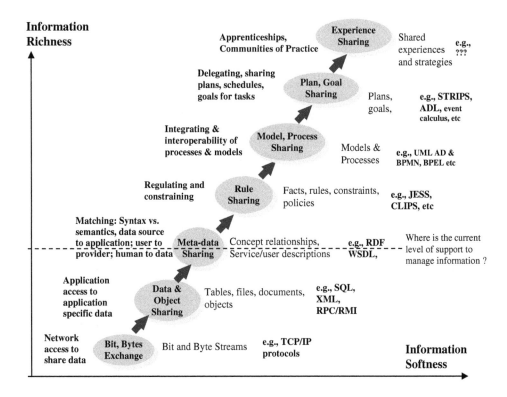

Figure 9.4 Multiple information representations are needed and need to be managed as we move to increasingly rich and soft information. The dotted line indicates our current ability in terms of robust system and tool support to manage these richer, softer types of information

---

[24] If the semantics are not specified in a machine-readable form, different users and applications are more likely to use, interpret the field differently and this can lead to interoperability problems.

needed to store and execute the process, parse the structure of a semantic message with respect to an ontology domain model, and store the results. Because message transport systems need high performance, interaction protocols used in general in distributed computing tend to be specified using a syntactical representation[25] rather than using a semantic one.

A specific type of KR sharing that has attracted much research interest is based upon *Belief, Desire and Intention (BDI)* models. A BDI model defines the internal organisation of an agent with respect to explicit commitment (through intentions) to goals while potentially having only partial information. Hence an agent's intention model must have a set of conditions under which it can revoke the commitment. The logic and semantics of BDI model have been controversial since their inception e.g., in terms of the way the 'belief' predicate is actually used. The use of belief is sometimes little different from a symbolic way of indicating what state information the agent has (Werner, 1996). Whether this state management includes a full BDI model will depend on the requirements of the agent architecture and on whether or not the design and support of a solution for a problem-solving domain can benefit from these notions or not and if a system really is intentional, whether or not we can coherently view it as such (Dennett, 1987). Dennett further proposes (quoted in Ferber, 1999, pp. 313–316) that there is not a single level of intention but multiple levels of intention which the observer or receiver may not be able to differentiate: reacting to an undirected message from a sender which may or may not affect any receiver e.g., a cry for help; (intentionality level 0); sender sends in order to create an effect in a receiver, e.g., to get specific response back such as to answer a specific query (intentionality level 1), etc.

Irrespective of whether or not an interaction protocol is represented semantically, the payload or content it conveys could still be represented semantically. This is fine because the semantic content only needs to be interpreted by the sender and receiver in the computation nodes and not by the network elements. This suggests a way to use semantic headers that are not seen by the core message transport system and hence reduce the performance of the core message transport system. Semantic message headers could be defined and attached to the semantic body, which define semantics for the communication context for the content, and are only interpreted at the sender and receiver.

9.3.3.2 Agent Communication Languages and Linguistic-based Protocols

One of the best-known interaction protocol models for intelligent systems is based upon *Agent Communication Languages* or *ACL* which is in turn based upon a type of linguistic protocol called *Speech Acts*, also called *Communicative Acts* (Figure 9.5) This focuses on the principle that models and meaning are attributed through interaction as much as standalone contemplation. The origin of speech act theories is usually traced to lectures given by the linguist John Austin in 1955 (Austin, 1975). Austin noticed that some speech utterances are like physical actions in that they appear to change the state of the world, e.g., pronouncing someone as 'man and wife' as part of a religious ceremony or sending a message to set a new fact that changes the state of a KB. Speech act theories are pragmatic theories of language, i.e., theories of language use: they attempt to account for how language is used by people every day to achieve their goals and intentions.

One of the potential *benefits* is that if a generic model of communicative acts could be specified, it could be used across all knowledge domains, enhancing service interoperability. Of course, some instantiation of the service actions would be needed in order to ground the semantics and this could vary across services and service domains. In contrast, currently, each application

---

[25] For exmaople, the syntax of protocols can be specified using the more abstract notation of context-free grammars such as Backus Naur Form (BNF) or in a more concrete computational format such as XML.

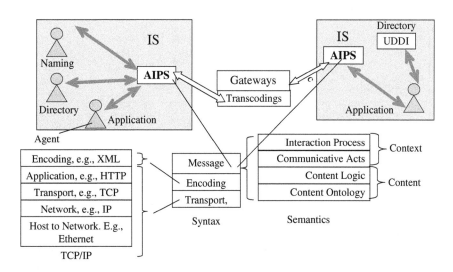

Figure 9.5 Multiple ISs designed as MAS interaction using an Agent Interaction Protocol Suite or Agent Language

domain and even multiple applications within that domain specify their own sets of service actions. This makes interoperability using service actions defined in heterogeneous service models complex.

The basic structure of a communicative act follows the basic structure of an action as used in planning (Section 8.7.2). Actions can be represented in terms of (pre-)conditions which, if true, enable actions to be triggered. Post-conditions of effects define what should now be true if the action was successfully executed. There are different specifications for communicative acts depending on how and what preconditions and post-conditions are defined and how they are represented and what range and classification for the types of speech acts are used. Poslad (2007) surveys the classifications of speech acts. The most useful types of communicative acts are assertive, akin to a set information action, directives, akin to task requests, queries, mediating actions such as propagate and phatics that seek to establish, check, prolong and interrupt, i.e., to help control communication.

The specification of the semantics of the speech acts themselves can be done in several ways. Austin (1975) originally interpreted the meaning in terms of a locutionary model in three parts: (1) the generation of speech (locution); (2) the choice of speech act (illocution); and (3) the intended effect of the speech act on the receiver (perlocution). FIPA has formally specified its ACL semantics focus in terms transferring the sender's mental attitude to one or more receivers (BDI model) but models of society or third parties are not considered (Singh, 1998). Because this type of BDI semantics has these problems in practice, Singh suggests, use of alternative (to BDI) semantics for FIPA-ACL. In addition, Poslad (2007) considers three other alternative semantic specifications for ACLs. *Contract programming model semantics*, e.g., KQML uses a type of programming by contract model to specify its semantics in terms of preconditions, post-conditions and completion conditions for each of its communicative acts. Semantic commitments based upon social conventions can be used. Finally, the IP context can be used as the semantics for the communicative act. For example, the FIPA Interaction Protocol model makes a rudimentary attempt at a social model in the sense that the interaction is related to the organisational roles of the interacting parties and the semantics of each CA in an IP is interpreted within the context of the IP. A critical look at

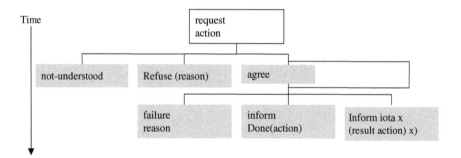

Figure 9.6   The FIPA request interaction protocol

speech act theory from a linguistic perspective is given by Allwood (1977), Junichiro (2003) and Flowerdew (2006).

Specifications of communication using speech acts can be specified in terms of how multiple communicative acts can be used as part of different interaction patterns and how a communicative act links to the message content (Section 9.3.3.1). FIPA defines ten or so interaction patterns which differ depending on whether or not they are information versus task sharing, push versus pull and one-to-one sender to receiver or one-to-many (Poslad, 2007) The request interaction pattern is an example of a task-sharing one-to-one pull type interaction (Figure 9.6). Although it seems similar to a client–server type request-response pattern, it is much richer in the sense that the responder can optionally choose to acknowledge the request and supply the result later. In addition, the responder also has different options to signal lack of understanding of the request, failure for some reason such as lack of sufficient ICT resources and refusal when although it could do the task for some reason, it chooses not to.

Several protocols are needed to support interaction for intelligent applications using communicative acts: an interaction protocol, a communicative act protocol and a content protocol. The content protocol may separate the content ontology from the content logic in order to support the flexibility to allow different logics to be used with the same content language. In addition, a message encoding protocol and a transport protocol are often defined in order for message middleware services to send the message.

In terms of the TCP/IP protocol suite, the ACL behaves as a suite of multi-layer protocols at the application level and for this reason has been termed an AIPS or Agent Interaction Protocol Suite or AIPS (Poslad, 2007). Note several agent communication language models have been specified such as a two-layer AIPS consisting of KQML (Knowledge Query Meta Language) and KIF (Knowledge Interchange Format) (Genesereth and Ketchpel, 1994; Finin and Fritzson, 1994) and the FIPA ACL (Poslad and Charlton, 2001) which define a six-layer AIPS consisting of interaction protocol, communicative act, content ontology, content logic, message encoding and message transport (see Figure 9.5). Several surveys have been performed on ACLs including Labrou *et al.* (1999), Singh (1998) and Poslad (2007).

9.3.4 *Further Examples of the Use of Interaction Protocols*

As an example of the benefits of using rich and flexible interaction protocols, a general scenario for resource access is considered, e.g., accessing and displaying or playing audio-video content. In the simple case, we invoke an AV-player (service) passing the details of the source of the AV content of our choice. But when we play the AV source, it fails, the system must then decide how to proceed. Typically the requester asks for assistance by searching for help in a well-known place, a directory.

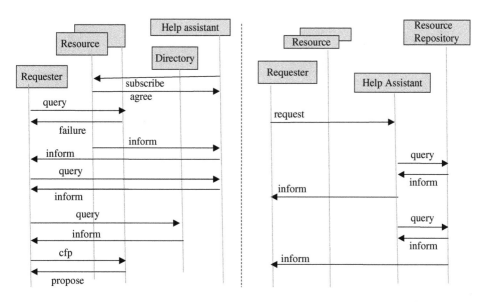

Figure 9.7   Part of the interaction for the plan given in Figure 9.10: locating help when access to resource fails (left) and delegating the task of resource access (right) to a help assistant

Once the requester finds assistance, providing some conditions have been fulfilled such as authentication and competency checks, the requester can choose to delegate the resource access task to the help assistant (see Figure 9.7 which illustrates part of the plan give in Figure 9.10). In more detail, a Help peer registers itself with a provider to be informed when it fails to fulfil a request from a requester (a subscriber). The Help peer then announces itself to the requester. The requester then queries Help about use of a resource X. Help advices A to ask resource depository D something. D tells the requester that peers E, F, G, and H have resource X. The requester issues a contract (a call for proposals or cfp) to E, F, G, and H to ask them to bid to supply the resource as A does not know which of these can provide the resource most favourably, etc.

There may be certain conditions under which a requester does not want to delegate a task to someone else for reasons of privacy, security. Instead in this case, the requester may just seek advice from the assistant rather than delegating the task to the assistant (Figure 9.8). Then requester can issue a call for proposals to different resource providers because the requester may want to compare multiple bids from providers in order to select the most favourable one.

9.3.5 Multi-Agent Systems

Multi-agent Systems (MAS) can be classified both in terms of the type of individual agents they contain, and in terms of the types of interaction they support. Generally, if an IS is represented by an agent, then a MAS represents multiple interacting IS. When MASs interact with other MASs, they represent systems of systems interacting. Singh and Huhn (2005) characterise the properties of MAS as a degree of dynamism, degree of scale (numbers of agents), type of (organisational) control, homogeneous versus heterogeneous types of individual agent, and the type of agent interaction (e.g., goal exchange, belief exchange, etc.). Use of an appropriate ACL (Agent Communication Language) can support these MAS properties.

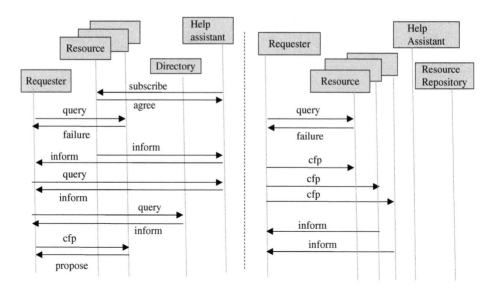

Figure 9.8 Part of the interaction for the plan given in Figure 9.10: asking for advice (left) and negotiating (right) resource access from multiple resource providers

9.3.5.1 ACL and Agent Platform Design

The main type of MAS considered here is one in which individual agents use an AIPS or ACL to interact (Genesereth and Ketchpel, 1994). Of the ACLs proposed, the most commonly used one and the one that has been standardised is the FIPA-ACL.[26] The FIPA AIPS defines about ten different interactive protocols, over twenty different communicative acts. It also has the flexibility to support different content logic representation and different transport encodings and transport protocols that can be oriented to different environments such as lower bandwidth wireless network use versus higher bandwidth wired network use.

An agent platform is a concrete, computational form or reification of a multi-agent specification that defines a set of agent middleware services to facilitate the interaction of agents, including agent management, accessed through some API. The core service is to support interaction using an ACL. FIPA has specified additional agent middleware services, typically including agent name and agent life-cycle management services called the AMS, a directory facilitator service called the DF. An example of two agents situated on two different heterogeneous platforms interacting is given in Figure 9.5.

A design issue is whether or not to model these service processes as agents.[27] If they are agents, then the benefits of the specific internal system design can be leveraged, i.e., a communications agent can reason about how best to communicate. However, this must be balanced against the additional complexity and potential performance drop when realising services as agents (Poslad

---

[26] FIPA, the Foundation for Intelligent Physical Agents, became an IEEE standards activity in 2005. It can be found at http://www.fipa.org, accessed June 2007..

[27] An agent is a specific means of accessing service actions. If agents interact via an ACL, then the services which agents offer must be accessed via that ACL. If agents use goal-based IS designs, then agent interaction is goal-based, etc.

and Charlton, 2001). Several agent toolkits have been developed which support the FIPA ACL and agent specifications, e.g., JADE (Bellifemine *et al.*, 1999), FIPA-OS (Poslad *et al.*, 2000), etc. Generally, such toolkits support homogeneous internal agent designs. For example, the FIPA agent toolkits mentioned support a core design of a reactive agent. Extensions to agent toolkits can be defined to support interaction using more specific model representations such as BDI.

9.3.5.2 Multi-Agent System Application Design

There are many examples of IS or *Agent-Oriented Software Engineering* (*AOSE*) methodologies (also called Agent-Oriented Development or AOD) proposed by different developers but these can be classified into two types: those which extend or adapt non-IS system methodologies, e.g., object-oriented based AOSE; and those based upon AI methodologies. These are starting to converge although some proponents would argue you cannot simply enhance a non-IS method of development to support AI. Applying a simplified AOSE methodology consists of developing two main models: specifying the individual interacting entity roles and the their properties such as their interaction protocols and knowledge-sharing model (organisational view model); specifying the nature of the internal IS computation for each organisational entity or role, e.g., determine the plans to achieve a goal or reasoning to check the validity of a new fact.

AOSE design is captured in two main model views: an *organisational view* (Figure 9.9) which specifies the types of agents and roles, and an *operational view* (Figure 9.10) which specifies the interaction constrained by goals and plans of actions to achieve those goals. Modelling active entities in the system as agents versus roles is often a matter of preference or style. However, (organisational) roles support a more dynamic approach. A role can be used: to separate responsibility for service access from identity; to enable agents to combine multiple roles; to enable several agents to play the same role (organisational role redundancy); to enable agents to change roles at run time; to enhance re-usable structures and patterns; and to provide direction to support management policies. Plans determine the system operation and are expressed at a simple level in

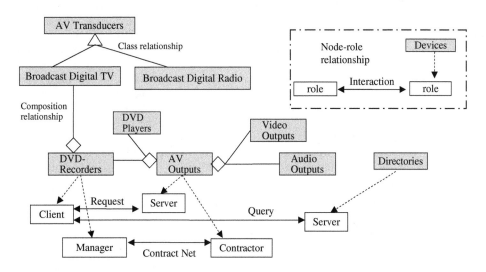

Figure 9.9 Organisational entities (agents) can play multiple roles. Organisational roles constrain the type of interaction

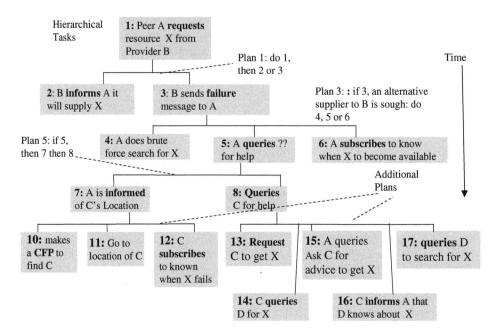

Figure 9.10 A simple planning model to achieve a goal which defines redundant paths through tasks (redundant sequences of tasks) which can be enacted to reach the goal and which can use redundant peers to enact tasks

terms of sets of interactions, and in turn interactions that determine the roles of entities in the organisation (organisational-interaction duality, Section 9.2.3.3).

An example of an operational view is a simple *task-oriented* description of a problem is given in Figure 9.10. The goal is to display images from the digital camera peer on some (visual) display, i.e., part of the personal memories scenario from Chapter 1. The goal is normally achieved using a very simple default plan which in this case consists of a sequence of two actions: the AV source peer such as a digital camera or its storage media requests the use of the image reading functions of an AV player which is connected to a display, the provider peer. The provider then responds by signalling to the AV source that the display is ready. However, the default plan fails because the display is being used by another user application so it reports a failure, display not available. Various other alternative plans could then be evaluated involving cooperative interaction with other peers to see if they can help the requester achieve the goal. The plan here is very simple, they are sequences of task with the outcome of the tasks being evaluated to decide how to proceed. More complex tasks can be decomposed into simple tasks using a planning decomposition method such as HTN or HTA (Section 8.7.3). Many applications and projects have used this type of MAS model for the system design for IS interaction.

9.4 Some Generic Intelligent Interaction Applications

There are many intelligent interaction applications which can be classified into: CCI, e.g., autonomous systems; HCI, system–environment interaction such as (human) interactive systems where IS can act as mediators to enhanced human interaction which would be far less rich without it; and CPI. This section focuses on social type HCI interaction models and their application to CCI and CPI.

9.4.1 Social Networking and Media Exchange

Providing content for public and private access is now cheap relative to the average standard of living. This, combined with our instinctive need to communicate, means that we are not only prepared to adapt to using these new communication methods but, in general, it is a fundamental desire. However, the new means of delivering content leave us open to the potential hazards that exist in the physical world such as fraud, vulnerability, theft of our identity, or being used to exploit others known and trusted by us.

The scale and potential for such exploitation are possible because computers can be automated to track and monitor certain types of communication and automated to create certain types of automated response. As one of the main benefits of knowledge engineering is to enable higher-level semantic interoperability through making representations formal and explicit, it opens a particular concern of creating more opportunity for semantic-based attacks. That is, the same formal semantics that are used to enable semantic interoperability, e.g. use of the device profile to enable the automated integration of a new application, can also be used inappropriately to deliberately sabotage or infiltrate an application. This misuse can potentially benefit from the semantics because it can be more targeted in its approach and appear more authentic because it may have access to 'personal' and organisational information, which hence provides a false sense of authenticity to the user.

Social media experience can take many different forms, including text, images, audio, and video. The most common kinds of social media experiences are blogs, social networks, content communities (sometimes called *folksonomies*), wikis, podcasts and forums.

- *Blogs*: perhaps the best-known form of sharing personal content, blogs are online journals, with the most recent entries appearing first. Now there are *microblogs* for sharing small brief text updates, suitable for access on resource-constrained devices such as mobile user devices.
- *Social networks*: these websites allow people to build personal websites and then connect with friends to share content and communication. The best-known example of a social network is MySpace,[28] which had over one hundred million members.
- *Content communities*: communities which organise and share particular kinds of content. The most popular kinds of content communities tend to be around sharing photos, e.g., Flickr,[29] sharing bookmarked links, e.g., del.icio.us[30] and sharing videos, e.g., YouTube.[31]
- *Wikis*: these websites allow people to add content to or edit the information on them, acting as a communal document or database, e.g., Wikipedia.[32]
- *Podcasts*: audio and video files which are made available by subscription through many content providers. Content be downloaded automatically when new content is added.

In general, the trend is that large amounts of content are created and shared among the users and this creates a stronger move to the Web as a user-driven application platform.[33] It is not feasible to

---

[28] Social networking website, http://www.myspace.com/, accessed Nov. 2007.

[29] A photograph sharing community, e.g., http://www.flickr.com/, accessed Nov. 2007.

[30] A social bookmarks manager. Users can add bookmarks to their list and categorise them. Website example, del.icio.us/,/, accessed April 2008.

[31] Video sharing community, e.g., http://www.youtube,com/, accessed Nov. 2007.

[32] The best-known wiki is the online encyclopaedia Wikipedia, which started in 2001 and has over 2.5 million articles published in English alone in 2008, e.g., http://en.wikipedia.org/wiki/Main_Page, accessed Jan. 2007, and overall has more than 10 million articles in more than 250 languages.

[33] This has loosely been referred to as Web 2.0, a term first coined by Tim O'Reilly in 2004. This was originally intended as a business revolution, it has currently no new specifications associated with it.

expect even diligent users to annotate and add details to their content and organise this in a methodical and consistent. To help the users to manage and organise their content requires certain contextual knowledge that comes from a number of places e.g. content annotations, semantic metadata, contact lists, the way the user organises contact lists as family and friends, etc. To organise content for users requires that a system has certain pertinent and significant knowledge about the users that comes from different places. The challenge in any solution is in the way the knowledge is aggregated to provide a contextual filter that can be applied to the organisation of the content. So creation of this contextual knowledge and the use of this contextual knowledge can help the user in managing and sharing their content:

- *Personalisation* provides the user with social ranking preferences and contextual grouping of content when presenting and searching social content.
- *Self-organisation* uses a device-oriented profile and user profile knowledge to organise and move social content around
- *Self-governance* uses users' preferences and policies to attach access rules to social content when sharing and managing content.

9.4.2 Recommender and Referral Systems

Recommender and referral systems enable people or systems to rely on the ratings of others or the ratings of similar things in order to make and simplify choices such as service choices. Such systems are also often used to simplify personalisation.

9.4.2.1 Recommender Systems

Recommenders are types of personalisation software that make personalised recommendations of goods, services, people based upon some small inputs from the user. This is viewed as crucial for e-commerce sites. Some online stores make millions of recommendations per day because recommenders have been shown to substantially increase sales at on-line stores. There are two main types of recommender system: content-based filtering that finds things similar to ones you like, and collaborative-filtering that finds things liked by people who are similar to you. Most recommender systems inherently hide the identity of the sources of the recommendations, in part because they are often an aggregation that does not maintain a member list of recommendations. In many e-communities and forums, the norm is to use fictitious identities and to deliberately hide true identities. Anonymous opinions may be okay to choose a movie or CD but most of us would not bet our job, or manage a sensitive operation, based upon anonymous recommendations. Examples of the use of recommender system for UbiCom use include the following. The AmbieSense Project (Göker and Myrhaug, 2002) situates each user task within a use-case. This uses case-based reasoning and location-awareness in order to make user recommendations. RECO (Pignotti *et al.*, 2004) situates each user task within a sequence and learns users' preferences over time, in order to make user recommendations.

9.4.2.2 Content-based Recommendations

Content-based recommendations are based on information on the content of items rather than on other users' opinions. This uses a machine learning algorithm to induce a profile of the users' preferences from examples based on a feature-based description of content. There are several potential advantages to a content-based recommendation approach. There is no need for data on other users. There are no cold-start or sparsity problems. It is able to recommend to users with unique tastes. It is able to recommend new and unpopular items. It can provide explanations of recommended items by listing content-features that caused an item to be recommended.

The challenges (disadvantages to overcome) of content-based recommendations are as follows. It requires content that can be encoded as meaningful features. Currently, much content-based matching is in terms of browsing or searching and uses syntax rather than semantics or context. Its users' tastes must be represented as a learnable function of these content features. It is unable to exploit quality judgments of other users unless these are somehow included in the content features.

9.4.2.3 Collaborative Filtering

Collaborative filtering presumes that a database of many users' ratings of a variety of items is maintained. Then for a given user, other similar users whose ratings strongly correlate with the current user can be found. Items can be rated by these similar users, but not actually rated by the current user. Many existing commercial recommenders use this approach, e.g. Amazon.

A typical collaborative filtering method is to weight all users with respect to similarity with the active user. Typically, the Pearson correlation coefficient is used between ratings for active user and another user. A subset of the users (neighbours) which act as predictors is selected. Ratings are normalised and a prediction from a weighted combination of the selected neighbours' ratings is computed. The items with the highest predicted ratings are presented as recommendations. Challenges for collaborative filtering are as follows. There needs to be enough other users already in the system to find a match (*cold start*). If there are many items to be recommended, even if there are many users, the user ratings matrix is sparse, and it is hard to find users that have rated the same items (*sparsity*). It can be difficult to recommend items that are considered to be very different from existing ones (*popularity bias*).

Referrals are the trusted recommendations by known people, in contrast to recommendations that are anonymous. For serious life and business decisions, people often value the opinion of a trusted expert more, rather than an anonymous decision. If an expert is not personally known, then a reference to one can be found via a chain of friends and colleagues using a *referral chain*. A referral chain provides a way to judge the quality of an expert's advice and a reason for the expert to respond in a trustworthy manner. Finding good referral chains is slow, time-consuming but necessary.

9.4.3 Pervasive Work Flow Management for People

Whereas physical distance is much less of a barrier for communication and workflow, virtual distance between employees in terms of differing beliefs, systems and experiences is still a barrier. Restructuring and lack of trust between different layers of organization can reduce the effectiveness of the cooperation between them. Maxwell's solution (2000) is to provide more informal communication channels. Shepherdson *et al.* (2003) have developed and applied a workflow management framework, mPower, which supports more decentralised worker-oriented teamwork coordination, enabling workers to schedule work requests, to trade work requests and work-shifts, to make collective decisions, to extend or reduce work-hours and to call on additional expertise. More flexible and utility-based travel can also be planned and re-planned. In this workflow management framework for mobile workers, there is scope for knowledge-based management of jobs, allowing jobs to be composed based upon expertise, and the ability to trade and update the expertise needed to do a job. There is also a mechanism by which workers can initiate interaction with a mentor.

9.4.4 Trust Management

Trust is an inherent property in UbiCom systems in which one autonomous component cannot completely control another autonomous component but which may nevertheless need to rely or one another or require some cooperation from it. *Trust*[34] in social organisations is a general expectation, explicitly evaluated, that one autonomous component, the *truster*, can rely on another autonomous component, the *trustee*, in order to share information, tasks, goals, etc., with them. There several characteristics inherent in trusting. There is some expectation that the trustee has some degree of reliability, competency within a specific context and that the trustee has the honesty and commitment to act on the truster's behalf, e.g., to carry out a task on behalf of someone else. Second, inherent in trusting a trustee is although there is a likelihood of success, there is some greater risk that the trustee may fail. This *trust risk* in turn depends on other factors such as the degree of loss if the trustee fails and the likelihood of the trustee failing. Third, the trustee is not under the direct control of the truster. The truster is often not able to monitor the trustee, nor directly control the trustee. There are several dimensions or metrics to specify trust for use in open UbiCom systems such as personal trust or impersonal trust, or with respect to the disposition of the truster to trust, which ranges from being averse to trust to being eager to trust or if *distrust* is modelled as the complete absence of trust (zero trust) or is considered in terms of negative metrics for trust (Marsh, 1994; McKnight *et al.*, 1998).

Personal trust arises from our own subjective direct experiences with a trustee. The trustee is trusted because they have acted at least satisfactorily in the past for the truster and the expectation is that they will do so again. In contrast, *interpersonal trust*, relying on others, without any necessarily having any direct experience of the trustees but perhaps trusting them because they are part of some normative institution (Section 9.2.3.2) or because others have good experiences of them and can recommend them, recommender systems (Section 9.4.2). An example of interpersonal trust is the use of a trustee of a trusted third party such as certificate authority or issuer who attests via a digital signature of a certificate that a particular identity (subject) is associated with set of credentials and a particular policy to use credentials to do something such as access a particular resource (Section 12.2.5). Another example of interpersonal trust is a trustee that is a service provider who provides some service on behalf of a truster, the requester of the service.

In many computer systems, although a notion of trust may be implied, such as the use of a trusted platform, or a trusted third party, there is often no explicit computation model of trust incorporated. Trust is a more useful issue for external interaction rather than internal interaction. Internal interaction is often designed to use well-defined notions of control which can obviate the need for trust. In external interactions between one autonomous system and its environment or another autonomous system, the use of a centralised control mechanism is not possible by design. Sometimes there are concepts which are akin to trust[35] used in distributed systems, for example, a system may define a quality of service for another peer to provide, or a requester can examine the collective reputation of another peer before deciding to interact with it. Peers are also often defined as eager to trust, ready to blindly trust, e.g., if the provider has a service description in well-known directory, then the provider must be trustworthy,[36] etc.

The use of explicit social trust models has at least one major benefit for use in UbiCom systems: it can be used to evaluate which peers to select to interact with in open, dynamic, non-deterministic, distributed environments where peers are not previously known to each other. There are several

---

[34] Social control based upon trust is sometimes referred to as a *soft security* in contrast to *hard security* which is control based on encryption algorithms. Soft security is viewed as a more effective mechanism for security, in terms of robustness, scalability, and adaptability, in pervasive environments such as information-sharing communities that support inter-organisational interactions (Hexmoor *et al.*, 2006). The relationship or dependency of the truster on the trustee is referred as a *trust relation*.

[35] This is referred to as an *ad hoc trust* model.

[36] Of course, the directory may offer no control or checks about whether or not malicious providers can offer services or malicious clients can search for services, hence blind trust in the directory service can be associated with an unknown trust relation between one peer and another.

ways in which social concepts trust can be incorporated into computation form and used in UbiCom systems. The design issues are also discussed by Langheinrich (2003). Authentication-based policy systems are based upon PKI. Authorisation-based policy systems such as SPKI and the PolicyMaker system (Blaze *et al.*, 1996) model security credentials, security policies and define trust delegation relations between trusters and trustees which may involve chains of trustees. These trust models define trust metrics that can express different degrees of risk of failure and likelihoods of trust succeeding and can take into account multi-valued metrics for trust.[37] Shankar and Arbaugh (2002) propose an attribute vector theoretical model for modelling trust relationships between entities because this captures both the identity-based and context-based trust relationships in a simple and expressive manner. Yin *et al.* (2006) propose a vector model to express the multiple attributes of trust and show how condition-action rules can be used to adapt this model to a healthcare pervasive computing application.

There are numerous applications of multi-agent-system models that can use collaborative filtering mechanisms based upon recommendations and reputations as well as policy-based MAS models to support impersonal trust. Ramchurn *et al.* (2004) consider two main aspects of design trust for MAS. First, to allow agents to trust each other, there is a need to endow them with the ability to reason about the reciprocative nature, reliability, or honesty of their counterparts The second main approach to trust used in MAS is to design protocols and mechanisms of interactions, i.e., the rules of encounter so that participants will find no better option than telling the truth and interacting honestly with each other. *Institutional trust* can be used when a peer utilises the control measures of institutions and norms. In practice, many trust decisions are still performed by the human operators of the UbiCom systems. The issue of developing fully automated systems that perform trust evaluation is still very challenging and is still a matter for further research. This includes the range of parameters that can influence decisions to trust, how to validate these systems, and the psychological and legal consequences and accountability when delegating trust decision to clever machines.

EXERCISES

1. What is meant by interaction multiplicity? Discuss the types of multiplicity when multiple dumb peers interact, when cooperative intelligent peers interact and when competitive intelligent peers interact.

2. Define the key characteristics of a cooperative system. Discuss two basic designs to support cooperation based upon perfect coordination and explicit communication.

3. Discuss designs for coordination of multiple cooperative systems based upon: service composition models, interaction protocols with inbuilt coordination mechanisms, joint planning and on joint intentions.

4. Discuss how the uses of norms can act as a design for a perfect coordination model.

5. Define and apply the following aspects of hierarchical organizational models: hierarchical containment, organisational roles and missions, organisational interaction and boundary spanner.

6. Compare and contrast the types of mediator identified in Section 9.2.2 and discuss which types are in use in different service domains. Why do patterns of mediation seem more

---

[37] McKnight *et al.* (1998), for example, define five trust dimensions for trust, interpersonal trust, trusting beliefs, system trust, dispositional trust and decision to trust.

EXERCISES (continued)

common in practice than others? Also analyse the classification given in Decker *et al.* (1997) that includes front agents, bodyguards, annonymisers, recommender, and discuss their current use. For the types of mediator tagged as '???', in Figure 9.3, discuss if their use can be justified.

7. Compare and contrast the use of negotiation versus consensus when allocating limiting resources to multiple requestor, e.g., multiple ICT functions or applications within a device such as mobile phone, which has limited energy resources and who try to agree how best to use the energy available.

8. Discuss three designs to support competitive interaction market-place agreements between service and resource providers and requesters, negotiation and voting.

9. Define the characteristics that make speech act communication richer than conventional network communication.

10. What are recommender systems? Why are they useful? What are the benefits of using referrals rather than recommenders?

References

Allwood, J. (1977) A critical look at speech act theory. In P. Dahl (ed.) *Logic, Pragmatics and Grammar*, Lund, Studentlitteratur, pp. 53–69. Available from http://www.ling.gu.se/~jens/publications/docs001-050/012.pdf, accessed April 2008.

Austin, J.L. (1975) *How to Do Things with Words*. 2nd edn. Ed. J. OUrmson and M. Sbisà. Cambridge, MA: Harvard University Press.

Bellifemine, F., Poggi, A. and Rimassa, G. (1999) JADE – A FIPA-compliant agent framework. In *Proceedings of 4th International Conference on the Practical Application of Intelligent Agents and Multi-Agents*, PAAM'99, pp. 97–108.

Blaze, M., Feigenbaum, J. and Lacy J. (1996) Decentralized trust management. In *Proceedings of IEEE Symposium on Security and Privacy*, pp. 164–173.

Boella, G., van der Torre, L. and Verhagen, H. (2005) Introduction to normative multiagent systems. 1st International Symposium on Normative Multiagent Systems (NorMAS2005), at AISB'05: Social Intelligence and Interaction in Animals, Robots and Agents: 1–7.

Connelly, K. and Khalil, A. (2004) On negotiating automatic device configuration in smart environments. In *Proceedings of 2nd IEEE Annual Conference on Pervasive Computing*, PerCom 2002, Pervasive Computing and Communications Workshop, pp. 213–219.

Davis, R. and Smith, R.G. (1983) Negotiation as a metaphor for distributed problem solving. *Artificial Intelligence*, 20(1): 63–109.

Decker, K., Sycara, K. and Williamson, M. (1997) Middle-agents for the Internet. In *Proceedings of the 15th International Joint Conference on Artificial Intelligence*, IJCAI 1997, pp. 578–583.

Dellarocas, C. and Klein, M. (2000) Civil agent societies: tools for inventing open agent-mediated electronic marketplaces. *Lecture Notes in Computer Science*, 1788: 24–39.

Dennett, D.C. (1987) *The Intentional Stance*. Cambridge, MA: MIT Press.

Di Caro, G. and Dorigo, M. (1998) Ant colonies for adaptive routing in packet-switched communications networks. In Proceedings of 5th International Conference on Parallel Problem Solving from Nature. *Lecture Notes in Computer Science*, 1498: 673–682.

Durfee, E.H. (2001) Distributed Problem Solving and Planning. *Lecture Notes in Computer Science* (LNCS), 2086: 118–149.

Durfee, E.H., Lesser, V.R. and Corkill, D.D. (1989) Trends in cooperative distributed problem solving. *IEEE Transactions on Knowledge and Data Engineering*, 1(1): 63–83.

Fasli, M. (2007) *Agent Technology for e-Commerce*. Chichester: John Wiley & Sons, Ltd.

Federrath, H. (2005) Privacy enhanced technologies: methods – markets – misuse. In *Proceedings of 2nd International Conference on Trust, Privacy, and Security in Digital Business* (TrustBus '05), LNCS 3592, pp. 1–9.

Ferber, J. (1999) *Multi-Agent Systems: An Introduction to Distributed Artificial Intelligence*, Reading. MA: Addison-Wesley.

Finin, T. and Fritzson, R. (1994) KQML: A language and protocol for knowledge and information exchange. In *Proceedings of 19th International Distributed Artificial Intelligence Workshop*, pp. 127–136.

Flowerdew, J. (2006) Problems of speech act theory from an applied perspective. *Language Learning*, 40(1): 79–105.

Freeman, E. and Gelernter, D. (2007) Beyond Lifestreams: the inevitable demise of the desktop metaphor. In V. Kaptelinin and M. Czerwinski (eds) *Beyond the Desktop Metaphor: Designing Integrated Digital Work Environments*. Cambridge, MA: MIT Press, pp. 19–48.

Gasser, L. (2001) Perspectives on Organisations in Multi-agent Systems. In M. Luck *et al.* (eds) *Proceedings ACAI 2001, Lecture Notes in Artificial Intelligence*, 2086, pp. 1–16.

Genesereth, M.R. and Ketchpel, S.P. (1994) Software agents. *Communications of the ACM*, 37(7): 48–53.

Göker, A. and Myrhaug, H.I. (2002) User context and personalisation. Paper presented at European Conference on Case-Based Reasoning (ECCBR): 1–7.

Gorman, M.E., Groves, J.F. and Catalano, R.K. (2004) Societal dimensions of nanotechnology. *IEEE Technology and Society*, 23(4): 55–62.

Hexmoor, H., Wilson, S. and Bhattaram, S. (2006) A theoretical inter-organisational trust-based security model. *Knowledge Engineering Review*, 21(2): 127–161.

Huang, E.M. and Truong, K.N. (2008) Breaking the disposable technology paradigm: opportunities for sustainable interaction design for mobile phones. In *Proceedings 26th Annual SIGCHI Conference on Human Factors in Computing Systems*, pp. 323–332.

Huhns, M.N. and Singh, M.P. (1998) Agents and multiagent systems: themes, approaches, and challenges. In Huhns, M.N. and Singh, M.P (eds) *Readings in Agents*. San Francisco: Morgan Kaufmann Publishers, Inc, pp. 1–23.

Hung, L.X., Giang, P.D. and Zhung, Y. (2006) A trust-based security architecture for ubiquitous computing systems. *Lecture Notes in Computer Science*, 3975: 753–754.

Jennings, N.R. (1993) Specification and Implementation of a belief-desire-joint-intention architecture for collaborative problem solving. *International Journal of Cooperative Information Systems*, 2(3): 289–318.

Jennings, N.R. (1996) Coordination techniques for distributed artificial intelligence. In *Foundations of Distributed Artificial Intelligence*, Chichester: John Wiley & Sons, Ltd.

Jennings, N.R., Faratin, P., Lomuscio, A.R., *et al.* (2002) Automated negotiation: prospects, methods and challenges. *International Journal of Group Decision and Negotiation*, 10(2): 199–215.

Junichiro, M. (2003) The fine-grained theory of acts and the speech act theory. In *Proceedings 1st International Workshop on Language Understanding and Agents for Real World Interaction*, pp. 63–69.

Kennedy, J., Eberhart, R.C. and Shi, Y. (2001) *Swarm Intelligence*. San Francisco: Morgan Kaufmann.

Kotsis, G. (2002) Performance management in ubiquitous computing environments. In *Proceedings 15th International Conference on Computer Communication*, pp. 988–997.

Krauss, S. (2001) Automated negotiation and decision making in multiagent environments. In Luck, M. *et al.* (eds) *Proceedings Advanced Course on Artificial Intelligence, ACAI 2001, Lecture Notes in AI*, 2086: 150–172.

Kräuchi, Ph., Wäger, P.A., Eugster, M. Grossmann and G., Hilty, L. (2005) End-of-life impacts of pervasive computing. *IEEE Technology and Society*, 24(1): 45–53.

Kulik, J., Heinzelman, W.R. and Balakrishnan, H. (2002) Negotiation-based protocols for disseminating information in wireless sensor networks. *Wireless Networks*, 8: 169–185.

Kurzweil, R. (2001) Promise and peril – the deeply intertwined poles of 21st century technology. *Communications of the ACM*, 44(3): 88–91.

Labrou, Y., Finin, T. and Peng, Y (1999) The current landscape of agent communication languages. *Intelligent Systems*, 14(2), 45–52.

Langheinrich, M. (2003) When trust does not compute – the role of trust in ubiquitous computing. Workshop on Privacy at the 5th International Conference on Ubiquitous Computing, UbiCom 2003., Available online from http://citeseer.ist.psu.edu/691072.html, accessed April 2008.

Maes, P. (1994) Agents that reduce work and information overload. *Communications of the ACM*, 37(7): 30–40.

Marsá, I., Velasco, J.R. López, M.A. *et al.* (2005) A fully-distributed, multiagent approach to negotiation in mobile ad-hoc networks. International Association for Development of the Information Society conference, IADIS'05, 253–260.

Marsh, S. (1994) Formalising trust as a computational concept. PhD thesis, University of Stirling. Available on line from http://www.cs.stir.ac.uk/research/publications/techreps/pdf/TR133.pdf, retrieved June 2007.

Maxwell, C. (2000) The future of work – understanding the role of technology. *BT Technology Journal*, 18(1): 55–56.

McKnight, H.D., Cummings, L.L. and Chervany, N.L. (1998) Initial trust formation in new organisational relationships. *Academy of Management Review*, 23(3): 473–490.

Millonas, M.M. (1994) Swarms, phase transitions, and collective intelligence. In *Proceedings Artificial Life III* ed. C.G. Langton, Santa Fe Institute, Addison-Wesley, pp. 417–445.

Nystrom, D., Tesanovic, A., Norstrom, C., *et al.* (2002) Data management issues in vehicle control systems: a case study. In *Proceedings 14th Euromicro Conference on Real-Time Systems*, pp. 249–256.

Oliver, J.Y., Amirtharajah, R., Akella, V. *et al.* (2007) Life cycle aware computing: reusing silicon *Technology* 40(12): 56–61.

Parashar, M. (2007) Autonomic grid computing: concepts, requirements, and infrastructure. In M. Parashar and S. Hariri (eds) *Autonomic Computing: Concepts, Infrastructure, and Applications*. Boca Raton, FL: CRC Press, pp. 49–70.

Park, S. and Yanga, S-B. (2008) An efficient multilateral negotiation system for pervasive computing environments. *Engineering Applications of Artificial Intelligence*, 21(4): 633–643.

Pignotti E., Edwards, P. and Grimnes, G.A. (2004) Context-aware personalised service delivery. In *European Conference on Artificial Intelligence, ECAI 2004*, pp. 1077–1078.

Poslad, S. (2007) Specifying multi-agent system interaction. *ACM Transactions on Autonomous and Adaptive Systems* (TAAS), 2(4), article 15: 1–24.

Poslad, S., Buckle, P. and Hadingham, R.G. (2000) The FIPA-OS agent platform: Open Source for Open Standards. *Proceedings of the 5th International Conference on the Practical Application of Intelligent Agents and Multi-Agent Technology* (PAAM) 2000, Manchester, UK, 2000, pp. 355–368.

Poslad, S. and Charlton, P. (2001) Standardizing agent interoperability: the FIPA approach. In M. Luck, V. Marík, O., Stepánková and R. Trappl (eds) Multi-Agent Systems and Applications, *Lecture Notes in Computer Science* (LNCS), 2086: 98–117.

Poslad, S., Laamanen, H., Malaka, R., *et al.* (2001) CRUMPET: Creation of User-friendly Mobile services PErsonalised for Tourism. In *Proceedings 3G2001 Mobile Communication Technologies*, London, 2001, pp. 28–32.

Poslad, S., Zuo, L. and Huang, X. (2007) Multi-agent system technology in distributed database systems. In P. Haastrup and J. Wurtz (eds) *Environmental Data Exchange Network for Inland Water*. Oxford: Elsevier, pp. 97–122.

Ramchurn, S.D., Huynh, D. and Jennings N.R. (2004) Trust in multi-agent systems. *The Knowledge Engineering Review*, Vol. 19:1, 1–25.

Rosenschein, J.S. and Zlotkin, G. (1994) Designing conventions for automated negotiation. *AI Magazine* 15(3): 29–46.

Sakamura, K. and Koshizuka, N. (2001) The eTRON wide-area distributed-system architecture for e-commerce. *IEEE Micro*, 21(6): 7–12.

Sandholm, T. (1999) Distributed rational decision making. In G. Weiß (ed.) *Multiagent Systems: A Modern Introduction to Distributed Artificial Intelligence*. Cambridge, MA: MIT Press, pp. 201–258.

Searle, J. (1995) *The Construction of Social Reality*. New York: Free Press.

Shankar, N. and Arbaugh, W. (2002) On trust for ubiquitous computing. Workshop on Security in ubiquitous computing, at UbiCom 2002, Available on line from http://www.teco.edu/~philip/UbiCom2002ws/index.htm, accessed April 2008.

Shepherdson, J.W., Lee, H. and Mihailescu, P. (2003) mPower – a component-based development framework for multi-agent systems to support business processes. *BT Technology Journal* 21(4): 92–103.

Smith, R.G. (1979) A framework for distributed problem solving. In *Proceedings 6th International Joint Conference Artificial Intelligence*, pp. 836–841.

Singh, M. (1998) Agent communication languages: rethinking the principles. *IEEE Computer* 13(12): 40–47.

Singh, M.P. and Huhn, M.N. (2005) *Service-Oriented Computing: Semantics, Processes, Agents.* New York: John Wiley & Sons, Ltd.

Watson, R.T. (2006) *Data Management, Databases and Organizations,* 5th edn. Chichester: John Wiley & Sons, Ltd.

Werner, E. (1996) Logical foundations of distributed artificial intelligence. In G.M.P. O'Hare and N.R. Jennings (eds) *Foundations of Distributed Artificial Intelligence.* New York: Wiley Interscience, pp. 57–117.

White, C.D., Masanetb, E., Rosenc, C.M., *et al.* (2003) Product recovery with some byte: an overview of management challenges and environmental consequences in reverse manufacturing for the computer industry. *Journal of Cleaner Production,* 11(4): 445–458.

Wooldridge, M. (2001). *An Introduction to MultiAgent Systems.* Chichester: John Wiley & Sons, Ltd.

Wooldridge, M. and Jennings, N.R. (1998) Pitfalls of agent-oriented development. In *Proceedings 2nd International Conference on Autonomous Agents,* pp. 385–391.

Wong, H.C. and Sycara, K. (2000) A taxonomy of middle-agents for the Internet. *Proceedings 4th International Conference on MultiAgent Systems,* pp.465–466.

Wurman, P.R., Wellman, M.P. and Walsh, W.E. (2001) A parameterisation of the auction design space. *Games and Economic Behavior,* 35: 304–338.

Yin, S., Ray, I and Ray, I. (2006) A trust model for pervasive computing environments. International Conference on Collaborative Computing: Networking, Applications and Worksharing, CollaborateCom, pp. 1–6.

Yong, C.Y., Bing, Q., Wilson, D.J. *et al.* (2007) Co-ordinated management of intelligent pervasive spaces. Paper presented at 5th IEEE International Conference on Industrial Informatics, pp. 529–534.

10

Autonomous Systems and Artificial Life

10.1 Introduction

Autonomy is considered to be a core property of UbiCom systems, enabling systems to operate independently, without external intervention. Autonomous systems operate at the opposite end of the spectrum to completely manual, interactive, HCI systems. Without autonomous systems, the sheer number and variety of tasks in an advanced technological society that require human interaction would overwhelm us and make system operation unmanageable. An automatic system is a specific type of autonomous system, designed to act without human intervention in its normal operating mode, and to execute specific preset processes defined as sequences of actions. Automatic systems are often designed to work in deterministic, possibly dynamic, environments, to incorporate simple models of environment behaviour, and possibly to incorporate algorithms to control the environment (closed-loop control systems). More general types of autonomous systems in contrast to automatic systems are needed to support self-operation in open, heterogeneous, and dynamic world environments. Autonomous systems[1] are designed to operate to achieve internal goals, independently, without any external control (execution autonomy), supporting some of the self-star properties (Section 10.2.2).

10.1.1 Chapter Overview

The next section (Section 10.2) describes the main types of autonomous system and the properties of a more general type of autonomous system model called the self-star model. Next, reflective type self-aware systems are discussed (Section 10.3). This is followed by a description of autonomic or self-management system models (Section 10.4). The remaining sections

---

[1] A good discussion of the autonomy properties for multi-agent design is given by Singh and Huhns (2005) who discuss pure autonomy, social autonomy, interface autonomy, execution autonomy and design autonomy.

Ubiquitous Computing: Smart Devices, Environments and Interactions   Stefan Poslad
© 2009 John Wiley & Sons, Ltd

cover self-organising systems as complex systems (Section 10.5) and systems based upon artificial life (Section 10.6).

10.2 Basic Autonomous Intra-Acting Systems

10.2.1 Types of Autonomous System

There are four major types of design for more general autonomous systems. First, autonomous systems can be designed to consist of dynamically *reusable* and *extensible components* to enable new tasks to be assigned (*design autonomy*) to existing components. Otherwise there is an overhead and disruption in dealing with legacy systems that perform which are no longer useful but nevertheless still consume environment resources. Simpler autonomous systems can be composed into more complex systems by defining interfaces (*interface autonomy*) and by using orchestration or choreography to interact with individual systems. Components can use service-oriented computing techniques. Component functions could be reprogrammable e.g., using mobile codes. These types of systems tend to have a strong notion of ICT environment autonomy but not of their physical environment and not necessarily of their human environment.

Second, there are *event-driven architectures* (EDA) and *context-aware* system designs for autonomous systems. Autonomous systems which operate in dynamic environments need to be aware of how dynamic environment states changes – *context-awareness*. These systems can be designed to sense events, to filter them, then to select actions and also to modify this selection if the context changes – context adaptation. EDA and context-aware systems can be passive (open loop with a human in the loop) or closed loop. They may consider only the current context, i.e., are reactive or consider only episodic events or they may consider past events, i.e., model (past) sequential input, in order to act predictively. These systems tend to focus on physical world awareness and user-awareness as the system decides how to adapt to such context changes based upon how they affect the current, active, user goals (Section 7.2).

Third, autonomous systems can be designed to be hybrid goal-based (Section 8.3.4) and environment model-based IS systems (Section 8.3.3). Rather than specifying a specific set or sequence of actions for a system to do which can easily fail if any one of the actions fails, goal-based planning systems can dynamically plan or self-plan their own actions to achieve a goal, enabling users to just specify high-level goals for an IS and for the IS to re-plan tasks if an existing plan fails. Goals are constrained by service-level policies. These systems typically incorporate a model of their environment's behaviour and how this is affected by their own goal-directed behaviour. These systems can also be designed to take into account the goals of other autonomous systems that follow their own self-interested goals or which can be co-opted to achieve joint goals such as organisational or societal goals. This type of design is referred to as the *distributed AI* or *multi-agent system* design for autonomous systems. Generally the focus of such systems is on supporting a notion of social autonomy or self-interested behaviour rather than on supporting ICT environment autonomy or physical environment autonomy. These goal-based, policy-constrained systems can also be designed to self-configure, optimise, heal, etc, to achieve their goals – *autonomic and self-managing* systems. An autonomic system that supports self-healing behaves similarly to a fault-tolerant system (Section 12.2.6).

Fourth, autonomous systems can be designed to be pre-configured with inbuilt local goals to define their execution that may or may not meet some global constraints. Once their operation starts, they are self-regulating without any global control. Key challenges here are, first, how to define the goals, plans and policies for execution and, second, what to do if these need to be changed or maintained as pure self-operating systems cannot be externally controlled to change these. The remainder of this section focuses on different types of local versus global self-star design models.

10.2.1.1 Autonomous Intelligent Systems

Autonomy[2] is often described as one of the key properties of an IS and agent-based systems. An IS may have a physical embodiment or software embodiment which may be free to be executed in any part of an internetworked virtual computer. Gouaich (2004) differentiates between two main kinds of autonomy: *self-governance* and *independence*. This self-governance type of autonomy requires intelligent entities in order to formulate the laws to control the system.

Steels (1995) differentiates the term 'autonomous' from the term 'automatic'. Unlike the term 'autonomous', automatic refers to a system being self-steering but under whose laws control originates from some external source such as a human designer. To be self-steering requires a system to sense its environment and act upon it based upon what it has sensed: an event-driven systems or reactive type IS. Autonomous are generally first automatic systems but which are extended so that they become self-governed. Steels describes a (cooperative) autonomous agent as a self-contained system which not only has to perform functions, perhaps on the behalf of others to fulfil its organisational role but also has its own self-interest.

Independence refers to the characteristic that one controls its own actions and resources and is independent of those of other ISs. To make one system independent of another means hiding the internal processes from external user access, they are kept private only allowing access to some functions via a controlled (public) interface. There is a danger in supporting reflection (Section 10.3) on a system, to reveal how a system's internal state changes to others such as diagnostic processing, because this can lead to a loss of independence.

Those familiar with object-oriented software modelling will recognise that this type of independence can be offered by systems which support abstraction and information hiding. Hence, there is a lack of differentiation for autonomy as independence between a conventional object-oriented system and an intelligent system. The differentiation between an intelligent system design and an object-oriented type design comes from the self-governance type of autonomy which is supported more in the IS but less so in a generic object-oriented system.

In addition, there is the concept of social autonomy, the organisational autonomy of multiple interacting IS. However, frequently one IS will delegate actions to another one, making it dependent on the actions of another IS, at least during that session of operation. As an IS even within an organisation is to an extent independent of another, there is greater risk of failure for the delegated actions because of misunderstandings, disagreements and conflicts, error, unknown private utilities and self-interests may operate. There are different designs for IS to cope with the risk of delegation of actions, for example, a goal-based IS design could have multiple redundant plans which also allow it to utilise the actions of yet other IS if one IS causes it problems. Falcone and Castelfranchi (2001) propose a model for social autonomy that supports a dynamic level of control that is based upon an explicit theory of delegation (and trust) and which specifies different dimensions and levels of delegation and which relates delegation to the notion and the levels of autonomy

10.2.1.2 Limitation of Autonomous Systems

Generally, systems may be designed to operate autonomously but the design and commissioning of the system such as system hardware and some software installation and system repair and maintenance are manual. There are several general critiques in using operational systems which

---

[2] The term 'autonomous' originates from the Greek words *autos* meaning self and *nomos* meaning rule or law. It was first applied to a Greek city-state where its citizens made their own laws, as opposed to living under those of an external governing power.

support full autonomy (Alterman, 2000). These must be balanced against the benefits of using such systems (Section 10.3). Criticisms include the difference between the developer's area of expertise being the system, whereas the user's area of expertise is the task environment. It may take considerable time and experience and many iterations of use from a pool of users, i.e., a community of practice, to understand and to optimise a system with respect to a task environment.

Human users are also able to mitigate the imperfections in the design of the system.[3] It may not be possible to determine and fix all the operational aspects of design for a given application in advance. Systems and their environments may evolve piecemeal over time, hence it may become important to alter the system operation over time. One AI approach to deal with a dynamic task environment is to imbue the system with some machine learning capability (Section 8.3.6). Alterman (2000) also considers a bilateral learning design in which not only does the system learn about the user's emerging practice but the user also learns about applying the system tool to a task environment. This is referred to as a *joint runtime learner*. Joint runtime learners include systems in which users either implicitly (type 1) or explicitly (type 2) learn about the system's processing of data and vice versa. Type 1 systems require systems having to understand users' model of it and some form of mediation between users' (mental) model of the system and the system model itself, which are often at quite different levels of abstraction.

A final argument against supporting individual system autonomy concerns the challenge in making a system able to act to achieve its own goals, independently of other systems, and of its environment. This often leads to the use of a self-contained monolithic design for a system. Often a more practical level of design can be to support a more limited level of autonomy and to combine this with a level of social intelligence (Section 10.2.1.1). See also the discussion about designing systems to support human versus machine intelligence (Section 13.7).

10.2.2 Self- Properties of Intra-Action*

Self-star properties, also called the *self-* self-x* or *auto-** properties, refer to a set or properties such as self-optimising, self-healing, self-protecting, self-organising, etc. (Table 10.1). Which of these needs to be supported depends upon the application domain and system design. The essence of the self-star properties is that application processes are not influenced by external behaviour or control but are dependent on, or driven by, their internal behaviour and by internal control. Attempts to externally upset the system will be resisted and moderated in self-star systems. Some system behaviour is decentralised and local,[4] e.g., optimisation, however, other system behaviour is community-wide, global, e.g., self-protection.

A variety of so-called self-star properties have been proposed to model complex systems whose components have some autonomy and propensity to maintain and improve their own operation in the faces of external environment perturbations. Kephart and Chess (2003) have highlighted the group of self-* properties of self-configuration, self-optimisation, self-healing and self-protection to realise the original Horn (2001) vision of autonomic computing. Nami and Bertels (2007), in their survey of autonomic computing, include many different self-* properties, considering the four properties mentioned in Kephart and Chess (2003) as major characteristics and all others as minor characteristics. Organic Computing (Müller-Schloer, 2004) is a system that dynamically adapts to

---

[3] It is noted that although users can improve the performance of a system, they could also denigrate the performance of the system and inadvertently or maliciously cause the system to fail.

[4] Local properties are also referred as microscopic properties; properties which act over a community of autonomous components are referred to as global or macroscopic properties.

Table 10.1   Types of self-star properties for UbiCom Systems

Self-star property	Description	Example systems
Self-configuring	Automated system configuration of specific components based upon high-level policies, with others that are affected automatically adjusting their configuration to maintain a service level	Automated Negotiation MAS
Self-regulating	A system that operates to maintain some parameter, e.g., QoS, within a reset range without external control	*Closed-loop control* systems, *Normative* MAS
Self-optimising, self-tuning	System continually monitors itself in order to optimise or improve its own performance and efficiency	Load-balancing (Grid); P2P distributed lookups, content propagation
Self-learning	Systems use machine learning techniques such as unsupervised learning which does not require external control	Unsupervised machine learning, leaning agents
Self-healing, Self-recovery	System automatically detects, (self-)diagnoses, and repairs localized software and hardware problem	Fault-Tolerant system, Volatile service models
Self-protecting	System detects and defends itself against malicious attacks and cascading failures. It can use an anticipatory approach to try to prevent system-wide failures	Soft security, social trust MAS
Self-aware Self-inspection Self-decision	System must know itself. It must know the extent of its own resources and the resources it links to. A system must be aware of its internal components and external links in order to control and manage them	Reflective systems; goal-based, plan-based, utility-based MAS
Self-interested	System that is oriented to pursue its own goals and ignore the goals of others	Competitive MAS
Self-organising	System structure driven by physics-type models without explicit pressure or involvement from outside the system	Swarm Computing, Amorphous Computing
Self-creating, Self-assembly, Self-replicating	System driven by ecological and social-type models without explicit pressure or involvement from outside the system System's members are self-motivated and self-driven, generating complexity and order in a creative response to a continuously changing strategic demand	Genetic algorithms, Amorphous computing
Self-evolution, Emergence	Coherent processes for emergence, at the macro-level or global level that dynamically arise from the interactions between the parts at the micro-level	
Self-managing,[1] or self-governing	A system that manages itself without external intervention. What is being managed can vary dependent on the system and application. Self -management also refers to a set of self-star processes such as autonomic computing rather than a single self-star process	Normative MAS, cooperative MAS, competitive MAS, Models based upon SNMP

(*continued overleaf*)

Table 10.1 (*continued*)

Self-star property	Description	Example systems
Self-describing, self-explaining Self-representing	A system explains itself. It is capable of being understood (by humans) without further explanation.	Well-designed HCI systems, Reflective systems

Note: [1]There is a range of self-managing system definitions in use. It can refer to a system that manages itself without human intervention. It can refer to a system that is self-configuring or self-optimising or to a system that supports a group of self-star properties such as autonomic computing.

its environment using self-organisation, self-configuration, self-optimisation, self-healing, self-protection, self-explaining, and context-awareness. Context-awareness is complementary to self-awareness, whereas context-awareness focuses on a system optimising its behaviour through monitoring its external environment and self-awareness focuses on monitoring its internal behaviour in order to optimise its behaviour.

Examples of applications of self-star systems are discussed in the following sections. One of the difficulties with the self-star model if it merely means doing things self-contained or doing things internally is that it is quite a general concept and that the range of members for the self-star model is potentially very large. There is no reason why the list of types of self-star system given in Table 10.1 could not be expanded a great deal further to incorporate self-referencing, self-interaction (intranets), etc. For some researchers, the essence of the self-star model is the autonomic computing model or self-management model in terms of the four core major self-configuration, self-optimisation, self-healing and self-protection properties with all other self-star properties being regarded as minor or supporting self-star properties.

10.3 Reflective and Self-Aware Systems

10.3.1 Self-Awareness

Context-awareness (Chapter 7) focuses on an awareness of a system's external environment context, the user of the system, the physical environment and the ICT infrastructure, periodically sensing this, automatically detecting significant change, and using this to adapt the internal system's behaviour to external environment behaviour. An associated design issue is what degree, if any, of awareness the system needs of its own internal behaviour. The basic design is that an ICT system does not process itself or is not aware of its own actions.[5] Generally, an important trait of intelligent machines is that it knows what it is doing and knows what to do next. Much of the focus is on recognising what a system's environment is doing and (thinking about) selecting the appropriate action. Equally important is the system knowing what its internal condition is and how it acts. Reflection is the process by which a system can observe its own structure and behaviour, reason about these and possibly modify these.

---

[5] In his original design for a computer architecture in 1945, Von Neumann made certain assumptions to ensure valid processing, one of which is minimal self-processing, i.e., a system does not monitor or change itself (Whitworth and Ryu, 2009).

A simple application of self-awareness is to monitor use of the online resources to check their status and to modify the internal behaviour to conserve resources more as resources reach minimal levels, For example, such a system could decide to increase the compression at which it stores content because it assesses that the recording of the current program may exceed the storage to complete the recording in uncompressed mode. Another core application of self-awareness is robots, e.g., robot arms must be aware of their inertial force and braking force in order to move and position accurately in the physical world (Section 6.7). More complex self-awareness can involve re-planning actions through analysing and predicting any current plans of actions that will likely fail.

Reflection has several benefits for UbiCom such as: facilitating debugging of operational errors in systems as a precursor to improved system fault-detection and autonomy; improving (external) context-awareness by supplementing it with a system's self-awareness; enabling greater system adaptation of run-time system behaviour possibly as a precursor to fault correction and supporting self-explanation to external environments. Reflection is considered in more detail in Section 10.3.3.

10.3.2 Self-Describing and Self-Explaining Systems

A further use of reflection is to enable a system to support external explanation to an external environment including humans. A *self-describing* system, also referred to as self-representing system, is able to describe itself from the perspective of what it does. This could be based upon its actions and state being represented declaratively as knowledge. A *self-explaining* system describes itself from the perspective of how and why. It is capable of being understood (by humans) without further explanation. These are often referred to as characteristics of a *meta-level* of the system as opposed to the operational level of a system. Self-explaining could be designed by reasoning internally about a system's state and actions. There are several levels at which self-descriptions and self-explanations can be supported in a system (Table 10.2).

UbiCom systems which support some degree of interaction with human users often need to explain their operation because otherwise human operators may override the system operation and operate the system incorrectly. Explanations or descriptions of systems are also useful when different independent (cooperative and self-interested) systems interoperate and need to resolve conflicting operations. Explanations can include describing a system's current ICT context in terms

Table 10.2 Increasing levels of support for an evolution of systems from self-describing (level 1), through to basic self-awareness (level 2), to self-explaining (levels 3 and 4) and to self-empowerment and autonomic behaviour (level 5)

Level	Design characteristics	Current design	Ideal design
1	Systems can describe themselves: State, tasks, plans, goals, etc. Descriptions may be internal, external, on-line or off-line	Yes	Yes
2a	Devices are self-aware of their internal condition	No	Yes
2b	Devices able to self-diagnose and report error conditions	Sometimes	Yes
	Use of single versus multiple input and output modes of interaction	Single	Multiple
3	Devices are designed to justify their actions	No	Yes
4	Devices are able to orientate the level of explanation to the user	No	Yes
5	Device is autonomic and empowered to plan its own actions to fulfil its goals	No	Yes

of its operational state and active tasks, describing why it is doing what it is doing (which plan it is following), and describing and sharing its current goals.

A common reason why semi-automatic systems are less usable and therefore are less used is the lack of attention paid to good HCI design, coupled with the reliance that human operators can understand the operation through using a paper-based or electronic external operator's manual, in natural language form, as the explanation of the system. There are several inherent disadvantages in using an external operator's manual as follows. The manual is not always available as it is not bound and located with the device it explains. The instructions for operation depend on the system state, the operation actions are modal, but users can't tell which state the system is in, so don't know which action should be used. Users cannot understand the instructions or understand how the instructions apply to the current user context. Instructions manuals define a sequence of operation but do not define the possible errors which lead to undocumented deviations from the sequences explained. The manual explains multiple device models but users cannot identify the model they have, possibly because the model numbering has changed. The manual may be incorrect or out of date.[6]

Self-describing systems are able to provide much richer descriptions of themselves, sharing not just their state and actions, but also their plans and goals (Table 10.2). These descriptions can be short, structured lists of properties in the form of *annotations* or *tags* (Section 6.2) or they be much richer, fuller, descriptions. Descriptions can be represented syntactically using some grammar to define a simple syntax, e.g., as an ordering of data fields – a byte orbit stream, or it can be represented semantically to define complex structures of concepts, properties, relations and constraints. These descriptions may be internal versus external and on-line versus off-line, e.g., a paper manual is external and off-line; electronic manuals are external and on-line. They may also be externally co-located versus externally not co-located (Section 6.2).

To some extent, intentional violation of a self-star property can occur. For example, imagine each device supports self-descriptions, these would require the device to have internal resources to store these descriptions, and to output and display these descriptions. It would also mean that they may need to have network resources available in case these descriptions need to be upgraded. In practice, what happens is the meta-level description of a system is often external to the system. Descriptions from multiple devices may also be stored centrally externally, i.e., in a directory. This means that the devices do not have to be designed to have the internal resources to support self-descriptions. Descriptions can be added later to legacy devices which have no descriptions and which have no internal resources for electronic descriptions. A hybrid technique is where the device can be queried for the address, e.g., URL, of its description, but which resides elsewhere in its environment, e.g., the Cooltown and Semacode projects.

At level two in Table 10.2, devices are aware of the status of operation. This status can be output periodically or only on request. Systems may be aware of error states and log and report these if they cannot be internally documented. Systems can also provide mechanisms and interfaces to output their external status for other meta-level processes or diagnostic management processes to analyse, e.g., the JTAG interface (Section 6.6.1). The system may also be able to trigger searches for external advice to diagnosis problems and manually display it and even execute the advice. At level three in Table 10.2, devices know their state in relation to their plans and goals and explain the relation between the state and the plan and goals thus to explain why they are doing what they are doing. They are able to use multi-modal channels to communicate so that if one channel fails,

---

[6] This list of manual explanation limitations is far from complete. It is left as an exercise for you to add your own reasons and experiences to this list.

another can be used. They are aware of why they are currently acting the way they are because they can relate their current action to a current plan. One of the limitations with internalising descriptions and explanations, of having pure self-descriptions and having self-explanations, is that they are normally only oriented to one viewpoint of use of the system. To support self-explanations, systems need to be able to allow policies and preferences to be expressed at different levels of abstraction and use-oriented terminology (Poslad and Zuo, 2008). Systems may need to link to multiple information sources that contain user experiences and reconcile these, i.e., refer to *communities of practice*.

10.3.3 Self-Modifying Systems Based Upon Reflective Computation

There are three elements to the reflection process: (1) the exposure and observation of the internal behaviour to support observation (*instrumentation*); (2) reflective computation about the operational computation (*introspection*); and (3) possible modification to the operational computation (*adaptation*).[7] To support reflection, reflective computation is done by a system at *a meta-level about* its own operation at the application or base level of the system (Figure 10.1). Often, a reflective system designed so that generic functions to support reflection at the meta-level are supported in reflective system middleware.

To enable this meta-level processing requires meta-level descriptions. Reflection can be implemented in a range of programming languages to interface between the meta-level and base-level (Maes, 1987). In procedural and object-oriented reflective languages, the self-representation of the system is the implementation of the system. This can lead to compromises in choosing a programming language for efficient execution versus for efficient reflection,[8] e.g., in terms of what can be reflected and what can be modified. In declarative reflections models using knowledge-based semantic languages (Section 8.4) and logic languages (Section 8.5), the system reflection and the system operation are implemented in separate models and the self-representation of the system is not the implementation of the system. Design issues concern how the reflection is represented, what triggers the reflection and what parts of the system can be reflected upon and the performance and security related management aspects of using reflection.

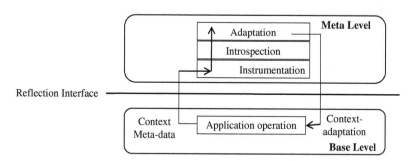

Figure 10.1   Reflective system architecture

---

[7] Generally, this seems similar to the life-cycle of machine-learning system operation. The main differences lie in the detailed mechanisms of how they work.

[8] At the current time, two of the main programming languages for distributed computing, Sun's Java and Microsoft's .NET both support reflective APIs. These both allow O-O programs to examine instances of classes and members at run-time.

Reflection is often seen as an extension to the middleware model (Section 3.2.3), called reflective middleware, in addition to being purely a computation model. Rather than making the environment context transparent or hidden from applications, reflective middleware enables the environment context to be exposed to applications so that they consider from an application perspective how to deal with dynamic application contexts. Hence, a reflective model is often applied as a design for context-aware systems. Here are a few examples. Venkatasubramanian *et al.* (2001) describe how reflective middleware enables applications to adapt to the QoS of the underlying infrastructure. Capra *et al.* (2003) discuss a reflective context-aware application that adapts system behaviour for mobile contexts. Here, open distributed services may define constraints or policies to regulate context-awareness. However, multiple contexts which need to be composed and their associated policies may conflict. Introspection based upon auctions can enable policy conflicts to be resolved. Another example is given by Meng and Poslad (2008) who use reflection about a multi-valued objective routing algorithm to enable spatial-routing to adapt to dynamic environment changes in a more flexible and reliable way.

10.4 Self-Management and Autonomic Computing

Autonomic computing is one of the most well-known types of self-* systems. *Autonomic computing* was inspired by analogy with the human body's nervous system. In the same way that the autonomic nervous system acts and reacts to stimuli to regulate and protect the human body, independent of the individual's conscious inputs, autonomic computing can support analogous behaviours in virtual computing systems (Horn, 2001). Autonomic systems are constructed as a group of locally interacting autonomous entities that cooperate to maintain system-wide behaviour without any external control. The motivation for autonomic systems was to deal with the obstacle of IT system complexity: 'The growing complexity of the IT infrastructure threatens to undermine the very benefits information technology aims to provide' (Horn, 2001). If elements of autonomic computing designs are incorporated into concrete distributed computing architectures such as WS SOC and Grid Computing, autonomic computing can enable these to better manage complexity, to support more adaptive allocations of resources and to simplify the modelling, assembly and deployment of components.

Kephart and Chess (2003) have identified the group of key self-star properties that characterise autonomic computing to be self-configuration, self-optimisation, self-healing and self-protection. Autonomic computing is also referred to as *self-managing systems* or *self-governing systems*. According to Ganek (2007), the core enabling properties for autonomic systems are self-awareness of one's own capabilities and those of other autonomic components, context-aware adaptation and use of planning to control one's own behaviour constrained by system policies. Kephart (2005) has reviewed the research challenges of autonomic computing in general.

There are three basic designs for self-* computing in terms of where the control loop and policies are applied within the system (Figure 10.2). First, policies can reside at the global or macro level and then control can reside at the local or micro level. In the first case, local policies are set and regulated in each resource which may, or may not, be designed to map to a global policy. Second, both the control and policies can reside at the global level. In both these cases, a global policy is set and multiple autonomic systems have to self-organise themselves to adhere to this policy.[9] The

---

[9] Some consider the global autonomic control to be the main or sole mode of operation of autonomic systems e.g., Zambonelli (2006), however, the autonomic system blueprint proposed by IBM also covers the local autonomic control (Section 10.4.1).

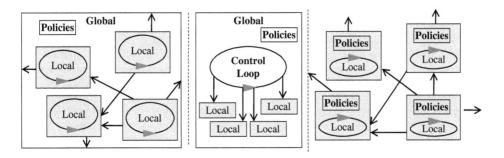

Figure 10.2   Three major types of internal self-* system control of resources: global policies driven local self-* control (left) global policies driven global self-* control (middle), local policies driven local self-* control (right)

third option is where both control and policies can reside at the local level – the latter is reminiscent of an emergent system Examples of the use of local control to adhere to global behaviour are first, the simple *flock model* or *herd model* for mobile entities proposed by Reynolds (1987). This is based upon three simple local rules: *separation*: steer to avoid crowding local flock-mates; *alignment*: steer towards the average heading of local flock-mates; *cohesion*: steer to move toward the average position of local flock-mates. A second example is the use of swarm intelligence. For the global policies driven by a local autonomic control (loops), a key issue is how multiple local autonomous participants derive the local rules which will support the global policy.

De Wolf and Holvoet (2007) have proposed a taxonomy for basic self-star properties of decentralised autonomic systems. They analyse autonomic system properties with respect to generic versus specific properties,[10] type of coordination used (Section 10.4.1) and how the system can be validated. Their classification of generic properties of autonomic systems is quite similar to the classification of environments for agent-based systems given by Russell and Norvig (2003), e.g., continuous versus discrete, episodic versus sequential but also includes global versus local[11] (Figure 10.2) and single-shot versus multiple-shot[12] behaviours. Hence, agent designs oriented to these environments can form the basis of designs of autonomic systems which support these properties. Their specific properties relate to the type of organisation: spatially dependent, role-based, group-based, resource access control and self-protecting. Methods for coordination of autonomic systems are dependent on the type of organisation and are discussed in a later section (Section 10.4.1).

---

[10] The self-* properties proposed De Wolf and Holvoet (2007) are much lower-level characteristics compared to the higher-level given in Table 10.1.

[11] Tuning servers individually (also called locally or on a microscopic scale) may be beneficial, e.g., modifying the individual performance to reflect the local energy supply. However, in other applications, tuning servers individually, which appear to be functioning well on their own, may not, in fact, contribute to optimal end-to-end (also called global or macroscopic scale) performance.

[12] The definition of single shot versus multiple shot property, whether something is achieved once or needs to be achieved several times depends upon the definition of a user session. Single shot actions that are repeated across sessions equate to a multi-shot property.

10.4.1 Autonomic Computing Design

High-level conceptual architectural models for autonomic computing, described by Ganek (2007) and Sweitzer and Draper (2007), consist of five components (Figure 10.3): a user interface (task manager); an autonomic manager with an autonomic control loop; a knowledge base about the managed resources including management policies; a standardised interface to access managed resources (TouchPoint) and a service-based communications network, the ESB (Section 3.3.3.8). The task manager supports high-level user policy-based management of user goals. The human user is regarded as being external to the system, also referred to as *non-self*. The control loop may be situated in the manager components themselves, alternatively the control loops can be designed to be situated in the resources themselves (Figure 10.3). The perspective of what constitutes the system, where the system boundary is and where the system is viewed from, can introduce contradictions. From the viewpoint of the resources, autonomic resource managers are external to them and hence the resources are inherently not self-managed There is an inherent performance degradation issue when managers are distant from the resources they manage. From the point of view of the resource managers, the resources are external to them but the management is internal.

Several current distributed system designs support self-star properties as indicated in Table 10.1. Simple Network Management Protocol management information base, SNMP and the MIB can be used to implement the TouchPoint and part of the knowledge base respectively. Effectors in the TouchPoint can issue set instructions to change the state of resources. Semantic and syntactical metadata wrappers can be used to describe the structures of resources in richer ways. Sensors can poll resources or receive notifications about events in resources. Designs for autonomic control loop can be based on feedback control algorithms (Section 6.6) and based upon the action selection loop design of intelligent systems (Section 8.3, Figure 10.4).

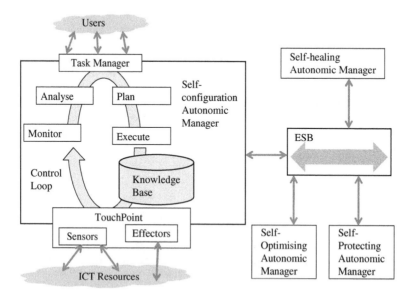

Figure 10.3 A high-level schematic architecture for an autonomic computer system that uses managers as opposed to resources to implement the control loop

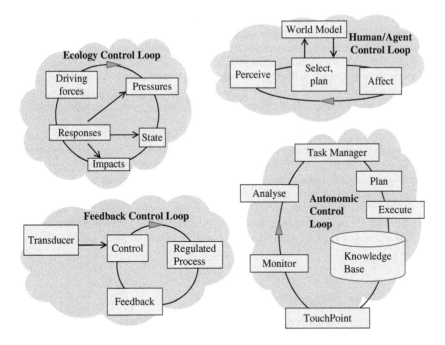

Figure 10.4   Control loops to support self-management in different kinds of natural and artificial systems

There are several practical underlying designs and technology to reify autonomic computing systems rather than to reify the individual components such as the autonomic control loop in the autonomic manager or the TouchPoint (see also Figure 10.4) these include: *event-drive architecture* or EDA (Section 3.3.3.6) and context-aware computing (Section 7.2), feedback-control systems, ecology systems and Multi-Agent Systems (Section 9.3.5). In addition, autonomic computing can itself be used as part of other types of systems in order to endow those other systems with the properties of autonomic computing, e.g., Grid Computing and Web Services.

An example of the use of an event-driven architecture to design autonomous computing systems is described by Wile and Egyed (2004). This consists of three layers: a bottom layer of sensors or probes and effectors to interface with resources. A middle layer of gauges transforms the low-level events from sensors into higher level application contexts. These context events are then processed in the upper control layer which contains simple event condition action control loops to enable events which meet predefined conditions to trigger actions. Their application is an email virus checker in which sensor events are gauged to detect the occurrence of events that could be considered suspicious, including the creation of new processes and the destruction of existing processes. This EDA model is similar in principle to a basic context-aware computing system model given in Section 7.2.

Both autonomic computing and MAS (Section 9.3.5.2) seem to exhibit many similar properties. Both are based upon interacting autonomous, proactive and goal-based components. Autonomic computing focuses more on autonomous components automating and maintaining their operation, i.e., on safety management, via local views that combine self-awareness and local models of external actions. MAS tend to focus more on cooperative and competitive social interaction models, on orchestrating interactions to achieve goals and on some rich sharing of context and expertise with others to give them more knowledge to interact.

Examples of specific support for self-star properties in MAS are as follows. The use of self-agreements using MAS which support negotiation between two or more parties without recourse to a third (external) party which can be used to implement self-configuration. MAS can support self-organising and self-regulation using normative modes of operation. MAS systems that are goal-based, plan-based and utility-based, naturally support an awareness of how their choice of actions determine how they act. This can equally support context-awareness of their actions on their environment and self-awareness. Self-optimising and self-tuning systems can be supported using learning agents and machine-learning techniques.

Self-star systems can be designed to support different maturity levels[13] of self-star properties: basic, managed, predictive, adaptive, autonomic. At the basic level, each resource is managed in isolation. At the managed level, multiple resources can be managed from a common point. At the predictive level, data mining and data correlation techniques are used to recognise patterns, to predict the optimal configuration and to provide advice on what course of action the administrator should take. At the adaptive level, the system can automatically take actions to effect resources based on the sensed information and predictive actions. At the autonomic level, the control of resources is governed by high-level user level policies and goals which users are able to modify.

10.4.2 Autonomic Computing Applications

The emerging complexity in some computing grids requires more adaptive models of design and autonomic computing is a way to enable this adaptation. Parashar (2007) describes AutoMate which is designed to support self-managing Grid applications. An example of a self-optimising application considers how channels are allocated to meet peak demand for calls in different mobile phone cells (Shackleton et al., 2004). Adjacent cell base stations may try to use the same channel. Rather than use centralised optimisations which cannot handle local peak demand fluctuation well, a decentralised 'mutual inhibition' technique, inspired by studying cell specialisation during fruit fly creation, is used which is based upon local rules is used by base-stations for interacting with their neighbours. In essence, each base-station send signals to its neighbours attempting to stop them from using its 'favourite' channels, and it must respond to such signals from its neighbours by reducing its 'preference' for their favourite channels. When there is a 'clash' because two base-stations both want a particular channel, this will be resolved with one base-station emerging victorious and the other relinquishing that channel. Another example of self-optimising is load-balancing in servers to meet a designated QoS under variable external processor loads (Bennani and Menasce, 2004). This is an example of using a centralised adaptive feedback controller (Section 6.6.3) to self-regulate a variable load in relation to the target QoS based upon actual service workload and predicted future workloads based upon statistical analysis. The actual and predicted workloads are fed into a performance model based upon a network queuing model to determine the expected QoS.

A core type of self-protecting application concerns intrusion detection. The nodes in the network could be designed to mimic a neighbourhood watch scheme in which each peer is empowered to monitor and report suspicious activity[14] rather than to use specialised peers to police the community (Shackleton et al., 2004). Individual devices compute an 'alert level' on the basis of locally

---

[13] An architectural blueprint for autonomic computing, IBM. 2003. Available from www-306.ibm.com/autonomic/pdfs/ACwpFinal.pdf, accessed July 2007.

[14] For example, it is said that late-night pizza deliveries to the Pentagon, the US Department of Defense, in Washington, DC, prior to the run-up to the Gulf War, led journalists to infer some intense activity was afoot by linking task to intent and that US involvement in the war was imminent.

detectable network activity and then exchange beacon signals, which define the alert level, but also identity, internal state, etc., with other nodes. The collected beacon signals are used by every member to update their own alert level. A key issue is how to filter those levels of anomalous events which should be reported as abnormal versus those that are normal variation versus those events where it is still uncertain about the normal versus uncertain state. Bigham *et al.* (2003) use an equation-based model to help classify normal versus abnormal behaviours because the application is a SCADA system for electricity control which has well-known equations for normal electricity flow. Qu and Hariri (2007) also used an equation-based model to classify behaviour into normal or abnormal. Self-protection designs focus most on distributed detection rather than on adapting the current protection. Note autonomic computing is often designed to enable homogeneous systems to self-regulate. It is not yet clear how it can enable heterogeneous competitive systems to cooperate, to support symbiotic systems that have independent yet to a degree interdependent goals.

10.4.3 Modelling and Management Self-Star Systems

A key challenge with managing autonomic is how to model and manage systems which are dynamically decentralised and to an extent, non-deterministic. De Wolf and Holvoet (2007) consider methods for determining the performance of autonomic computing components but these approaches also apply to modelling these systems in general. Of these methods, *unit testing* and *formal proofs* are deemed to be too limited because they can deal with only checking the behaviour of microscopic and static, closed systems respectively. This leaves four remaining candidate methods which can be used to validate the behaviour of self-star systems, statistical methods, equation-based versus equation-free macroscopic methods and time-series chaos theory. *Statistical methods* involve determining the behaviour of a representative sample and using that to determine and predict the behaviour of the wider system population. *Equation-based computation methods* analyse system dynamics, e.g., in the form of possibly partial differential equations, while *equation-free computation methods* simulate simplifying behaviour to be equivalent to evaluating the outcome of the equations. *Time series chaos theory* analysis describes non-linear behaviour, e.g., how sensitive the evolution of a system is to changes in its initial behaviour.

Current research that models the behaviour and interaction in simple life-forms forms an important contribution to improve models of complexity. Woolfram[15] (2002) states that mathematical equations do not capture many of nature's most essential mechanisms. He thinks of complex systems in terms of processes that move them from a start state via intermediates states to a goal state rather than in terms of equations. Even extremely simple programs can produce behaviour of immense complexity. Snooks (2007) argues, however, that although modelling systems purely based upon physics interaction may lead to useful self-organising system models, models must take into account higher-level social science interaction models, to account for higher social life forces which cannot be explained using physics-type attractors or repulsion models alone. When models of services and system complexity are driven and controlled using human social and organisational interaction, models of social self-creation need to be combined with simple reactive and rule-based models for self-organising models. So-called hybrid architectures and design based on agents, called hybrid agents, have been proposed to combine low-level reactive behaviour with higher-level reasoning and policy-driven behaviour.

---

[15] Stephen Wolfram played a pivotal role in creating the field of *computational physics* – the use of computers to model problems in basic physics, by designing the Mathematica program.

10.5 Complex Systems

A *complex system* is defined as a system whose properties are not fully explained by an understanding of its component parts (Goldenfeld and Kadanoff, 1999). In computation, a complex system often represents a hard problem that cannot be solved within polynomial time. Complex ubiquitous systems consist of systems of many systems such as systems of billions of networked elements structured into millions of interacting networks (complex networks), systems of many sensors or many MEMS and nano devices (amorphous computing).

Complex systems can arise when there are many possible combinations of interactions because of many interrelationships and interdependencies between them. Complexity can also arise through simple interactions and repeated applications of simple rules, e.g., chaotic behaviour can be caused by small perturbations in a system; *positive feedback* can lead to recruitment and reinforcement; *negative feedback* can lead to saturation, exhaustion, or competition. Simple behaviours interacting in a manner can produce a range of interesting complex behaviours for designing complex systems. Complex situations can also arise out of actions that obey simple laws, e.g., the laws of physics.

Kelly (1995), quoted in Modis (2003), asks whether different types of system complexity are comparable. 'How do we know one thing or process is more complex than another? Is a cucumber more complex than a Cadillac? Is a meadow more complex than a mammal brain? Is a zebra more complex than a national economy?' Kelly (1995) considers which types of mathematical models of complexity can be used to compare whether one type of system is more complex than another. It is not clear how accurate or equivalent the complexity models of one phenomenon, e.g., biological, when used as a model for another phenomenon, e.g., an artificial one such as a computer network. Modis (2003) considers a model of relative complexity, in which some evolutionary step or change in a system is proportional to the length of the ensuing status. Modus also discusses the growth in complexity and whether or not this growth is exponential or bell-shaped (also referred to as a logistic fit).

The conventional technique to modelling complex systems is the *divide-and-conquer* or *reductionist* approach in which more complex behaviour is functionally decomposed into simpler atomic component parts which function without side-effects, i.e., the outputs or actions depend only on the inputs and on the functional model of the component. Some systems, however, have macro properties which are almost impossible to predict from knowledge of the micro level properties of the individual parts of the system, e.g., emergent systems, non-linear, far-from-equilibrium thermodynamics, non-deterministic, probabilistic adaptive and interactive types systems and many types of physical world and biological systems. One of the key design challenges is whether or not complex system that are designed by specifying local interactions can be controlled, constrained or can be coherent to enable the separate lower level components to act at a higher level in some unified way.

10.5.1 Self-Organization and Interaction

Self-organization is a set of dynamical processes whereby stable or transient structures or order appears at a higher or global level of a system from the interactions between the lower-level or local entities. The things that are organised are the active peers themselves and possibly the resources in the environment that they access because these are often interdependent. The rules underlying the behaviour that specify the interactions among the entities are implemented on the basis of local information, without any reference to the global pattern. Self-organised behaviour can be characterised by key properties such as: the *creation of spatiotemporal structures* in an initially homogeneous medium, e.g. nest architectures, foraging trails, or social organisation; the existence of *multiple equilibrium* and possible coexistence of several stable states; the existence of *bifurcations*, divisions into two branches with further divisions following when some parameters are varied.

De Wolf and Holvoet (2005) make a distinction between emergent and self-organising systems. *Emergence* is global behaviour that dynamically and coherently arises from the interactions of the local[16] part and cannot be traced back to the individual parts. There is an increase in order without external control and adaptability. Such emergents are novel with respect to the individual parts of the system. A benefit of emergents is that the macro level is insensitive to changes at the micro level.

Self-organisation is an adaptable behaviour that autonomously acquires and maintains an increased order, statistical complexity, structure, etc. The focus of self-organisation is on an organisation that is not externally controlled. Emergence can have a micro–macro effect, but may not be self-organising, hence one can have emergence without self-organising. A system can also be self-organising without emergence. But these are often combined, either emergence is the result of a self-organising process or emergence results in self-organisation.[17]

The constraints on an organisation, internal to the system, result from the interactions within the system which are independent of the physical nature of those components. Hence there is a duality between interactions determining the identity and behaviour of the organisation, and vice versa, the interaction themselves being constrained by the social organisation (role) in which they occur. Interaction mechanisms for organising and coordinating autonomic system components are as follows.

Using *digital stigmergy*, individual autonomic peer components interact by modifying, including marking their local environment. The context of where and what in the environment is marked or modified is significant. Digital stigmergy is based upon a blackboard-style coordination technique but the technique is much more dynamic than that (see Exercises). The term *Stigmergy* was first used by Pierre-Paul Grasse in the 1950s to describe the indirect communication taking place among individuals in social insects, e.g., termites, ants, bees, wasps, etc, societies based upon his studies of the reconstruction of termite nests (Bonabeau *et al.*, 1999). Grasse noted that the coordination of tasks and the regulation of constructions did not depend directly on the workers but on the constructions or modifications to the environment. It is the modifications[18] themselves which direct and guide the workers. Digital Stigmergy can be used to tag the environment to indicate and hence discover use patterns. It can be used to support adaptivity computations to compute responses to the environment, e.g., to compute different energy utility functions for the proximity of food to the nest (*proximity principle*). Groups should consider not only time and space factors but also quality factors, e.g., the safety of a resource location or a path to a resource (*quality principle*). Groups should not allocate all of their resources in very limited ways but as insurance against the sudden resource changes due to environmental fluctuations by diversifying (*diverse response*). A group should not shift its behaviour in response to each fluctuation of the environment as the gains may outweigh the energy expended in reconfiguring access (*stability principle*). However, when there are significant gains in reconfiguring a group to adapt to environment fluctuations, this should be done (*adaptability principle*).

---

[16] The *emergent* or process of macro behaviour emerging from the micro or local behaviour is also called the *micro-macro effect*. The *macro level* is also referred to as the *global level* while the *micro level* is also referred to as the *local level* or *level of the individual*. Emergent properties cannot be studied by *reductionism*, physically taking a system apart and looking at the individual parts in isolation. *Coherence* refers to a logical and consistent correlation of parts.

[17] Hence, when Biskupski *et al.* (2007) define self-organising systems as systems in which global behaviour can emerge from specified local interactions that are constrained using local rules without recourse to global knowledge, they are really combining the properties of self-organisation and emergence.

[18] A modification to the environment can be the form of a marker is for others, e.g., ants leave a chemical scent or pheromone to mark a trail, *information-related stigmergy*, when foraging for food to indicate the path from a nest to the food source. Another form of stigmergy, called *task-related stigmergy*, alters the environment to cause further similar action by others, e.g., ants can leave sand grains at random locations to cause others to leave grains at the same locations leading to ants nests being formed.

It is important to optimise the level of randomness in the group in order to strike a balance between complete order and total chaos. Comprehensive insights to the use of swarm intelligence are given by Bonabeau *et al.* (1999) and Kennedy *et al.* (2001). Applications of digital stigmergy are also regarded by some researchers as applications of swarm intelligence, e.g., Kassabalidis *et al.* (2001). They consider stigmergy as the fundamental principle for network routing applications of swarm intelligence. Bai and Zhao (2006) survey the application of swarm intelligence for power distribution and energy regulation systems.

Co-field coordination, also called *gradient-based coordination* and *wave propagation coordination* are inspired by physics forces in nature. Autonomous entities can spread out a computation field or co-field, throughout the local environment, perhaps until the field meets some boundary. This can be used to provide context information to other entities to follow the field. Token-based coordination resource access controls can be circulated among resource users, the current token holder has exclusive access to these resources until it releases the token or a time limit expires (Section 11.7.1). Co-field coordination based upon time and space could be used by the user to signal that they want priority use of specific resources. One example of co-fields already mentioned is that of Reynolds (1987) which uses field for repulsion (separation of individuals) and attraction so that individuals stick together (cohesion and alignment). Mamei *et al.* (2004) give an example of co-fields to support tourists in planning their activities, such as scheduling attendance at specific exhibitions at specific times, having a group of students split up in the museum according to teacher-specific rules, helping a tourist avoid crowds across a large and unfamiliar museum, and in coordinating such movements with other unknown tourists. In some types of communication centres, peers can set acts to flexibly behave as adaptive routers, transceivers, cooperative networks (Section 11.7.8.6), this is a form of physical world electromagnetic wave propagation coordination.

In *tag based coordination*, observable labels are attached to peers. These can be used to make control decisions, to allow peer self-organisation according to tag types. An example of a labelling scheme is the MPLS protocol used to allow multiple kinds of network media packets to be handled according to packet type. *Token-based coordination* is similar to tag-based coordination but here the token is used to tag who can access a resource and is circulated around the peers. It can be used to control access to shared devices in social spaces. Some researchers have applied the ant model to resource allocation in networks to enable them to adapt to continuous node failure and to the addition of new nodes and resources and changes in traffic conditions.

Candidate designs for self-organising systems can be based upon multi-agent system (MAS) models, cellular computers and amorphous computing. Some multi-agent system architectures are designed to support relatively fixed organisational models based upon quite complex agents and relatively complex cooperative and competitive interaction, but these often lack the flexibility to dynamically reorganise and to self-organise. An alternative MAS architecture for self-organising systems is to base it upon simpler reactive agents and agent interaction, e.g., Brooks's subsumption architecture (Brooks, 1986). Cellular computing is based upon simple units such as finite state automata, vast parallelism, and locality of connection patterns between cells.[19] Amorphous computers are based upon co-field interaction (Nagpal, 2002).

Stigmergy, tag-based, gradient-field and token-based assignment tend to be applied to organisations of homogeneous peers and resources. They are models of inanimate things, tend to be single-dimensional and focus on supply-side driven interactions. These also need some security, as fields, tags and tokens may act as unattended resources and indicators, making it potentially easier for

---

[19] Cellular computing is similar to the massively parallel subtype of parallel computing, rather than to the supercomputer type of parallel computing (see Section 3.2.2.2).

rogue peers to modify stigmergic tags, co-fields,[20] tags and tokens, to their advantage. Additional methods for organisation can be based upon social needs, norms, policies and organisational roles and can employ state-based coordination, utility-based coordination, (Section 9.2.4). Market-based (self-interested peers) can be used as an efficient way for providers to allocate resources to customers when these act rationally (Section 9.2.4).

10.5.2 Self-Creation and Self-Replication

The self-organising models discussed so far focus on how existing peers and resources are optimised through self-organisation, they do not say much about how peers can self-create to replace faulty nodes or how they can replicate and expand to take full advantage of the environment and resources. Self-replication is considered to be a hallmark of living systems. Freitas and Merkle (2004) consider that serious scientific study of artificial self-replicating structures or machines has now been underway for more than 70 years, after first being anticipated by Bernal in 1929. Von Neuman and Ulman proposed the idea of self-producing systems based upon cellular automata in the 1950s (Wolfram, 2002). Evolution of self-producing systems requires strategies that lead to cumulative selection of traits occurring over multiple cycles of reproduction otherwise systems would be stuck with their fixed traits that were set at design time. Genetic programming can be used to search through the space of traits in order to select the best traits (Banzhaf *et al.*, 1998).

One of the most well-known examples of artificial self-replicating mechanisms is a *computer virus*,[21] analogous to a biological virus. A computer virus can be defined as software that contains self-replicating mechanisms, analogous to a biological virus which uses cellular mechanisms and materials to reproduce itself and which cannot exist by itself. It requires a host (Guinier, 1989). A virus is often introduced into a host system and hidden within other application software that performs some useful non-malicious function. A virus contains a trigger that activates the self-replication mechanism when certain conditions occur such as a particular date and contains a signature. In contrast to a virus, a worm can exist independently from a host. Models of *artificial immune systems* are being researched and developed as biologically inspired self-organising anti-virus systems (Dasgupta, 2006).

There are concerns that macro and micro devices designed for one purpose could reconfigure themselves and function to fulfil some other undesirable goal, depending on the degree of intelligence and local autonomy. This could be handled by designing blocking codes into these devices. Second, such devices could possibly self-replicate and overrun physical environments like biological environments.[22] It is hard to perceive that this could happen with MEMS type devices as these require highly specialised manufacturing equipment and are based upon specialised artificial materials. However, if nano devices are manufactured using more common natural materials and if they contain an innate ability to rearrange and self-organise these materials, then the risk for self-replicating nano-sized physical viruses increases.

---

[20] For example, because of the availability of low cost, micro fabricated components, it is getting easier to generate local electromagnetic fields. It is easier to set up false transmitters to lure wireless clients away from bone fide transmitters.

[21] A computer virus is also referred to as a *self-reproducing* or *self-replicating program* (*SRP*). A computer virus is a software virus rather than a hardware virus.

[22] Such device based embodied viruses are referred to as *hardware viruses* in order to differentiate them from (virtual computing) *software viruses*. The term computer virus, to date, generally refers to software viruses.

10.6 Artificial Life

UbiCom systems based upon *Artificial Life* are systems which mimic natural life and are characterised as follows. They have a finite lifetime from birth to death. They use selective reproduction. Their offspring inherit some of traits of the parents. They use survival of the fittest (evolution). They support the ability to expand in numbers to command a space or habitat. They can respond to stimuli in a habitat, acting to maintain it and adapting to it. Individuals can act autonomously as well as collectively to survive by foraging for energy and other needs. This represents another type of model, as an alternative to learning-based IS, that promotes system adaptation and fault-tolerance. Even relatively simple living organisms show a surprising propensity to live and survive. Whereas some self-organising mechanisms use single-dimension optimisation techniques such as gradient alignment techniques and single-dimension goal-directed systems, some types of artificial life computing model, e.g., evolutionary computing, can perform multi-dimensional optimisations. Living organisms are consummate problem solvers. Hence the motivation for the use of artificial life models is that they are a good model to solve complex optimisation problems and are a good model for self-optimisation, e.g., Sutcliffe *et al.* (2007) describe the use of evolutionary programming to derive optimal design for a naval command and control system in which designs are optimised along three dimensions: reliability, performance time and cost.

10.6.1 Finite State Automata Models

The basic units for *Finite State Automata* (*FSA*), also called *Finite State Machine* (*FSM*) models and cellular automata, are *automata*, something that acts on a set of inputs and computes a set of outputs. An FSA is similar to a reactive type of IS (Section 8.3.2) in that it senses what is in its environment and responds to it based upon a set of inbuilt rules. The main characteristics for an FSM are that this type of system has a finite number of states to represent its state of processing, its actions depend upon its internal state and any inputs adhere to a specific syntax, i.e., data received sequentially. There are different types of FSM: FSMs which can accept or sense inputs, which can generate or emit outputs using fixed rules in response to an input or which can transform inputs into outputs. A FSM can be represented in several ways, mathematically, as tables, or graphically, e.g., as a directed graph such as a Markov diagram or Markov State diagram (Figure 10.5). FSMs can be used to model devices with a finite set of states such as off, on and standby.

(Multi) cellular automata models, originally proposed by John von Neumann in the 1950s as formal models of self-reproducing organisms (Sarkar, 2000), can be designed in the form of a grid, where each cell is represented as an FSM, exists in one of two states, dead or alive and which use basic transformation rules to transform a cell into dead or alive. Example rules proposed by

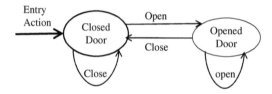

Figure 10.5   A finite state machine represented as a Markov graph for a door control device

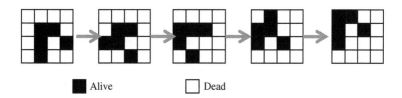

■ Alive   □ Dead

Figure 10.6   Five successive generations of Conway's game of life show how a gliding pattern in which a shape shifts position

Conway in his *Game of Life* (Gardner, 1970) are first, any dead cell with exactly three live neighbours comes to life. Second, any live cell with two or three live neighbours is happy and remains unchanged in the next generation. Third, any live cell with fewer than two live neighbours dies of loneliness. Fourth,, any live cell with more than three live neighbours dies from overcrowding. This example illustrates the principle that reactive-type behaviours combined with a set of simple transformation rules, when repeatedly applied, can model more complex behaviours, e.g., the gliding pattern (Figure 10.6). Other more complex rules of life can also be formulated. Another example of simple rules governing collective behaviour is that of Reynolds model for flocks of animals (Section 10.4). Multiple individual FSMs can be interconnected to form device networks.

10.6.2 Evolutionary Computing

Evolutionary computing involves computer algorithms which make cumulative selections from a population of entities to solve a problem. The behaviour of entities in the system is governed by implicit behaviours and goals, e.g., entities need to acquire resources to continue to live and entities have to survive a defined time period, rather than act through explicit behaviours. Cumulative means that in each generation or step of evolution, existing entities reproduce to form a new generation of entities.

Although much of work in evolutionary computing originated from research in cell automata, the ideas are also changed somewhat. New generations are based upon the best traits of the previous generation rather than on rules which define how an existing generation transforms itself into the next one. There is a set of possible outcomes for reproduction determined by natural selection. Only the survivors, the fittest, can reproduce. Reproduction of a child combines the traits of its parents. Reproduction can involve mutation, in which an offspring differs slightly from its parents. According to Jain and Karr (1995), the main types of evolutionary computing techniques include Genetic Algorithms (GA) and Genetic Programming (GP), Evolutionary Strategies (ES) and Evolutionary Programming (EP).

The work on GA began in the 1960s (Holland, 1975). GA starts with the generation of a random initial population of possible solutions to a problem. Solutions are analogous to genetic chromosomes and are represented declaratively as sets of strings where each string is analogous to a gene. The values the strings can take are analogous to alleles. Each solution is assigned a fitness score according to how good the solution is. GA is based upon two key traits: natural selection of solutions using the selection operation and reproduction using the cross-over and mutation operator. Natural selection determines which members of the population (representing potential solutions to a problem) survive and reproduce, i.e., the ones with the best fitness score, producing a new generation with a greater number of solutions with higher fitness functions. Reproduction ensures mixing and recombination among the genes of their offspring. The cross-over operator allows the gene strings of two parents to be randomly mixed to explore variations of existing solutions. Mutation allows genes to be selected at random and the alleles to be changed at random or changed deterministically leading to completely new solutions being introduced.

The main advantage of a GA is that it is able to manipulate numerous potential solutions to a problem simultaneously, reducing the possibility of the GA getting stuck in local minima, unlike many other search algorithms, because the whole space of possible solutions is searched simultaneously. John Koza developed genetic GP techniques in the 1990s as an implementation of GA representing the space of potential solutions as a graph which are then searched using GP. ES is another variation which seems similar to GA but has two main differences: a cross-over operator is not used in ES. ES uses more problem-specific strings representing problem specific solutions whereas GA uses more or less problem-specific string values which often may just be binary values. Hence GA is a more generic optimisation method than ES.

Briscoe and de Wilde (2006) propose a new distributed optimization architecture they call an *Ecosystem-Orientated Architecture* (EOA) created by extending a Service-Oriented Architecture (SOA) with *Distributed Evolutionary Computing* (DEC). Evolutionary computing is used locally within an ecosystem[23] to find individual members of the ecosystem to satisfy locally relevant constraints. Individuals within the Digital Ecosystem represent applications (groups of services), created in response to user requests by using evolutionary optimization to aggregate services. Local searches for individuals are accelerated and will yield better local optima, because a prior distributed optimisation stage already provides a good sampling of the search space by making use of computations already performed in other peers with similar constraints. Individuals can migrate through the digital ecosystem and adapt to find niches where they are useful in fulfilling other user requests.

EXERCISES

1. Discuss what types of artificial intelligence, if any, can be inspired from biological models of: human brain, human nervous system, collective behaviour of social insects and collective cellular automata

2. What is an automatic system? What is an autonomic system? What is a self-star system property? Explain how these are related.

3. Compare and contrast autonomous systems versus IS systems and MAS.

4. For the four main scenarios given in Section 1.1.2 of this text or for your own scenarios, outline designs for these scenarios based upon autonomic computing and define which self-star properties the systems should have.

5. Define what the motivation is for a self-explaining system. Discuss one or two systems from your own experience with regard to their degree of support for self-explaining. Are there any additional causes of self-explanation not covered in Section 10.3.2?

6. Compare and contrast the idea of digital stigmergy (Section 10.5.1) with the idea of a *local shared data repository* such as a *local blackboard* or *message board* in terms of the mobility of the message transmitter and the context where messages are left.

7. Debate the application of stigmergy techniques which allow the users of ICT devices to re-organise the device to optimise the performance for them. Are there any disadvantages?

---

[23] A digital ecosystem is analogous to a biological ecosystem which consists of a community, a set of organisms from different species interacting together and which interacts with and lives in harmony with the part of their physical environment where the species live called the habitat.

EXERCISES (continued)

8. What is a reflective system? Describe in principle how it works, how it can be used to support a self-explaining system and how it can be used to support an adaptive context-aware system.
9. Discuss how complex systems can arise out of relatively simple system interaction models. Give two different examples for rules of interaction.

References

Alterman, R. (2000) Rethinking autonomy. *Minds and Machines*, 10(1): 15–30.

Bai, H. and Zhao, B. (2006) A survey on application of swarm intelligence computation to electric power System. 6th World Congress on Intelligent Control and Automation, 2006. *WCICA 2006*, 2: 7587–7591.

Banzhaf, W., Nordin, P., Keller, R.E, *et al.* (1998) *Genetic Programming: An Introduction: On the Automatic Evolution of Computer Programs and Its Applications.* San Francisco: Morgan Kaufmann.

Bennani, M.N. and Menasce, D.A. (2004) Assessing the robustness of self-managing computer systems under highly variable workloads. In *Proceedings International Conference on Autonomic Computing*, pp. 62–69.

Bigham, J., Gamez, D. and Lu, N. (2003) Safeguarding SCADA system with anomaly detection. In Proceedings 2nd International Workshop on Mathematical Methods, Models and Architectures for Computer Networks Security (MMM-ACNS'03), *Lecture Notes in Computer Science*, 2776: 171–182.

Biskupski, B., Dowling J. and Sacha, J. 2007. Properties and mechanisms of self-organizing MANET and P2P systems. *ACM Transactions of Autonomous Adaptive Systems*, 2(1): 1–34.

Bonabeau, E., Dorigo, M. and Theraulaz, G. (1999) *Swarm Intelligence: From Natural to Artificial Systems.* Oxford: Oxford University Press.

Briscoe, G. and DeWilde, P. (2006) Digital ecosystems: evolving service-orientated architectures. In *Proceedings 1st International Conference on Bio-inspired Models of Network, Information and Computing Systems.* ACM International Conference Proceedings Series, 275, Article No. 17.

Brooks, R.A. (1986) A robust layered control system for a mobile robot. *IEEE Journal of Robotics and Automation*, 2(1): 14–23.

Capra, L., Emmerich, W. and Mascolo, C. (2003) CARISMA: Context-Aware Reflective mIddleware System for Mobile Applications. *IEEE Transactions on Software Engineering*, 29(10): 929–945.

Dasgupta, D. (2006) Advances in artificial immune systems. *IEEE Computational Intelligence*, 1(4): 40–49.

De Wolf, T. and Holvoet, T. (2005) Emergence versus self-organisation: different concepts but promising when combined. In S. Brueckner, G. Di Marzo Serugendo, A., Karageorgos, *et al.* (eds) *Engineering Self Organising Systems: Methodologies and Applications. Lecture Notes in Computer Science*, 3464: 1–15.

De Wolf, T. and Holvoet, T. (2007) A taxonomy for self-* properties in decentralized autonomic computing. In M. Parashar, and S. Hariri (eds) *Autonomic Computing: Concepts, Infrastructure, and Applications.* Boca Raton, FL: CRC Press, pp. 101–120.

Falcone, R. and Castelfranchi, C. (2001) The human in the loop of a delegated agent: the theory of adjustable social autonomy. *IEEE Transactions on Systems, Man, and Cybernetics, Part A*, 31(5): 406–418

Freitas, Jr., R.A. and Merkle, R.C. (2004) *Kinematic, Self-Replicating Machines.* Landes Bioscience.

Ganek, A. (2007) Overview of autonomic computing: origins, evaluation, direction. In M. Parashar, and S. Hariri (eds) *Autonomic Computing: Concepts, Infrastructure, and Applications.* Boca Raton, FL: CRC Press, pp. 3–18.

Gardner, M. (1970) Mathematical games: the fantastic combinations of John Conway's new solitaire game 'life'. *Scientific American*, 223(10): 120–123.

Goldenfeld, N. and Kadanoff, L.P. (1999) Simple lessons from complexity. *Science*, 284(5411): 87–89.

Gouaich, A. (2004) Requirements for achieving software agents' autonomy and defining their responsibility. *Lecture Notes in Computer Science (LNCS)*, 2969: 128–139.

Guinier D. (1989) Biological versus computer viruses. *ACM Special Interest Group on Security, Audit, and Control (SIGSAC) Review*, 7(2): 1–15.

Holland, J.H. (1975) *Adaptation in Natural and Artificial Systems*. Ann Arbor, MI: University of Michigan Press.

Horn P. (2001) Autonomic computing: IBM's perspective on the state of information technology', also known as IBM's Autonomic Computing Manifesto. Retrieved from http://www.research.ibm.com/autonomic/ manifesto/autonomic_computing.pdf on Nov. 2007.

Jain, L.C. and Karr, C.L. (1995) Introduction to evolutionary computing techniques. In *Proceedings Electronic Technology Directions*, pp. 122–127.

Kassabalidis, I., El-Sharkawi, M.A., Marks, R.J. et al. (2001) Swarm intelligence for routing in communication networks. *IEEE Global Telecommunications Conference, GLOBECOM '01*, 6: 3613–3617.

Kelly, K. (1995) Out of Control: The New Biology of Machines, Social Systems and the Economic World. New York: Perseus Press.

Kennedy, J., Eberhart, R C. and Shi, Y. (2001) *Swarm Intelligence*. San Francisco: Morgan Kaufmann.

Kephart, J.O. (2005) Research challenges of autonomic computing. 2005. In *Proceedings 27th International Conference on Software Engineering (ICSE 2005)*: 15–22.

Kephart, J.0. and Chess, D.M. (2003) The vision of autonomic computing. *Computer*, 36(1): 41–52.

Maes, P. (1987) Concepts and experiments in computational reflection. In *Proceedings OOPSLA, ACM SIGPLAN Notices*, 22(12): 147–155.

Mamei, M., Zambonelli, F. and Leonardi, L. (2004) Cofields: a physically inspired approach to motion coordination. *IEEE Pervasive Computing*, 3(2): 52–61.

Meng, D. and Poslad, S. (2008) A reflective context-aware system for spatial routing applications. In Proceedings 6th International Workshop on Middleware for Pervasive and Ad-Hoc Computing (MPAC'08), December 2nd, 2008, Leuven, Belgium (Accepted).

Modis, T. (2003) The limits of complexity and change, *Futurist* (May–June): 26–32.

Müller-Schloer, C. (2004) Organic computing – on the feasibility of controlled emergence. In *Proceedings of the Int Conference on Hardware/Software Codesign and System Synthesis (CODES 2004)*, pp. 2–5.

Nagpal, R. (2002) Programmable self-assembly using biologically-inspired multiagent control. In *Proceedings 1st International Conference on Autonomous Agents and Multiagent Systems*, pp. 418–425.

Nami, M.R. and Bertels, K. (2007) A survey of autonomic computing systems. In *Proceedings 3rd International Conference on Autonomic and Autonomous Systems*, pp. 26–30.

Parashar, M. (2007) Autonomic grid computing: concepts, requirements, and infrastructure. In M. Parashar, and S. Hariri (eds) *Autonomic Computing: Concepts, Infrastructure, and Applications*. Boca Raton, FL: CRC Press, pp. 49–70.

Poslad S. and Zuo, L. (2008) An adaptive semantic framework to support multiple user viewpoints over multiple databases. In M. Wallace, M. Angelides and P. Mylonas (eds) *Advances in Semantic Media Adaptation and Personalisation*, Series: Studies in Computational Intelligence, Vol. 93, pp. 261–284.

Qu, G. and Hariri, S. (2007) Anomaly-based self protection against network attacks. In M. Parashar, and S. Hariri (eds) *Autonomic Computing: Concepts, Infrastructure, and Applications*. Boca Raton, FL: CRC Press, pp. 493–531.

Reynolds, C.W. (1987) Flocks, herds, and schools: a distributed behavioral model. In: *SIGGRAPH '87 Conference Proceedings, Computer Graphics*, 21(4): 25–34.

Russell, S. and Norvig, P. (2003) *Artificial Intelligence: A Modern Approach*, 2nd edn. Upper Saddle River, NJ: Prentice Hall, pp. 32–58.

Sarkar, P. (2000) A brief history of cellular automata. *ACM Computing Surveys*, 32(1): 81–107.

Shackleton, M., Saffre, F., Tateson, R., Bonsma, E. and Roadknight, C. (2004) Autonomic computing for pervasive ICT - a whole-system perspective. *BT Technology Journal*, 22(3): 191–199.

Singh M.P. and Huhns M.N. (2005) *Service-Oriented Computing*: Semantics, Processes, Agents. Chichester: John Wiley & Sons, Ltd.

Snooks, G.D. (2007). Self-organisation or Selfcreation? From social physics to realist dynamics. Discussion Paper 546, Centre for Economic Policy Research, Research School of Social Sciences. Australian National University. Available on-line from http://ideas.repec.org/p/auu/dpaper/546.html, accessed March 2008.

Steels, L. (1995) When are robots intelligent autonomous agents? *Robotics and Autonomous Systems*, 15: 3–9.

Sutcliffe, A., Chang, W-C. and Neville, R.S. (2007) Applying evolutionary computing to complex systems design. *IEEE Transactions on Systems, Man and Cybernetics*, Part A, 37(5): 770–779.

Sweitzer, J.W. and Draper, C. (2007) Architecture overview for autonomic computing. In M. Parashar, and S. Hariri (eds) *Autonomic Computing: Concepts, Infrastructure, and Applications*. Boca Raton, FL: CRC Press, pp. 71–98.

Theraulaz, G. and Bonabeau, E. (1999) A brief history of stigmergy. *MIT Artificial Life*, 5(2): 97–116.

Venkatasubramanian, N., Deshpande, M., Mohapatra, S, *et al.* (2001) Design and implementation of a composable reflective middleware framework. *21st International Conference Distributed Computing Systems*: 1644–1653.

Whitworth, B. and Ryu, H. (2009, forthcoming) A comparison of human and computer information processing. In M. Pagani (ed,) *Encyclopedia of Multimedia Technology and Networking*, 2nd edn, vol. 1, pp. 230–239,

Wile, D.S. and Egyed, A. (2004) An externalized infrastructure for self-healing systems. In *Proceedings 4th Working IEEE/IFIP Conference on Software Architecture*, pp. 285–298.

Wolfram, S. (2002) *A New Kind of Science*. Wolfram Media. Book extracts and notes are available on-line from http://www.wolframscience.com/thebook.html, accessed April 2008.

Zambonelli, F. (2006) Self-management and the many facets of 'nonself'. *IEEE Intelligent Systems*, 21(2): 53–55.

11

Ubiquitous Communication

11.1 Introduction

Ubiquitous applications use communication networks to access relevant remote external information and tasks, anywhere and anytime. Although, communication access can be modelled as part of the internal system, the core of the communication network infrastructure is considered to be external to the UbiCom system and part of the system's virtual computing environment. Different applications require different combinations of network functions and services, e.g., data streaming, minimal jitter, type of media access control, etc. Different networks support different sets of communication functions in different ways. Key design issues concern, first, whether or not these communication functions are largely transparent to services (network-oriented) or whether or not communication is exposed via some interfaces and able to be configured and controlled by services (service-oriented). Second, there is the issue of whether or not to make all networked services ubiquitous, attached anywhere and accessible from anywhere just in case these services may be needed, versus selectively accessing networked services, e.g., some services may be limited to a locality.

Many general and introductory texts and descriptions about networking are oriented towards specific types of networks. For example, data[1] communication, traditionally focuses most on the communication of alpha-numeric data. Telecoms focuses on voice communication and its use as an *underlay*[2] *network* for data and audio-video over telecoms. Broadcast audio-video networks use separate radio and TV networks or wireless networks. Because one interpretation of UbiCom, is

---

[1] The term data has an ambiguous meaning. It can refer to a specific type of content such as alphanumeric or text data. It can refer to any type of content including audio, video and text, etc. The context of usage of the term data should determine its meaning

[2] An underlay network refers to the underlying physical network. In contrast, an *overlay network* is some logical network topology that overlays the physical network topology. An application which is underlay-aware is aware of some of the underlying characteristics of the network. This is useful, because for example, an application can reduce the resolution of audio-video content if it knows that there is a low-bandwidth link. Underlay aware, networks Tang (2005) are a subtype of ICT-awareness (Section 6.7.2) which is a sub-type of context-awareness (Section 6.2).

Ubiquitous Computing: Smart Devices, Environments and Interactions   Stefan Poslad
© 2009 John Wiley & Sons, Ltd

any content on any network, anytime, anywhere, the complete range of different media networks is treated holistically in this chapter.

11.1.1 Chapter Overview

This first section continues giving an overview of communication networks. The following sections then look at each of the major kinds of network which were historically designed and managed as separate networks: audio unicast or voice networks, audio broadcast networks (radio), (fixed) data networks and wireless data networks. Video broadcasting networks are discussed as part of the move to offer integrated multimedia service networks over a common network infrastructure (Section 11.5). The next section (Section 11.6) focuses on the use of communication networks to support pervasive services based upon wireless networks, electricity grids, people and mobile users. The use of communication by devices pervasively embedded or scattered in the physical world and the use of human type and social networking interaction are discussed elsewhere. The final section (Section 11.7), examines some outstanding network design issues.

11.2 Audio Networks

Audio networks were the first types of pervasive communications Networks. There are two basic types: audio unicast networks (PSTN) and an audio broadcast (radio) networks.

11.2.1 PSTN Voice Networks

Networks based upon *Public Switched Telephone Network* (*PSTN*) are designed to support voice communication.[3] This originally used analogue transmissions but this has been replaced by digital transmission in the core network as this is more cost-effective. However, the edge of the network that connects to homes and business, the *local loop*, still remains analogue in most regions. Many workplaces today still use separate networks for voice and data although there is a progression towards combined voice, data and audio-video networks. Telephones act as access devices to the PSTN and are typically connected to a private circuit switched network or *Private Branch Exchange* (*PBX*) at work. The PBX controls access to a smaller number of external connections. Individual home users tend to be connected to a PSTN using a single-line local loop to a local switching station. The PSTN can also be accessed using fixed or mobile phones. Many users also use the PSTN as an access network to data exchange networks, i.e., the Internet. Computer systems at work tend to connect to separate Local Area Networks, and then use a router to connect to an external network managed by an *Internet Service Provider* (*ISP*).

The first PSTNs used circuit-switching to interlink different network links to form an end-to-end connection. Here, network links were used exclusively by pairs of callers when they are online. PSTNs were designed to be very resilient. The circuit switching used in telecoms networks typically uses a hierarchy of about five levels of switching offices that handle particular number ranges (Tanenbaum, 1996). A call made from one number to another telephone number in a different range is passed up the tree of switching stations until a higher-level switch is found that can switch between the number ranges and associated phones. In some cases some switching offices on the

---

[3] Telecoms, also called PSTN, was the first global electronics network to be established, it preceded the data Internet. It is referred to as a unicast network because voice communication is often one (initiating) sender to one receiver.

same level may also be cross-linked to prevent the need to pass calls up to a higher-level switching and to provide greater resilience through alternate paths.

Unlike packet-switched networks, circuit-switched networks were originally designed to first set up a dedicated circuit of links between switching offices during the call initiation, starting when the first number is dialled. An end-to-end connection through switching offices must be completed before the actual voice call starts. It is then used exclusively between two parties for the duration of the call. This has the important benefit is that it is naturally easier to maintain a higher QoS but at the expense of non-optimal use of the channel, e.g., no information is transferred during a pause in the voice call at the network level, although the pause may be meaningful at the application layer. Allowing multiple data streams to be interleaved enables one application to transmit data during the time another one has paused. Later with the advent of digital telecoms networks, multiple voice packets could also be multiplexed over shared links. Telecoms networks are driven to support global interoperability and standards that allow users in one world region to (voice) call users in other regions.

11.2.2 Intelligent Networks and IP Multimedia Subsystems

The earliest types of digital telecommunication networks were designed to support specific services using specialised logic contained in specialised switching network elements. Any new features or services proposed have to be added and implemented directly in the core switch systems which led to very long introduction times for new services. The *Intelligent*[4] *Networks* (*IN*) network service model, standardised by the ITU-T enables Telcos to offer new value added and customised voice services such as toll free calls, e.g., '0800' numbers. This supports independent component-based services in general purpose computer nodes rather in special switching nodes. This enables service providers to drive new services rather than network providers and allows them to use these to form flexible *overlay networks* (Section 11.7.8.4).

Active development in new IN services has declined in recent years although there are still many systems across the world which use this technology, e.g., to support toll-free calls. The emphasis is now more on the development of telecom services and APIs rather than on developing new telecom network protocols. Although there seems to be a clear move to IP-based networks, in the shorter term, *hybrid IN* and Internet service architectures for mobile users are being proposed such as *IP Multimedia Subsystems* (*IMS*).

IMS was originally developed for 3G wireless networks but WLAN and fixed network support has also been added (Crespi, 2005). Users access IMS using IP. A key challenge is application-layer control (signalling) protocol for controlling voice/video session, multimedia conference, messaging and Presence over IP. This type of multiple media transmission control can be performed using the IETF SIP (Session Initiation Protocol) (Schulzrinne and Rosenberg, 2000) which seems to have superseded ITU's earlier H.323 protocol. The basic entities in a typical SIP system involve: a series of mediators where users sign in and are authenticated, location servers are used to track user locations, presence servers detect if users are active, proxy servers and redirect servers assist in call forwarding and an MCU or multi-point control unit mixes multimedia streams. SIP can use three different types of MCU: full mesh, mixer and multicast. In a full mesh, every participant builds a signalling path with every other participant and sends an individual copy of the media stream to the others – this only scales to very small groups. A mixer or bridge takes several media streams and

---

[4] N.B. the term intelligence here seems to refer to a weaker form of intelligence, to the use of more general programmable logic to promote a faster time to market for new telecom services than the use of any stronger form of artificial intelligence.

replicates them to all participants. Neither full mesh nor mixers scale to large conferences, hence a network layer multicast is used to support this.

11.2.3 ADLS Broadband

Asynchronous Digital Subscriber Line (*ADSL*) transmission technology can be used to increase the transmission capability over existing physical, e.g., copper-wire PSTN type, access networks. It does this by exploiting the fact that audio telephony signals require only about a 3 kHz bandwidth but a typical line can transmit usable signals up to approximately 1 MHz. High-frequency signals, however, can be attenuated and are subject to more electrical interference. The signal to noise ratio for transmissions is more dependent on distance. However, Digital Signal Processing (DSP) can be used to support signal modulation, commonly based upon *Discrete Multi-Tone* (*DMT*) an international standard, and *Carrierless Amplitude modulation Phase modulation* (*CAP*) to encode signals for improved transmission and then to recover the original signals. At both the exchange and the customer premises, devices called splitters separate/combine the existing telephony signals from the ADSL signal. Higher transmission rate services such as a higher-speed Internet access and video on demand are also available depending on the performance characteristics of the underlying physical network.

11.2.4 Wireless Telecoms Networks

To support access anywhere for mobile or cell phone users, wide-area wireless telecoms networks have been established. There are different networks depending on geographic region and on the Generation (G) of the wireless network such as 1G analogue and 2G digital. These differ primarily in the way they are designed to share access to the wireless network among different users. *Global System for Mobile Communications* (*GSM*) is a 2G, Time Division Multiple Access (TDMA) network prevalent in Europe, parts of Asia, Africa and Australia but not in the USA and the Far East. *Code Division Multiple Access* (*CDMA*) is another 2G digital cellular network system mainly used for cell phones in North America. These networks can interoperate via gateways to allow CDMA phones to call GSM phones but a CDMA phone can't be used directly on GSM network and vice versa. Phones can also be designed to be used on both networks. Wireless transmitters or base stations have a limited range and are positioned so that they cover an area or wireless cell, which overlap each other to a degree. This is designed so that when a user moves between cells, one base station can hand over communication to an adjacent one, often transparently without the user being aware of this.

DECT (*Digital Enhanced Cordless Telecommunications*)[5] is has been deployed in over 100 countries worldwide to access wireless voice communication within a local area. It uses both TDMA and Time Division Duplex algorithms to avoid interference from other DECT system typically giving about 120 duplex channels in a device when operating at 1.88–1.9 GHz. It has also been specified to operate in the licence-free 2.4 GHz ISM frequency band. However, the latter frequency band is commonly used by many household appliances and devices such as microwave ovens so good design to deal with common RF interference problems is needed. DECT also supports data exchange using the *DECT Packet Radio Service* (*DPRS*) at data rates of up to

---

[5] The DECT Forum, http://www.dectweb.com/, accessed Jan. 2007. Following the success of DECT in Europe, Africa and South America, a variant of DECT has been developed for North America called Worldwide Digital Cordless Telecommunications (WDCT).

2Mbit/s using demodulation and it supports Multimedia Access using the *DECT Multimedia Access Profile* (*DMAP*). A low cost digital system based upon DECT, called corDECT, has also been used to provide wireless voice and data services in rural areas (Section 11.7.6). In terms of applications and market, DECT and Bluetooth are similar. In the UK, the majority of LAN cordless household telephones use DECT. However, the majority of WAN mobile phones, games consoles, etc. use Bluetooth. Hence Bluetooth rather than DECT seems to becoming more pervasive for a greater range of devices. Although DECT seems well established to access wireless telecoms networks services in many countries, it is also faces strong competition here from WAN mobile phones.

11.2.5 Audio Broadcast (Radio Entertainment) Networks

There are several benefits in using audio broadcasting or radio.[6] It supports one sender to many receivers. It is more ubiquitous than video as it supports a multi-modal interface that allows humans to listen to music and voice while engaged in many everyday physical world activities that require visual concentration, e.g., driving a vehicle. Third, radio (and TV) receivers, unlike wireless data networks, are inherently designed to handle and tune into receive a wide spectrum of RF broadcasts on multiple channels.

Digital radio has been introduced as a replacement for analogue radio. For digital radio, the Eureka 147 Digital Audio Broadcast (DAB) standard is most commonly used and is coordinated by the World DMB Forum.[7] DAB uses the MPEG-1 Audio Layer 2 audio (MP2) codec for audio broadcasting while personal players use the MP3 codec. The main original objectives of DAB were: (1) to provide radio at CD-quality; (2) to provide better in-car reception quality than using FM analogue radio;[8] (3) to use the spectrum more efficiently; (4) to allow tuning by the name of the station rather than by frequency; and (5) to allow data to be transmitted.

11.3 Data Networks

According to Naughton (1999), the early Internet in the1960s was based upon several innovations. First, it was based upon the shift from (single-tasking) *batch computers* to (multi-tasking) *time-shared computers*. Second, the original intention was to directly connect each computer node to each other in the form of a *peer-to-peer* (*P2P*) *mesh network*. It was reasoned that this would not

---

[6] Analogue radio was developed in the late nineteenth century, for example, Marconi established the world's first radio station on the Isle of Wight in 1897. A second key development was the superhet receiver circuit, invented by Edwin Armstrong in 1918, that improved the ability of radio circuits to tune into different frequencies, channels or stations. The superhet receiver circuit exploits the physics of mixing two RFs together, produces four frequencies, one of which is the difference between the two frequencies and which is lower and more easily filtered to remove noise and can be amplified.

[7] The world DAB forum, http://www.worlddab.org/, accessed May 2007.

[8] DAB broadcasters tend to use data transmission rates of 128 Kbps rather than 192 Kbps in order to cram lots of radio channels or stations into a limited spectrum. Studies conclude that 256kbit/s has been judged to provide a high quality stereo broadcast signal and even with 192 kbit/s, it is relatively easy to hear imperfections in critical audio material. The use of a 128 Kbps channel rate leads to signal quality that is worse than FM for stationary access. DAB has also weak error rate correction. As a result, the DAB+ standard with a better and more efficient transmission codec has been proposed but is not yet widely used. DAB+ uses the HE-AAC version 2 audio codec, commonly known as AAC+ which is about three times more efficient than MP2, but this makes DAB+ non compatible with the previous DAB standard.

scale up[9] and hence, it was decided that computers would not connect directly to each other but would be connected via intermediate nodes, dedicated network computers originally known as *Interface Message Processors*. Unlike the very expensive time-shared mainframe computers that cost millions of dollars at the time, these computers were simpler. They had no persistence to support permanent records of the data stored but simply stored and forwarded data. Third, a shift from analogue to digital communication was needed in order to avoid signal degradation across multi-node networks. Fourth, a network is needed to support high capacity and resilient network paths. Fifth, large data was split into fixed size data packets and, sixth, there was a shift from a circuit switched Telecoms network model to a packet-switched data model. The latter allows different packets from the same source to be switched to travel different paths and to allow multiple packets from different sources to be multiplexed along the same path to maximise the utility of expensive low capacity network links.

11.3.1 Network Protocol Suites

Network protocols generally define a fixed length for pieces of data to be transmitted, called the *data packet* size. This makes it easier to store and forward data packets when sending and receiving and to transmit data more reliably and efficiency. Data messages that exceed the fixed length of the data packet size can be split into smaller related individual data packets. This is called *data segmentation*.

To send and receive content data packets, additional control packets are used such as synchronisation and acknowledge data packets. The types of data and control packets are defined in a network communication protocol. Each data packet is labelled with the address of an end receiver computer node in a particular network. This enables packets from multiple messages to be interleaved or multiplexed to use the same part of the network. Each type of physical network link protocol defines its own type of data packet and multiplexing. Data packets defined by one data protocol can be tunnelled or encapsulated in other types of data packet. This is often used to build or layer more functionality for more complex communication protocols on top of another one and to hide the complexity of a lower-level communication protocol from a higher-level one, see Figure 11.1. Hence, users and applications are not exposed to the intricacies of the lower-level internet and to host-to-network protocols.

11.3.2 Addressing

Before communication can occur between network elements such as computers, they need to be allocated network addresses. In computer networks, network addresses are allocated to networks in logical ranges called subnets. To transfer messages across networks requires the use of a special network computer called a router (see below). Network addresses consist of two parts: a (sub) network address and a host computer address. *IPv4* supports 32 bit or 2^{32} (about 4.3 billion) addresses. This is insufficient for giving even one address to every living person, let *al*one supporting multiple embedded and portable devices per person. *IPv6*, however, supports 128 bit or 2^{128} addresses. There are varying estimates to how many IPv6 addresses are available for person depending on different address allocation schemes but somewhere in the range 10^{24}–10^{28} addresses for each of the roughly 6.5 billion people alive today. However, the majority of addresses used today are still IPv4 – there is no consensus when the IPv4 address space will run out as various techniques exist to

---

[9] The number of connections C is given by $C = N(N\text{-}1)/2$, i.e., connectivity increases in proportion to the square of the number of computers connected.

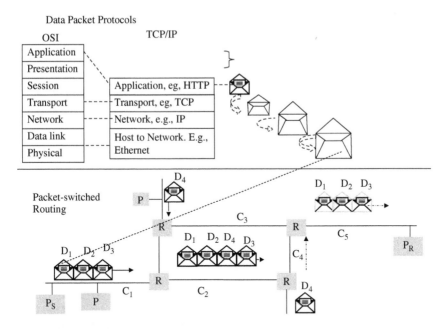

Figure 11.1 Data messages for an application are fragmented into packets D_1 to D_3 for delivery across distinct communication networks C_1 to C_5. Data protocols can be combined to encapsulate data corresponding to a higher-level more complex protocol and to map data into a simpler lower-level one

reuse addresses. It may be before 2010, before 2020 or after 2020. The possible explosion of networked enabled, smaller embedded, devices and sensors may be a driver for the use of IPv6 addresses.

11.3.3 Routing and Internetworking

Multiple paths may be available between sender and receiver nodes located on different networks which may be interlinked to form an Internet. Data may be too large to be transmitted in one go and may need to be split into multiple packets, that can get transmitted along multiple routes, referred to as *data routing*. This is normally performed at the network level without applications being aware of this (routing is transparent to applications).

Routers examine the addresses of data packets to decide whether or not packets should be forwarded to another particular network. The information to decide where to route packets is defined in routing tables held in the routers. Routers communicate with each other using specialised routing protocols that provide updates to the routing tables. One of the most common used routing algorithms is *Open Shortest Path First (OSPF)*. Each OSPF router broadcasts its routes and in turn requests information about routes that it can connect to. Generally, dynamic routing is aimed at fixed wired infrastructure networks, where senders and receivers are at fixed locations and where the routing is kept hidden. Packets transmitted along multiple routes may arrive out of order or get lost and need to be transmitted and reassembled in order at the destination.

Data networks are commonly designed to use a type of simple unreliable *Packet Switched Data Network (PSDN)* protocol, e.g., the IP or Internet Protocol. IP data packets are mapped or encapsulated into host-to-network layer data packets. A PSDN is unreliable because data can get lost or related data packets can be received out of order compared to the transmission order.

Hence, a PSDN protocol is often used in conjunction with an additional transport layer protocol that provides reliable and sequenced data delivery. One example is the *Transmission Control Protocol or TCP* that operates between the application end-points in the access networks. Reliability is accomplished with a system of synchronisation and acknowledgement control packets, timeouts and retries. When data loss occurs in the network, the data transmission stalls while the protocol detects the loss and retransmits the missing data. Applications can mask the effect of this by buffering data (Section 3.3.3) although this increases message latency.

TCP can also adjust the transmission rate to avoid overwhelming the receiver (flow-control) or the network in between (congestion control). However, TCP requires extra resources. The use of a three-way synchronisation or hand shake is used in order to set up the connection and this takes time. Further, TCP can only be used to connect a single sender and receiver. If we wish to support multi-cast communication, we need to use a different transport protocol such as the *UDP, the User Datagram Protocol*. Unlike TCP, UDP is far simpler to implement, it can be used for multi-cast, it does not need to synchronise itself at the start and end of a session and it does not need time-out, or retransmit or use any kind of flow-control or congestion control.

Multimedia streaming refers to data content that is continuously received by, and normally displayed to, the end-user as it is being delivered by the provider. Unreliable transport protocols, such as UDP, can be used to send media streams. This is simple and efficient but packets can be lost or corrupted in transit. Depending on the protocol and the extent of the loss, receivers may be able to recover the data with error correction techniques, may interpolate over the missing data, or may just suffer a dropout (a missing data part). The *Real-time Streaming Protocol* (*RTSP*), *Real-time Transport Protocol* (*RTP*) and the Real-time Transport Control Protocol (RTCP) were specifically designed to stream media over networks. The latter two are built on top of UDP. Reliable protocols, such as the Transmission Control Protocol (TCP) can also be used for more reliable media streaming. However, these accomplish this with a system of time-outs, retries and by retransmitting missing data, which makes this more complex to implement. It also means that when there is data loss on the network, media streams may freeze while the protocol handlers detect the loss. Clients can minimise the effect of this by buffering data for display.

11.4 Wireless Data Networks

There are several benefits to using wireless networks for UbiCom:

- *Anywhere*: In contrast to wired networks which can only be accessed at a fixed number of network junctions, wireless networks give users the freedom to access them anywhere where they are still in range of a wireless transmitter or hub that they have access permission for.
- *Mobility*: Wireless communication networks can be accessed while moving. The cost of installing wireless transmitters and receivers typically is much cheaper than a wired network.
- *Less disruptive*: Wireless networks can be used in areas where wired networks would be considered too inconvenient, disruptive or expensive to install, e.g., in old historical buildings and in emergency situations.
- *Adaptivity*: Wireless networks are also considered more adaptive in terms of their ability to expand or shrink the coverage of the network and to vary the density of coverage, installing more transmitters and capacity in high-populated in contrast to more rural areas.

11.4.1 Types of Wireless Network

A wide variety of wireless networks available is given in Table 11.1. Networks vary according to the type of infrastructure, the network range, frequencies used, the type of signal modulation to

Table 11.1 A comparison of the characteristics of wireless networks used for different kinds of services

Device / Service	Frequency (Hz)	Transmitter ange (M)	Bit rate (bps)	Energy (W) & other factors, e.g., attenuation
Dust:				
Smart Dust (Berkeley)		1–20 KM	1 M (burst)	0.1 nJ/bit
Sensor				
Radio:				
AM	0.5–1.6 M		Analogue	20 KW transmitter
Short wave:	5.9–26.1 M		Analogue	
Citizens Band (CB):	26.9–27.4 M		Analogue	
FM	88–108 M	100 k	Analogue	20 KW transmitter
DAB	174–240 M	80–160 k	128 K	
Television				
Analogue TV	75, 200 M	100 K	Analogue	
Cable TV / channel	6 M	6M / channel	0.5–10 M	
Satellite TV	10.9–14.5 G	40000 k		
Telecoms				
Mobile phones	0.8,0.9,1.8 G	0.1–5 k	50–400 k	
DECT	40–50,900 M	100	56 K	
PSTN	0.3–3.3 k	N/A	15–64 k	
ADSL	0.003–8 M	5 K	0.5–10 M	
Data networks				
Ethernet	20,100,250,600 M	100	10,100,1000 M	
WiFi	2.4 G	100	2–54 M	500 mW
WiMax	2–10 G	8 K	70 M	
Consumer electronics:				
Garage doors, alarms	40 M	10	Analogue	
Baby monitors	49 M	10	Analogue	
Radio-control cars	72 M	10	Analogue	
Microwave ovens	2.4 G	0.2	Analogue	Shielded
TV remote control IR	35 K	0.2	Analogue	Line of sight
Other short-range:				
RFID	12,13,900 M	0–5	0.1	
Bluetooth	125–135 K	30	1 M	60 mW (active mode)
ZigBee	0.9, 2.4 G	100	20–250 k	1–100 mW (active mode)
UWB	2.45 G	10	0.1–5 G	250 mW
Other long-range:				
Wildlife tracking	215–220 M			
Air traffic control radar	960–1215 M	50–100 K	Analogue	
GPS	1.2–1.6 G	40 M		Line of sight
MIR space station	145–437 M	40 M		
Deep space radio	2.29–2.3 G	>400000 K		20 KW transmitter

increase channel efficiency and channel sharing, the bandwidth available and power consumption. For example, most global, wide area and local area wireless networks are infrastructure dependent and use fixed transmitters, e.g., satellite, mobile phone, *WLAN* or *WiFi, WiMax*. In contrast, in an ad hoc wireless network the transmitters and routers are dynamic, e.g., packet radios and sensor nets. Mobile wireless networks can vary in the range they cover. They can be global, e.g., satellite; wide area covering 100s to 1000s of km, e.g., mobile phone networks such as GSM, TDMA, CDMA; local area and metropolitan networks covering 100m to 5 km, e.g. WLAN; or personal networks covering 1–10 m, e.g., Bluetooth, ZigBee and InfraRed.

The range of potential access depends upon the frequency of transmission, the strength of the transmitter and on factors such as the attenuation of the signal, for example by moisture, water and different kinds of solid objects. There is also a relationship between frequency and the maximum data ranges that can be transmitted. Generally, the higher the frequency, the greater the data transmission rate but the greater the attenuation.

Spatial Efficiency or SE in Bits per Second per unit area is considered a useful metric to describe the data rates that are available within a local area. In addition, as power consumption is a particular constraint for mobile wireless transceivers, a power efficiency metric, Bits per Second per Watt, may be useful.[10]

There is also a proliferation of new wireless services being offered over a multiple networks such as satellites, cellular networks, and over wireless LANs (WLANs). This is fuelling concern over how to allocate or to reallocate scarce radio frequency (RF) spectrum resources and support multi-protocol wireless networks. New techniques are needed to allow the spectrum to be used more flexibly and efficiently. These include smart antennas, smart modulation and digital signal processing, multi-user detection, ad hoc networking and *software radio* which moves the radio functionality from hardware into software (Buracchini, 2000). Software radio alters traditional radio designs in three main ways. It moves analogue/digital (A/D) conversion as close to the receiving antenna as possible. It substitutes software for hardware processing. It facilitates a transition from dedicated to general-purpose hardware. Each of these changes has important implications for the economics of wireless services (Lehr *et al.*, 2002).

11.4.2 WLAN and WiMAX

Wireless LANs, also called *WLANs* are local area wireless networks that adhere to the IEEE 802.11 set of standards that can operate at different frequency ranges and support different message transfer rates. In June 2003, the 802.11g standard was ratified. This works in the 2.4 GHz band (like 802.11b) and operates to a maximum raw data rate of 54 Mbps. A typical WLAN network consists of computers with WLAN cards that connect to WLAN access nodes that have a wired connection to an internet. WLAN currently tends to be inbuilt into laptop computers but not currently in many mobile phones. *Wi-Fi* is a registered trademark of the Wi-Fi Alliance, a trade organisation that tests and certifies equipment compliance with the 802.11x standards.

WiMAX, the *Worldwide Interoperability for Microwave Access*, from the WiMAX Forum , is proposed as wireless wide-area broadband access technology, based upon the IEEE 802.16 standard, typically offering 10 Mb/s over 10 KM although speeds up to 70 Mb/s are achievable over 10 KM. (Nuaymi, 2007). The IEEE 802.11 and 802.16 technologies are distinguished by their type of medium access control (MAC) whereas 802.11 MAC uses CSMA ('listen before talk') and is

---

[10] Spatial and power efficiency could also be combined into as a signal metric for wireless transmission Bits / S / M / Watt.

connectionless. The 802.16 MAC supports full QoS, bandwidth-on-demand, is connection-oriented; it supports centralised control and scheduling and offers multimedia support.

11.4.3 Bluetooth

Bluetooth[11] is a standard for short-range wireless communication over about 1–100 m depending on the class of device and power. Bluetooth applications include both local communication and increasingly local control. Unlike IR, Bluetooth does not require a line of sight between the transmitter and receiver. Some current popular Bluetooth devices and applications include hands-free mobile phone headsets and car kits to support hands-free communication when driving and wireless controllers of game consoles.[12]

Bluetooth uses the same radio frequencies as WLAN but with higher power consumption resulting in a stronger connection. WLAN requires more set-up. It uses access nodes and is better suited for operating full-scale networks because it enables a faster connection, a better range from the base station, and better security than Bluetooth. Unlike WLAN, Bluetooth does not require much configuration to set up shared resources, transmit files or to set up audio links (for example, headsets and hands-free devices). Bluetooth devices advertise all services they actually provide; this makes the utility of the service much more accessible, without the need to worry about network addresses, permissions, etc. Bluetooth devices use an auto-discovery mode to enable devices to discover each other within range. Bluetooth access devices tend to form small ad hoc networks, *Piconets*, where two or more Bluetooth units share the same channel, one device acts as a master and the devices connected to it act as slaves. A set of two or more interconnected Piconets form a *Scatternet* using a slave node as a gateway between the two and use a TDM MAC scheme to support shared access.

11.4.4 ZigBee

ZigBee is a specification for a suite of communication protocols from the ZigBee alliance formed in 2002 (Geer, 2005). It uses small, low-power digital radios based on the IEEE 802.15.4 standard for *Wireless Personal Area Networks* (*WPAN*) as a beaconing technique in which a node continuously transmits small packets to advertise its presence to other mobile units. This then tries to establish a connection to start networks, then letting other devices join in. ZigBee operates in the industrial, scientific and medical (ISM) radio bands; 868 MHz in Europe, 915 MHz in the USA and 2.4 GHz in most jurisdictions worldwide.

ZigBee protocols are intended for use in embedded applications requiring low data rates and low power consumption and low latency. ZigBee's current focus is to define a general-purpose, inexpensive, self-organising, mesh network that can be used for industrial control, embedded sensing, medical data collection, smoke and intruder warning, building automation and home automation. The resulting network will use very small amounts of power so individual devices might run for a year or two using the original battery. The technology is also intended to be simpler and cheaper and has a smaller code footprint than other WPANs such as Bluetooth. Currently, Bluetooth still dominates the local communication. Bluetooth consumer devices can be used in a sleep mode to reduce power consumption. However, when a ZigBee device is powered, it can wake up and get a packet across a network connection in around 15 milliseconds. In contrast, a Bluetooth device in a similar state would take around 3 seconds to wake up and respond.

---

[11] See http://www.bluetooth.com/, accessed Jan. 2007.
[12] Nintendo Wii, Sony PlayStation 3 and Xbox 360, use Bluetooth in their wireless controllers.

There are three different types of ZigBee device. A *ZigBee coordinator* (ZC) forms the root of the network tree and can bridge to other networks. A *ZigBee Router* (ZR) acts as an intermediate router, passing data from other devices. A *ZigBee End Device* (ZED) contains just enough functionality to talk to its parent node (either the coordinator or a router) It cannot relay data from other devices and requires the least amount of memory. ZigBee protocols use an ad-hoc, on-demand, distance vector to automatically construct a low-speed ad-hoc network of nodes. In most large network instances, the network will be a cluster of clusters. It can also form a mesh or a single cluster.

11.4.5 Infrared

Infrared (*IR*) is a short-range low bandwidth data communication used to communicate between computers and between computer peripherals. Remote controls and IrDA devices use infrared light-emitting diodes (LEDs) to emit infrared radiation which is focused by a plastic lens into a narrow beam. The beam is modulated, i.e. switched on and off, to encode the data. The receiver uses a silicon photodiode to convert the infrared radiation to an electric current. It responds only to the rapidly pulsing signal created by the transmitter, and filters out slowly changing infrared radiation from ambient light. Infrared communication[13] is more useful for indoor use. IR requires a line of sight from the transmitter to the receiver. It does not penetrate walls and so does not interfere with other devices in adjoining rooms. Infrared is the most common way for remote controls to command appliances.

11.4.6 UWB

Ultra-Wideband (*UWB*) is a technology for transmitting information at data rates exceeding 100 m bits/s, spread over a large bandwidth (>500 MHz) at a low power range, over short distances. The FCC has authorised the unlicensed use of UWB in the 3.1–10.6 GHz frequency range. This is intended to provide an efficient use of scarce radio bandwidth while enabling both high data rate wireless connectivity within BANs, PANs and within buildings and at longer-range, low data rate applications, as well as radar, collision obstacle avoidance, precision altimetry and imaging systems (Kalghatgil, 2007). In contrast to conventional wireless systems that need to use baseband signals to modulate radio frequency (RF) carrier signals, UWB can be used to directly transmit signals at baseband frequencies. In pulse-based UWB, the transmitter only needs to operate during the pulse transmission, producing a strong duty cycle on the radio and minimising baseline power consumption. As most of the complexity of UWB communication is in the receiver, simple, low power transmitters can be supported.

11.4.7 Satellite and Microwave Communication

Satellite communication has the potential for truly ubiquitous global communication. Commercial satellites use parts of the microwave[14] range frequencies for transmission (Fiedziuszko, 2002). The

---

[13] RF devices usually conform to standards published by *IrDA*, the *Infrared Data Association*, http://www.irda.org/, accessed 2007-10.

[14] Equipment may be described qualitatively as '*microwave*' when its signal wavelength is about the same as the dimensions of the transmission equipment, i.e., about 1 mm to 300 mm wavelengths with frequencies corresponding between 300 megahertz and 300 gigahertz. Above 300 GHz, electromagnetic signals get absorbed too much by the Earth's atmosphere.

first communication satellites in the early 1960s, e.g., *TELSTAR 1*, did not provide global coverage but acted as mobile relays to store-and-forward transmissions between ground stations in different locations, hence ground stations could only communicate for part of the day (Evans, 1995). In the mid-1960s, so-called *geostationary satellites*, which remain in a fixed location and operate in the 4 and 6 GHz range, the C-Band, were launched.[15] Geostationary satellites have the advantage of much simpler antennae design and configuration and a relatively small number of these satellites can be interlinked to provide global coverage. Satellite design must contend with station-keeping design, e.g., the influence of the sun and moon in causing orbit perturbations, communication payload design, handling the large round trip delay for transmitting and receiving signals and efficient multiple channel utilisation.

For two decades, satellites were superior to undersea RF cables in terms of both bandwidth and cost for long-distance communication. With the advent of long-distance fibre optic cables[16] and the installation of terrestrial VHF transmitters for mobile phones, satellite communication is no longer considered superior. In the 2000s, a major use of satellite communication is for wide-area TV video broadcasts although is also being challenged by cable networks. An important use of satellite communication is to serve very large numbers of thin routes, which cannot economically be served by cable (Evans, 1995).

In the mid-1990s, new satellite systems were developed to provide two-way interactive services operating in the *Ka-band spectrum* at 30 and 20 GHz (Yen, 2000). These offer several advantages such as higher data rates, in the range from 1–20 G bps per satellite, a smaller equipment size and high gain spot beams to sequentially interconnect specific uplinks to specific downlinks using FDM or TDMA to access multiple channels. However, the Ka-band frequencies are more susceptible to propagation impairments and are increasingly affected by the Earth's atmosphere compared to satellite transmissions using lower frequencies.

11.4.8 Roaming between Local Wireless LANs

There are several reasons for using Internet wireless networks: to support roaming between short-range wireless networks, to select an optimum network when several overlap, e.g., based upon bandwidth, security etc, and to allow short-range services to interconnect with longer range ones, e.g., to notify someone remotely that something of interest remote is happening locally. For example, logistics companies may use WLAN access points in warehouses to interlink scanners that audit tagged goods that are being loaded, unloaded and being stored. Global positioning system capabilities built into the device can help dispatchers know the exact location of any truck and warn a driver about to deliver a package to the wrong location. Customers can thus get up-to-date information about the location of their package within seconds of its location being updated.

Some other useful combinations are to use a WLAN to access the Internet or for a phone to use WLAN and VoIP to connect to an access-node when in range of a WLAN transmitter otherwise one would have to use a more costly and slower WWAN network to transfer data, if out of range. Generally, operating system support is required to support pipelines across heterogeneous networks.

Generic Access Network (GAN), also known as *Unlicensed Mobile Access (UMA)*, is a telecommunication system allowing seamless roaming and handover between local area networks

---

[15] The first satellite was the Russian *SPUTNIK* satellite launched in 1957. An early geostationary satellite COMSTAT 1 was able to provide live television coverage for the USof the 1964 Olympic Games which were hosted in Japan.
[16] The first *long-distance fibre-optic* (TAT-8) cable was laid under the Atlantic in 1988.

and wide area networks using dual-mode mobile phones. The aim of GAN is to enable GSM mobile operators to offer fully converged connectivity using their existing core network. Subscribers could then seamlessly roam from one cellular network to a WLAN, maintaining the call as they move from one to the other. As cellular operators increase the variety of services and applications they offer to their customers, the issue of in-building coverage increases in significance. Multi-mode handsets with base-stations acting as gateway can be used by operators to enable allow seamless access to services on different networks. Although GAN technology is already available and deployed by a number of cellular carriers, it is not yet used in connection with 3G wireless telecoms networks.

In contrast, *Femtocells* are small cellular access points which provide enhanced coverage specifically in residential environments, enabling operators to provide fixed-mobile converged voice, data and video services such as IPTV (Ho and Claussen, 2007). Femtocells can work with all UMTS terminals. Applications specific to femtocells will initially relate to presence-based activation. Femtocells can be used with standards beyond High Speed Packet Access (HSPA) which uses modulation improvements to enhance the performance of UMTS. Although Femtocells may not significantly affect the dual-mode handset market, both technologies may survive with new handover techniques based on SIP, e.g., *VCC* (*Voice Call Continuity*), emerging.

11.5 Universal and Transparent Audio, Video and Alphanumeric Data Network Access

Traditionally, different content media are delivered over different types of network because they needed to use networks with different properties and because they were developed by separate technical groups. Current distributed services are often still driven by economics and availability, first selecting networks based upon availability and cost, and then by deciding what services to subscribe to over them. Often the cost is still prohibitive for many users to access multiple redundant networks and then to mix and match services from different providers. However, services are increasingly being delivered over a common network such as IP or cable and multiple services are being offered as a bundle by network service providers.

Audio and Video (AV) broadcast *Content Based Networks* (*CBN*) have different drivers compared to telecoms and network networks. The first obvious difference is that digital AV CBN transmits streamed multimedia audio and video. The content represents time-based data streams containing individual elements or frames that need to be generated and accessed at some fixed rate. If timing is not maintained or cannot be masked, distortions such as jitter occur.

Second, broadcast networks are designed more for simplex or one-way, one to many, synchronised transmissions. In contrast, telecoms networks are developed to support duplex or two-way, one-to-one communication. Whereas the Internet was developed, at least initially, to support asynchronous communication. In contrast to telecoms networks that require global interoperability, video broadcast networks are often oriented to a regional rather than a global customer base. Video content is richer and is more likely to be tailored to a specific region in terms of language and culture. As a result, there are many national competitive and incompatible video broadcast systems. Unlike telephone calls that are usually one-to-one, there is a one-to-many relationship between the receiver and the channels of content the receiver can select. There are various degrees of interaction allowing receivers to view, pause, record and review or skip parts of audio video content. The presentation of video synchronised with audio is more complex than the presentation of voice alone.

The Internet has focused most on alphanumeric data transmission and support to manage reliable and unreliable data streams, mainly for paired senders and receiver. The support for scalable audio and video content streamed broadcasts over the Internet is still maturing, whereas the scalable broadcast of audio entertainment, voice and video over dedicated networks has matured. The adoption of compatible standards for the triple-play (audio, video and alphanumeric data) will facilitate their integration. It could be envisaged for example, that common codecs could be reused across multiple audio-video services, e.g. across digital radio, TV, voice broadcasts, audio players and the Internet, but sometimes applications have different requirements and different characteristics are needed.[17]

11.5.1 Combined Voice and Data Networks

In residential buildings, often because of cost considerations, a single external communication line exists and multiple services, such as voice and text applications, and video, are accessed over this single line using a modem. There are different types of modem, e.g., Cable, *Digital Subscriber Lines* (*DSL*) and its many variants, such as *Asynchronous Digital Subscriber Line* (*ASDL*) over a fixed telecoms line, and air DSL.

These can integrate voice and text data and transmit them over a single external telephone network connection. These are replacing the use of the older *ISDN, Integrated Services Digital Network*, systems. These are also replacing the use of dial-up modems that take over the voice local loop connections. In order to send both data and voice over the same network link, the analogue phone voice signal frequency of approximately 3 KHz is modulated at a higher frequency to send data. The lower frequency voice signal and the higher frequency voice signal are split to prevent these different frequency signals from interfering with each other in part of the access network.

Cable TV modems are in some ways similar to ADSL modems and provide another possibility for local access to data networks in residential areas. Some of the bandwidth from unused analogue TV channels is allocated for data use. Again a splitter is used to split the TV signal from the data signal and numbers of individual cable users are multiplexed together at the cable TV provider. DSL and cable modem use essentially a CDMA or frequency division technique to allow text and voice data to be used concurrently on the same access network but use separate core networks (Figure 11.2).

Voice over IP (*VoIP*) refers to the use of an IP, packet-switched data network, to interleave text data and voice to be transmitted over the same network rather than being split. In transmitting voice as data packets over IP, there are three important communication requirements: to minimise delays, jitter and packet-loss. Delays can be caused by IP networks storing and buffering data during transfer that leads either the sender or receiver to think that the other party has paused the communication when they have not. Both parties can them send and receive at the same time. Jitter refers to the variability of delay of transmission of data through the network. Sometimes packets will get lost if the network becomes congested of if the receiver becomes overloaded. There are several ways to minimise delays, jitter and packet-loss: by adding more network capacity, or by network routers being designed to *prioritise data* packets according to different classes, gold, silver, bronze, etc. For example, routers can detect the data type priority and then move voice data ahead of text data in the store and forward router buffers when the networks become congested, if it is labelled as higher priority.

---

[17] Freeview, digital satellite and digital cable video broadcasting, DAB, CD and DVD players tend to use the same audio codec (MP2), whereas many consumers use the MP3 audio code.

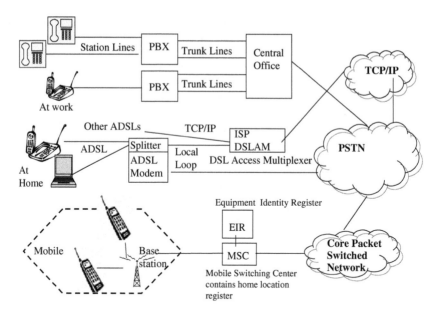

Figure 11.2   A typical telecoms network that can support voice and data over fixed and wireless links

PSTNs typically use a standardised codec (64 kbps) to digitise voice whereas VoIP digitisation can vary from 4 to 64 Kbps, with varying compression and providing variable sound quality. PSTN users have experience of using a range of basic to more sophisticated voice services, e.g., to signal the sender ID before voice calls are connected and to support multi-party conference calls. There are two basic protocols for VoIP, signalling IETF's SIP and ITU's H.323. Both of these use the *Realtime Transport Protocol* or *RTP* to transfer time-sensitive messages. Of these two, the *Session Initiation Protocol* or *SIP* is the more widely used. VoIP can be accessed either using a software phone application, on a computer or via a dedicated VoIP hardware phone linked using a serial line such as USB or via a wireless link.

11.5.2 Combined Audio-Video and Data Content Distribution Networks

In contrast to voice and radio, the delivery of audio-video streams is more complex because video needs to be synchronised to audio. Users can interact more with live video streams to pause, and replay video and users often record and edit video, etc. Video networks can be modelled at different levels as viewed by different stakeholders such as content creators, network providers that act as content program distributors for selections of content, TV receiver/player/recorder/display device manufacturers and users. Users select services and networks based upon cost, choice of content and availability of access within a region. Traditionally, the three different types of conventional networks for broadcasting audio-video entertainment content are: VHF TV, satellite TV and cable TV.

An analogue television broadcast signal can be augmented with digital data by embedding this data in the *Vertical Blanking Interval* or *VBI* part of the television signal. The TV VBI signal has been used for Teletext data transmission in Europe (Damouny, 1984) and closed-captioning in the USA. *Teletext*[18] offers a range of text-based information including national, international news, sports results, weather and TV schedules. Subtitles (or closed caption) information for video broadcasts can also be transmitted in the Teletext signal. Teletext organises its text information into numbered pages and then broadcasts these in sequence. When a viewer keys in a page number, the receiver waits until that information is broadcast again, typically a few seconds, before the requested Teletext signal can be received, decoded and displayed. This response time can be improved by buffering some or all of the Teletext pages to improve performance.

Video is increasingly being broadcast in digital encoded form rather than analogue form. There are multiple standards for digital video broadcasting such as the European DVB system (Reimers, 2004), the US ATSC system, the Japanese ISDB system, plus some proprietary versions of these. Of these, the DVB system appears to be the most widely used. DVB is modelled in a similar way to TCP/IP at an abstract level in that content can be transmitted over multiple types of physical link such as satellite (DVB-S, DVB-S2 and DVB-SH) cable (DVB-C); terrestrial television (DVB-T) and terrestrial television for hand-helds (DVB-H). Data is transmitted as MPEG-2 transport streams. The core link standards of ISDB are similar to those of DVB and are all based on the MPEG-2 video and audio coding and transport streams.

An illustrative video broadcast network that uses cable, based upon DVB-C is shown in Figure 11.3. The cable provider's central office called a *Head End*, multiplexes and encodes the individual audio, video, and data components into an *MPEG-2 program stream*. Program streams may also originate from content providers and other broadcasters. Multiple program streams are multiplexed into an *MPEG-2 Transport Stream* at the Head End and are then passed over fibre optical physical links to the *Cable Modem Termination Systems (CMTSs)*.

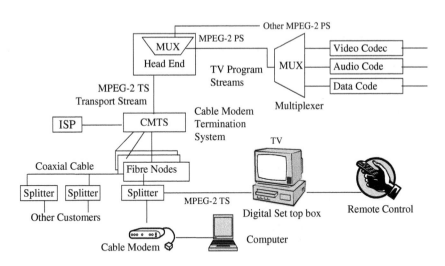

Figure 11.3   A video broadcasting network over cable that also supports the cable provider operating as an ISP

---

[18] Teletext is regarded by some as a predecessor of the Web in the sense that users select the content, page numbers for viewing. Unlike the Web, teletext is not hyperlinked and is broadcast. This makes the system reliable and it performs well as the number of users increase. In contrast, Websites can become inaccessible due to a high unexpected demand for information when major events happen.

Each CMTS can connect about 1000 or so customer premises. Splitters can be used at Customers' premises to multiplex video with text and voice data.

An example digital video broadcast system based upon open standards used for UK digital terrestrial video broadcasts, called Freeview, is considered. This represents the first horizontal market in digital television in the world, with several multiplex operators and an open market in receivers. At this time, it is unique in using an open standard API as its operating system. It uses the *ISO MHEG 5* standard based on the work of *DAVIC* (Buford and Gopal, 1997).

Digital TV broadcast both standard-definition television (SDTV) and *High-Definition Television (HDTV)* content. All early SDTV television standards were analogue in nature, and SDTV digital television systems derive much of their structure from the need to be compatible with analogue television, e.g., its interlaced scan. No agreement was reached on a single standard.

An additional challenge is to use the same network to support communication by multimedia applications simultaneously, e.g., a home network may be used for Web document access, VoIP and video streaming. The concept of transmitting audio-video and alphanumeric data content over fixed and mobile networks and has been called *quad-play networks*. There is a need to be able to multiplex heterogeneous packets from multiple applications which have different sensitivities to time delays and jitter. There are several ways to do this: protocols such as MPLS, Diffserv and RSVP can be added to existing IPv4 networks while IPv6 has more inbuilt support for this. *MultiProtocol Label Switching (MPLS)* was designed to provide a unified data-carrying service for both circuit-based and packet-switching networks. It can be used to carry many different kinds of traffic, including IP packets over different types of underlying physical and logical networks. Whereas MPLS supports faster packet handling by routing packets as part of a stream, *Differentiated Services (Diffserv)* tags each packet individually so that individual packets can get priority routing. The Resource Reservation Protocol (RSVP) can be used to reserve paths for different priority data but it cannot handle variable delays to individual packets hence it is used less.

There are a range of consumer devices that can access multimedia content over multiple networks. Types of single-, dual-, tri-, quad-, pentad-, sextet- and septet-play service bundling were considered in Section 2.3.1. To date, most service access is serialised, the user switches between single offerings rather than orchestrating interoperable services.

11.5.3 On-demand, Interactive and Distributed Content

Conventional AV content was expensive to create, was created to be popular across a large diverse audience and was delivered over a small number of broadcast channels to a largely passive audience with limited audience interaction. Audience interaction was simple, allowing audiences to decide which programs to view. In contrast, *Video-On-demand (VoD)* dedicates a single channel to each user and enables the user to interact with the video to pause, rewind, fast-forward, etc (Thouin and Coates, 2007). The deployment of TV services over IP-based network (IPTV) uses MPEG-2 as the content format and DVB technologies to interface between an IP network and a digital TV receiver to enable live media broadcast and user interaction to pause and continue content on demand. Stienstra (2006) describes a basic DVB architecture to support this.

11.6 Ubiquitous Networks

11.6.1 Wireless Networks

It may be supposed that the mere use of wireless can support ubiquitous access. However, there are several issues that make this challenging. From the signal attenuation characteristics also affect ubiquitous access. For example, the frequency range used by satellite *Global Positioning Systems or GPS* leads to significant attenuation by buildings, thus preventing its use for position determination in doors. Thus, full ubiquitous access for some services is just not possible. Instead the focus is more maximum ubiquitous access within theoretical and practical constraints. Multipath effects can cause analogue media signals to fade but can be masked in digital broadcasts by using error correction techniques. Second, many wireless networks use transmitters that have a limited range. Adjacent wireless networks and transmitters that support the same service can use different frequency, code or time division based multiplexing to prevent different sources at the same frequency from interfering with each other. This means that users who roam between adjacent transmitters need to use receivers that are designed to transparently switch frequencies as one signal fades and another one becomes stronger. Often frequencies may be allocated to service providers at a regional level so it may not be possible for a provider to use the same transmission frequency in multiple regions. This makes the switch-over from one wireless network to another less transparent. Third power consumption must be considered. Wireless terminals may be able to be used anywhere where wireless transmissions can be sent and received but if a mobile wireless terminal runs out of power, it cannot send or receive – it must be plugged into a power outlet, if available for later use.

Depending on whether transmitters are active or need to be polled, access can be designed such that transmitters poll for any receivers in range or vice versa. This enables local services to be automatically accessed only when receivers are in range.

Unlike wired networks, wireless network are not so easily contained within a defined boundary, it is inherently easier to eavesdrop on, or for freeloaders to access, unless the network has security. Wireless networks, particularly WLAN, use unlicensed frequency bands such as 802.1, as the efficient use of the RF spectrum is a limiting factor. Interfering radio sources, such as microwave food heating devices, may appear at any time and are unavoidable. This creates contention problems and reduces throughput. Channel agility and routing agility are the main ways to overcome these problems. If wireless networks use a single RF channel in each node, that channel is used both for peer-to-peer access and for 'backbone' access to a wireless access point or gateway to a wired network. To overcome these problems, MESH networks have been proposed, these use multiple RF transceivers and channels, together with collision detection and avoidance.

11.6.2 Power Line Communication (PLC)

Power Line Communication (PLC) is an alternative way to ubiquitously access data and audio-video content. Wherever there is an electricity power line connection, the same network that conducts electricity to deliver energy can modulate electricity as a signal and be used as a channel to communicate data and audio-video content. PLC describes a range of systems for using electricity distribution wires for simultaneous distribution of data. The carrier can communicate voice and data by superimposing a modulating signal over the standard 50 or 60 Hz alternating current (AC). This can be used in home automation for remote control of lighting and appliances without the installation of additional control wiring, e.g., *X10* and the *HomePlug* powerline alliance (HomePlug, 2006). An example of an X10 project is the use of a gateway programmed in Erlang (Aurell, 2005) that connects to X10 devices. HomePlug defines a number of standards for connecting devices via power lines in the home, for transmitting HDTV and VoIP around the home and for a to-the-home connection. Other situations where the electricity network can be used to provide Internet content include electric vehicles that maintain a permanent contact to an overhead or underfoot electricity supply, e.g., electric trains, buses and trams.[19]

---

[19] And of course your underground city metro where mobile communication suffers because signals cannot penetrate ground sufficiently to reach underground tunnels.

11.6.3 Personal Area Networks

Wireless Personal Area Networks (*WPAN*) have been specified by the IEEE P802.15 working group (Braley, *et al.*, 2000). A *Personal Area Network* (*PAN*) is normally confined to a person or object that typically extends up to 10 meters in all directions and envelops two or more objects or persons whether stationary or in motion. The WLAN standard was initially considered for extension but was dropped because of its higher power requirements and its higher network management over-head. Bluetooth, Zigbee and IR are examples of systems that can be used to implement a PAN. Typical applications include: multi-network phone depending on network access (intercom, cord-less phone and cell phone); mobile Internet access on the move; interactive local conferencing to instantly exchange information; hands-free head-set to communicate and (voice) control and automatic synchronisation of information for mobile users.

11.6.4 Body Area Networks

A *Body Area Network or BAN* consists of a set of mobile and compact intercommunicating sensors that are either wearable or implanted into the human body. A typical BAN application can monitor vital body parameters and movements and either store them in some device on the body for later upload and analysis[20] or periodically transmit data in real-time via some external network interface. Gyselinckx *et al.* (2006) describe the use of a system for the simultaneous acquisition of EEG, (electrical brain activity), ECG (electrical cardiac activity and EMG (electrical muscle activity) signals and transmission of these signals wirelessly to a base-station. Gyselinckx *et al* (2006) contend that both Bluetooth and Zigbee cannot meet the combined Wireless BAN power and data requirements. Typical chipsets for these radios consume in the order of 10 to 100mW for data rates of 100 to 1000kbps, leading to a power efficiency of roughly 100 to 1000mW/Mbps or nJ/bit. Wireless transmitters need to be one to two orders more power-efficient and hence UWB (Section 11.4 11.6) with a data rate of 10kbps and with 5µW power consumption, is used.

BANs and PANs[21] seem quite similar. A BAN is often referred to as more personal, a specific subset of a PAN which use the body as the electrical conduit between devices. PANs can extend the scope of the network into metre-sized personal space and even to larger social spaces. Because of the nature of electromagnetism, where electric signals can cause both capacitive and inductive fields, and moving magnetic fields that can cause electrical currents, electrical conduction through the body can cause inductive and capacitive fields to protrude outside the body. Zimmerman (1996) distinguishes *near field* (capacitive and inductive electric field effects) and *far field* (radio) communication in PANs. Far-field (radio) communication is more suscep-tible to eavesdropping and interference than near-field communication due to the former's propagation properties. An isotropic radio transmitter propagates energy with a signal strength that decreases with distance squared in far-field transmissions, whereas near-field strength decreases with distance cubed. Near-field communication can also operate at much lower frequencies (0.1 to 1 megahertz) than far field which can be generated directly from inexpensive microcontrollers. The design issues in using near-field versus far-field effects for physical media communication in PANs is also a design issue for passive RFID tags attached to physical objects (Section 6.2.4).

---

[20] This is often referred to as a Holter monitor within the field of heart electrical signal, ECG, monitoring.

[21] The development of the Personal Area Network (PAN) grew out of a cross-disciplinary exchange between two groups at the MIT Media Laboratory, Mike Hawley's Personal Information Architecture Group and Neil Gershenfeld's Physics and Media Group (Zimmerman, 1996).

Zimmerman has proposed the use of electronic devices connected as part of near-field BANs which can exchange digital information by capacitively coupling picoamp currents through the body. A low-frequency carrier, less than one megahertz, was used so that no energy was propagated, minimising remote eavesdropping and interference by neighbouring BANs. He has demonstrated a near-field BAN system to exchange business cards via a handshake.

11.6.5 Mobile Users Networks

Not all network access by mobile users, applications and devices need be via wireless networks and vice versa (Figure 11.4). Wireless access devices can be static and mobile users can move in between wired or wireless hotspots such as in Internet cafes.

Perkins (2002) makes a distinction between mobility and portability network support. *Mobility* refers to (the roaming type of) mobile applications in which users are free to continuously move anywhere within range of a wireless network. *Portability* refers to a mobile device that is stopped or suspended during the roaming itself but attached at discrete access points, intermittently. This is useful for mobile office-type scenarios when users work at multiple locations, e.g., at a central business premises, at customer premises and at home. There are several design issues for networks to support mobile users including how to locate and address mobile users, how to route data to mobile receivers, channel allocation for multiple users whether or not there is a fixed network topology or infrastructure or an ad hoc one (Varshney, 1999). Thus we can also classify mobile network support in terms of whether or not the network infrastructure is fixed or dynamic; whether or not the user is assigned a new ('care of') address every time the user changes location or keeps the same address independent of location. The advantage of mobile user support at the network level of the network protocol stack means that mobility, at least to some extent, is transparent to applications.

11.6.5.1 Mobile Addresses

The network location or address for a mobile user needs to be determined in order for a user to receive data. It is easier to send as the user just has to locate the nearest access network base station. There are two basic approaches to mobile user addressing: keeping the address the same in different locations or giving a new address for every change of location. If the user keeps the same address, the network must keep track of users' nearest base station or access point address, e.g., mobile phone and LAN. For the portability type mobility, a new address is acquired typically using

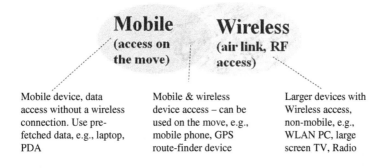

Figure 11.4 The difference between mobile and wireless

DHCP, the *Dynamic Host Configuration Protocol* (Droms, 1993). The user changes address depending on location, e.g., mobile IP (Perkins, 2002), then routing becomes more complex. Either the user's address can be updated by a base station paging mobile users periodically, e.g., mobile phones, or the mobile user can notify services of updates whenever there is a significant move, e.g., moving between LAN hot-spots and using DHCP to advertise their new location.

11.6.5.2 Single-Path Routing

There are two types of routing for mobile users: single-path routes, e.g., mobile phone or cellular networks, and wireless ad hoc, multi-path route, networks that can be used for battlefield, disaster zone and ad hoc Personal Area Network applications. An example of a single-path route is *Mobile IP* consisting of Mobile Node, Home agent and foreign agent (Perkins, 2002). A *mobile node* is a host computer or router that changes its point of attachment from one network or subnetwork to another, without changing its home IP address. A *Home Agent*[22] is a router on a mobile node's home network that delivers data to mobile nodes, and maintains current location information for each. A foreign agent is a router on a mobile node's visited network that cooperates with the home agent to complete the delivery of data to the mobile node while it is away from home. Mobile IP performs three main functions: (1) discovery: mobility agents advertise their availability on each link for which they provide a service and mobile nodes can use DHCP to receive a care-of-address; (2) registration: when the mobile node is away from home, it registers its care-of-address with its home agent; and (3) tunnelling: is used for data to be delivered to the mobile node when it is away from home, the home agent has to tunnel the data to the care-of-address.

11.6.5.3 Multi-Path Routing in Mobile Ad hoc Networks (MANETs)

In the networks we have discussed so far, it is assumed that there is a fixed network link that computers are connected to and that computers have a fixed network address. If the sender and receiver computer are on the same network subnet, the sender and receiver communicate without the need for routing. However, if computers are on different subnets, fixed dedicated network nodes such as switches, routers and base-stations are used by the processing elements in the network, to manage messages within a network domain such as a subnet, and to forward messages bound for other network domains.

In contrast to fixed computer networks, *ad hoc networks* use connections that are established for the duration of one session and require no base station or fixed router. Ad hoc networks that support mobile nodes are called *Mobile Ad hoc Networks* (*MANETs*).[23] Rather than used dedicated router nodes, each node is willing to forward data for other nodes, and so the determination of which nodes forward data is made dynamically based on the network connectivity, hence the name ad hoc (Figure 11.5). Instead, devices discover others within range to form a network for those computers. Devices may search for target nodes that are out of range by *flooding* the network with broadcasts that are forwarded by each node. Connections can be made over multiple nodes (multi-hop ad hoc network). Routing protocols then provide stable connections even if nodes are moving

---

[22] The term agent within networking generally refers to a distributed computing node that communicates over a network rather than to some specific sub-type of artificial intelligence.

[23] Wireless multiplayer gaming such as Sony's PlayStation and the Nintendo DS game consoles can use ad hoc connections and MANETs.

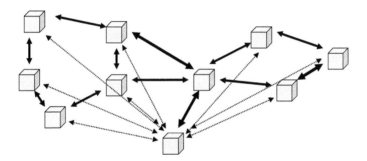

Figure 11.5 An ad hoc network has no dedicated router nodes. Instead each computer can act as a router to forward messages. The full lines indicate the hop by hop data transfer. Each node is not necessarily connected to every other node. It is not a full mesh network. The dotted lines indicate some examples of end-to-end data transfer from one example node to another node

around. MANETs can also be used in situations where a useful network infrastructure is not already in place, in natural disasters and in armed conflict situations.

Ad hoc routing is often more complex than single-path routing although applications do not need to be aware of the complexity of such routing. Biradar and Patil (2006) have classified and compared routing techniques for use in wireless ad hoc multi-hop networks based upon whether the routing is predetermined or on-demand, uses periodic updates versus event-driven updates to routes, uses flat versus hierarchical structured of network nodes, uses source routing versus hop-by-hop routing and whether single-paths or multiple paths are used. Routing which depends on the type of content is not considered in this survey.

11.7 Further Network Design Issues

11.7.1 Network Access Control

Different networks use a range of access control techniques to handle network resource allocation problems and to allow multiple users to access network media that has a limited capacity.

GSM is designed to use *Time Division Multiple Access* (*TDMA*) to share access among multiple users. In TDMA, several users can split the transmission time into a slot per channel. Code Division Multiple Access or CDMA, is used by many US mobile carriers, in which voice signals are spread across a band of frequencies or channels using a spread spectrum. Later generations of wireless networks such as 3G are designed to use much more efficient multi-access methods such as *Wideband Code Division Multiple Access* (*WCDMA*) that can support higher bandwidths. WLANs are typically based on sharing frequencies between several active users. Because many simultaneous users may cause packet collisions, and hence waste channel bandwidth, and because it can be difficult to detect some nodes, i.e., hidden nodes, it is important that packet collisions be avoided. Hence WLANs typically use a *Multiple Access with Collision Avoidance* (*MACA*) type transmission protocol. Here a wireless network node makes an announcement by sending a control packet, a *Request-To-Send* (*RTS*) with the length of the data frame to send before the packet is sent. This is sent before a sender sends any actual packets and informs other nodes to keep silent. If the receiver allows the transmission, it replies to the sender in a signal called *Clear-To-Send* (*CTS*) with the length of the frame that is about to receive. Any other nodes hearing a RTS signal refrain from sending until they hear a CTS signal.

Other active nodes wait or back off the transmission duration estimated from the CTS, plus an extra random time, before they attempt to transmit.

Another option is to use a *Carrier Sense Multiple Access with Collision Detection* (*CSMA/CD*) network control protocol in which a carrier sensing scheme is used. Here if a sender detects another signal while transmitting data, it stops transmitting its data, transmits a *jam signal*,[24] and then waits for a random time interval before trying to send its data again. An example use of CDMA/CD is the Ethernet[25] LAN.

Token-based systems are used to control access to local networks using special control messages, tokens, which continuously circulate throughout a system, e.g., structured as a *token ring*[26] topology. A network token can be set to states which indicate if the network is free or busy. A network node which wishes to use the network to transmit data must wait to receive a token in a free state. It can then change the token to a busy sending state and then start to transmit its data. This data circulates with the busy token, signalling to all other network attached nodes that they must wait to transmit. When the data reaches the receiver, it changes the token status to indicate a busy received state, the token then circulates back to the sender, which sees the busy received state as an acknowledgement that the data has been received and releases the token for others to access the network. The protocol is that network nodes which have the token and are busy must release it within a time limit. Network nodes can be designated to be an active monitor or standby monitor to manage the creation of tokens and to deal with lost tokens. Tokens-based systems are also a useful design model as part of more general resource access control and security management (Section 12.2.5).

11.7.2 Ubiquitous Versus Localised Access

Networks may also be designed to support only local access by organisations, to serve only particular populations of users and applications and not to be accessed remotely across a wide area. The driver for this is the need to tailor services for local needs. Some services that only serve a local area, e.g., local restaurants that only serve customers that visit it, may advertise only to local customers. The restaurant may decide to advertise and broadcast to any users in the vicinity as they have a surplus of food that will not keep for users who need to travel from longer distances. Services can be restricted to local access because they are only available on wired local networks that are situated in buildings or because they are offered on wireless local area networks and because they need to be kept internal by an organisation. Local services can also be designed to have access control to enable them to be accessed remotely and securely by a closed group of users, operating as *Virtual Private Networks* or *VPNs*.

For networks to support ubiquitous access, it seems logical to use Wide Area Networks or WAN. Even if these are actually constructed out of large numbers of distinct networks that are internetworked, they can be designed to make the routing or hand-over between different

---

[24] This also known as a *backoff delay* and is typically determined using a truncated *Binary Exponential Backoff* (*BEB*) algorithm (Christensen et al., 1998).

[25] However, CSMA/CD is no longer used in faster transmission specifications such as 10 Gigabit Ethernet because alternatives such as the use of *Full-Duplex Repeater* (*FDR*) offer superior performance over CSMA/CD (Christensen et al., 1998).

[26] The Token Ring network was proposed by IBM and promoted in the early 1980s but Ethernet seemed to start dominate it as a network protocol for LANs in the early 1990s, because in part Ethernet performed better at bulk information exchange whereas Token Ring seems better at real-time performance with lower loads. These days Token Ring is no longer used.

networks transparent to users – users experience one virtual network. There are at least three major types of WAN: the wired and wireless-based Internet, wireless satellite networks and the wired and wireless telephone networks. These use different protocols and manage internetworking differently.

11.7.3 Controlling Network Access: Firewalls, NATs and VPNs

Many ICT resources connected to the Internet are protected to control access to specific resources by specific users or closed user group. If access is not restricted, then freeloaders could gain access to ICT resources that they have no stake in and denial of service would increase as unknown users overload the use of resources at non-deterministic times. There are several methods to protect access to local networks based upon restricting access to computers with designated media access control or IP network addresses or based upon certificate-based authentication. Other methods include disconnecting some network elements from remote access and to use firewalls, NATs and VPNs (Dennis, 2002).

A *firewall* is a router or special purpose computer that monitors all packets entering and leaving and network and filters packets, e.g., based upon network address, thus restricting access to packets with a designated network address. Firewalls can be designed according to which level of the network they work: packet-level and application level firewalls. *Packet-level firewalls* examine and filter data packets according to the IP or Transport layer address. This type of firewall can be susceptible to packet address *spoofing*, in which malicious processes try to replace real IP addresses that lack network access with other IP addresses that have better network access. *Application level firewalls* require users to log on in to the firewall computer before they can access any other network elements.

Network Address Translation (NAT) is a computer that acts as a proxy[27] to translate IP addresses inside a network, keeping them private, to other IP addresses that are visible outside the network. Firewalls may also be designed to support multiple levels of network interfaces called *multi-homed* hosts[28] rather than using the same network interface for external connection and internal connection. Some organisations use multiple levels of different types and combinations of firewall, NAT and multi-homed host computers to protect key network resources. Because of the growing number of IP devices in residential buildings, the need for strong security such as firewalls to prevent unauthorised access to control is becoming a necessity for home use.

Equally important to restricting access to ICT resources is the ability to restrict the use of resources on remote networks to specific users that are accessed over a public Internet. A common technique to achieve this is to use a *Virtual Private Network (VPN)*. VPNs essentially allow users to establish a lower layer virtual channel or virtual circuit for exchanging data packets, but hidden from, high-level applications. This is called *tunnelling*. Multiple heterogeneous applications may use the same tunnel. These lower layer packets are said to be tunnelled through or to overlay the public network. Packets are usually encrypted when they are tunnelled to prevent the content (data packets) from being intercepted and examined by unauthorised

---

[27] Sometimes a distinction is made that *proxy host* services are more transient and session-based, whereas a *bastion host* retains the state of its services permanently. Typically, firewalls are designed as bastions rather than proxies (Rhee, 2003).

[28] Host computers with one, two, or three network interfaces are called *single-homed*, *dual-homed* and *tri-homed* hosts. Often a tri-homed firewall is used to create a *De-militarised Zone (DMZ)*, a separate perimeter network that lies between the external network and the internal network to offer more protection for the internal network (Rhee, 2003).

users. Users normally authenticate themselves at a VPN client or access devices in order to gain access to remote resources via the *VPN tunnel*. There are several types of VPN tunnel used, depending on the type of network layer packets are tunnelled through, the type of encryption used and the type of access control to the VPN used.

11.7.4 Group Communication: Transmissions for Multiple Receivers

Sending the same message from a single source to a defined group of multiple receivers, *multicast communication* or *group communication*, can be useful in order to provide some fault-tolerance for the content, or in order to share information within a group, e.g., a conference, or when a sender cannot or does not want to limit themselves to interaction with a particular receiver. There may be hardware support for multicast so that large group messages can be sent efficiently. If there is no hardware support, i.e., in routers, messages need to be sent sequentially in practice, although this may be hidden. To avoid the overhead in managing large groups, groups can be split into hierarchies that can cascade group transmissions to lower members. Group membership may or may not be visible to the members depending upon the design. In *unreliable multicast*, messages are transmitted to all members without acknowledgement. In *reliable multicast*, the message transmission makes the best effort to deliver to all receivers in the group: it may be delivered to some but not all nodes. In *atomic multicast*, a message is either received by all receivers or none of them. Messages can be tagged with a sequenced identifier to indicate the ordering. Acknowledgements can be used to support more reliable group communications.

11.7.5 Internetworking Heterogeneous Networks

Ideally, universal access means that any type of data may be accessed simultaneously anywhere over any kind of network. Historically, many separate types of communication network exist that are not interlinked. Networks are heterogeneous in terms of the physical media that electromagnetic signals propagate through. For example, signals may propagate through wired copper or optical fibre networks or through wireless or *Over-The-Air* (*OTA*) networks. These different types of physical links of the network have different signal capacities and have different signal attenuations and hence different requirements for signal amplification and repeaters. Each type of physical media network, e.g., Ethernet, *Point to Point Protocol* (*PPP*) etc., defines its own protocols for partitioning and for structuring data into packets for transmission. Networks are heterogeneous in terms of the types of content or media they exchange, such as video, audio and (alphanumeric) data. Currently, separate networks are still predominantly used to distribute audio (voice and music), video and data. Networks are heterogeneous in terms of the types of applications that use them, in terms of their architecture or topology and in terms of how the individual networks are interlinked and managed.

It seems useful to design and standardise specifications to interlink a heterogeneous internetwork in such a way that it acts as a universal network for applications and users and in which the individual network heterogeneity characteristics are hidden. Universal networks have been specified for Internets based upon the *Transmission Control Protocol / Internet protocol* suite or TCP/IP and for wireless and mobile use, *Universal Mobile Transmission Service* or *UMTS*[29] based upon the 3rd generation mobile phone technology.

---

[29] Currently, the most common form of UMTS uses *W-CDMA* for the underlying wireless network access. This is specified by the *3GPP forum*, http://www.3gpp.org, whose specifications are in turn based upon the *ITU IMT-2000* requirements for 3G cellular radio systems which specifies W-CDMA as one of its five radio interfaces.

11.7.6 Global Use: Low-Cost Access Networks for Rural Use

In theory, wireless networks could be ubiquitous but in practice they aren't in many regions. This is often because communication and content-delivery networks are built and operated as commercial businesses that have to generate sufficient revenue to exist. Currently, the total worldwide Internet usage penetration was only about 18% but only about 4% in Africa (Miniwatts, 2007). This is contrast to 29% of the global population using GSM-type mobile phone technology (GSMWorld, 2007), more if other types of mobile phone are also included. People in some rural areas may not be able to pay either to install the local (access) loop or to subscribe to services across the communications and CDNs so low-cost networks and access terminals are needed. Typically, a wireless local access loop is used to minimise the installation costs.

In rural India, for example, the *CorDECT* system (Jhunjhunwala *et al.*, 1998) has been proposed to support local access. It is based on the DECT standard which initially was designed for use with cordless telephones. This uses a MAC protocol called *MC-TDMA* which performs both time and frequency division in order to accommodate multiple channels. It typically operates over distances of up to 10 km with data rates supporting data rates of 35 and 70kbps. The conventional listen-before-you-talk-type MAC is problematic when used in low bandwidth transmission over several km (Raman and Chebrolu, 2007).

A second method of providing low-cost interactive information services is to leverage broadcast TV as a service access network. The *PrintCast* system augments television programmes, specifically the TV VBI signal, with print-related information to give end-users additional material that they can retain on paper (Gupta *et al.*, 2006). PrintCast combines print data from a PC with the program AV signal using a standard device called an inserter that encodes the signal and transmits it with traditional broadcast equipment. Viewers who receive the signal then use a PrintCast decoder to decode the data from an analog television signal and relay data to a printer. PrintCast applications include voter information guides for elections, coupons associated with advertisements, homework exercises for distance learning, and printing recipes for cooking shows.

A third method to provide wireless access networks includes *Very Small Aperture Terminal* (*VSAT*), a product range based on principles for designing small ground terminals enabling cheaper satellite access, and Spread Spectrum, a communications technique for the transmission of signal over some band of radio spectrum designed primarily to enhance performance in the presence of interference. A fourth option is to use mesh networks (Section 11.7.8.5).

11.7.7 Separating Management and Control from Usage

There are different options for designing application use versus control and management of networks (Figure 11.6). The aim of this architectural model is to separate concerns about media access, internetworking, applications from the concerns about data flows, control of the communication and management of the communication.[30] Management concerns system-wide, global, regulation functions, the FCAPS functions. Control concerns local functions such as synchronisation, error correction, network congestion and receiver overload, forwarding, routing and media access control.[31]

---

[30] These are alternatively referred to as the data, control and management planes because they are orthogonal concerns to network media access, internetworking and applications. There is an overlap between control and management but generally control involves a specific type of operational management whereas management is more generic including, for example, the FCAPS functions.

[31] Examples of control signalling between a telephone user and the telephone network include dialling digits, providing the dial tone to tell you that you are online, accessing a voice mailbox, sending a call-waiting tone, dialling *66 to retry a busy number, etc.

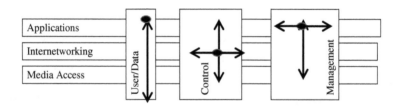

Figure 11.6 Data, control and management flows across the different layers in a simplified network model

In an application-centric model, the application defines the user data flow; the control and management are application-driven and application-specific and all three take place over a single communication channel.

With *in-band signalling*, part of the data or bearer channel capacity must be reserved for control signalling. Alternatively, an additional channel needs to be used – *out-of-band signalling*. One advantage of using a separate, out-of-band channel for signalling is that with a single channel, the in-band or control messages may be held up by the very content messages they seek to control. In addition, it may be easier for users to accidentally or intentionally fake control signals to disrupt services. Hence, PSTN networks use an out-of-band signalling system commonly referred to as SS7 to control telephone content messages. The TCP/IP FTP File Transfer Protocol is another example of a protocol that uses separate channels for content exchange and for control.

In some systems, each major application uses its own dedicated network, e.g., video broadcast network, audio broadcast network, voice network, data network, etc., hence management is application (network) specific. As multimedia content applications are becoming integrated into single networks, management needs to be factored out of the applications and networks so they can be applied across networks and applications. This leads to the idea of a separate management plane and the design of application neutral management services.

11.7.8 Service-Oriented Networks

Traditionally, different application services were coupled to specific networks because different applications need different levels of support for data transmission functions, such as *latency*,[32] sequencing, performance and reliability, channel sharing, data control and security. It is simpler to design networks to support one specific set of application requirements rather than to support multiple applications that have different messaging requirements because these may conflict. However, this has the disadvantage that more complex multi-service networks need to be provided and maintained and it is harder to integrate data from applications that exist on different networks. These days, services are oriented to be coupled less to specific networks and to be available across heterogeneous networks (Figure 11.7). Architectures were network oriented. To use a service, users must subscribe to a particular network and service configuration on the network, e.g., voice calls via a telecoms network and audio-video content via an audio-video wireless broadcast network. This section starts by first considering the major types of homogeneous networks and their services and then looks at network models that can provide universal access to heterogeneous services.

---

[32] Message latency is the time taken to transfer an empty message.

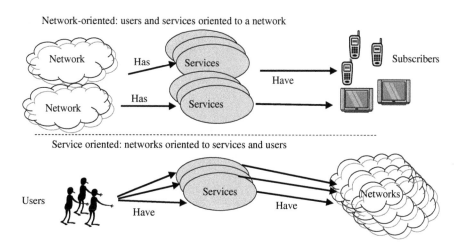

Figure 11.7   From network oriented service models to service-oriented network models

11.7.8.1 Service-Orientation at the Network Edge

A simple abstraction of a network topology to support universality is to partition a communications network into two parts: an *access network* or *edge network* and a *core network* and to design these networks to appear as a universal network. Access networks, as the name implies, is the network that users and application use to access the network. The core network comprises the interlinked networks that then link to the access networks. An important design decision is whether or not to put the complexity or intelligence into the core network, e.g., PSTN, versus in the edge network or in both, e.g., IP networks. There is an argument for end-to-end or edge-based complexity and to keep the core network relatively simple to optimise the core network only to forward and route data packets (Saltzer *et al.*, 1984). The main argument is that 'functions placed at low levels of a system may be redundant or of little value when compared with the cost of providing them at that low level.' This implies that a simple and as neutral as possible network should be used, with the intelligence, complexity, and functionality supported in the edge network by applications.

Widespread adoption of IP in the core network has given the Internet a nearly universal interoperability that allows all end users to access Internet applications and content on a non-discriminatory basis. A network neutrality vision for the communications and content delivery world in which every end user can obtain access to every available application and piece of information is quite compelling. However, it has led to some content providers resisting more open access to the edge network as they will lose market share.

Network homogeneity can reduce the utility of the network both by reducing diversity and by biasing the market against certain types of applications (Yoo, 2005). Changes to the core network and standardisation can take some considerable time to reach fruition, e.g., more network addresses (IPv6, 1991), security (IPSEC 1993); multicast (IP multicast 1990). It can have the perverse effect of reinforcing the sources of market failure used and justifying regulatory intervention in the first place. It can further entrench monopoly power by dampening incentives to invest in alternative network neutrality. Instead, Yoo proposes that network service providers should embrace network diversity. The core network should support multiple core network functions not just those biased towards to individual packet-switched data type transmission. They should, for example, also support streaming audio-video transmission. Nikolaidis (2007) makes a slightly different argument against network neutrality in the sense that the set of universal services that users expect from the network is now getting more diverse. Whereas, previously, the universal services were based upon data packet forwarding network type services such as email and the Web, essential services in the future may include fast search engine access, A-V content on demand and support for real-time 3D virtual communities.

11.7.8.2 Content-based Networks

A *content-based network* is a communication infrastructure in which the flow of messages through the network is driven by the content of the messages, rather than by linking specific senders to specific receivers. With this communication pattern, receivers subscribe to the types of content that are of interest to them without regard to any specific source (unless that is one of the selection criteria). Senders simply publish information without addressing it to any specific destination (Carzaniga and Wolf, 2001). Conventional solutions to this include, first, allocating bands of addresses to each type of content and interest and then to multi-cast content to this band of addresses rather than to specific receivers. A second approach is to use a content-based publish/ subscribe middleware service, e.g., an application-level information broker that supports rich information selection capabilities. However, both of these would not scale up to support many different kinds of content and interests combined with high performance.

In a content-based network, nodes are not assigned unique network addresses. Data is not addressed to any specific node or node group. Instead, each node advertises a receiver predicate (or r-predicate) that defines data of interest for that node that the node intends to receive. Nodes can also send out datagrams, which the network will forward to all the nodes with matching r-predicates. Similar to a traditional network, the semantics of a content-based addressing scheme is realised by the forwarding and routing functions in a router. The router computes its output based on the datagram content and on its internal forwarding table that is maintained by gathering, combining, and exchanging predicates and other routing information with adjacent nodes (Carzaniga and Wolf, 2001).

11.7.8.3 Programmable Networks

Typically, service providers do not have access to the router, e.g., the router controls environments algorithms and router states, in order to optimise network use for different applications. This makes the deployment of new network services, which could be far more flexible than proprietary control systems, impossible due to the closed nature of network nodes. *Programmable networks* allow some of the network elements to be reprogrammed dynamically, perhaps by injecting executable codes that could support application-specific services, into network elements such as routers and switches. The downside to this is that complexity increases and there is an increasing risk of instability and for malicious code to take control of core network elements. Campbell *et al.* (1999) identified two main initiatives to establish programmable networks: DARAPA's *Active Networks (AN)* program and the *Open Signalling (Opensig)* community. The main difference between the two is that the Opensig approach seeks to establish open programmable network interfaces, whereas the AN approach is to support injecting executable code into network elements to reprogram them. An example of open network APIs is the Parlay/OSA initiative (Moerdijk and Klostermann, 2003). Tennenhouse *et al.* (1997) provide a good survey of AN research.

11.7.8.4 Overlay Networks

An *overlay network* is a virtual network built on top of a physical network that provides an infrastructure to one or more applications. It handles the forwarding and handling of application data in ways that are different from or in competition with the basic underlying physical network such as the Internet or PSTN. It can be operated in an organised and coherent way by third parties, which may include collections of end-users. Many standard applications such as email, VoIP, Web search engines, etc. can benefit from providing services in special intermediate service nodes which can, for example, cache information. Overlay networks have the major benefit in that new communication services can be introduced which might otherwise require modifying the current Internet infrastructure, increasing costs and the need for coordination of the Internet service providers all over the world to make this kind of global updates infeasible[33] (Clark *et al.*, 2006). Another issue is that different applications may need different levels of reliability, performance and latency and security and access control. Nodes in the overlay network are connected by virtual or logical links. Application-specific overlay networks can be incrementally deployed on end-hosts running an overlay protocol, e.g., based upon a *Distributed Hash Table*[34] with modified (virtual node) routing without requiring changes to the physical routers. One of the earliest types of overlay networks was used by Telcos to offer non-geographical phone number services such as free phone (0800) numbers, local rate phone and premium phone rate numbers, which are charged at the same rate no matter where they are dialled from.

11.7.8.5 Mesh Networks

In a *full mesh network* topology, every network node is connected using point-to-point connections to every other one but connecting every node to every other node is costly to wire and costly power-wise to transmit to each other. Hence, in practice, mesh networks are usually *partial mesh networks*, in which each node is not connected to every other node. Partial mesh networks tend to combine ring and star-based network topologies. *Wireless Mesh Networks* (*WMN*s) are partial mesh, ad hoc, networks that can significantly improve the performance, at a lower cost and at a lower power output compared to other types of WLAN such as Wireless Local Area Networks (WLANs), Wireless Personal Area Networks (WPANs), and Wireless Metropolitan Area Networks (WMANs) (Akyildiza *et al.*, 2005). WMN is lower power because it uses a set of lower power multi-hop transmissions[35] rather than needing a single more powerful transmission to base-station. A WNM may be a suitable solution in rural areas where conventional base-station wireless type network support or DSL support maybe patchy. However, each WMN receiver is now more complex and more costly as it must also act as a relay.

Instead of using a sophisticated and costly, centralised base-station, each wireless receiver in a WMN can act as a relay point or node for other receivers within range, enabling a data signal to pass through several nodes as it progresses from the core network to its destination. To this end, the WMN acts as a kind of *cooperative network* for its users.

WMNs can also be used to support interoperability between heterogeneous wireless networks, to improve their performance and to avoid the dead zones that can occur in some types of wireless networks. In WMNs, each node operates not only as a host but also as a router (mesh clients), forwarding packets on behalf of other nodes that may not be within direct wireless transmission

---

[33] This may be one of the reasons why IPv6 has not taken off as much as it could have because of the reluctance to upgrade so many IPv4 network elements, e.g., routers, everywhere.

[34] DHT performs the functions of a hash table (HT) in which stored values can be looked up if you have the key, a simple (hash) function converts the key (name) to a storage address to achieve the value. A DHT distributes and stores the HT over multiple computer nodes. A common design for a DHT is to base it on a circular, double-linked list.

[35] This same concept of the use of multi-hop routing to reduce power transmission costs is used by sensor nets (Section 6.3).

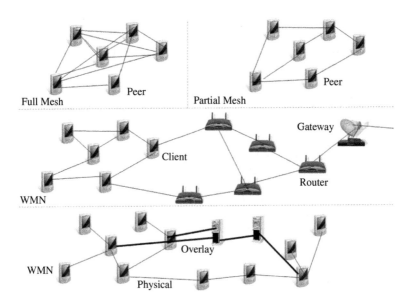

Figure 11.8   Mesh networks, wireless mesh networks and overlay networks

range of their destinations (Figure 11.8). In addition, dedicated mesh routers contain additional routing capabilities and bridging and gateway function to other networks. A WMN can also dynamically self-organise and self-configure mesh connectivity to support ad hoc multi-hop networking. Akyildiza *et al.* (2005) summarise the main characteristics and benefits of WMN as:

- Multi-hop wireless network: extends coverage range without sacrificing the channel capacity, without a direct *line-of-sight* (*LOS*) link.
- Ad hoc networking: supports self-forming, self-healing, and self-organisation for multipoint-to-multipoint communications enabling the network to grow gradually as needed.
- Mobility dependence on the type of mesh nodes: mesh routers usually have minimal mobility, while mesh clients can be stationary or mobile nodes.
- Multiple types of network access: both backhaul access to the Internet and peer-to-peer (P2P) communications are supported.
- Mesh routers usually do not have strict power consumption constraints and may not be appropriate for some types of mesh clients, those that act as sensors in a sensor network where power consumption is the primary concern.
- WMNs are compatible and can be integrated with multiple IEEE 802.11 (Wifi) type networks.
- WMNs aim to diversify the capabilities of ad hoc networks.

Consequently, ad hoc networks can actually be considered as a subset of WMNs. WMNs consist of a wireless backbone with mesh routers to provide large coverage, connectivity, and robustness in the wireless domain but which may not be reliable. Unlike ad hoc networks, where end-user devices also perform routing and configuration, WMNs contain mesh routers for these functionalities. Hence, the load, energy consumption, and application load on end-user devices is significantly decreased. Mesh routers can be equipped with multiple radios to perform peer-to-peer routing and backbone access functionalities. This enables a separation of two main types of traffic in the wireless domain and increases performance compared to ad hoc networks. Ad hoc networks provide routing using the end-user devices, the network topology and connectivity depend on the movement of users, making routing protocols, network configuration and deployment, more complex.

11.7.8.6 Cooperative Networks

Some network access devices cannot access multiple networks in order to communicate, they just have access to one network connection – this is often in order to simplify design, to reduce costs or because no redundancy is necessary. Some other network access devices have inbuilt support for heterogeneous network access, e.g., to Bluetooth, to infrared, to WiFi and to GSM networks. Each of these networks must be used in isolation, they do not interoperate. A third case is that multiple types of the same type of physical and network layer may exist because multiple independent users and providers may offer overlapping wireless networks within the same vicinity but yet again these do not interoperate. These can overlap and the coincidence of multiple overlapping networks will increase as more networks are installed but yet again these networks do not interoperate.

Cooperative communication is designed to enable single-antenna mobile access devices to reap some of the benefits of being *Multiple Input Multiple Outputs* (MIMO) systems (Nosratinia *et al.*, 2004). A specific problem that cooperative communication can solve at the physical media layer concerns signal *fading* because thermal noise, shadowing due to fixed obstacles and due to signal attenuation can vary significantly over the course of a given transmission. Transmitting independent copies of the signal that will face independent fading generates diversity and can effectively combat the deleterious effects of fading through combining these multiple signals. Cooperative communication can be used to generate this diversity. In essence, cooperative communication treats communication nodes as it they are part of a WMN but multicasts the same message over multiple routes with different fading effects to the same receiver. The other benefit, apart from improving the recovery of the signal in the face of fading, is that power consumption can be reduced for transmission even when it ought to increase as multiple users are now using each transmitter. Overall less power can then be used to cover for fading.

EXERCISES

1. Describe mobile network types and characteristics in terms of: what is mobile? What types of mobile terminal or service access device is used? What factors characterise mobile terminals? What types of wireless networks are used? What factors characterise wireless networks?
2. Mobile device communication over wireless networks is enough to satisfy all of our ubiquitous computing needs. Discuss.
3. Discuss the pros and cons of choosing short-range versus long-range communication if both are available when a group of people are situated at a common location or a common meeting place. Consider the energy usage and the message latency, etc. If these are advantageous for local interaction, why is it that long-range communication is often more commonly used in practice?
4. Discuss convergence of many network services into a single network in the home. How can this be designed? What are the design issues? Do you have any initial experiences of this?
5. Consider the interference and access control issues as more diverse devices and networks are used within a locality such as the home.

EXERCISES (continued)

6. Discuss what is meant by, and the need for, service-oriented networks, content networks, programmable networks and overlay networks.
7. Compare and contrast cooperative networks (Section 11.7.8.6) with social interaction models (Chapter 9).
8. You are asked to plan a communications infrastructure for a rural village in some developing country that has a limited electricity grid connection but no communication network within 30 km versus developing a communications infrastructure in some rural village in a developed country that has electricity grid, a PSTN network and a 2G+ mobile phone network available. Compare and contrast your designs.
9. Discuss the use of a network-oriented model based upon IP everywhere to underpin pervasive computing services.

References

Akyildiza, I F., Wangb, X., Wang, W. (2005) Wireless mesh networks: a survey. *Computer Networks.* 47(4): 445–487.

Aurell S. (2005) Applications: remote controlling devices using instant messaging: building an intelligent gateway in Erlang/OTP. In *Proceedings of the 2005 ACM SIGPLAN Workshop on Erlang ERLANG '05*, pp. 46–51.

Beyda, W.J. (2005) *Data Communications from Basics to Broadband*, 4th edn. Upper Saddle River, NJ: Pearson Prentice Hall.

Biradar, R.V. and Patil, V.C. (2006) Classification and comparison of routing techniques in wireless ad hoc networks. International Symposium on Ad Hoc and Ubiquitous Computing, ISAUHC'06: 7–12.

Braley, R.C., Gifford, I.C. and Heile R.F. (2000) Wireless personal area networks: an overview of the IEEE P802.15 working group. *ACM SIGMOBILE Mobile Computing and Communications Review*, 4(1): 26–33.

Buford, J. and Gopal, C . (1997) Delivery of MHEG-5 in a DAVIC ADSL network. In *Proceedings of 5th ACM International Conference on Multimedia:*, pp. 75–85.

Buracchini, E. (2000) The software radio concept. *IEEE Communications*, 38(9): 138–143.

Campbell, A.T., De Meer, H.G., Kounavis, M.E. *et al.* (1999) A survey of programmable networks. *ACM SIGCOMM Computer Communication Review*, 29(2): 7–23.

Carzaniga, A. and Wolf, A.L. (2001) Content-based networking: a new communication infrastructure. Proceedings NSF Workshop on an Infrastructure for Mobile and Wireless Systems. *Lecture Notes in Computer Science*, 2538: 59–68.

Christensen, K.J., Molle, M. and Li, S. (1998) Comparison of the Gigabit Ethernet Full-Duplex Repeater, CSMA/CD, and 1000/100-Mbps Switched Ethernet. In *Proceedings of 23rd Annual IEEE Conference on Local Computer Networks*, pp. 336–344.

Clark, D., Lehr, B, Bauer, S. *et al.* (2006) Overlay networks and future of the Internet. *Communications and Strategies*, 63: 1–21.

Crespi, N. (2005) A new architecture for wireline access to the 3G IP Multimedia Subsystem. 5th International Conference on Information, Communications and Signal Processing, 493–497.

Damouny, N.G. (1984) Teletext decoders – keeping up with the latest technology advances. *IEEE Transactions on Consumer Electronics*, 30(3): 429–436.

Dennis, A. (2002) *Networking in the Internet Age*. New York: John Wiley & Sons, Inc.

Droms, R. (1993) Dynamic host configuration protocol. *RFC 1541*. http://www.rfc.net/, accessed Dec. 2006.

Evans, J.V. (1995) Twenty years of international satellite communication. International Conference on 100 Years of Radio, pp. 239–245.

Fiedziuszko, S.J. (2002) Satellites and microwaves. *14th International Conference on Microwaves, Radar and Wireless Communications*, MIKON-2002, 3: 937–953.

Geer, D. (2005) Users make a beeline for ZigBee sensor technology. *Computer*, 38(12): 16–19.

GSMWorld (2007) GSM Facts and Figures. Available from http://www.gsmworld.com/news/statistics/index.shtml, accessed Sept. 2007.

Gupta, A., Ranganathan, P., Sarin, *et al.* (2006) IT Infrastructure in emerging markets: arguing for an end-to-end perspective. *IEEE Pervasive Computing*, 5(2): 24–31.

Gyselinckx, B., Vullers, R., Hoof, C.V., *et al.* (2006) Human ++ : emerging technology for body area networks. *IFIP International Conference on Very Large Scale Integration*, pp. 175–180.

Ho, L.T.W. and Claussen, H. (2007) Effects of user-deployed, co-channel femtocells on the call drop probability in a residential scenario. In *Proceedings of 18th Annual IEEE International Symposium on Personal, Indoor and Mobile Radio Communications (PIMRC'07)*, pp. 1–5.

HomePlug 1.0 Technology White Paper. Available from http://www.homeplug.org/en/products/whitepapers.asp, accessed Sept. 2006.

Jhunjhunwala, A., Ramamurthi, B. and Gonzalves, T.A. (1998) The role of technology in telecom expansion in India, *IEEE Communications*, 36(11): 88–94.

Kalghatgil, A.T. (2007) Challenges in the design of an impulse radio based ultra wide band transceiver. *IEEE International Conference on Signal Processing, Communications and Networking*: 1–5.

Lehr, W., Merino, F. and Gillett, S.E. (2002) Software radio: implications for wireless services, industry structure, and public policy. International Telecommunications Society Conference.

Magedanz, T. and Popescu-Zeletin, R. (1996) *Intelligent Networks: Basic Technology, Standards and Evolution*, Thompson Computer Press.

Miniwatts (2007) Internet world stats: usage and population statistics. Available from http://www.internetworldstats.com/stats.htm, accessed Sept. 2007.

Moerdijk, A-J. and Klostermann, L. (2003) Opening the networks with Parlay/OSA: Standards and aspects behind the APIs. *IEEE Network*, 17(3): 58–64.

Naughton, J. (1999) *A Brief History of the Future: The Origins of the Internet*. Phoenix.

Nikolaidis I. (2007) Editor's note – the unbearable simplicity of being neutral. *IEEE Network*, 21(2): 2–3.

Nosratinia, A., Hunter, T.E. and Hedayat, A. (2004) Cooperative communication in wireless networks. *IEEE Communication*, 42(10): 74–80.

Nuaymi, L. (2007) *WiMAX: Technology for Broadband Wireless Access*. John Wiley & Sons, Ltd.

Ojanperä, T. (2006) Convergence transforms Internet. *Wireless Personal Communications*, 37(3–4): 167–185.

Perkins, C.E. (2002) Mobile IP – Updated. *IEEE Communications*, 40(5): 66–82.

Raman, B. and Chebrolu K. (2007) Experiences in using WiFi for rural internet in India. *IEEE Communications*, 45(1): 104–110.

Reimers, U. (ed.) (2004) *DVB: The Family of International Standards for Digital Video Broadcasting*, 2nd edn. Berlin: Springer-Verlag.

Rhee, M.Y. (2003) *Internet Security: Cryptographic Principles, Algorithms and Protocols*. Chichester: John Wiley & Sons, pp. 339–353.

Saltzer, J.H., Reed, D.P. and Clark, D.D. (1984) End-to-end arguments in system design. *ACM Transactions on Communications*, 2 (4): 277–288.

Schulzrinne, H. and Rosenberg, J. (2000) The session initiation protocol: Internet-centric signalling. *IEEE Communications*, 38(10): 134–141.

Stienstra, A.J. (2006) Technologies for DVB services on the Internet. In *Proceedings of the IEEE*, 94(1): 228–236.

Tang, C. (2005) Underlay-aware overlay networks. PhD thesis, Michigan State University. Available from http://www.cse.msu.edu/~mckinley/.

Tanenbaum A.S. (1996) *Computer Networks*. 3rd edn. Upper Saddle River, NJ: Prentice Hall.

Tennenhouse, L., Smith, J.M., Sincoskie, W.D. *et al.* (1997) A survey of active network research. *IEEE Communications*, 35(1): 80–86.

Thouin, F. and Coates, M.J. (2007) Video-on-demand networks: design approaches and future challenges. *IEEE Network*, 21(2): 42–48.

Varshney, U. (1999) Networking support for mobile computing. *Communications of the Association for Information Systems, AIS*, 1 (1): 1–29.

Vaxevanakis, K., Zahariadis, Th. and Vogiatzis N. (2003) A review on wireless home network technologies. *ACM SIGMOBILE Mobile Computing and Communications Review*, 7(2), 59–68.

Yen, L. (2000) Satellite communications for the millennium. *International Symposium on Antennas and Propagation*, 2: 530–533.

Yoo, C.S. (2005) Beyond network neutrality. *Harvard Journal of Law and Technology*, 19(1): 1–78.

Zimmerman, T.G (1996) Personal area networks: near-field intrabody communication. *IBM Systems Journal*, 5 (3–4): 609–617.

12

Management of Smart Devices

12.1 Introduction

System management concerns collecting information about the operation of a system and making operational and strategic decisions to actively maintain or modify the system operation. Operational system management and its sub-types such as security management generally concern three main management activities to maintain the operation of a system: (1) *monitoring* to detect management change events; (2) *prevention* to control and handle management change events through the use of configurations and policies; and (3) *correction*: to handle disruptive causes of management changes and faults that result from management events.

The most mature models for operational management UbiCom systems consider UbiCom systems as subtypes of distributed ICT systems and adopt the techniques for managing distributed ICT systems used for managing UbiCom systems. In addition, because ICT services and resources are owned and operated by individual humans and human organisations, these tend to incorporate human-inspired system operational models e.g., services can be regulated according to service-level agreements.

There are further management challenges introduced by UbiCom systems that are not adequately covered using conventional distributed ICT system management alone. For example, more diverse smart ICT devices are used, e.g., devices can be micro or nano-sized and can be embedded in non-ICT objects. ICT devices can be used in diverse physical and human, smart environments and environment devices. Devices in smart environments need to be more aware of the human and physical context. In addition, UbiCom systems may themselves provide new management solutions for ICT systems.

In addition to managing system operation, UbiCom system management also covers the *creation* of systems and the *dissolution* of systems. Rather than just considering dissolving services from the virtual computing environment so that they are no longer advertised and available, UbiCom systems management must also consider their dissolution in a sustainable way from the physical environment and their dissolution from their human environment. This dissolution is closely tied to the way they are created, how they are planned, designed and whether their design leads to a usable, useful and used system.

Ubiquitous Computing: Smart Devices, Environments and Interactions   Stefan Poslad
© 2009 John Wiley & Sons, Ltd

12.1.1 Chapter Overview

The following sections analyse management of UbiCom systems in terms of the three core designs of smart devices, smart environments and smart interaction. The management of smart devices in virtual environments (Section 12,2) covers the core management of the distributed property of UbiCom systems in terms of the use of dedicated remote service access points the use of RSAP and service policies and in terms of the core management functions such as safety and security. Next, management models for user-centred design of smart devices in human environments are considered. The next section (Section 12.3) expands the set of service management models beyond the (RSAP) and service contract models given in the previous section. In this section, one of the great perils of using smart devices is considered, that of privacy invasion. The final section considers the management of smart physical environments section (Section 12.4). The focus is on the management of context-aware UbiCom systems, the management of smart dust systems and the use of devices which are likely to be untethered or left unattended.

12.2 Managing Smart Devices in Virtual Environments

Management of smart devices in virtual (ICT) environments is first considered from different viewpoints such as the data processing and the network viewpoint of ICT systems. Then device management is considered with respect to which system functions are managed and the type of challenges (Table 12.1).

12.2.1 Process and Application Management

Once applications are installed and registered with the device operating system, the operating system (MTOS) manages the system to enable the user to select any task or any group of tasks to be executed, even when the number of tasks ready to execute exceeds the number of CPUs available. A MTOS supports the execution of multiple concurrent process applications on a range of high resource smart devices by managing memory, process control and communication transparently to applications and users (Section 3.4.3). Operating systems have some inbuilt support to prevent processes from hogging the CPU and I/O resources and from over-writing the memory used by someone else. In energy-constrained portable devices, the operating system also manages the use of the limited energy available using a range of power-saving techniques (Section 4.3.4). In mobile devices, the operating system and network virtual environment also supports the ability for mobile receivers and transmitters to send and receive messages from different parts of a global network (Section 11.7.6). Different strategies for managing the download and installation of applications into smart devices are discussed in Section 12.3.1.

12.2.2 Network-Oriented Management

The network viewpoint of ICT systems defines two main components: computer nodes and network elements to interlink them (Section 3.1.2). Traditionally, a distinction is made between *network management* and *system management* although these two are inherently linked.[1] Network management concerns monitoring communication services running on network elements

---

[1] One of the first computer companies to really highlight the close interplay between the network and computer was Sun with their slogan 'the network is the computer' in the 1990s. The pervasive use of Web and Internet-based applications which automatically access remote network elements reinforces this idea.

Table 12.1 Management requirements for smart devices

UbiCom component	Management challenges
Smart Devices in virtual environments	Manage distributed open, dynamic, heterogeneous volatile services and resources
	Manage FCAPS functions of systems
	Manage multiple processes within one ICT devices versus distributed across multiple devices
	Manage communication between mobile pad and tab type devices
	Manage information metadata linked to information
Smart Devices in user-centred environments	Manage implicit human computer interaction with individual users
	Low maintenance of devices by end users and owners. Manage devices as part of human centred activities that may be life-long, exceeding the life-time of service access devices many times over
Devices in Smart (Physical) Environments	Manage single dedicated tasks on embedded system using ASOS
	Manage dust-sized devices that may get lost, move outside a prescribed management domain
	Manage computer device interaction with physical environment
	Manage interactions devices sited in unsupervised, shared physical spaces, e.g., wall mounted devices
	Managing devices throughout their whole environment life-cycle including disposal
Smart Interaction, between humans, their devices and physical environment	Manage multiple interactions in multiple activities
	Manage interaction in open systems and environments
	Manage individual, versus social organisational (cooperative) concerns versus economic market (self-interested) concerns
	Manage local versus global self-* management (autonomic systems)
	Support self-explaining systems
	Support emergent versus self-organising systems
	Support self-creating versus self-organising systems

such as hubs, switches, routers, gateways and modems in the core of the network, e.g., a SNMP model. These elements tend to run on ASOS computers. System management concerns maintaining the operation of application services running on MTOS computer systems at the edge or end of the network and is strongly influenced by network management initiatives in the telecommunications domain, e.g., the FCAPS model.

Many fundamental communications functions, services, especially in single media content networks, are still managed at the network level in the core network, even although there seems to be a trend to service-oriented network management for the physical signal distribution, data encoding, data channel sharing and efficiency, error checking and correction and data transfer control. There seem to be two approaches with respect to managing heterogeneous content and applications with different requirements for jitter, delays and packet loss. One approach is keep the core functions and network routing elements simple and to support these at the edges of the network in applications. A second approach is to increase support in the core network to better manage the different requirements of heterogeneous media using IPv4 protocols such as MPLS, Diffserv and RSVP or using IPv6 (Section 11.5.2).

12.2.2.1 FCAPS

Without good system management, the probability of (partial) system failure increases, system re-configuration is harder, the likelihood of service non-availability increases and security violations increase. Good management is a particular concern for network service providers as network downtime leads to lost revenue, opportunities and productivity. This was particularly a concern for telecoms systems operators that tended to charge users per packet or on the duration of the connection session. This led to standardisation of common network management functions for *TNM (Telecommunication Network Management)* by ISO/IEC (1989) and by ITU-T (ITU-T, 2000) and is often referred to as *FCAPS*, an acronym for *Fault, Configuration, Accounting, Performance and Security* (Table 12.1)

The TMN model for managing Open Systems within a telecommunications network defines four logical layers: Business Management (BM), Service Management (SM), Network Management (NM) and Element Management (EM) (Figure 12.1). FCAPS are the management functions spread across these logical layers.

The FCAPS model functions define the basic requirements for managing distributed computers and hence UbiCom systems. Many computer service vendors offer client–server implementations[2] of the FCAPS functions via various APIs, typically the Simple Network Management Protocol SNMP. SNMP traps can be used to support fault management. Configuration management can be supported via SNMP events or using UPnP and other device discovery protocols and then stored in a database. Stored events can be filtered and exported to other applications such as billing systems to support accounting management. ICMP, SNMP and HTTP can be typically used to exchange events to support performance management. SNMPv3 security functions can be used to support security management.

The FCAPS properties, although conceived for telecommunications networks, can also be applied to other types of network such as data networks, which may use a flat-rate network service usage

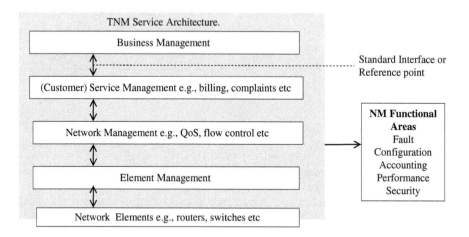

Figure 12.1 Telecommunication Network Management (TMN) Services and Network Management (NM) functional areas

---

[2] Open source implementations of FCAPS are also available such as OpenNMS, an enterprise grade open source network management system, see http://www.opennms.org, accessed Nov. 2007.

Table 12.2 FCAPS network management functions

Management type	Description
Monitoring	Log persistent and transient events
Fault/Safety	Detect fault type of events that lead to system failure
	Organise and manage fault cascades in which a root fault leads to numerous child faults, generally the child faults should be suppressed
	Report faults to an appropriate authority or manager
	Automatic correction and handling of some faults
Configuration	Set or modify parameters that control routine operation
	Track resources defined by their resource descriptions
	Track changes in status of resources such as failures
	Manage activation and deactivation of resources
Accounting/Auditing	Track service usage and inform authorities about usage and usage costs
	Set limits on resource usage
	Automatic handling when usage exceeds limits
Performance	Collect network statistics using polling or event push
	Evaluate performance under normal and degraded conditions
	Monitor events that exceed thresholds etc
Security	Manage access control, confidentiality, data, etc
	Monitor and detect security events that pertain to these
	Automatic correction of selected security events

charge. Accounting can be replaced with more general auditing. Fault, configuration, and security management are often integrated and accessed via the operating system in Personal Computers (PCs) nodes. Accounting and performance operational management are of less interest to most users. The FCAP functional areas are outlined in Table 12.2. Each of these is described in turn. FCAPS management can be applied to smart devices, smart environments and smart interaction.

12.2.3 Monitoring and Accounting

Monitoring obtains the information required to support management functions. Typical information includes usage, current status snapshots, status changes and unusual event reports. Monitoring can be configured as follows in terms of: how often to save the system status and what the system status saved is, e.g., the difference from a previously saved state or as a stand-alone state. Monitoring involves three separate processes, analysis, filtering and auditing. Analysis involves gathering and correlating distributed information. This may need conversion into a common format and to compare events with a past history and the use of case-based patterns. Filtering involves reducing and selecting events based upon predefined rules in order to reduce the information overload and storage, e.g., notify something if anyone accesses a particular system between 18:00 and 08:00 hours. Different information is needed for different purposes, e.g., to support the different FCAPS management functions and to support different application requirements. Filters need to support selective archiving by end-users and should be user configurable. Two main methods or protocols to monitor networked devices or hosts are to use ICMP (Section 12.2.3.1) or SNMP (Section 12.2.3.2).

Accounting Management records the use of resources and services by individuals and by groups, e.g. by department. This can also be used to calculate the variable cost of usage and to support itemised billing. However, this is also useful even when there is a fixed, e.g., monthly, cost for usage. Accounting can be used to trigger rules to govern fair access, to set usage quotas, to assess longer-term trends and patterns of system use beyond that derived purely from monitoring and so help to plan system evolution and re-configuration. Accounting can also be used to identify sources of events and to capture the effect of events as they unfold.

12.2.3.1 ICMP

The *Internet Control Message Protocol* or *ICMP*[3] is an integral part of TCP/IP-based networked systems and can be used to send messages about unusual events and error conditions in an IP network. A common use of ICMP is to check if a computer node is reachable on a network. A sender process on one computer node sends an *ICMP echo request* message to a receiver process on another node which should respond with an *ICMP echo response* message. An ICMP timestamp request and response message is similar to the echo message but also indicates the message Round Trip Time and is used to measure network performance. Applications which use this protocol are often referred to as a Ping application. Although, ICMP is useful for occasional use, it is an unreliable network protocol, it does not scale well and floods large networks if many network elements are periodically pinged. Hence other network protocols such as SNMP have been proposed to provide a more systematic way to monitor and manage large networks.

12.2.3.2 SNMP

The main components of the *Simple Network Management Protocol* or *SNMP* model (Figure 12.2) consist of network elements, agents, proxies, managers, network management information database and the network management protocol. Network elements or network objects[4] are the things to be managed or that can be left unmanaged. Agents[5] are the management processes that execute as processes in the network element and are used to monitor and filter events, such as to dynamically assign an IP address, buffer overflow and the status of one or more managed network elements in their domain. Agents execute as processes in network elements that are managed. Managers query information from agents, process and store the information. In the SNMP model, the functional complexity is in the managers, keeping the agents simple. Proxies or proxy agents can be used as wrapper agents for unmanaged network objects that are unable to have network agent processes running in them.

The Management Database (MDB) or Management Information Base (MIB) is used to store information about the network elements for use by managers. This syntax and semantics of the information are defined by a SMI (Structure of Management Information) specification.[6] SNMP supports communication between agents and managers and between managers. Managers can either regularly poll agents to pull management information or agents can sent push management information to managers in response to trigger events.

In the SNMP model, a manager may manage multiple agents. Managers and agents can be organised hierarchically. Agent and manager roles can be combined into one node and unmanaged

---

[3] ICMP is specified in RFC 792 (1981), retrieved from http://www.ietf.org/rfc/rfc.html. Several extensions have been proposed to RFC 792. The use of ICMP echo or timestamp request and response messages is commonly referred to as a *ping*.

[4] This type of network object is not an object in terms of an object-oriented system because these network objects have state data but no methods other than the implicit methods to get and set this state data.

[5] This type of agent is different from the intelligent type of agents (Chapter 8.).

[6] SMI is represented in a syntax that is a superset of ASN.1, the Abstract Syntax Notation One taken from the OSI network model, See Tanenbaum (2002) for a good description of SNMP and computer networks in general.

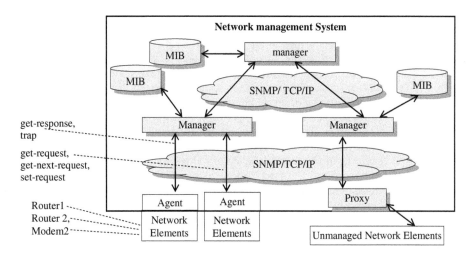

Figure 12.2   Basic architecture for network management

elements can be managed via proxies. SNMP (version 1) defines just five types of messages, get-request, get-next-request (a variation of the get-request to get parts of a data-set in sequence such as a routing table), set-request (to set some parameters for a network object), get-response (to get a response to a request) and trap (to send unsolicited management data by an agent to a manager). Three versions[7] of SNMP have been developed (Subramanian, 2000).

SNMP has the potential to be used for UbiCom. It is well supported in many network elements. It can support a range of FCAPS management functions (Section 12.2.1). It supports an asynchronous application protocol that can handle volatile communication and it supports low resource agent processes, however, resource managers are often resource-hungry and are complex to maintain.

There are several practical issues in using SNMP based UbiCom. SNMP MIB data structures are complex and need to be specified at a low level of detail. MIBs can be converted to other representations such as XML to make the MIBs more accessible by heterogeneous applications (Soares and Thiry, 2002). SNMP is not inbuilt into many devices but these can be treated as unmanaged and managed by proxies. Agent proxies can act as application gateways to allow SNMP events to be mapped to non-SNMP events such as UPnP events (Murtaza1 *et al.*, 2006). There is a range of standards for event-driven architectures to support management events in addition to SNMP events (Section 3.3.3.6). Systems, e.g., those based upon SNMP, can produce a huge set of network events that need to be managed, stored and filtered. SNMP is oriented to a single user, managed devices, the use of a few SNMP manager processes, a single physical space and to static rather than dynamic managed system elements. SNMP requires a high degree of technical ability and is not suitable for many end-users without usable SNMP tools.

---

[7] SNMPv1 was released in 1990 (RFC 1157), see http://www.ietf.org/rfc/rfc.html. SNMPv2 formalised the manager-to-manager communication by adding a manager-to-manager message type; P2P manager communication is also supported. Support for bulk data message type was added. In addition, SMI was re-specified and hence SNMPv2 is not compatible with SNMPv1. SNMPv3, released in 1998, added security and improved the modularisation of the architecture and supports interoperability with SNMPv1 and SNMPv2.

12.2.4 Configuration Management

A *configuration* is used as a specification of the settings of the attributes that can be modified and any invariant attributes, such as an identification code, that characterise a user device. *Configuration management*[8] involves four main management functions: identification of the configuration, change control (or change management), accounting and verification and auditing. Drivers to change the configuration or reconfiguration and reorganisation are often driven by the need to maintain performance and a quality of service. Reconfiguration may be triggered by ICT resource contention, e.g., processing, storage, communication and limitations. Many embedded systems and personal systems need to interrupt their service and reboot themselves for reconfiguration changes to take effect. The more user and system processes that have access to change the configuration, the more frequently this interruption will occur. Configuration conflicts may also become common in multi-domain managed UbiCom systems when different parts of a system use different versions of a resource such as a graphics adaptor, e.g., an operating system may upgrade the graphics adaptor which then becomes incompatible for use with particular applications.

In user-centred virtual ICT environments such as home networks, users may lack the expertise to manually maintain configuration information in ad hoc device environments as new devices are added and as some old devices become obsolete. Ideally, zero manual configuration of devices by users should be necessary. This is achieved to some extent in some parts of the UbiCom system such as zeroconf networks (Section 3.3.1) Consequently, virtual home ICT environments should be designed to support device discovery and hence device *plug and play* (Section 3.3.2), together with device *unplug and play*[9] and *unplug and stop play*. The demand for a flexible topology to organise the interplay between heterogeneous devices requires that the configuration of the network is re-discovered and mapped automatically at regular intervals.

Another challenge is how to deal with multiple independent configurations or how multiple control devices can control a common appliance, occurring concurrently, e.g., several people at the same location want the lighting or volume set to different levels; or users at different locations try to configure the same thing differently. This leads to configuration conflict. Solutions to solve configuration conflicts can use similar techniques to solve operational conflicts (Section 12.2.8.4).

12.2.5 Security Management

Security management concerns the assessment of the risk of *threats* which cause some loss of value to system assets, heightened through any system vulnerabilities or weaknesses and developing and maintaining appropriate *safeguards* or security *controls* to protect assets against threats (Tan *et al.*, 2006). There are three basic types of safeguard:

* *Detection*, e.g., periodic system audits and scans; integrity checks and checksums are used to identify modifications to stored information, messages and code; digital signatures can be used to attest who has written what.

---

[8] Configuration management was first proposed as a technical management discipline by the United States Department of Defense in the 1950s. See http://en.wikipedia.org/wiki/Configuration_management, accessed July 2007.
[9] Unplug and play: handling and possible maintaining services where desired, in spite of devices being inadvertently unplugged because of the presence of redundant services being offered by multiple devices. Of course unplug may also be deliberate in that case it should be unplug and stop play.

- *Prevention*, e.g., encrypting confidential data to prevent eavesdropping; authentication based upon strong passwords and certificates; authorisation granted on access control policies based upon user identity and user roles.
- *Correction*: handle what has happened after a threat has occurred, e.g., countermeasures such as blacklisting and blocking access from users and networks that were sources of previous threats; use of a disaster recovery plan.

Both detection and correction offer *a priori protection* while correction offers *a posterior protection*. UbiCom System security can be modelled in terms of Viewpoints of sets of Safeguards that protect the system assets (the items of value in a system) against Threats (the actions that decrease that actively the value of assets) – the V-SAT model of security (Tan *et al.*, 2004), see Figure 12.3.

Security management depends upon the specification of two main relationships: of threats against assets and of safeguards against assets. *Risk assessment* is used to model the assets of value in a system and their loss in value in relation to the probability of the threat happening and the probability of the threat succeeding. A *security policy* specifies the assets of value, which safeguards protect them and which threats affect which assets. Risk assessment and security policy may need to be specified with respect to different viewpoints of the system as threats may depend upon the applied use of the system and the model of how the system relates to its environment.

Table 12.3 shows relationships between threats to assets and their safeguards. This high-level analysis does not include quantitative estimates of the probability of threats occurring, succeeding and the loss in value of assets. It also includes representative threats rather than including all the main threats, assets and safeguards. It is in natural language form, security policies are often expressed in natural language form, rather than in a machine-readable form for direct operational use. Information on which to quantify the risk of threats can be taken from system audits and from reports of threats by others and on valuations of assets and of the cost of operating and maintaining assets.

The use of conventional security safeguards are covered in any standard book on computer security, e.g., Mel and Baker (2001), or Gollmann (2005). Several specific security safeguards are dealt with elsewhere, e.g., the use of firewalls, NAT and VPNs type safeguards to protect network assets (Section 11.7.3); the use of data integrity management to detect and prevent inconsistent lean hard data (Section 12.2.9).

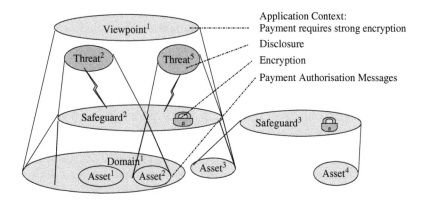

Figure 12.3 V-SAT model of security: viewpoints of safeguards that protect the assets of the systems against threats

Table 12.3 Relation between threats, assets and safeguards from the viewpoint of the user of a smart mobile device

Threat	Asset	Safeguards
Sender masquerade	Information about real sender shared with fake sender. Actions requested by a fake sender are performed by receiver	Authenticate caller identity Call back real sender Strong password-based access control
Unauthorised access: to read from or write to device;	Unwanted received or sent messages, message costs, unauthorised access to remote services such as owner's bank accounts	Access control at access device, at message server
Denial of Service:	No access to local device services such as music player, camera etc	Physical device security
Communication access device lost, stolen or damaged	Cost and delay in using devices for acquiring a replacement access device and contract	Use of decentralised access device model; fault-tolerance and data back-up
Network loss	Can't send or receive	Server side backup, access device redundancy. Use pre-cache, delayed write
Execute viruses	Data can be corrupted; normal access to services can be interrupted,	Virus-checker to check
Loss of confidentiality	Personal details, social and business relations and activities are made public	Encrypt messages sent and stored data
	Track access device and hence owner	Legal safeguards, service policies
Local data corruption	Local information loss, e.g., address book	Data Integrity Management
Message corruption	Message loss	Access control
Repudiation of sent or received messages	Loss and delay in message requests or information exchange	Audit to detect and verify messages sent in access devices and in the network

12.2.5.1 Encryption Support for Confidentiality, Authentication and Authorisation

The core techniques in distributed systems concern *encryption* that uses symmetric or asymmetric keys to support confidentiality, authentication and authorisation or access control. An encryption algorithm or cipher is used to transform clear text or plain text into encrypted text using an encryption key. A corresponding decryption key converts the encrypted text back into plain text using a corresponding decryption key. In *symmetric encryption*, the same key is used for encryption and decryption and needs to be kept secret – a *secret key*. A key challenge in using symmetric encryption is how to distribute secret keys while keeping them confidential in transit. A common solution is to use some further key, even a public key, to encrypt a secret key in order to share it.

In *asymmetric encryption* or *public key encryption* algorithms, the public key is made available in an unrestricted fashion and used for encryption by the sender, whereas a corresponding private key, kept secret by the receiver, is used to decrypt a message in the receiver. The private key cannot be derived from the public key. This eases the problem of the sender and receiver having to somehow share the same secret key. Public key encryption enables the public keys used for encryption to be made public. Public key encryption can also be used to support authentication and authorisation of documents, messages and user credentials, as well as supporting the confidentiality of these using digital signatures and certificates. The content of a document is converted to a unique value using an algorithm such as a hash function so that this value changes if the document is changed. The unique value can then be encrypted using a private encryption key, the *signature key*, and this is called a *digital signature*. Anyone knowing someone's public key and knowing the hash function used to characterise the document can verify the digital signature corresponds to someone's private key – the verification key.

Authorisation often involves user identities being validated, e.g. the resource user wants to ascertain that he or she divulges their security credentials to the right party, in this case to a valid holder of the resource access rights. Although public key encryption can be used for authentication, a key issue is how does someone know that the public key, sometimes used as a verification key in digital signatures, actually belongs to a particular identity? Anyone could claim that they hold a particular identity. There needs to be something or someone to attest that a particular identity is bound to a particular identity. This is often an identity certificate, a document that someone else, another, trusted, third party, attests that a particular security credential, e.g., a public key, belongs to a particular unique identity. This attestation uses another digital signature, this time belonging to the trusted third party. Certificates can also define the access control rules associated with a particular identity.

Identity certificate authorities are used to create these certificates.[10] Often these authorities act as trusted third parties on behalf of the interacting parties such as resource providers and users. This approach requires the use of a complex middleware infrastructure which often involves the use of identity certificate chains of authorities. Users must be able to identify certificate authorities and authorities, must be accessible and be trusted. Certificates also need to be attested as being signed by valid keys rather than by expired or revoked keys.

In *Identity-Based Access Control (IBAC)* or authorisation, users need to a present security credentials or evidence, e.g., a user identity associated with a key e.g., shared secret key such as a password or a public key. These user credentials are then compared against the set of credentials that allow access to a resource, e.g., specified in an access control list or access control policies. If the presented user credentials can be matched to the credentials that allow access, access is granted, otherwise it is withheld.

In an open environment, greater flexibility is useful in order for one party to authorise another party to act on their behalf. Several restrictions can be removed: the need for globally unique identities,[11] the need for access control to be associated with globally unique identifiers, the need for issuers to directly define access rights. An example of such a scheme is the *Simple Public Key Infrastructure (SPKI)*.[12] This supports the use of local rather than global identities, associations of local names with access control privileges and the ability for an issuer to allow a subject to delegate its access control privileges to someone else.

An analysis of the use of smart mobile devices reveals an increased risk to secure middleware, to secure access devices and to secure content, compared to desktop computers.

12.2.5.2 Securing the System and its Middleware

The use of *seamless networks*[13] of IP networks everywhere, pervasive wireless communication, access by static and mobile devices, allows even remote peers to access local area network

---

[10] Certificate authorities are also referred to as *issuers* as they issue access control privileges to users, *subjects* that access specific resources.

[11] Another example of security being possible and beneficial without identification is when privacy is needed (Section 12.1.1.1.).

[12] SPKI is defined in RFC 2692 and 2693. IETF RFC, Request For Comments specifications are available from www.ietf.org/rfc.html, accessed Aug. 2007.

[13] Seamless connectivity covers both a static access node that changes its remote service access points and a mobile access node that moves, changing its local access points. In theory, it should be easy to set policies to differentiate between local versus remote access.

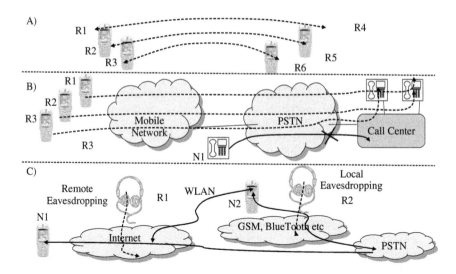

Figure 12.4 Some examples of threats through the use of seamless (wireless) networks, where R indicates a Rogue User and N indicates a Normal User: (a) compromised phones can overload a network, as free-loader use a local network; (b) remote users can overload a network, preventing access by a local user; (c) local and remote users eavesdrop on a normal user.

resources and has been termed the disappearing perimeter or disappearing security boundary. The security in local wireless networks connected to the Internet can potentially allow remote users to eavesdrop and read someone else's transmitted data[14] (Figure 12.4). Mobile users are more likely to operate within unfamiliar ICT infrastructures which have a greater potential for the use of unauthenticated and unverified mediators. This increases the threat of unauthorised access, modifications and disclosure. It may also be easy to synthesise rogue wireless access nodes so that instead of service clients connecting to genuine service providers, they instead mistakenly bind to fake service providers. Mobility can also make it easier for one mobile malicious peer or group, if blocked from resource access in one domain, to simply move to another domain where they just restart their disruptive activities. To an extent this can be combated by conventional safeguards such as the use of cryptographic data encoding and the use of VPNs of public networks (Section 11.8.3) but these often lack flexibility and responsiveness to work in dynamic and distributed systems.

DoS threats can also seek to jam specific frequencies or flood a range of radio frequencies used by a wireless access node with RF interference. In this case, the RF protocols used by an access node can be designed to make the threat of *jamming* and *flooding* more difficult.[15] Two well-known techniques to impede jamming and flooding are to use *Frequency Hopping Spread spectrum*

---

[14] Note that the remote user, in contrast to the local user, must filter many more data packets to try to detect those of interest. This requires a substantial filtering capability and one that is efficient in a large high performance Internet.

[15] Wireless communication is generally defined to operate assuming there is interference present. This interference may be advertent or inadvertent. Hence some degree of support to protect against advertent jamming and flooding is often an inherent part of wireless communication design.

(FHSS) and *Direct Sequence Spread Spectrum (DSSS)*.[16] With FHSS, the RF is split into frequency bands and the transmitter randomly hops between different slots in a way known to valid receivers. DSSS encodes each bit into a special 8-bit or 11-bit code and spreads the transmission of the encoded bits across the frequency spectrum. Jamming and flooding are still possible but it is more expensive to do this as more frequencies need to be jammed with interference.

Attacks by intermediaries are well known, e.g., the *man-in-the-middle attack* allows a first party's public key to be substituted by a third party's public key so that the second party encrypts information with the third party's public key enabling the third party to essentially see all the second party's confidential information. The intermediary or third party is also able to spoof or fake the receiver addresses specified by a sender so that messages are exchanged with a fake receiver rather than a real intended receiver. Then if the third party presents a fake login form that the first party user thinks logs him or her onto a particular service provider, the third party simply collects the login details of the user.

Some preventative and corrective safeguards to handle denial of service threats to mobile devices have also been dealt with elsewhere (Section 4.2). First, back-ups, occasional wireless-synchronization or wired synchronisation of data can be used. Second, a remote-access model can be used to support a virtual distributed UVE type desktop on the mobile device so that data that appears to be local is actually managed remotely. Third, mobile devices can be designed to deal with the threat of volatile remote service access using the techniques described in Section 3.3.3.9. Finally, techniques based upon self-healing and self-protecting (Section 10.4) can also be used.

12.2.5.3 Securing Access Devices

Devices may be left unsecured because their owners expect that they will remain in their physical control, however, if they leave the physical control of their owners, they are open for use by anyone. In the eTRON system (Sakamura and Koshizuka, 2001), any information to be protected is stored in *tamper-proof*[17] hardware of eTRON nodes which mutually authenticate each other, restrict communication to legitimate parties and where all communication is also encrypted. Devices that reach their end of the lifetime either because they fail or because their owners give them up may also leave trails of valuable information behind. Often, devices with inbuilt network security are supplied in a wide open access mode without any security as otherwise the device cannot connect to anything without users being able to understand how to customise the security configuration. In this case, users may maintain the device operation in a wide-open mode.

As the density of devices that operate within a locality increases, devices may pair up or connect in a session with the wrong service because multiple services and devices available at any pairing could be randomly chosen. Centralised trusted third party authentication server infrastructures are often used in fixed networks as a common safeguard. However, authentication servers may be absent or unreachable because network access is volatile. Hence, there is a need for secure transient association between two entities, e.g., owner and device.

---

[16] Both DHSS and FHSS are specified as different variants of the IEEE 802.11b (WiFi) specifications. DHSS is also used in CDMA mobile phone systems. DHSS tends to dominate the wireless WAN and LAN marketplace, rather than FHSS, as it can support higher data bandwidths (Denis, 2002).

[17] Tampering covers reading, writing and deleting. As Stajano (2002) points out, *tamper-proof* devices often only prevent tampering up to a threshold level, hence, the term *tamper-resistant* is more accurate. If that level is exceeded, the system should prevent anything from being removed intact. Tamper-proofing may only protect of the system, e.g., the CPU. This may still allow other parts of the system, e.g., the peripherals, to be tampered with. A cheaper solution may be to use a *tamper-evident* device that makes it impossible for a device or its resources to be tampered with, without detection.

A policy model to support secure transient association called the Resurrecting Duckling security policy model has been proposed by Stajano (2002) as follows. The two state principles are as follows. When the duckling entity, e.g., a smart device, is created, it begins in the *imprintable state*. Anyone can take it over by executing some set protocol such as sending it the private key of the its principal, the mother duck. Once this happens, it transitions into the *imprinted state* and only obeys its mother duck. A transition back from the imprinted state to an imprintable state can be triggered by a death transition. Death can be ordered by the mother or can occur after a predefined time interval or after the completion of some task. The duckling must be designed in such a way that an unnatural death caused by an assailant is uneconomical for the assailant to gain control of the duckling.

Low-resource devices may lack the resources in terms of CPU power, stored energy, wireless bandwidth, etc. to support confidentiality and access control-based prevent safeguards such as encryption or decryption within a reasonable amount of time. This may lead to the selective use of such safeguards in lightweight devices during particular parts of an application session. In addition, the use of less process-hungry cryptographic operations such as symmetric rather than asymmetric key cryptography and shorter keys is encouraged. Several schemes have been proposed to address some of the resource limitations of securing communication for low-bandwidth, low-resource smart devices, e.g., the *Guy Fawkes protocol* (Anderson *et al.*, 1998). The Guy Fawkes protocol is proposed as a less expensive authentication protocol than public key type digital signatures. It signs messages using only two computations of a hash function and one reference to a time-stamping service. It is aimed at protecting message sequences relying on an initial authenticated start message in the sequence and use of the previous message to help compute the hash for the next message in the sequence.

A general type of attack on devices with limited energy reserves is to cause their energy reserves to be unnecessarily expended. For example, a common strategy to conserve power is for devices to enter various power-saving modes, e.g., various sleep and hibernation modes. A *sleep deprivation attack* or threat makes just enough legitimate requests to prevent a device from entering its energy-saving mode. A *barrage attack* bombards victim nodes with legitimate requests. Of these two threats, the barrage threat requires more energy by the attacker and is easier to detect (Pirretti *et al.*, 2005).

12.2.5.4 Securing Information

Access to information is simpler to manage if it can be secured in the static sense at the point of access or where it resides, using access control systems based upon policy management. However, the design of highly distributed systems such as P2P systems (Section 3.2.6), inherently supports decentralised file sharing, allowing peer users a greater degree of autonomous control over their data and resources. P2P system designs can also be designed to make file sharing anonymous so that a receiver may not know from which other peer computer its information originated from.

Securing data distributed from creators and producers via publishers, who may own the copyright to the content, and distributors to consumers or owners of a copy of the data, so that the creators and publishers retain some rights to restrict or allow access to the data is called Digital Rights Management[18] (DRM). The aim of DRM is to restrict copying or conversion by consumers and owners, to balance between owners making several copies for personal use on several devices versus someone distributing content for free or selling on content illegally.

UbiCom and the use of converged multi-play networks (Section 2.3.1) in theory allow any content to be accessed via any network and any device. According to Merabti and Llewellyn-

---

[18] Some people think that DRM is a strange form of security because the data is protected from the owners of the (copy of the) data. Others think that the use of the term *rights* is legally misleading and that DRM should refer to *Digital Restriction Management*.

Jones (2006), there are three major aspects of current DRM techniques used in Trusted Computing Platform (Oppliger and Rytz, 2005), Digital watermarking (Kirovski *et al.*, 2004) and Fingerprinting (Clausen and Kurth, 2004), which preclude their use in a ubiquitous setting. These characteristics are their: inflexibility with regard to fluid data flow; a centralised nature that relies on enforcement from a central point, usually the content producer or the publisher, and considerable computing power requirements, for example, for watermark checking. Hence, it is proposed that distributed trust models (Section 9.4.4) are used to supplement existing DRM techniques (Merabti and Llewellyn-Jones, 2006).

12.2.6 Fault Management

A *fault* is defined as the cause of one or more observed error, or abnormal, events. *Fault Management* or *Safety management* concerns maintaining core ICT service operations. Fault management overlaps with security management in that both cover preventive and corrective type safeguards to deal with (malicious) DoS attacks by both outsiders and insiders and by those that cause *inadvertent faults*(without intent) versus those that cause *malicious faults* (with intent). However, fault management differs from security management in that fault management also covers dealing with inadvertent human operator and design faults as causes which are not generally covered as part of DoS prevention in security management. Faults that cause systems to stop working are easier to diagnose than *Byzantine faults*[19] (Lamport *et al.*, 1982) in which component are still alive but operating incorrectly. Chetan *et al.* (2005) give an analysis of the range of faults in pervasive computing, separating these into device, application, network and service failures. They propose a fault tolerance technique that uses context information to tolerate application and device faults and the use of a fail-stop fault model to deal with Byzantine faults.

Fault management is crucial for maintaining the operation of critical infrastructures used for energy distribution, telecommunications, transportation, logistics, intelligent HVAC, banking and maintaining patient care through monitoring, detecting, preventing and anticipating anomaly events. Fault or safety management involves several basic processes: fault prevention, fault prediction, fault event monitoring, fault detection, fault diagnosis, fault handling and fault-tolerance.

Fault prevention or *fault-avoidance* concerns specifying a system to avoid internal faults or defects under the control of the system and to anticipate and handle those fault events that can be caused by external environment influences outside the control of the system. There are several established processes in software engineering (Pressman, 1997) to try to identify and prevent faults from occurring throughout different phases of the development process such as planning, risk analysis, quality assurance, design walkthroughs and testing and validation. Other techniques that support fault prevention include the use of norms and institutions in open decentralised systems (Section 9.2.3.2).

Fault detection often involves the use of a test or validation phase when a system is commissioned. This aims to test if the system is operating normally or not. Push or pull interaction is used to gather possible error events as follows. Error messages can be transmitted when a fault occurs (fault reporting) or system components can be regularly polled for a response. Recorded states can then be compared against defined error states.

---

[19] A *Byzantine Fault or Byzantine Failure* is also known as the *Byzantine Generals Problem*. This refers to conflicts during the Roman empire of the Middle Ages which, centred on its capital of Constantinople where multiple generals of the Byzantine Empire's army must decide unanimously whether or not to attack some enemy army. The problem is complicated by the geographic separation of the generals, who must communicate by sending messengers to each other over foreign territory and by the presence of traitors among the generals. Traitors can seek to trick the generals or confuse them into actions that are detrimental to winning the war.

Fault diagnosis is the analysis and classification of faults. Most fault detection and diagnosis are aimed at detecting point faults at run-time and then in handling the detected faults. Fault diagnosis can be used to identify and differentiate a root fault from a cascade of further faults that result and propagate from the root fault. Fault diagnosis can involve system testing.

Faults may be random or non-deterministic such as random electromagnetic interference caused by high sunspot activity or gradual system deterioration or wear, e.g., a rechargeable battery not recharging fully, mechanical wear of the moveable hard-disk read/write arm. Faults may be due to incorrect internal design. Faults can be caused by a system's environment causing it to operate outside its safe operating conditions, e.g., a fire can cause electrical components to overheat which may overload other components causing them to overheat and catch fire. In *Supervisory Control And Data Acquisition* or *SCADA* systems, human operators who watch near real-time data can issue commands in an open-loop control system that can inadvertently or maliciously, cause faults. Faults can lead to sustained system termination. There are many different types of system failure such as a fail-stop, fail-silent, timing-fail, write-fail, read-fail and network or communications failure leading to temporary network partitioning. Byzantine failures can be solved by treating other general orders as votes and modelling the solution as a majority voting agreement (Lamport et al., 1982) and by using game theory techniques.

Fault prediction concerns analysing patterns of activity with a view to pre-empting anticipated faults by deploying corrective measures before they occur. To achieve this, *simulations* can be used to model, to try to predict events and pre-emptively handle them by recognising the patterns that lead up to the fault. For example, Woolf et al. (2007) have simulated cascade-type catastrophes in complex networks where an overload in one part leads to subsequent overloading in neighbouring parts. A further fault approach is to model the normal data flows and control operations within a system and to detect anomalies caused by attempts to change or damage the system. This has the advantage that it can detect unknown attacks and the actions of malicious insiders, but unless this is handled carefully, this can generate many false alarms (Bigham et al., 2003). *Symbiotic simulations* are types of simulations which can also be used on-line, rather than used off-line. The symbiotic simulation interacts with the physical system, driven by real-time data collected from the physical system under control, the simulation's environment, enabling the results from the 'what-if' experiments performed by the simulator to be used to control the dynamic behaviour of the physical system (Aydt et al., 2008).

Fault handling concerns the correction of a fault system to a state acceptable for continued operation, thus preventing the system reaching a permanent failure state. Some basic strategies for fault-handling include: masking such as a quick restart, dynamic correction of an error; containment, prevention of error propagation across defined boundaries; repair involving reconfiguration and fault tolerance. *Fault tolerance* concerns the use of redundant elements to aid recovery from detected faults, e.g., critical systems of businesses can use a hot-standby or by Redundant Array of Inexpensive Drives or RAID disks.

12.2.7 Performance Management

Sometimes it may not be possible to specify absolute single point boundaries for system behaviour that distinguish behaviour as being correct or as being acceptable or not.[20] *Performance management* often involves maintaining the operation of the system, such as data throughput and

---

[20] Consider the start-up or boot-up of an electronic device, there is no global rule for a target time to declare that by this time either the system has started operating normally or not.

minimising data loss, i.e., its performance within agreed limits. There are several specific ways to manage performance such as best effort, QoS and SLA.

In a *best effort* system, the system is not managed to guarantee a Quality of Service or QoS or a defined level of performance. The performance will vary depending on the varying workload and system operational capability. This is often a reasonable low-cost solution to performance management providing there is sufficient spare capacity that can be used to support peak demand performance. A world with excessive or 'infinite' resources such as network bandwidth, however, contradicts the basic economic notion that all commodities are inherently scarce (Yoo, 2005). In an open distributed UbiCom system in which the environment load on the system is variable and in which the different system resources, in different environments, belong to different management domains, best effort performance management is somewhat hit and miss.

System performance may be maintained with respect to a QoS. In computer networks, this refers to the use of Resource Reservation Control (RRC) mechanisms, such as RSVP, MPLS and DiffServ (Section 11.5.2) rather than achieving a certain service quality or service level. This is useful for UbiCom because this takes into account that resources may be heterogeneously used and managed. However, it is unclear how RRC mechanisms can operate and adapt in environments where there are highly variable workloads with highly variable QoS requirements. To this end much more accurate models of workload are needed for UbiCom systems that incorporate user behaviour-oriented workload modelling, hierarchical workload modelling and adaptive workload modelling (Kotsis, 2002). Often the QoS is set from the service providers' perspective using targets that they can quantify. In contrast, the Quality of Experience or QoE is a more subjective measure of a customer's experiences of services because often the usage experience criteria are very different.[21] General criteria that characterise user experiences have been discussed elsewhere (Chapter 5). Services can also be managed to achieve a target performance level, using a SLA, agreed between providers and customers (Section 12.2.8.3).

12.2.8 Service-Oriented Computer Management

Any individual component or group of UbiCom system functions can be modelled as services (Section 3.2.4) and can then be managed as services. Common service functions, such as communication or network, data processing, data storage, human computer interaction and various types of operational management including security, and their associated management functions can be factored out of individual applications into middleware and then be managed there on behalf of multiple applications and users. Increasingly, systems management occurs at a human social organizational level using policies and SLAs, to complement management at the ICT level of the system in terms of data throughput and transaction rate.

12.2.8.1 Metrics for Evaluating the Use of SOA

Kalasapur *et al.* (2006) have devised categories of metrics to evaluate the use of SOAs in pervasive environments: service density, service availability,[22] service potential, service impact and service redundancy.[23]

---

[21] Service providers may be more concerned with uptime, throughput, processing power, etc., whereas users may be more concerned with the start-up time, the number of key strokes to active functions in different modes, etc.

[22] Service availability was called *degree of support* by the authors but service availability is a more descriptive name and is used here.

[23] Service redundancy was called *service reconfiguration* by the authors but service redundancy is a more descriptive name and is used here.

Two metrics for service composition are proposed, service composition length and service composition sustainability:

- *Service Density* is a quantitative measure of the ability of the ICT environment to support user tasks, defined as the ratio of the total number of services to the number of requests; the higher the value of the ratio, the higher the probability of success for a request.
- *Service Potential* is the ability of an individual service to take part in multiple user tasks defined as the ratio of the number of tasks a service can be part of, to the total number of user tasks that can be supported in the environment.
- *Service Availability* is the ability of the ICT environment to support multiple user tasks defined as the ratio of the number of matches available for user requests to the number of unique user requests. An environment with a high value is said to be highly available.
- *Service Impact*: when services are added or removed, this leads to change in the state of the ICT environment. This is defined as the difference in the service availability, with and without the service.
- *Service Redundancy* is a measure of the replication in functionality of a service in the ICT environment. Services can fail due to power limitations or mobility. Then the environment needs to identify an alternative service, which matches the original functionality. For each service, the redundancy can be measured as the ratio of the number of alternative services available to the number of user requests that use the service as part of a composed service or as a whole service.
- *Service Composition Length* is defined as the number of services used to compose a solution to satisfy each service request. The average length of composition for a particular request is defined as the ratio of total number of service elements used in the composition to the number of services specified in the request. It is beneficial to have as small a value of this ratio as possible.
- *Service Composition Sustainability* is a measure of the environment's ability to sustain a composition when one or more of the employed services fail. This depends on the number of alternatives available for the failed service. For each user task, the composition sustainability is defined as the ratio of the sum of alternative compositions available for each requested service to the number of services specified within each request.

For a given set of tasks, it is advisable to support services that can guarantee higher service availability, and services with higher service potential to maximise resource utilisation. In highly dynamic environments where the state of resources can change frequently, it is necessary to introduce redundant services, thereby limiting the impact of a single service. Service composition is an effective mechanism to support reuse of services, QoS assurances, mobility and fault tolerance within SOAs (Kalasapur *et al.*, 2006).

12.2.8.2 Distributed Resource Management and the Grid

A *Grid* is a distributed ICT system model of heterogeneous ICT systems that agree to share their local resources[24] such as processing and data storage, with each other, behaving as a virtual computer for its users (Section 3.2.4). The Grid resource management system (RMS) is a core component of the Grid that supports adaptability, extensibility, and scalability, allowing systems

---

[24] Some authors view data process and workflow type resources as synonymous. Others make the distinction that workflow is a particular type of process involving the coordination of action between multiple autonomous parties.

with different administrative policies: to inter-operate while preserving site autonomy, to co-allocate resources, to support load-balancing, to set and maintain quality of service and to meet computational cost constraints. The RMS manages the pool of resources available to the Grid, i.e. the scheduling of processors, network bandwidth and disk storage with respect to policies that govern how the resources should be used by the Grid so that resources can still meet their local resource demands. It may be necessary to employ a federation of RMSs instead of a single RMS because of different administrative policies and resource heterogeneity. In general, requiring the RMS to support multiple policies requires scheduling mechanisms to solve a multi-criteria optimisation problem (Krauter *et al.*, 2002).

There are several challenges in using the Grid model for managing UbiCom resources. As mentioned previously, UbiCom system resources are more heterogeneous than Grid computing resources (Section 3.2.5). Berman (1999) notes that managing resources must be considered at two different levels, the system level and the application level, and that management at both levels simultaneously may be challenging because they have different performance goals. At the system or global level, the concern is scheduling resources to optimise *fairness* to ensure all resources so that requests are satisfied and to optimise resource *utilization, e.g.*, the amount of time a resource is used. These goals may conflict with the need to optimise the performance of individual applications with respect to minimum executions time, resolution and performance. Some solutions to optimising the scheduling different applications with different performance parameters can be to adopt resource management models from telecoms networks.

12.2.8.3 SLA Management of Services

In some SOA models, e.g., OMG SOA (Section 3.2.4), services are modelled more specifically as specifications of sets of operations that can be offered as part of a contract or a Service Level Agreement (SLA) between providers and users. The contract specifies quantifiable operational targets or outcomes for service levels. The operation of the services can then be monitored to see if any deviations from these targets occur. Important application domains for the use of SLA management are telecom network and help-desk provision. SLAs can be established to maintain levels of services with respect to: minimum levels for the percentage of calls that are abandoned while waiting to be answered; the maximum time for the time it takes for a call to be answered by the service desk; the minimum percentage of help requests that get answered within a definite time-frame, the minimum percentage of help requests that can be resolved by the first responder without having to call back the customer later with new information. SLAs are useful for operational management if rules for maintaining quantifiable target levels of service can be established and can be monitored.

SLAs require the performance of two types of behaviour to be modelled: the load performance and the system performance. The load model models the workload applied to the system and incorporates the behaviour of the users, e.g., in an ecommerce application, the number of requests that are made and how many users leave the site prematurely because of poor service. The system model models the performance of services as these process user requests. SLAs used in UbiCom systems also require models of how SLAs for individual services can be aggregated when individual services are combined into composite services (Daly *et al.*, 2002).

12.2.8.4 Policy-based Service Management

A more general rule-based system for managing system operation than using service level agreements is to use policy-based service management. Policies are operating rules that can be used as a means for maintaining the order, consistency, security, and organizational goals or mission. For example, system operation policies are rules governing the choices in the behaviour of a system. Policy-based management within a domain involves having explicit declarative representation that can be dynamically manipulated in order to manage the operational configuration of systems. Core application domains for policy-based management of distributed systems include security-based policy-based management, particularly access control[25] and user privacy management (Duflos et al., 2002; Sloman and Lupu, 2002), and network-based management (Stone et al., 2001). Policy-based management can also be used in general for service management, mobility and context-aware management. User level policy management can be used as part of a vision of iHCI (Section 5) and autonomic computing (Section 10.4). Instead of managing the detailed low level (re)configuration of parts of the system, the user specifies high-level policies for the system using some policy model. Policy-based management can be used to manage context-aware media streaming to a desktop system or mobile device. An example policy is User A's favourite music is automatically started at the user's desktop machine as soon as the user enters room A. Once the system detects that the user has walked out of the room, the system will then pause or stop the music (Syukur et al., 2004).

Policy-based systems can be designed in a range of ways. Policies can be modelled using a range of semantic models from weak semantic models such as XML (Syukur et al., 2004) to strong ones based on expressive ontologies combined with logic reasoning such as KIF and OWL. For example, Tan et al. (2006) represent security policies for managing open distributed systems using KIF, the Knowledge Interchange Format.

Policy conflicts can arise because multiple policies may be triggered during the same point of an executing process. They can also arise during open system interaction when two previously separate systems overlap or need to be orchestrated into a composite system. Policy conflicts can cause policies to be triggered in a non-deterministic and inconsistent way, e.g., switch the heating off because the motion detector detects no one is in the room but there is an elderly person sleeping who may then get too cold. There are several ways to resolve policy conflicts. If policy rules have different priorities, higher priority rule can take precedence of lower policy rules.. Another method is that when multiple policies apply but they are of variable flexibility, the application of less flexible rule takes precedence. Further techniques to resolve policy conflicts include: analysis of policies with possibilities to merge policies, use of negotiation between parties to find a new agreement between different policies (Tan et al., 2006) and use of voting such as majority voting to agree on the policy of the majority.

12.2.8.5 Pervasive Work Flow Management for Services

Workflow as a means to compose and orchestrate services is discussed in Section 3.3.4. Montagut and Molva (2005) propose managing pervasive workflows in terms of distributed control and distributed task assignment. Here each device to be managed is assigned a role for the workflow and needs to have the local resources (fat client) to execute a local workflow engine. Devices and the services or work offered can be dynamically discovered and scheduled. It is not clear if and how distributed advertisements to describe device capabilities can be made. They do not synchronise tasks or handle faults in tasks. Ontologies can be used to define metadata descriptions of heterogeneous devices, to enable them to be interlinked and to allow devices to be related to user or application contexts and compose multiple devices and services according to context constraints and policies e.g., SMS over a phone call if the user is in a conference room (Maffioletti et al., 2004).

---

[25] One of the earliest access control models was the Bell-La Padula model proposed in 1973 (Bell, 2005).

12.2.9 Information Management

Information management is central to UbiCom system management. Any aspect of the system that needs to be managed, including application-specific operations and generic configuration, performance and security operations can be modelled as information and then managed using information management techniques. There are several information characteristics that need to be managed such as information volume, persistence, integrity, distribution, discovery, namespace and interoperability. These in turn are affected by information characteristics such as *hardness,*[26] *richness*[27] and *structure*[28] (Watson, 2006).

12.2.9.1 Information Applications

Data can be considered as the raw input into data processes whose output, the processed data, is called information. In practice, these simple definitions of data and information overlap. Although the focus on data management often seems to be about information storage, in reality it usefully focuses on information retrieval.[29] Data stored on physical storage resources is organised by the computer operating system file system and manager. Information applications such as hypertext information systems, email, news or work group systems, electronic calendars, and Relational Database Management System (RDBMS) applications such as transaction processing systems (TPS), use the OS file system to support data storage and retrieval.

The most basic type of information file structure or syntax is to just store data as a linear sequence of bits in the order they are created, e.g., audio and video streams. Each type of information application and information provider uses different data structures for storage and retrieval. In addition, the meaning, semantics or interpretation of the information is often implicit and it varies according to the particular usage or pragmatics. These heterogeneities in terms of syntax, semantics and pragmatics make it more complex to retrieve information and to use information. Generally the data content, even within the basic type of file structure, also includes various syntactical indices and metadata to aid data retrieval and management.

The individual information files from multiple applications are organised hierarchically into folders or directories by a MTOS according to user or application-specific categories, e.g., music, work projects, home projects, etc., and hierarchies across one or more devices. Files are created and retrieved via the File Manager application. Design issues in organising user information using the MTOS file system and file manager are described further in Section 12.3.3.

12.2.9.2 Rich Versus Lean and Soft Versus Hard Information

In terms of current ICT system support for managing data, this is more oriented to managing lean, hard data used for specific tactical tasks, using RDBMSs, HTML/XML byte sequenced and

---

[26] Hardness is a subjective measure of the accuracy and reliability of information. Information is hard if there is no ambiguity and information can be measured accurately, e.g., the price of gold in a particular stock market. The opposite of hardness is softness, e.g., Art.

[27] Richness refers to the ability of information and media to change human understanding, overcome differing conceptual frames of reference, or clarify ambiguous situations in a timely manner (Markus, 1994). The opposite of richness is leanness, e.g., mathematical data.

[28] Common data structures include linear, hierarchical, graphical and matrix.

[29] Data that is stored but never retrieved and has no explicit retrieval strategy seems to have little value for the amount of storage resources it consumes.

random-access files. However, Individual human use still often uses softer data, stored in hard paper formats for daily activities such as calendars, address books and to do lists. Organizations' information management has several focuses. Operational data such as business transactions and inventories of organizational assets such as employees, equipment, buildings are represented as lean, hard data, acquired in TPSs and stored in spread-sheets in file systems or RDBMSs. There is a variety of other operational data such as images of building plans, products and audio visual streamed data such as phone calls and video surveillance. The operational data is analysed to make decisions about how well different parts of the business are operating. Automated tools such as decision support systems and data mining are used to analyse the hard lean data. However, the softer, richer data often has to be manually analysed.

Organizations tend to be more explicit goal and policy driven when operating strategically rather than day to day. However, there is often an information gap between the desired goal and the present operational performance data and this is in part due to the difference in the higher softness and richness of the data needed for strategic goals compared to what is available for daily operational activities. In terms of the use of information in daily activities of individuals, there is a similar gap between the operational information available to support daily tasks versus the information needed to support long-term user activities and user goals. There are several key challenges here. First, soft and rich information needs to be represented so that it is machine-readable and machine-understandable and in order to automate more of the information processing needed to support individual activities and individual and organisation strategies and goals. A second key challenge is to support the full life-cycle of information processes for rich soft data to acquire and maintain such data. Third, the orchestration and choreography of multiple heterogeneous information structures (Section 3.3.4) need to be supported.

Often, because of the maturity, integrity management and great performance of the RDBMS model for information storage and retrieval, other types of, perhaps richer and softer information, representation such as knowledge or semantic information models (Section 8.4) are mapped to the RDBMS model to leverage these strengths of the RDBMS model for these other information. This requires a mapping of the other model to the RDBMS model versa for retrieval. Providing the cost and performance of these mappings are manageable, the superior data management support using RDBMS models will be beneficial.

12.2.9.3 Managing the Information Explosion

Operating increasing numbers of UbiCom system applications will require the ability to leverage and to cope with the data explosion from the increasing range and number of interactive devices that can sense the analogue physical world and can read and record multi-channel, multimedia content. For example, in the My e-Director 2012 project[30] (Poslad et al., 2009), the aim is to make more audio-video recorded information sources available from multiple camera angles and to allow users to select and orchestrate from which camera angles they will watch specific episodes in sports events, enabling them to see things from multiple perspectives. Although, this has the potential to enrich the viewing experience, it also generates huge amounts of data that need to be managed. This will also be affected by DRM and by data privacy issues if the interaction is personalised. Similarly, as more smart sensors become embedded in smart devices and smart environments, these can similarly generate increasing amounts of information that need to be

---

[30] The My-e-Director 2012, Real-Time Context-Aware and Personalised Media Streaming Environments for Large Scale Broadcasting Applications project concerns researching and developing more advanced interactive sports viewing for the 2012 Olympics, http://www.myedirector2012.eu/, accessed April 2008.

managed. Calculations of the volume of data must also take into account the volume of metadata used to describe the data, e.g., often audio information streams are recorded as metadata to describe video content streams.

Several studies have attempted to estimate the amount of information to be managed. Lyman and Varian (2003) estimate that the world produces new data at the rate of two to three Exabytes per year. If the world population is about 6.5 billion[31] people, each individual on average produces about a third of a billion bytes per year. Gantz *et al.* (2007) have estimated that in 2006 the digital universe was 161 Exabytes[32] and that by 2010. 70% of this information will be generated by individuals as opposed to organizations. Note also that only a tiny fraction of this is currently Web-based hypertext data, e.g., HTML and XML.

Another interesting estimate is the calculation of how much information is needed to record the personal experiences and sensory inputs of an individual throughout a typical lifetime of about eighty years – the personal memories scenario given in chapter 1. One estimate is that the data portion per individual over a lifetime is about 0.03 terabytes, but this not an estimate of continuously recording multi-sensory input (Lyman and Varian, 2003). Gordon Bell, with the assistance of others at Microsoft, is attempting to capture a lifetime's worth of personal information (Gemmell *et al.*, 2006). They estimate that one terabyte will hold a text-audio lifetime at twentieth-century resolutions and quantities. Dix (2002) estimated[33] that were someone to wear a 100 kbits/s video camera, recording continuously over a lifetime, they would record about 27.5 terabytes of AV data. Want and Pering (2003) have also estimated the capacity to record an individual lifetime of data. Assuming an average lifetime of 80 years, with a fraction of life awake of two-thirds, a compressed video sample rate of 512kbps, the storage capacity required is about 100 terabytes and that this could be stored in a 2.5 cm^2 IC by about 2012. It is not just the sheer volume of information that is a management challenge it is also the multitude of information channels which if each one required human decision-making would overwhelm our human ability to utilise this beneficially. In addition, the huge volume of metadata generated is also a huge management challenge.

12.2.9.4 Managing Multimedia Content

Multimedia[34] content management concerns: feature extraction from various media such as text, speech, music and video; feature integration into metadata media streams to enrich the interaction with multimedia streamed content that they are synchronised to; information retrieval of stored multimedia using multimedia metadata indices and management of multimedia metadata and data. There are several issues which make single non-alphanumeric text media data and multimedia content harder to manage compared to alphanumeric text management. Many stages of pre-processing of larger volumes of data are often needed to extract the media features of interest, such as the main colours in the background of an image and the average pitch of the introduction to a piece of music, to represent these as metadata indices.

---

[31] A billion here is taken to be a thousand million not a million million.

[32] An Exabyte is 10 bytes or one million Terabytes (10bytes).

[33] Each of these estimates is different in part because they make different assumptions of what information is accrued by an individual. Dix's estimate includes continual recording even when sleeping. Gemmell *et al.*'s (2006) estimate is about more active selective recording of lower resolution AV information by humans. No estimate is made of the capacity to store other human senses such as smell, taste and touch and to store the physical and emotional context in which these occur.

[34] The term multimedia sometimes refers to the use of single non-alphanumeric text content such as audio, image or video content or can refer to the simultaneous or combined use of multiple media content.

Multimedia content is implicitly rich and soft in nature but the automated analysis and classi-fication of multimedia data often generate leaner and harder indices or metrics – the so-called semantic content to syntactical classifier gap. Techniques to address this gap include mapping the low-level media features to high-level semantic concepts under human supervision. If two instances of multimedia content are compared and are classified to have similar low-level features, the high-level semantic annotation of one instance could also be assigned to the other instance. Another technique is to combine use of multiple multimedia indices which have different levels and degrees of semantics e.g., to combine the features extracted by language processing and semantic analysis of the text caption associated with the image with the extracted visual features of the image (Kesorn and Poslad, 2008).

In addition, because of the digital nature of multimedia content systems and because most multimedia metadata used to simplify queries is alphanumeric, there is a gap in mediating between non-alphanumeric multimedia content and the alphanumeric multimedia metadata or indices. Techniques to handle this gap include allowing queries to be expressed in the form of the content. So rather than users having to type the name of a song or artist in order to retrieve a piece of music, which has the limitation that the user may not know these or know them correctly, the user can, for example, try to hum or sing part of the piece of music (Ghias *et al.*, 1995). The system aims to match one segment of a piece of the music, to another piece of (stored) music and to retrieve music that is similar. The advantage for the user is that the query is expressed in the same type of media as the content itself. Vision and audio recognition enables humans to interact much more naturally with the physical world. For example, recording images of buildings and signposts we are facing enables us to use wireless mobile devices to input pictures as queries to locate our orientation and position and to identify and characterise some physical world object. Recording a sound bite can enable us to identify a type of animal or a piece of music.

In order to improve the performance of multimedia content matching, the content can be converted to alphanumeric content because alphanumeric searching and processing are much more efficient, in a way that is transparent to the user. Thus the pitch and rhythm dimensions of music could be mapped to text characters, enabling the musical words generated to be indexed using existing text search engines (Doraisamy and Rüger, 2004).

12.2.9.5 Managing Lean and Hard Data Using RDBMSs

RDBMSs are the system of choice for managing the integrity of hard, lean, attribute-based, application dependent, data in organisations. Data retrieval focuses on specifying known patterns and then trying to match the retrieved data to the pattern. RDBMSs are oriented to storing and querying factual instances of individual data entities stored as data tables, e.g., what kinds of printer are installed on a particular network, and to query cardinality relationships between the attributes of different data entities, e.g., which of all the printers can print images from the set of all the digital cameras at the full image resolution of the camera?

The main information management requirements for lean, hard data is to maintain the integrity of the data. Data integrity requirements can be subdivided into protecting the existence of data, maintaining the quality of data and ensuring data confidentiality. The existence of data can be protected using preventive measures which isolate the data and curative measures that support data backups and recovery. The quality of data can be maintained using access control techniques, integrity constraints, data validation and concurrency update control transactions mechanisms to manage access and changes by multiple users. Confidentiality can be handled using access control and encryption.

However, data in each RDBMS source is held and organised to support and maintain integrity in a local application-centric way. The RDBMS model is not designed to correlate data across multiple applications stored in multiple databases and to reuse data outside the applications that created their data. RDBMS system extensions such as distributed databases and data warehouses are needed to support the aggregation of data from multiple RDBMS. Additional processes such as data export, filtering, cleaning and transcoding are used so that the integrity of data sourced from multiple stores can be supported. In addition, because of the complexity of maintaining application processes and their interdependencies on their computation environment, data often has a longevity that could often usefully exceed the longevity of the type of application that created it. Hence, there is an increasing trend to move from network-centric and application-centric to data-centric management which is network and application agnostic.

RDBMSs are designed to enable multiple applications to organise their alphanumeric data entities into flat organisations of table data entities whose attributes are linked with cardinality relationships. Managing data the RDBMS way maintains data integrity for an application's data but loses the richer, e.g., hierarchical composition and class relationships between data entities. The RDBMS-type data organisation also does not support powerful searching across and interlinking multiple distributed application RDBMS data. Many RDBMS applications such as TPS automatically input data into RDBMSs. As far as the MTOS is concerned, RDBMS behaves as just another MTOS file system user, although the MTOS file system is largely hidden by the RDBMS from applications. Many applications store their information in their own data structures in a proprietary way not based upon RDBMS data structures and lacking its strong data integrity support.

12.2.9.6 Managing Metadata

Metadata, also referred to as *annotations*, is information to self-describe data. Some common types of metadata include: data file attributes[35] such as the file name, file size, date of creation or modification and extension; MPEG-2/4/7/21[36] video streams; RDBMS data schema. Ideally metadata should be stored with and bound to the data it describes. The limitations in using metadata are, first, that additional resources such as storage are required to compute the metadata, and to store the metadata. Second, additional write operations are needed to store the metadata along with the data. Third, metadata may be managed in a different way and become separated and disassociated with the data that they refer to, although the file system should maintain *syntactical bindings* between the data and metadata.

The main benefits of the use of metadata are, first, that it eases data discovery and data retrieval. Instead of reading and searching the whole of the data, the system can read and search the smaller metadata, e.g., Web search engines first search the metadata. Second, the availability of metadata aids the operational management of systems, e.g., if intermediate integrity check-sums are used with data transmission, any errors or interference can simple roll back data transmission to the last uncorrupted part rather than retransmitting from the start. Third, metadata can be used to promote interoperability because data is explicitly described, differences in data structures and semantics can be more easily analysed and resolved.

---

[35] In most OSs, core file attribute metadata for each file such as file size, etc. is stored in a separate special file called something like the Master File Table or MFT. In the Mac OS operating system, a resource fork file stores metadata about icons, the shapes of windows, etc., alongside data stored within the data fork file. MacOS metadata can be user-defined.

[36] MPEG, the ISO/IEC Moving Picture Experts Group specifies a range of specifications: MPEG-2 metadata supports encoding, decoding information; MPEG-4 metadata supports scene descriptions. MPEG-7 is metadata to describe the multimedia content in XML; MPEG-21 metadata supports machine-readable licence information i.e., to support DRM.

Metadata can be described by a range of attributes, some of which may be application-specific rather than generic, and can be classified in several ways. Two early classifications for metadata are the *ANSI 3-schema architecture* for database management systems and the intensional data versus extensional data classification (Mark and Roussopoulos, 1986). The ANSI 3-schema architecture consists of: the information meaning described in the conceptual schema or logical schema (upper layer), e.g., the relational data model; the external data representations described in external schema (middle layer), e.g., application-specific views of data such as using application-specific views or virtual tables in the relational model; the internal physical data structure layout described in the internal schema (bottom layer), e.g., indices that point to addresses of data in disk tracks. The various schema are often time-invariant or intensional data whereas the data instances of the data schema vary in time and are referred to as the extensional data component. The main benefit of this kind of layered metadata modelling for UbiCom systems is a separation of concerns. It separates and makes independent the descriptions of the data structure from the actual data instances, thus enabling many different extensional data and internal data schema to be used with a common intensional data and conceptual data schema. Mark and Roussopoulos (1986) describe a system architecture to support the use of explicit meta data models for RDBMSs.

Missier *et al.* (2007) discuss managing semantic, external schema, type metadata and in particular managing the semantic binding rather than syntactic binding of metadata to data. They describe a Semantic Binding Framework (SBF) whose design requirements are to support a uniform way to maintain correct associations among Grid resources, metadata, and knowledge entities whenever they change, as they evolve and to support access control to metadata. Their SBF does not, however, support dealing with the common interoperability problems of UbiCom systems when distributed metadata with heterogeneous semantics need to be harmonised. There are several examples of SBFs designed to support semantic interoperability. For example, Stjernholm *et al.* (2007) describe the use of an SBF with semantic interoperability to support distributed queries to heterogeneous environment databases. In an extension to this model, Poslad and Zuo (2008) show how a SBF can support multi-lateral user viewpoints (multiple external schema) of heterogeneous environment data.

12.3 Managing Smart Devices in Human User-Centred Environments

12.3.1 Managing Richer and Softer Data

Richer, softer data arises, for example, from natural language-based verbal and written conversations and monologues, from activities and experiences and from human expression in various forms of art. Much of this form of human knowledge, experiences and expertise is not in a form that is well defined or structured for digital storage and retrieval. The main way to manage this data in ICT systems is to digitally record and convert human actions and communication using various audio-video devices into sequences of bytes which are then stored in data files and managed by the operating system file system. These files are categorised and organised locally by different individual users.

12.3.2 Service Management Models for Human User and Physical Environments

The dominant service management model used in smart devices is that they use a Remote Service Access Points (RSAP) model. This is designed to advertise services in static and centralised directories and to bundle the services on offer at multi-service access portals that can be downloaded and installed locally in order to maintain services on the device. Other service management

Table 12.4 Seven different models for user-centred service management

Type of service management	Description	Who controls, Administers	Type of Architecture
Monolithic, stand-alone or off-line	Services installed offline	End-user, owner	Monolithic, unconnected
Services as Appliances	Dedicated device performs preset task. May get updated on demand	User, service provider	Monolithic, P2P
Remote Service Access Points (RSAP)	Services are delegated to providers who reside remotely and are accessed over a network	Service provider	Thin client-server SOA
Service Contract	Services are specified by service contracts	Service provider	Thin client-server SOA
User Service Pool	Services are entrusted to a community of users to be managed	User, third-party	Grid, MAS and P2P Model
Software as a Service (SaaS)	Services reside locally and are updated on-demand	Third-party	Fat client-server, SOA
Self-managing	Service components have more local autonomy to act and interact	Self-managing	Autonomic, self-organising

models exist which differ with respect to how services are accessed and how service changes are managed (Table 12.4). Each of these is considered in turn.

In the off-line model, *monolithic service* or *stand-alone service* model, services are installed on a general-purpose software platform off-line. Services are installed when the platform is first configured. If it needs to be reconfigured and if additional software is supplied on removal media and input locally. Such a service model may still be useful today in situations where strict access-control is needed.

A *Service Appliance*[37] is a type of appliance that specialises in information to perform a specific activity and that has the ability to share information with other appliances. A service appliance is easier to learn to use, operate and maintain because it is a single task device that comes pre-installed, hence its UI can be tailored for a specific task. This avoids the complication of multiple-service devices where one service can inadvertently interfere with another one because it locks access to a shared resource such as an IO channel when it is active. Examples of service appliances, that are essentially special-purpose computers, include video broadcast set-top box receivers, printers, digital cameras and vacuum cleaners.

Some of these can only be networked to specific servers such as a vacuum cleaner reporting faults to a service centres. Others service appliance devices such as printers, have displays that have heterogeneous network capabilities including wireless and are starting to emerge as service hubs, operating peer-to-peer services and taking on services that were traditionally the domain of the PC. Printers can connect directly to cameras and perform simple picture editing such as cropping photographs. Cameras can connect directly to larger display devices such as TVs to view recordings in high definition and high fidelity. As computers become much more widely used, they may become much more of a commodity. The focus then shifts to the use of peripheral devices and

---

[37] Norman (1999) refers to information (service) appliances, a term coined earlier in 1978 by Raskin. Service appliances and stand-alone service device model are similar, both are designed to perform specialist tasks, however, the former is also designed to share information whereas the latter is not.

appliances as the most used devices to access network services, rather than using MTOS-based devices as a hub. One of the key challenges with this model is to consider who controls and coordinates this peer to peer service appliance interaction.

In a remote service access point (RSAP) model, also called the Application Service Provider or ASP Model (Tao, 2001), service access is delegated by a consumer to a provider. A thin-client access device connects to a remote provider's server. The provider may be discovered once and remain throughout all user sessions, or a new provider may be discovered each time. A simple client-interface is used to invoke and pull services, which are maintained on a remote server. The motivation for this is often to shift the system complexity and management away from the user, towards the service provider. The downside is if service access is volatile because of intermittent network access or server overload, user access will freeze. In some cases it may not be apparent which particular service instance the user is able to bind to and should select, while in other cases it may not even need to be apparent to the user. There are three variations of the RSAP service: the *service contract* model, the *delegated service* model and the *service pool* model.

In the *service contract* model, the maintenance of services is more formally delegated to other ICT service providers in a market-place: this can be achieved using SLAs which use a contract to specify target levels or guaranteed levels of service (Section 12.2.8.3). Customers can specify redundant access to alternative suppliers if one fails but such SLAs and redundancy may be too costly for SME and home users.

A user *service pool* is where users have resources they only partially use and provide these to some community of service users, interacting with them at a peer-to-peer level rather than using client–server interaction. These providers are often just experienced users with a social desire to give something back to the community. This level of service often has no guarantees. For example, in *communities of practice* (CoP), a pool of services may be built up among friends and families to which they informally delegate service management. This helps maintain ICT facilities in the home, thus avoiding the cost of using commercial services. A final key benefit of the user service pool is that users, as the resource and service providers, have a wealth of experience of the wider practical issues that providers who are not users cannot so easily accrue.

In a *Software as a Service (SaaS) model*,[38] (part of) the application can run on a local service infrastructure and parts run remotely. When new service updates become available, service clients are notified of these and can be configured to either automatically or manually install the latest service updates over the network. This is a push-type interaction. This may also be performed over a wireless link while the user is mobile and then it is referred to as an Over-The-Air (OTA) installation. For example, it is common for many types of network device to get a periodic service update such as the mobile phone BIOS upgrades or on a computer: office software, security software, publishing software, Web software, OS software, BIOS software, etc. These can be performed in the background to be transparent but they can trigger ill-timed system reboots and in some cases cause data loss from applications as applications are not terminated properly. Users can also configure the use of a manual mode to control updates to prevent the problems of the automatic mode but this is more complex to administer.

In a *self-maintenance services* model, services may be self-upgrading but this occurs at the level of the individual application, e.g., anti-virus software, or at the operating system level. Services may often not be able to correct a fault because it is outside their domain of management. Generally, autonomic infrastructures in the home are not self-optimising and self-healing.

These service management models can be combined so that a laptop comes preconfigured with services, can operate off-line but can also operate in SaaS mode on the move. In the Monolithic

---

[38] SaaS is also called a Application Service Provider (ASP) or on-Demand Service model

SaaS and remote-access models, a general purpose software platform such as a MTOS computer is used and the user or owner of the computer can decide which services can be downloaded. Here, the flexibility needs to be traded off against the maintenance complexity. For some users, the complexity of learning to use, to operate and to manage such systems is overwhelming because users may attempt to install services that are incompatible or too fragile to use with a particular service infrastructure. Two types of service management model are: (1) to put the know-how and 'intelligence' to manage services into a generic service infrastructure, e.g., the Grid model; and (2) to use biologically inspired management. These are dealt with elsewhere in this chapter.

12.3.3 User Task and Activity-Based Management

Much of what is termed personal computers and personal computing is more suited towards office workers who work on single fixed tasks, in a relatively uninterrupted manner, for long periods of time. In contrast there are other types of worker activity, that are prone to be interrupted, nomadic, of short duration and where multiple user activities are likely to be interleaved and used to achieve multiple user goals. In user-centred services, users' context for ICT events and service reconfiguration can be expressed at multiple knowledge viewpoints, e.g., using the mental model of different users. HCI support for this is described in Section 5.5. Knowledge-based models to support cross-device use and cross-activity use of a user context is described in Chapter 9.

12.3.4 Privacy Management

Violation of individual privacy is an oft quoted peril of UbiCom. Privacy is a type of state in which a person's identity and personal information are kept confidential from others. In order to interact in society and in business in several specific ways, such as ecommerce or voting to elect representatives, partial privacy rather than complete privacy is used. Here a person consents and entrust others with their identity and personal information. Full privacy in society is regarded by some as equally perilous as a lack of privacy. People are more likely to perform in a less responsible manner if they are able to escape the consequences of their actions. Hence, shared public and private virtual spaces and physical spaces are often designed and operated to support surveillance by authorised and responsible representatives[39] for legal purposes. There are generally several privacy concerns:

- *Anonymity versus Authentication. Anonymity* means other users are unable to determine the identity of a user bound to a subject or operation. Often, the user consents to release their identity to specified others, to be authenticated, e.g., for others to link something to them such as a payment transfer. In between identity and anonymity lies *Pseudonymity*[40] in which users and or subjects are unable to determine the identity of a user bound to a subject or operation, but this

---

[39] There are many societal issues here concerning the individual's right to privacy and use of consent to give up some privacy. One issue concerns which authorities such as law enforcement agencies have a legal right to invade personal privacy and with whom they share this information. The UK government in 2008 proposed that it would be beneficial to share personal information about its lowest income citizens with utility companies to enable the utility companies to match the most affordable tariffs to its customers. Another issue concerns *trust*, and preventing abuse of privacy by authorities.

[40] Goldberg (2000) distinguished a range of four levels of *nymity*: *anonymity*, *non-reversible pseudonymity* (ID not tied to a true identity), *reversible pseudonymity* (ID masked but tied to a true identity, also referred to as a *partial identity*) and identity.

user is still held accountable for its actions. *Identity management* concerns managing various (pseudonyms) of an individual person, i.e., administration of identity attributes including the development and choice of the partial identity and pseudonym to be (re-)used within a specific context or role.

- *Unlinkability*: Users are unable to determine whether a specific user caused certain specific operations in the system.
- *Unobservability*: Users cannot determine whether or not an operation is being performed.
- *Notifications, rights and consent*: Users have a right to be notified about the personal information collected by them and to give consent for its use.

The profusion of smart environment devices means that humans can be identified, tracked and profiled to a greater degree throughout the physical environment, e.g., location tracking of devices without the owner's consent. As human use smart mobile personal devices in more interactions with UbiCom systems and in more environments, there is more scope to leave identifying virtual trails that either singly or through being amassed and data mined can identify who people are and what are their behaviours referenced in time and space.

More interactions occur over shared physical networks and shared service and social spaces. It is also possible to sense smaller amounts of physical trails with a greater degree of sensitivity and accuracy, e.g., DNA profiling. Human behaviour can be tracked and observed without the human subjects knowing that their behavioural patterns can be analysed and can predict their behaviour. Making multiple nested selective queries can be made to compile *minority reports*.[41] Data mining techniques can be used to analyse sales data; it is routinely used to make recommendations to buy books, music, movies and toys (Han and Kamber, 2006). Businesses in the interest of reducing business wastage could predict and pre-empt when an employee will leave (Kasanoff, 2001). Tracking, or even stalking, someone is easier if you have access to their personal model. Psychological profiling of suspects can be used to identify and catch criminals. Businesses realise that to lure and enrich customer service experience, they must provide personalisation as part of their *customer relationship management* (*CRM*) and to support *one-to-one marketing*.

There are several potential safeguards to protect personal privacy. Privacy Enhanced Technologies (PET) can be used but it is complex and supports anonymity that is not suitable for all situations. Platforms based upon trust in service providers should adhere to the policies for the access control they advertise. Privacy legislation can in addition offer some protection but this is often bound to particular geographical regions. Titkov *et al.* (2006) and Price *et al.* (2005) have identified that a multilateral approach to personal privacy is needed that specifies: the regulatory regime they are currently in, the type of UbiCom service required, the type of data being disclosed, and their personal privacy policy. Each of these is discussed in more detail below.

12.3.4.1 Biometric User Identification

Biometric identification systems identify people by a unique biological characteristic that is inherent and bound to a individual or make us who we are, not by what we know (e.g., passwords) or by what we possess (e.g., a certificate, token, or key). Biometric systems can be based upon behaviour such as gait, signature, voice or keyboard typing, and on *physiological traits* such as retina pattern, fingerprint, DNA, palm and face (Table 12.5). Biometric identification is a useful component for

---

[41] A minority report here refers to a selection or series of one or more queries for information that can be used to identify a unique group or individual, in some cases circumventing access controls specifically designed to prevent this. A neat example of this is given in Watson (2006, p. 557).

Table 12.5 Different types of biometric identification

Type of ID or recognition	Characteristic
Physiological	
Face	Non-intrusive, capture using image or video camera, then extract features, and then match against stored records; can be used in crowds; sometimes low accuracy $\sim 50\%$ Face ID is affected by posture and face expression. Faces are not unique, e.g., identical twins, ID could be cloned
Fingerprint:	Set of ridges on the finger tip: Intrusive: requires special scanner. Left by contact between finger and a firm surface. ID is unique but could be cloned
Palm print:	Pattern of lines from veins, wrinkles, ridges on the inner surface of a hand between wrist and fingers). Intrusive: requires special scanner
Retina:	Patterns of blood vessels in thin tissue at back of eye. Intrusive: requires special scanner ID is unique even for identical twins
DNA:	Highly variable repeating sequences of DNA called mini-satellites at specific loci Intrusive: DNA can be extracted from blood, semen, skin, saliva or hair left behind by a person. A fairly unique ID
Prints of other body parts	Teeth and bite marks, footprint, lip prints, ear prints can also be used. ID uniqueness can be less, particularly in large groups, depending on the type of print
Behavioural	
Signature:	Name is written by a fluid gesture of hand. Intrusive; signatures can vary; ID check is often manual and has very variable accuracy. ID can be cloned with practice
Voice	Non-intrusive; use in one-to-one interaction. ID is affected by variations in voice over time, with diseases that affect the vocal cords, with social environment and encoder distortion when using electronic mediated voice
Other:	Indirect, pattern-based behavioural profiling. Requires access to user's environment and a recording of some duration of the user interaction Inexpensive capture; just log key-strokes
Keystrokes Walk	Requires kinematic capture: cameras to capture movement of the body in space (*kinematics*) and force plates to measure the forces involved in producing these movements (*kinetics*). ID uniqueness within large groups of people not clear

context-aware systems as it represents a potentially less obtrusive and natural way to identify a person and hence as a focus to automatically adapt devices' functions. Signature-based identification has been used routinely in business such as banking for many years but its accuracy is variable. Fingerprint identification has been routinely used in the fight against crime for many years. This is being complemented by the increasing use of *DNA profiling*[42] to identify people (Butler, 2005). The more accurate biometric identification techniques such as fingerprint scans to access devices and retinal scans at airports are becoming much more routinely used for access control.

Biometric identification has several main advantages including difficulty of ID cloning or forging, thus hindering identify theft and also the credentials cannot be forgotten by the individual. The disadvantages of such a system are that the individual can get inadvertently physically or

---

[42] DNA samples are being collected from convicted offenders and suspects, even if later found innocent, to retain samples in a database for cross-comparison against DNA profiles of biological evidence collected at any new crime scene. Some people even argue that we should all be DNA profiled, if we are innocent, it can help rule us out of being a suspect. DNA has been used in legal courts since 1988.

physiologically damaged, rendering the identification unusable. In some cases, accuracy is variable as the identifying characteristics vary over time, with a human's physical condition, stress, pose and social environment. Some biometric prints, e.g., DNA, can be relatively easily taken and used without permission of the owner.

Pattern-based observations of users' behaviours to construct user profiles were discussed in Chapter 5 on HCI and in Chapter 8 on context-awareness. The use of pattern-based behaviour for identification is an extension of this. The design issues for pattern-based biometric techniques, focusing on face and voice identification and keystroke analysis, are discussed by Kung *et al.* (2004) and Monrose and Rubin (1997) respectively.

Biometric identification can also involve *content-based feature extraction* and classification (Figure 12.5). This typically involves processing a biometric print to extract a multi-dimensional set of features or vectors represented by a mapping from the biometric print space to the feature space, for example, face recognition may involve identifying multiple feature dimensions involving head, eyes, mouth and nose and inter-relations. Often to ease analysis, the set of feature dimensions needs to be reduced. Techniques for *feature reduction* can be classed into *unsupervised projection methods* such as PCA (Principle Component Analysis) and ICA (Independent Component Analysis) or *supervised projection methods* such as LDA (Linear Discrimination Analysis) and SVM (Support Vector Machines).

The design trade-offs for biometric identification (Kung *et al.*, 2004) include balancing the false rejection rate and false acceptance rate; balancing accuracy, convenience and intrusiveness (increases stress); combining recognition and verification techniques, e.g., to use a first security (token-based) identification techniques for identification and then to use a second (biometric) technique for verification of the identity; centralised versus distributed matching of ID instance to ID database; time to complete identification, data storage requirements and compatibility between extractor and classifier.

12.3.4.2 Privacy-Invasive Technologies versus Privacy-Enhanced Technologies

Smart devices can broadly be classified into whether or not they invade privacy, Privacy-Invasive Technologies (PIT), or enhance privacy, Privacy Enhanced Technologies (PET) (Clarke, 2001). However, in practice, there is more of a range from strong PET, through weak PET, to weak PIT, to strong PIT. Strong PET supports anonymity or non-reversible pseudonymity, unlinkability and unobservability. Unobservability can be supported by incorporating into the system some loss or statistical variation into the system, so that not every piece of information is directly accessible or recoverable.

Most current systems rely on the use of trusted third parties, e.g., the use of public key cryptography infrastructures, to protect the privacy of someone and in general the security of

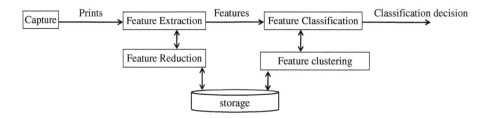

Figure 12.5 Block diagram for a content-based feature recognition and identification system

assets. Additional trusted third parties such as centralised client proxies are used as intermediaries to simplify access to services such as network providers, to access remote services. The idea of strong PET systems is to avoid the use of third party trust. One the earliest examples of the use of strong PET was to support email confidentiality (Chaum, 1981). A good overview of the use of PET is provided by Federrath (2005) which includes the use of MIXes, blind signature schemes and Onion Routing. Weak PETs are designed to support reversible pseudonymity, unlinkability and unobservability. Mobile phone communication is designed to support weak PET. The mobile phone ISP is trusted to reverse the pseudonym to an identity only for billing purposes and not to divulge these to others.

Saponas *et al.* (2007) analyse how three different types of UbiCom device such as the Nike + iPod Sport Kit, Slingbox Pro and Microsoft Zunes, designed to act as a weak PET, in practice end up being used as a weak PIT because they allow others to configure access to them and allow others to track what users are doing. The Nike + iPod Sport Kit consists of a wireless sensor that a user puts in one of his or her shoes and a receiver that she attaches to her iPod Nano. When the user walks or runs, the sensor wirelessly transmits information to the receiver, tagged using a unique identifier. This information can be eavesdropped and users' locations can be tracked. The Slingbox Pro allows users to remotely view, to sling or *place-shift*, the contents of their TV over the Internet. This content can be profiled even if message encryption is used in the case of the Slingbox. Microsoft's Zune is a portable digital media player with WiFi which supports ad hoc P2P type sharing of pictures and songs with others, even complete strangers. Zune devices can specify and block other Zune devices from accessing them but this can be circumvented.

Strong PIT is actively designed to support identity, linkability and observability of people within a service domain. A good example of strong PITs are many types of smart cards (Section 4.2) such as store cards and travel cards.

12.3.4.3 Entrusted Regulation of User Privacy to Service Providers

The Platform for Privacy Preferences Project, P3P, developed by the World Wide Web Consortium (W3C), has been recommended as an industry standard, providing a simple, automated way for users to gain more control over the use of personal information on websites that they visit. At its most basic level, P3P is a standardised set of multiple-choice questions, covering all the major aspects of a website's privacy policies. Taken together, they present a clear snapshot of how a site handles personal information about its users. P3P-enabled websites make this information available in a standard, machine-readable format. P3P-enabled browsers can 'read' this snapshot automatically and compare it to the consumer's own set of privacy preferences. P3P enhances user control by putting privacy policies where users can find them, in a form users can understand, and, most importantly, enables users to act on what they see. The P3P protocol has two parts: privacy policies and privacy preferences. Privacy policies is an XML format in which a website can encode its data-collection and data-use practices, and Privacy Preferences is a machine-readable specification of a user's preferences that can be programmatically compared against a privacy policy.

P3P has several weaknesses. The current P3P standard only provides a mechanism for websites to state their intentions regarding use of the personal information that they collect. Users need to trust providers to act responsibly concerning privacy. Mechanisms for enforcing that sites act according to their stated policies are beyond its scope. The process of sharing privacy information is server oriented. More specifically, in P3P, evaluation is done by matching user references against server provider policy, where the policy is managed by service providers. Users have no choice here but have to trust service provider completely. Hogben *et al.* (2002). however, discuss how to design a P3P system to comply with privacy legislation such as the EU privacy directive (Section 12.3.4.4) Since, P3P does not support authentication, there is no clear way to determine the legitimacy of the statements listed in service provider policy. The P3P specification only describes the meaning of a policy that restricts itself to the most primitive case. Complicated cases concerning privacy conflicts are not sufficiently addressed. The issue of pseudonymous and anonymous use of data is also largely unclear.

12.3.4.4 Legislative Approaches to Privacy

In a legislative approach to privacy, collectors of personal information are legally bound to provide a suitable means of notice and consent to users. However, there are differences in legislative approaches to privacy internationally, which complicates the legal transborder flow of information and which means that different designs for ICT systems to support this are needed. The European Data Privacy Directive incorporates a unique *opt-in* provision that presumes an expectation of data privacy as the default position in which users need to give consent for personal information to be collected. In this model, the approach is about deciding which personal information to collect and about interpreting and mapping the legislative rules in terms of service policies and building an appropriate service infrastructure to enforce these rules.

Whereas, for example, in the US, data collectors presume consent, and require an affirmative *opt-out* by the user not to collect personal information Under the US system, part of the focus is on how users perceive when they want to opt out because collectors of privacy information are invading their privacy. It has been shown, through work by Anne Adams, quoted in Lederer *et al.* (2002), that four interdependent factors determine an individual's perception of privacy: the information recipient, the intended usage of the information, the information sensitivity, and the context in which the disclosure occurs.

One of the challenges with the opt-out system, and to an extent with the opt-in system, is how to deal with the potential explosion of privacy notification and rights agreement or refusal events that could occur. These could easily overwhelm users of UbiCom systems. Lederer *et al.* (2002) propose using abstractions called faces to group together different sets of user-understandable privacy preferences, such as the secure shopper, cocktail party, hanging out with friends faces, etc, and then using these to select the level of privacy.

12.4 Managing Smart Devices in Physical Environments

Managing smart environments refers to physical environments which are smart because they are embedded with a range of smart devices or contain untethered smart devices that interlink into more pervasive virtual ICT environments and that enable humans to interact in richer ways with the physical environment. These devices in the physical environment are in turn smart because in part they can be designed to mimic the use of natural and physical world behaviours as useful and usable design models for UbiCom device behaviour. These are discussed in turn. The wider issues of whether designing too much active intelligence into smart environment devices makes them manifest physical problems into virtual problems and the rebound effect and change in behaviour in humans and the natural and physical world such as machines making people less human and machines changing social norms, are discussed elsewhere (Section 9.2.3.2).

12.4.1 Context-Awareness

Ubiquitous computing provides a variety of components and techniques to manage human behaviour within the physical world. Smart tags and sensors can provide finely-grained information to monitor people, with respect to their situations and to compile a history of usage, e.g., is a person driving safely or competent at parking their car? If systems were more aware of their operating context, they could elect to externalise more services and to reduce their local internal capacity for storing goods and services, when it is more cost-effective to do so but this makes a system more dependent and less autonomous. However, this is not necessarily a problem in a system that has an abundance of environment services. Systems could seek to dynamically self-adapt and to self-optimise their service provision, e.g., to charge more for transport when the weather is dark or wet.

There are two aspects of context management considered here: first, using context-awareness itself to improve management of systems used for physical world activities and for human world activities; second, the operational management of context-awareness throughout its lifecycle (Section 7.2).

12.4.1.1 Context-Aware Management of Physical and Human Activities

Several core applications of context-aware type systems to aid the management of the use of UbiCom systems in human and physical world environments have already been discussed such as location-aware management of mobile goods and users to improve distribution or to reach a destination (Section 7.4), service personalization (Section 5.7.4) and ICT system management such as managing communication based upon the recipient's ICT context (Section 7.6).

Context-aware Power Management (CAPM) aims to minimise the overall electricity consumption of a building, using context information such as location, while maintaining acceptable user-perceived device performance (Harris and Cahill, 2005). Their experimental trials revealed that location alone is not enough for effective power management. Composite contexts from multi-modal sensors are needed to determine finer grained user behaviour for effective power management. For example, a simple acoustic sensor could potentially differentiate different kinds of user behaviour: the office is quiet versus multiple people are talking versus the door is opening, etc. Processing this data, for example, by the use of Bayesian networks to differentiate and predict user behaviour patterns, will have an additional cost in both CPU cycles and hence energy consumed.

Context-aware Access Control aims to control access to resources based upon the context such as remote resource discovery enquiries, remote resource queries and remote resource invocation support connections with anyone, anywhere. In some cases, this is beneficial because it can allow local specialised resources to be sustained by a larger critical mass of users, the global base of all specialised users, rather than an otherwise small group of local specialised users. Alternatively, this can lead to otherwise local resources being flooded with remote access requests.

Location-aware access control or location-based access control is a means to filter and limit interaction and service enquiries and invocation to local senders and requesters. The motivation for this may be because the services on offer are only valid or fresh locally. There are a range of techniques to support location-aware access control. Intuitively, advertisements using local area (network) communication may seem a natural way to make services known only locally, by taking into account users' physical presence such as location when determining their access privileges. However, with the advent of application gateways, e.g., the Internet can propagate a local presence to far wider areas. Ardagna *et al.* (2006) have described the use of location-based credentials to limit access, while Bhatti *et al.* (2008) describe the use of a policy-based approach administrative tool that helps define access control related to user preference level and provider constraints. Person-based Access Control (PBAC) defines access control based upon a person's ID or personal preferences.

12.4.1.2 Management of Contexts and Events

A good analysis of the user and system management issues of context-aware is given by Van Bunningen *et al.* (2005). Context adaptation means the system proactively processes information on behalf of a user so that an action can be taken without requiring his or her attention – proactive or active context-awareness. Ideally this proactive decision should be understandable and controllable by users if they are present, i.e., is tractable, so that a user can see why something (proactive) happened. The minimal way to support this is to either continually display the changing context or to display it on request. The context can be displayed using visual cues in an abstraction that the user readily understands, e.g., display the position of moving goods or people on a map. The system may also need to give users the ability to modify the context when the system or user detects or suspects that it has been incorrectly determined.

Context management involves managing volatile connections to context sources and sinks. Flexible context representation mechanism is needed so as to provide conversion between different kinds of context information. Higher level rather than lower level contexts are stored. This has several advantages. First, to reduce storage space by, only storing the high level context (e.g. being in a meeting), instead of storing all low-level sensor information. A second advantage is that at a higher level, more computing power is available to do data compression. Third, context dependencies can be exploited to optimise storage by not storing derivable contexts. Designers of context-aware systems need to consider which contexts they should store and for how long and where it should be stored. Note the storage predictions for storing the visual context of users can lead to data storage requirements of the order of one to 100 terabytes if contexts are to be permanently kept (Section 12.1.1.1).

One of the main challenges in managing context-awareness is to perform context adaptation when faced with uncertainty, ambiguities, contradictions, and other logical inconsistencies in conflict during the context-awareness life-cycle (Section 7.2.8). Data mining techniques can be used to predict categorical (discrete, unordered) labels. Prediction models can be used to analyse past continuous data trends to predict unknown future data that follows that trend (Han and Kamber, 2006). Classification techniques can be based upon unsupervised learning, e.g., use of data clustering to automatically derive classifier labels, or on supervised learning, or reinforcement learning algorithms in which a supervisor derives the classifier labels. An example of the use of a supervised learning algorithm, decision tree induction to classify the user activity is shown in Figure 12.6 which illustrates the use of composite context. Design issues when classifying contexts are the ordering to apply the classifier decisions for the individual contexts, overlapping and conflicting decisions between classifiers and handling classifier uncertainty. Note that this assumes the use of proposition-type logic decisions which return true or false. If decisions are more conditional, then probabilistic classifier methods must be used such as a Bayesian classification.

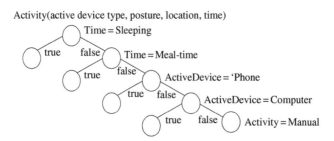

Figure 12.6 Classifying user activity as a composite context based upon a decision tree for individual contexts

12.4.2 Micro and Nano-Sized Devices

A critical part of the design of micro devices is very efficient power management in terms of possibly, renewable, power storage and power consumption for both sensing, processing and data transmission (Section 4.3.4). The complexity of design and verification is significant as the number of available transistors grows, exponentially increasing the productivity gap between the two. This has been historically tackled by employing larger and larger design and verification teams, but human resources are economically hard to scale. More automated tools will be needed for verification of micro and nano devices but these need to be linked directly to the implementation (fabrication) process rather than being separated as usual in most system designs (Goldstein, 2005). Another view is that defects may increase with micro-fabrication but because fabrication costs are low, multiple units may have enough units, providing the tools exist for defects that can be verified, isolated and ignored.

Micro and nano devices may be untethered, i.e., are airborne or can be fluid-borne. These depend on their physical environment for distribution and movement. Their movement cannot be controlled like macro-sized mobile devices,[43] instead they can be tracked (Römer, 2003). Again this needs to be low-energy on the device side otherwise the micro will expend all its energy in signalling for tracking. Some small devices are simply fixed in the environment or embedded within larger systems.

Micro and nano devices are harder to contain within a bounded physical space. There are some concerns about the environmental impact of such tiny devices that are not visible to the naked eye, how to track and control their spread and to manage the lifecycle from introduction, though operation, through retirement and reuse. The localised nature of these devices, the need to disperse a representative sample of these devices in a wider population of the physical environment and the anticipated failure rate of some individual devices mean that these devices are likely to be employed massively in parallel. The consequence of this is that it may be quite difficult to remove all instances of such tiny devices, leaving some or many in circulation that are dead or active. Such micro-devices could be more easily ingested by humans and then reside permanently in the body or absorbed into food chain. There are also concerns about the ability of these tiny devices to self-configure, self-replicate and to produce new unanticipated emergent behavioural norms (Section 10.5).

12.4.3 Unattended Embedded Devices

Embedded micro devices and macro devices often need to be left unattended for long periods, in relatively inaccessible environments, e.g., pace-makers that are implanted in the human body and remote sensors left in uninhabited physical environments. Unattended embedded devices[44] that are used for control, e.g., pace-maker implants, require stable timing to deliver control signals at set times, over time. The security design of devices to be tamper-resistant or at least tamper-evident is a key concern. As has already been mentioned, protecting specific parts of a device may be insufficient. Some threats can manipulate the local environment to cause the device to malfunction such as heating or freezing the environment. Hence a multi-lateral approach is needed to protect unattended devices, e.g., use of materials resistant to physical attacks, use of counter-measures and other corrective measures it tampering attacks are detected, and the use of *a priori* preventive measures such as encryption to lessen threats.

---

[43] We can refer to this as passive mobility, as opposed to active mobility where the mobility is activity controlled and managed by the owner of the device or some device controller such as a mobile robot.

[44] In addition to embedded devices being left unattended in the physical environment, other resources such as information, e.g., messages for others (stigmergy), can be left unattended in the physical environment (Section 13.5.1). This introduces some similar security concerns for unattended devices.

If embedded devices are tampered with and they fail, it is best if malicious or inadvertent failures are designed to be self-contained, so that if they fail safe, damage to their environment is minimised. This is also vital for smart environment devices which are safety critical. Inadvertent failures can also occur because although environment devices operate safely alone, they can still interfere with each other to fail. Embedded system devices should be designed to operate with high availability, high reliability and high stability.

EXERCISES

1. Compare and contrast the following management systems and protocols for a range of UbiCom systems such as dust, tab, pad and board-sized: ICMP-based, SNMP based, MTOS-based, RDBMS and Web/Internet protocols.
2. Compare and contrast security management versus safety management versus privacy management.
3. Compare and contrast security designs for different types of mobile devices such as mobile phone, SatNav device, AV remote control and smart cards. Could a common set of security requirements and design be used across this device range? What are the pros and cons of reusing the security design of one type of device in other types of device? E.g., using the current mobile phone security safeguards to protect SatNav devices?
4. Discuss how the imprinting model of Stajano (2002) can be combined with Body Area Networks (Section 11.7.4) to support different security support for personal mobile devices.
5. Discuss how to manage the safety of SCADA system using fault prediction techniques.
6. What is meant by service-oriented management? Discuss the following techniques for service-oriented management based upon grid resource management, policy-based service management, service-level agreements and pervasive workflows.
7. Differentiate between the management of hard and lean data versus soft and rich data versus the management.
8. Estimate how much information you personally have generated last year. How has it changed over the last 10 years? How will your data management techniques, if at all, change in the future?
9. Critique the design of MTOS of the leading vendors in terms of their suitability for use across devices, to support many interaction, to support the concept of a personal information cloud, to support a seamless shift between a personal and social system, to support participation in multiple activities.
10. Outline the design of an UbiCom system to maintain user privacy for smart dust, smart cards, smart phones, smart laptops and smart boards. Compare and contrast design for your different designs.

References

Anderson, R. Bergadano, F., Crispo, B. *et al.* (1998) A new family of authentication protocols. *ACM Operating Systems Review (SIGOPS)*, 32(4): 9–20.

Anisetti, M., Ardagna, C.A., Bellandi, V. *et al.* (2007) *OpenAmbient*: a pervasive access control architecture. In A.U. Schmidt, M. Kreutzer and R. Accorsi (eds) Long-term and dynamical aspects of information security: emerging trends. In *Information and Communication Security*. Nova Science Publisher Inc.

Ardagna, C.A., Cremonini, M., Damiani, E., *et al.* (2006) Supporting location-based conditions in access control policies. In *Proceedings* of *ACM Symposium on Information, Computer, and Communications Security*, pp. 212–222.

Aydt, H., Turner, S.J., Cai, W., *et al.* (2008) Symbiotic simulation systems: an extended definition motivated by symbiosis in biology. 22nd Workshop on Principles of Advanced and Distributed Simulation, PADS '08: 109–116.

Bell, D. (2005) Looking back at the Bell-La Padula model. In *Proceedings of 21st Annual Computer Security Applications Conference*, pp. 337–351.

Berman F. (1999) High-performance schedulers. In I. Foster and C. Kesselman (eds) *The Grid: Blueprint for a New Computing Infrastructure*. San Francisco: Morgan Kaufmann, pp. 279–310.

Bhatti, R., Damiani, M.L., Bettis, D.W., *et al.* (2008) Policy mapper: administering location-based access-control policies. *Internet Computing*, 12(2): 38–45.

Bigham, J., Gamez, D. and Lu, N. (2003) Safeguarding SCADA system with anomaly detection. In Proceedings of 2nd International Workshop on Mathematical Methods, Models and Architectures for Computer Networks Security (MMM-ACNS'03), *Lecture Notes in Computer Science*, 2776: 171–182.

Butler, J.M. (2005) *Forensic DNA Typing: Biology, Technology and Genetics of STR Markers*, 2nd edn. Elsevier Academic Press.

Capra, L., Emmerich, W. and Mascolo, C. (2005) CARISMA: Context-Aware Reflective mIddleware System for Mobile Applications. *IEEE Transactions On Software Engineering*, 29(10): 929–945.

Chaum, D. (1981) Untraceable electronic mail, return addresses, and digital pseudonyms. *Communications of the ACM*, 4(2): 84–88.

Chetan, S., Ranganathan, A. and Campbell, R. (2005) Towards fault tolerance pervasive computing. *IEEE Technology and Society*, 24(1): 38–44.

Clarke, R. (2001) Introducing PITs and PETs: technologies affecting privacy. *Privacy Law & Policy Reporter*, 7(9): 181–183. Available on-line from http://www.anu.edu.au/people/Roger.Clarke/DV/PITsPETs.html, accessed Jan. 2008.

Clausen, M. and Kurth, F. (2004) A unified approach to content-based and fault-tolerant music recognition. *IEEE Transactions on Multimedia*, 6(5): 717–731.

Daly, D., Kar, G. and Sanders, W.H. (2002) Modeling of service-level agreements for composed services. In M/Feridun *et al.* (eds) Proceedings of DSOM 2002, *Lecture Notes in Computer Science* (LNCS) 2506: 4–15.

Dennis, A. (2002) *Networking in the Internet Age*. Chichester: John Wiley & Sons, Inc, pp. 243–270.

Dingledine, R., Freedman, M. and Rubin, A. (2001) Free haven. In A. Oram (ed.) *Peer-to-Peer: Harnessing the Power of Disruptive Technologies*. New York: O'Reilly, pp. 159–187.

Dix, A. (2002) The ultimate interface and the sums of life? *Interfaces*. 50: 16.

Doraisamy, S. and Rüger, S. (2004) A polyphonic retrieval system using n-grams. *5th International Conference on Music Information Retrieval (ISMIR, 2004)*: 204–209.

Duflos, S., Diaz, G., Gay, V., *et al.* (2002) A comparative study of policy specification languages for secure distributed applications. *Lecture Notes in Computer Science (LNCS)*, 2506: 157–168.

Federrath, H. (2005) Privacy enhanced technologies: methods – markets – misuse. In *Proceedings 2nd International Conference on Trust, Privacy, and Security in Digital Business* (TrustBus '05), LNCS 3592, pp. 1–9.

Gantz, J.F., Reinsel, D., Chute, C., *et al.* (2007) The Expanding Digital Universe: A Forecast of Worldwide Information Growth Through 2010. IDC white paper. Retrieved from http://www.emc.com/about/destination/digital_universe, on 2007–09.

Gemmell, J., Bell, G. and Lueder, R. (2006) MyLifeBits: a personal database for everything. *Communications of the ACM*, 49(1): 88–95.

Ghias, A., Logan, J., Chamberlin, D., *et al.* (1995) Query by humming: musical information retrieval in an audio database. In *Proceedings of 3rd ACM Int Conference on Multimedia*, pp. 231–236.

Goldberg, I. (2000) A pseudonymous communications infrastructure for the Internet, PhD thesis, University of California at Berkeley.

Goldstein, S.C. (2005) The impact of the nanoscale on computing systems. In *Proceedings of IEEE/ACM International Conference on Computer-aided design*, ICCAD '05, pp. 654–660.

Gollmann, D. (2005) *Computer Security*. New York: John Wiley & Sons, Ltd.

Han, J. and Kamber M. (2006) *Data Mining: Concepts and Techniques*, 2nd edn. San Francisco: Morgan Kaufmann Publishers, pp. 285–382.

Harris, C. and Cahill, V. (2005) Exploiting user behaviour for context-aware power management. IEEE International Conference on Wireless and Mobile Computing, Networking and Communications (WiMob'2005), 4: 122–130.

Hogben, G., Jackson, T. and Wilikens M. (2002) A fully compliant research implementation of the P3P Standard for privacy protection: experiences and recommendations. *Lecture Notes in Computer Science*, 2502: 104–125.

ISO/IEC (1989) Open Systems Interconnect Basic Reference Model. Part 4 – management framework. Document No. ISO/IEC 7498-4. Retrieved from http://www.iso.org/iso/home.htm, Sept. 2007.

ITU-T, Telecommunication Standardisation Sector of the ITU (2000) M.3400 TMN management functions. Retrieved from http://www.itu.int/, Oct. 2007.

Kagal, L., Korolev, V., Avancha, S., *et al.* (2002) Centaurus: an infrastructure for service management in ubiquitous computing environments., *Wireless Networks*, 8: 619–635.

Kalasapur, S., Kumar, M. and Shirazi, B. (2006) Evaluating service oriented architectures (SOA) in pervasive computing. In *Proceedings of 4th Annual IEEE International Conference on Pervasive Computing and Communications* (PerCom 2006), pp. 276–285.

Kasanoff, B. (2001) *Making it Personal: How to Profit from Personalisation without Invading Privacy*. New York: Perseus Books.

Kesorn, K. and Poslad, S. (2008) Use of semantic enhancements to NLP of image captions to aid image retrieval. 3rd International Workshop on Semantic Media Adaptation and Personalisation, (SMAP 2008): accepted.

Kirovski, D., Malvar, H. and Yacobi, Y. (2004) A dual watermark-fingerprint system. *IEEE MultiMedia*, 11(3): 59–73.

Kotsis, G. (2002) Performance management in ubiquitous computing environments. In *Proceedings of 15th International Conference on Computer Communication*, pp. 988–997.

Krauter, K., Buyya, R. and Maheswaran, M. (2002) A taxonomy and survey of grid resource management systems for distributed computing. *Software Practice Experience*, 32: 135–164.

Kung, S.Y., Mak, M.W. and Lin, S.H. (2004) *Biometric Authentication: A Machine Learning Approach*. Upper Saddle River, NJ: Prentice-Hall.

Lamport, L., Shostak, R. and Pease, M. (1982) The Byzantine generals problem. *ACM Transactions on Programming Languages and System*, 4(3): 384–401.

Lansdale, M. (1988) The psychology of personal information management. *Applied Ergonomics*, 19(1): 55–66.

Lederer, S., Dey, A.K. and Mankoff, J. (2002) Everyday privacy in ubiquitous computing environments. In *Proceedings of UbiCom 2002 Workshop on Socially-informed Design of Privacy-enhancing Solutions in Ubiquitous Computing*.

Lyman, P. and Varian, H.R. (2003) How much information? Retrieved from http://www.sims.berkeley.edu/how-much-info-2003, Sept. 2007.

Maffioletti, S., Kouadri, S. and Hirsbrunner, M.B. (2004) Automatic resource and service management for ubiquitous computing environments. In *Proceedings 2nd IEEE Annual Conference on Pervasive Computing and Communications Workshops (PERCOMW'04)*.

Mark, L. and Roussopoulos, N. (1986) Metadata management. *IEEE Computer*, 19(12): 26–36.

Markus, M.L. (1994) Electronic mail as the medium of managerial choice. *Organisation Science*, 5(4): 502–527.

Mel, H.X. and Baker D. (2001) *Cryptography Decrypted*. Reading, MA: Addison-Wesley.

Merabti, M. and Llewellyn-Jones, D. (2006) Digital rights management in ubiquitous computing. *IEEE Multimedia*, 13(2): 32–42.

Missier, P., Alper, P., Corcho, O., *et al.* (2007) Requirements and services for metadata management. *IEEE Internet Computing*, 11(5): 17–25.

Monrose, F. and Rubin A. (1997) Authentication via keystroke dynamics. In *Proceedings of 4th ACM Conference on Computer and Communications Security*, Zurich, Switzerland, pp. 48–56.

Montagut, F., Molva, R. (2005) Enabling pervasive execution of workflows. *Proceedings International Conference on Collaborative Computing: Networking*, Applications and Worksharing.

Murtaza1, S.S., Amin, S.O. and Hong, C.S. (2006) Applications of SNMP in ubiquitous environment. *Korean Network Operations and Management (KNOM) Review*, 8(2): 14–19.

Norman, D.A. (1999) *The Invisible Computer*. Cambridge, MA: MIT Press.

Oppliger, R. and Rytz, R. (2005) Does trusted computing remedy computer security problems? *IEEE Security and Privacy*, 3(2): 16–19.

Park, I., Lee, D. and Hyun, S.J. (2005) A dynamic context-conflict management scheme for group-aware ubiquitous computing environments. In *Proceedings of 29th Annual International Computer Software and Applications Conference (COMPSAC'05)*.

Pirretti, M., Zhu, S., Vijaykrishnan, N., *et al.* (2006) The sleep deprivation attack in sensor networks: analysis and methods of defense. *International Journal of Distributed Sensor Networks*, 2(3): 267–287.

Poslad, S. and Zuo, L. (2008) An adaptive semantic framework to support multiple user viewpoints over multiple databases. In M. Wallace, M. Angelides and P. Mylonas (eds) *Advances in Semantic Media Adaptation and Personalisation*, Series: Studies in Computational Intelligence, Vol. 93, pp. 261–284.

Poslad, S., *et al.* (2009) Directing your own lives and Interactive Sports Channel. Paper presented at 10th International Workshop on Image Analysis for Multimedia Interactive Services, WIAIMS'09, Special Session on Event, Behaviour Video Analysis for Interactive Multimedia Services, 6–8 May, London, 2009.

Pressman, R.S. (1997) *Software Engineering: A Practitioner's Approach*, 4th edn. Maidenhead: McGraw-Hill.

Price, B.A., Adam, K. and Nuseibeh, B. (2005) Keeping ubiquitous computing to yourself: a practical model for user control of privacy. *International Journal of Human-Computer Studies*, 63(1–2): 228–253.

Römer, K. (2003) The lighthouse location system for smart dust. In *Proceedings of 1st International Conference on Mobile Systems, Applications and Services*, pp. 15–30.

Sakamura, K. and Koshizuka, N. (2001) The eTRON wide-area distributed-system architecture for E-commerce. *IEEE Micro*, 21(6): 7–12.

Saponas, T.S., Lester, J., Hartung, C., *et al.* (2007) Devices that tell on you: privacy trends in consumer ubiquitous computing. In *Proceedings of 16th USENIX Security Symposium*, pp. 55–70.

Singh, M.P. and Huhns, M.N. (2005) *Service-Oriented Computing: Semantics, Processes, Agents*. New York: John Wiley & Sons, Ltd, pp. 8–9.

Sloman, M. and Lupu, E. (2002) Security and management. *Policy Specification*, 16(2): 10–19.

Soares, A. and Thiry, M. (2002) Specification of a MIB XML for systems management. In *Proceedings of 27th Annual IEEE Conference on Local Computer Networks* (LCN'02), pp. 241–248.

Stajano, F. (2002) *Security for Ubiquitous Computing*. Chichester: John Wiley & Sons, Ltd.

Stjernholm, M., Poslad, S., Zuo, L. *et al.* (2007) An ontology-based approach for enhancing inland water information retrieval from heterogeneous databases. In P. Haastrup and J. Wurtz (eds) *Environmental Data Exchange Network for Inland Water*. Oxford: Elsevier, pp. 123–144.

Stone, G.N. Lundy, B. and Xie, G.G (2001) Network policy languages: a survey and a new approach. *IEEE Network*, 15(1): 10–21.

Subramanian, M. (2000) *Network Management: Principles and Practice*. Reading, MA: Addison-Wesley.

Syukur, E., Loke, S.W. and Stanski, P. (2004) A policy based framework for context aware ubiquitous services. *Lecture Notes in Computer Science*, 3207: 346–355.

Tan, J.J., Poslad, S. and Titkov, L. (2004) An ontological approach to harmonising security models for open services. In *Proceedings of 17th European Meeting on Cybernetics and Systems Research*, 2 594–599.

Tan, J.J., Poslad, S. and Titkov, L. (2006). A semantic approach to harmonising security models for open services. *Applied Artificial Intelligence Journal*, 20(2–4): 353–379.

Tanenbaum, A.S. (2002) *Computer Networks*. 4th edn. Harlow: Pearson Education.

Tao, L. (2001) Shifting paradigms with the application service provider model. *IEEE Computer*, 34(10): 32–39.

Titkov, L., Poslad, S. and Tan, J.J. (2006) An integrated approach to user-centered privacy for mobile information services. *Applied Artificial Intelligence Journal*, 20(2–4): 159–178.

Van Bunningen, A.H., Feng L. and Apers, P.M.G. (2005) Context for ubiquitous data management. *Proceedings International Workshop on Ubiquitous Data Management* (UDM'05), pp. 17–24.

Want, R. and Pering, T. (2003) New horizons for mobile computing. In *Proceedings of 1st IEEE International Conference on Pervasive Computing and Communications*, pp. 3–8.

Watson, R.T. (2006) *Data Management: Databases and Organisations*. 5th edn. Chichester: John Wiley & Sons, Ltd.

Woolf, M., Huang Z. and Mondragon, R.J. (2007) Building catastrophes: networks designed to fail by avalanche-like breakdown. *New Journal of Physics*, 9 (June). Available on-line http://www.iop.org/EJ/toc/1367-2630/9/6, accessed Jan. 2008.

Yoo, C.S. (2005) Beyond network neutrality. *Harvard Journal of Law & Technology*, 19(1): 1–77.

13

Ubiquitous System: Challenges and Outlook

13.1 Introduction

UbiCom defines a powerful vision for growing numbers of everyday human tasks to be automated by machines, enabling tasks and information to be accessed whenever and wherever they are needed but yet remaining hidden within the fabric of the physical world. There are many examples of UbiCom in everyday use, and there is pervasive use of sensors and controllers embedded into many everyday digital objects and electronic systems which can subsequently adapt their behaviour.

However, UbiCom brings with it its own set of challenges. More information and tasks may be more accessible but these can cognitively overload human users. While UbiCom systems can be designed to simplify access automating analogue tasks, the converse may also be true. For example, setting the time in an analogue clock, requires simply turning one or two knobs turned in a single direction to set the correct hours and minutes, in contrast, setting the time in a digital clock is often not so intuitive, it can require operating multiple unlabelled modal[1] buttons. Far from being models of hidden computing, most computers today are still quite obtrusive to access. ICT access often uses a keyboard and pointing device that can be difficult to use while performing other daily activities.

13.1.1 Chapter Overview

This final chapter continues with an overview of the challenges of UbiCom systems (Section 13.2) in terms of the support for the five core properties of UbiCom systems. The outlook begins by considering some trends for: developing smart devices (Section 13.3), smart interaction (Section 13.4), smart physical environment devices (Section 13.5) and smart human environment devices (Section 13.6). Next the interplay between human intelligence and machine intelligence is

---

[1] A modal user interface control triggers multiple actions depending on the mode or context. A single button (push and hold) could be used to increment and decrement (increment to maximum and then reset) hours, press again to access minutes and again to access hours, etc.

Ubiquitous Computing: Smart Devices, Environments and Interactions   Stefan Poslad
© 2009 John Wiley & Sons, Ltd

considered (Section 13.7), followed by a discussion of some of the social issues (Section 13.8). The final section ends with some final remarks about the smart DEI framework for UbiCom which has been proposed.

13.2 Overview of Challenges

13.2.1 Key Challenges

Several researchers have discussed the challenges of developing UbiCom systems. Weiser (1993), Demers (1994) and Satyanarayanan (2001) have given weighted insight into the design issues in developing resource-constrained devices, mobile devices and smart environments. Abowd and Mynatt (2000) have surveyed UbiCom applications along three themes: natural interfaces, context-awareness and automated capture and access and consider the main challenge to unify multiple interactions including social interactions. Edwards and Grinter (2001) have analysed the design issues in developing smart environments. Jessup and Robey (2002) discuss the social challenges. Davies and Gellersen (2002) give an overview of the technical and social challenges. Rogers (2006) has considered the design issues and challenges in using an iHCI type model for UbiCom. These challenges have been summarised in Table 13.1.

Table 13.1 Challenges in designing support for UbiCom system properties: distributed, iHCI, context-awareness, autonomous, intelligence

Property	Challenges
Distributed	Reliability can decrease as systems become more upgradeable and interoperable
	Openness increases incompatibilities, free-loader resource use, reduces responsiveness
	Less clearly defined system boundary
	Synchronising local cached data with remote, possibly centralised, data
	Bigger chance of unauthorised remote access, disclosure, decreased privacy & security
	Low-level events and composite events can flood and interrupt users
	Ad hoc interactions can be difficult to control and manage
	Undesired, inopportune or disruptive impromptu interaction disrupts system operation
	Overwhelming choice, multiple versions, heterogeneity
	Reduced cohesion, more complex trust interaction
	Distribution computation and communication costs outweigh gains
iHCI	Users get overloaded by interaction possibilities as the digital landscape expands
	Disappearing technology could lead to uncertainty about whether a system is working
	System takes away control from the user
	Disruptions can occur from unrecognized and nonsensical sources
	Ambiguous user intentions lead to incorrect system interpretation
	Loss of privacy & control because of an increasing indirect tracking capability
	Loss of presence in physical real-world because of continuous virtual interaction
Context-aware	No omnipresence − sense and act through layers of the environment that act as filters
	Uneven or patchy sensed environment events and context
	Incomplete, wrongly inferred, ambiguous, unwanted, context adaptation by system
	Ill-defined contexts, e.g., ICT, power, user etc, lead to non-optimal performance
	Localized scalability: density of interactions falls off as we moves away, otherwise, Users are overwhelmed by distant interactions of little relevance
	User goals and context often cannot be clearly defined and inferred
	Context adaptation leads to quicker, unconsidered, selection, decisions & commitments
	Balancing system versus application versus user control of context adaptation

Table 13.1   (*continued*)

Property	Challenges
Autonomous	Loss of high value macro mobile resources; Loss of many low value micro resources
	No-one wants to be an administrator, administration is more distributed and complex
	Undesired adaptation or unintelligible re- or self-configuration of the system
	Dependencies, e.g., of applications on system etc, limits autonomy
	Unanticipated, undesired and uncontrollable macro level behaviour emerges from micro level interaction
	Loss of control by user; greater system autonomy means system can say no
Intelligent	System infers state, knowledge, context, etc incorrectly
	Greater reliance and dependencies on systems of systems, interactions to operate
	Systems learn to operate outside its safe limits or conflicts with user intentions
	Systems exceed normal human behaviour limits, causing physical and mental damage
	Virtual organisation can masquerade as real organisations
	Byzantine, disruptive and malicious behaviour through collusion

13.2.2 Multi-Level Support for UbiCom Properties

Figure 13.1 proposes that there are graduated *levels of support* by UbiCom systems for each of the five core UbiCom system properties: level 1 (minimal), level 2 (basic), level 3 (medium), level 4 (high), level 5 (full). Levels of support could be used to indicate of levels of maturity. However, it is not necessary, nor necessarily desirable, to support the full level for each property: this depends upon the application and the situation. For example, a recommender, location-aware system application for a mobile user could be designed to support medium levels of support for the distributed (mobile ad hoc use), context-aware (presents location-based

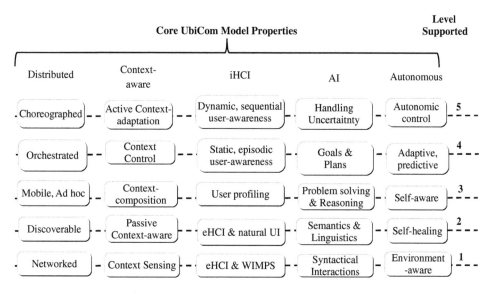

Figure 13.1   Graduated levels and system support for each of the five core UbiCom system properties

personal preferences to the user), iHCI (user can interact using natural interaction based upon personal profiles), AI (system models its environment and itself and can reason about it) and autonomous (system is aware of its environment, itself and can self-heal volatile communication) properties.

Some further comments about these levels of support are as follows. There do not need to be five graduated levels for each property, more or less levels could be defined. For each property, the level of support could be modelled using multi-dimensional hierarchies rather than single dimensional hierarchical models, e.g., see context-awareness (Section 7.2.2). One dimension for each property could reflect the degree of system proactivity while another dimension could reflect the number of independent or dependent factors which could be controlled by the system.

In Section 1.4, the term *smart* was defined to mean that the entity can be active, digital, networked, can operate to some extent autonomously, is reconfigurable and has local control of the resources it needs such as energy and data storage. A second way to define smart is in terms of the level of AI the system supports (Figure 13.1) so a minimal type of smartness is that a system has an explicit representation of its structure and state which it can share with others. A medium level of smartness is that systems support problem solving and reasoning. There are also multiple dimensions to intelligence in terms of the degree of human versus pure rational intelligence, the degree of individual versus collective intelligence used, the types of reasoning used and the types of environments intelligent entities can operate in (Chapter 8).

It may be fruitful to develop richer UbiCom systems that humans are able to trust by evolving them in particular ways, increasing the level of system control versus manual control unless the right balance is found for society and for the marketplace. For example, some intelligent and autonomous applications could preferable evolve from context-awareness and control systems along the path from unguided to single sensor-guided to multiple sensor-guided to multiple manual sensor-guided, to semi-automatic-guided, e.g., vehicle collision avoidance systems which operate under low speed conditions. Similarly, intelligent systems could evolve from systems which support minimal levels of AI until an appropriate level of intelligence for the application and its environment is supported.

13.2.3 Evolution Versus Revolution

Christensen (1997) categorised technology as either disruptive or sustaining. A *disruptive technology* is one that changes or replaces the accepted way of doing things. A *sustaining technology* enhances an existing product or service by refining it or making its creation and delivery more efficient. Patel and Pearson (2002) argue that many visions for future computing assume sustainable, incremental, evolutionary progress in technology, whereas history has shown repeatedly that markets are changed mostly by disruptive technologies, e.g., personal computer, MP3 music format, Web browser, etc. Norman (1999) gives some nice examples of disruptive technologies along with famous companies and people that rejected them. Technologies which were initially rejected included the radio, telephone, computers for business use, computers for personal use, the Polaroid camera, the Xerox copier, Mosaic Web Browser and the Apple Computer. Technology sometimes tries to drive use rather than use driving technology. Examples of the former are the use of the Internet to connect appliances such as the fridge, adding the ability to browse the Web to camcorders, making everything wireless.

13.2.4 Future Technologies

There are many sources of ideas for future technology. Many science fiction writers who may have trained or worked as scientists and engineers have described ideas which later turned into reality such as Arthur C. Clark who is credited with proposing the idea of orbiting satellites in 1945.[2] Many engineers and scientists in many different fields have proposed bold visions for the future use of ICT such as Van der Bush (1947), Weiser (1993), Pearson (2000), Kurzweil (2001), Greenfield (2006) and others.

Pearson (2000) has made about 500 detailed predictions with respect to a timeline up to 2050 and some of these deserve special mention. If we review many of the earlier predictions by such evidently smart people, by their very nature, these are hit and miss. In addition, there are other contributing factors. There may often be quite a lag between the feasibility of a prototype versus systems that actually lead to uptake by the masses. This may in turn in depend upon non-functional requirements such as efficiency, safety and reliability, ease of use for HCI and eco-friendly use in physical environments. Technological revolutions or evolutions are just one of the environments which must be affected in such a multi-disciplinary world which causes a hill-climbing effect for the social (including political, business and legal acceptance) and physical world to cause change. We need to understand the novel secondary effects of technology, to understand the complex interplay between systems and their environment. This argues for multi-disciplinary participation in developing new applications. The world continues to need more engineers, physicists, geographers, lawyers, sociologists, etc, and these need to talk to each other. Specific future technologies are considered in subsequent sections.

13.3 Smart Devices

Some of the main challenges for smart devices have been summarised in several sections such as services (Section 3.1.2), network communication (Section 11.7) and device management (Chapter 12). Here a few key challenges are revisited.

13.3.1 Smaller, More Functional Smart Devices

There is an evolutionary trend towards smaller, lower power, higher resourced devices (Figure 13.2). For example, consider mobile phones first produced in the 1950s, they weighed tens of kilos and needed to be carried in a suitcase or backpack. In the 1980s, phones became brick-sized and today mobile phones weigh tens to hundreds of grams. Phones can be manufactured to be much smaller, lighter and low-powered, if MEMS technology can be

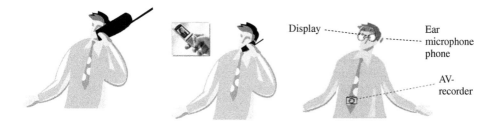

Figure 13.2   The trend towards smaller, low-powered, higher resources smart devices

---

[2] See http://www.clarkefoundation.org/acc/biography.phpm, accessed December 2007.

leveraged further, to realise hidden earpieces, microphones, cameras and displays in glasses. Many other applications of micro components are given elsewhere in this text.

The use of more flexible materials to act as ICT devices as discussed in the sections on Tangible UIs, Organic UIs and MEMS can lead to many more physical objects supporting dual or even multiple hidden virtual computing functions, e.g., clothes could sense human skin and reconfigure itself to offer more ventilation if it senses the skin is sweating. Clothes can also act as musical instruments or as communication conduits, to notify their wearers of incoming messages and calls.

Goldstein *et al* (2005) have proposed the concept of synthetic reality which combines self-organisation of multiple MEMS devices called catoms and tangible UIs. Catoms are individual MEMS components which can move in three dimensions in relation to other catoms and adhere to other catoms. Nano components can also act in a similar manner. Although this physical synthesis models can occur on demand offering greater flexibility of configuration, these could get out of control. This type of synthetic reality behaviour may mimic physical world behaviours, for example, MEMs-type computer viruses may exhibit the lethal potential of biological viruses (Kurzweil, 2001). Virtual world behaviour may even surpass the behaviours of their physical world counterparts. It is not clear if computer viruses can function exactly as biological viruses because the former are not embodied in the same way as biological viruses. If the latter get no nutrients from the environment, they may simply die. A MEMS type or nano computer virus may, however, remain passive, waiting forever to replenish its energy in order to restart its behaviour.

13.3.2 More Fluid Ensembles of Diverse Devices

A range of form factors for devices exist from dust-sized, tab-sized, pad-sized to board-sized. Flexible ensembles of micro and nano-sized devices were considered in the previous section (Section 13.3.1), flexible ensembles of macro-sized devices are discussed here. Digital technologies will continue to proliferate and become embedded and scattered in the physical environment. Devices may be planar or form more flexible two-dimensional skins or three-dimensional shapes. In the applications section (Section 2) and throughout this book, a range of macro-sized smart devices and smart environment applications have been described. In a smart office, specific lights, e.g., on the desk, can switch on when activity on the desk is sensed while other lights in the vicinity can remain deactivated. Many hidden and diverse devices with touchscreens or tangible and organic user interfaces can behave as hidden and pervasive computers enabling surfaces of many physical objects to function as sensors and displays (Figure 13.3).

In the bathroom, bathware such as smart mirrors can provide information about predicted conditions, e.g., the weather, that relate to activities and offer advice, e.g., what clothes to wear, how to set the heating and how to tend to the garden. Multi-sensor taps can automatically adjust the water flow and temperature depending on the personal preferences and anticipated activity such as shaving, washing hands and brushing teeth. Toilets could chemically analyse body waste and assess the well-being of the body. Toothbrushes could contain a micro-camera that projects an inside view of the mouth to the mirror to enable us to offer other views of our teeth in order to clean them more effectively. Sensors and controllers could be installed outside the home, in the garden, e.g., to inform its owners about local ground conditions. For example, sensing how much rain has fallen can inform gardeners whether there is a need to water plants or not. Multiple physical media networks such as electricity, audio and video broadcast networks, mobile and fixed voice networks and data over IP networks, can be active within the same physical space promoting ad hoc device interoperability.

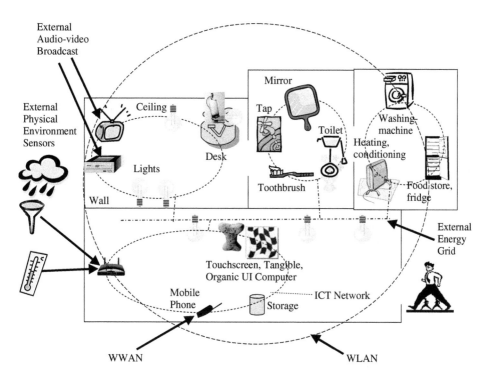

External
Audio-video
Broadcast

External
Physical
Environment
Sensors

Ceiling

Lights

Wall

Mirror

Tap

Toilet

Desk

Toothbrush

Washing
machine

Heating,
conditioning

Food store,
fridge

Touchscreen, Tangible,
Organic UI Computer

Mobile
Phone

Storage

ICT Network

External
Energy
Grid

WWAN   WLAN

Figure 13.3   The trend to embed and scatter numerous and even potentially overwhelming numbers of digital network devices into and bound to physical objects in the environment

13.3.3 Richer System Interaction and Interoperability

13.3.3.1 Migrating from Analogue to Digital Device Interaction

Many devices and systems exist which do not need to be digital or support any kind of flexible system to system interoperability. They can be designed to be stand-alone analogue single function appliance devices.. New versions of old appliances, which need to regulate their behaviour with respect to some external environment parameter, require sensing and control, can replace human sensing and control and incorporate digital electro mechanical devices or IC chips which support embedded computing. For example, modern fridge freezers include digital temperature controllers with multiple temperature sensors and an LED display. Washing machines include digital controllers to support multiple washing lifecycles for different kinds of clothes. Ovens can include internal analogue or digital temperature sensors and timers to regulate their temperature for set periods, according to different types of cooking such as defrosting and oven-roasting. They can include additional external digital temperature probes which can be inserted into food to indicate when the food is cooked.

The increasing use of IC chips and embedded computers in appliances and other devices situated in physical and human environments is motivated by more accurate, cheaper, sensing and control of processes situated in the physical environment. Digital components can be used to provide more accurate and reproducible recording and playing of AV content in digital form rather than analogue form which supports more flexible editing and annotation. However, many of these digital device functions may have little need to interact with other devices and external systems. The motivation for richer digital interaction is given in Section 13.3.3.2.

13.3.3.2 Richer Digital Device Interaction

Many individual digital systems, particularly those which are embedded systems, currently operate in isolation. There are several drivers for richer, flexible and dynamic ICT system to ICT system, and ICT system to environment interaction and interoperability: remote access, configuration and control; remote browsing and searching; multiple system integration; supporting multiple levels of system access and operational views of one system by other heterogeneous systems including humans and ICT systems; orchestration and choreography of individual ICT services and processes; sharing information, tasks, goals, processes, plans and experiences between systems and the environment.

An evolution pathway from less rich and soft information interaction to richer and softer (Section 12.3.1) information interaction is proposed. Application network protocols can be used to enable applications to be controlled remotely. Application data protocols enable different applications to exchange data structures. There is a range of rich and soft data structures and their relationship represented and exchanged, from simple single concept hierarchies to more complex multiple concept hierarchies. Richer interaction goes beyond sharing information to include sharing tasks, process, goals and experiences. However, although systems can potentially interoperate in such rich ways, support is needed to allow this system interaction to be managed within the constraints of their ICT, physical and human environments. In addition, the majority of this richer system interaction is C2C interaction with some limited HCI to support finely grained user configuration, rather than CPI or HPI (Section 1.3).

13.4 Smart Interaction

Smarter interaction between individual smart devices and smart environments is the key enabler to promote richer, more seamless, personal, social and public spaces. Abowd and Mynatt (2000) note that user activities rarely have a clear beginning or end and multi-session, interruptions are to be expected, multiple activities operate concurrently, contexts such as time are useful for filtering and adaptation and that associative models of information are needed. Johanson *et al.*'s (2002) analysis of interactive workspaces emphasises that designs can benefit from being location aware. They should, wherever practical, rely on social conventions to help make systems intelligible. They should support wide applicability (interoperability) and they need to be kept simple to be intelligible for users.

Some of the key challenges for interaction in smart environments have already been mentioned, earlier in this chapter. As the multiplicity of interactions increases, the physical world context can be used to tailor interaction and the user context can be further used to tailor iterations. However, contexts can be hard to determine and there is uncertainty in interacting with other systems in open dynamic environments. Edwards and Grinter (2001) have eloquently highlighted some of the key challenges for using ubiquitous computing applications in home-type smart environments but their analysis can be generalised to smart (physical world) environment interaction. Their seven challenges are: the 'accidentally' smart home, impromptu interoperability, no systems administrator, designing for domestic use, social implications of aware home technologies, reliability, and inference in the presence of ambiguity. Design challenges for smart environments are discussed in more detail below.

13.4.1 Unexpected Connectivity: Accidentally Smart Environments

Maintaining a smart ICT-enabled physical environment will occur iteratively. Piecemeal adoption and upgrades to an existing environment, e.g., using wireless speakers, doorbells,

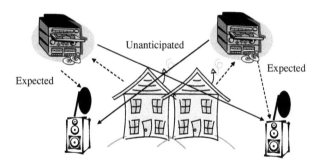

Figure 13.4 An example of unexpected connectivity: homeowners may not realise that their wireless speakers can actually connect themselves to sound sources in another house as easily as to sound sources within their own home

televisions and phones in a house, may lead to unpredictable level of complexity. Because of the changing form of the ICT devices in the physical world home environment, the intelligibility of how it functions for users can change over time. Users can be uncertain of the boundaries of their home, of how much privacy they have and of the locus of control in the system. For example, homeowners may not realise that their wireless speakers can actually connect themselves to sound sources in another house just as easily as to sound sources within their own home[3] (Figure 13.4). In conventional designs models, connectivity is explicit when physical wires are present, the 'range' of connectivity is more apparent,[4] connections are observable and connections don't change on their own. New design models of connectivity for wireless technologies are needed that are intelligible by users. These need to control the way in which inputs and outputs can be redirected in an ad hoc networks.

13.4.2 Impromptu Service Interoperability

Interoperability goes beyond simple connectivity. Whereas simple connectivity is concerned with redirecting outputs to different inputs, interoperability goes beyond this by considering more complex coordination and translations when diverse and heterogeneous devices interconnect. Many ICT devices in the physical environment are designed to be stand-alone, not to interoperate online, although these are sometimes designed to interoperate offline via removal media. Many devices are practically constrained, 'hard-coded', rather than theoretically constrained, to work with a sub-set of devices as islands of functionality rather than supporting pervasive interoperability. There are several reasons for this such as: cost, components are sourced from different vendors, are often acquired at different times, have different design constraints and issues, and would require the use of many diverse I/O hardware interfaces and channels.

---

[3] There are many other examples of this. For example, the introduction of new WLAN base-stations with stronger signals can cause WLAN clients to switch to them. Installation of wireless doorbells and remote controls for audio-video equipment can also lead to unexpected connectivity problems.

[4] The range of interaction seems narrower for a wired rather than wireless network but if the wired interaction accesses the Internet rather than remaining locally, data may flow over many networks that are not observable by senders outside their control and can be observable by remote others depending on the security levels supported.

Users' expectations are that systems should work together fluidly and flexibly. New design models of interoperability need to interoperate at multiple levels: at the I/O hardware, plug and play level, at the network level and at the service level. Raskin (2000) says that users ought to be able to plug and unplug devices and cables while devices are both on and off, without having to match up the variety of types of connector.[5] At the network level, wired and wireless network gateways can be used to convert transmissions between two or more heterogeneous networks. At the service level, systems need to be designed to support open, dynamic, models of service discovery that allow devices to acquire and negotiate new services interfaces at run-time, using both syntactic and semantic exchange. An example of a syntactic approach for fluid service interaction is to use the Web presence for physical and virtual resources model proposed in the Cooltown Project (Section 2.2.2.4). Examples of semantic approaches are discussed in Chapters 8 and 9.

13.5 Smart Physical Environment Device Interaction

Interaction with smart mobile devices and smart environment devices requires effective human–computer interaction design to make these systems useful but human interactions need to take into account that these are centred in physical world activities rather than just being centred in computer devices and in a virtual computing environment.

13.5.1 Context-Awareness: Ill-Defined Contexts Versus a Context-Free World

As has already been mentioned, according to many proponents, location or spatial type of context-awareness is one of the main drivers for future IT businesses and in particular mobile services (Marcussen, 2001). Beina *et al.* (2007) substantiate this claim. Business, transportation, health care, entertainment and public services will increasingly rely on the availability of location aware capabilities which will gradually be embedded in business and society. Location and sensor services will be dictated by business or safety needs in the early stages with minimum attention to the potential side effects. Major incidents, such as privacy violations, or social pressure, will likely be introduced in a second phase to address the full range of benefits and costs of these technologies which will drive their further adoption and acceptance. This may be true not just of context-specific UbiCom systems but of other types of computing too.

Context-aware systems are often expected to make decisions with limited context information about the world and with limited adaptation. The world may be only partially observable, e.g., the surface temperature may differ from the internal temperature. Adaptation, when performed, should be predictable, e.g., a user knows that a dropping temperature outside will cause the thermostat to turn on the heating, the system appears intelligible. Context-awareness may be ill-defined for several reasons, i.e., because the context is derived, e.g., a person's ID tag is in a room, therefore they must be present too. Contexts may be aggregated from several factors or indirectly inferred, e.g., someone's whereabouts is based upon their calendar. A system may try to anticipate and pre-empt actions based upon the context, e.g., a person who moves near a door, requires the door to open in order to pass through it. A system's

---

[5] Raskin points out that cable connectors are unnecessarily sexed into male and female type connectors that need to be matched differently in cable extensions and plugs. Far easier would be to design a single sex, hermaphroditic type of connector. Note cables may still be needed in a wireless environment because they can be shielded and lessen interference with other electromagnetic signals in unguided channels.

actions should make sense to the user, and be recoverable, enabling the system to roll back to a previous stable state.

Taylor *et al.* (2007) argue that active context-aware systems, e.g., systems that issue location-aware alerts to persons, are problematic on two counts: increased vulnerability to error and reduced human control. Replicating the complexity of the real world makes context-aware systems complex and vulnerable to error. Technology should be designed that empowers people to make decisions as they see fit. This is subjective, e.g., if someone cooking dinner is able to detect that some dinner guest is still at work. It means that they can choose either to delay dinner or to give the person still at work a call and urge them to hurry to the dinner location. A counter-example may be that it can sometimes overload humans and be unsafe to sometimes make decisions based upon their subjective preferences, e.g., to allow operators to speed up a maintenance cycle in order to avoid getting home late. It depends in part upon the social context whether or not active versus passive context-aware systems are preferable.

13.5.2 Lower Power and Sustainable Energy Usage

A range of power management techniques have been discussed for devices to reduce their power consumption. Passive electronic components can be used that do not require energy to maintain their state, e.g., electrophoretic displays. Tags can take energy from their interrogator, e.g., passive RFID displays. Active computation devices can adapt their power requirements based upon demand, e.g., dynamic voltage scaling control of CPUs. If power is supplied to multi-function, multi-components devices, components or functions not in use can be powered off. Smaller, micro-sized integrated circuit devices, e.g., MEMS, can be used, which require less energy to function. This in turn enables devices to operate more energy efficiently and untethered for longer without their energy store needing to be replenished or replaced. Multi-hop distributed lower power transmission can be used instead of higher power longer-range transmissions.

Most macro-sized mobile devices are currently powered by batteries, which need to be charged by attaching them to internal building energy grids. A battery has all of its chemicals stored inside, converting these chemicals into electricity. Batteries also need eventual replacement after a period of time because of the efficiency of charging decreases. Batteries suffer from hysteresis and degrade after a few thousand charge-discharge cycles. They operate over a relatively narrow temperature range.

Fuel cells are clean power sources that have much higher energy densities and lifetimes compared to batteries. With a fuel cell, providing there is a flow of chemicals into the cell, e.g., by replenishing the fuel, the electricity from the cell can supply devices without interruption. Most fuel cells in use today use hydrogen and oxygen as the chemicals. However, such fuel cells cannot respond quickly to changes in load: they cannot be efficiently used in isolation. Hybrid systems where a fuel cell based hybrid power source can be used, comprising a fuel cell operating as the primary power source and a Li-ion battery that has a better load response capability, operating as the secondary source. Hybrid fuel cells could enable computers to respond quickly enough to changing loads and to be powered for longer periods, of the order of ten hours (Zhuo *et al.*, 2006).

A second alternative to batteries are capacitors. In comparison to fuel cells, their capacity is much more limited. Because no chemical reactions are involved, capacitors are much more energy-efficient. They can be used over a wider temperature range, are much more responsive and can repeatedly store and release energy. Capacitors are better suited to complement batteries to supply quick bursts of energy, enabling the battery life to be extended. Some common uses are as a back-up energy supply for electronic circuits, camera zooms. They can be used in cranes to lift loads and they have a number of uses in cars to support door locks and power steering. The surface area together with the dielectric constant of the material separating them and the distance apart determine how much energy a capacitor can store. *Ultra capacitors*, also known as *super-capacitors* or *double-layer capacitors* are able to store much more charge than normal capacitors due to the type of materials used for the electrodes. These can be based upon carbon coating of electrodes and manufacturing holes in the carbon which substantially boost the surface area of the electrodes. The ability to grow many carbon nanotubes on to the surface of the electrodes may boost the energy stored even further.

Ambient renewable energy sources can enable devices to remain operational longer out in the physical world – this is referred to as *energy-harvesting*. Types of energy can be harvested include kinetic energy, mechanical vibrations, solar energy (Chalasani and Conrad, 2008) and bio-fuels. There are several ways kinetic energy can be harnessed from physical movement to generate power, e.g., wrist watches can be powered by hand and arm movement; laptops and radios and power generators can be powered using a wind-up handle or using pedal power. The wearer's movement moves magnets through wire coils generating an electric current which is used to power mobile devices. This same principle of *electromagnetic induction* is used in many type of power generator. *Piezoelectric* materials such as quartz, and lead zirconate titanate (PZT) can be used to convert mechanical energy from pressure, force, and vibrations into electricity. Electrostatic (capacitive) energy can be harvested, based upon the changing the separation and hence capacitance of vibration-dependent varactors or variable capacitors, converting mechanical energy into electrical energy. Atwood *et al.* (2001) have compared the energy generation capability of several devices and these are approximately $50\mu W/cm^3$ for thermoelectric, $116\mu W/cm^3$ for vibration, $350\mu W/cm^3$ for piezoelectric and $15mW/cm^3$ for solar cells. The kinetic energy of the physical environment, rather than human environment can also be used to generate power, including water flow such as tides and wind.

Energy optimisation in devices nevertheless faces several design challenges. A lack of feedback control leads to poor energy optimisation. For example, without the use of negative feedback to reduce the energy supply, e.g., food heaters generally cannot detect and regulate themselves not to unnecessarily overheat food. Energy can often be usefully converted into a form to be reused instead of being wasted, e.g., an object moves uphill increasing its potential energy but does not reuse this energy when going downhill or an object increases speed and then applies brakes to reduce speed. These sources of energy could be stored instead of being wasted.

Energy is used to provide services which are oriented to a human presence but humans may not be present. In this case, localised context-aware devices can be used, e.g., devices only switch on when they detect body movement or body heat, providing of course the sensing is sufficiently low power and does not outweigh the power saved. Energy needs to be optimised locally and globally across devices, e.g., if there are several lights in room, some of them are redundant wasting energy.

A ubiquitous home environment needs to be developed to support smart energy regulation to improve energy efficiency. Advanced electricity and gas meters could generate timely consumption data such as the energy consumption per device and the total consumption per unit time enabling customers to see when they are using energy, to manage that use more efficiently. In *demand response systems*, customers can choose to save money by adjusting energy use in response to dynamic price signals and policies. For example, during peak periods, when prices are higher, energy-consuming devices could be operated more frugally to save money. In direct load control systems, a form of demand response in which certain customer energy-consuming devices are controlled remotely by the electricity provider (or a third party) during peak demand periods. Configurable energy automation products such as 'smart home' technologies can enable customers to manage their energy use more efficiently through pre-programming or remote controls. Context-aware energy devices could switch themselves on in particular way or off when not in use but how this is specified needs analysis. It can be preset or configurable by the user. User activity awareness: e.g., the heating system could also be aware of the presence (or not) of the inhabitants in the building when regulating the heat. Self-organising applications allow these types of products to function together as resources within the electricity delivery system, e.g., not all the lights switch on when someone is near, just selected ones.

13.5.3 ECO-Friendly UbiCom Devices

Environmentally friendly or *eco-friendly* devices are devices which cause minimum or no harm to the environment. This requires considering device use throughout the whole of its life-cycle, from extraction of raw materials, through manufacture, through operation, through disposal. As more of the physical world is being annotated and augmented with digital systems, it is vital that devices behave as part of sustainable digital ecosystems. Otherwise, we will end up with an ever increasing collection of unused yet still usable electronics products to dispose off whose high cost of production and disposal is not offset sufficiently.

According to environment studies quoted by Huang and Truong (2008), consumer electronics account for about 1–4% of the municipal waste stream in Europe and the United States, but is responsible for 40% of the lead in this waste stream. Although the small size of a hand-held device means that its disposal yields less waste than that of a traditional desktop computer, its size also makes it more likely to be thrown away.[6] This also depends on how current the model is, how expensive it is and whether or not there is any sell on value. The more affordable ICT device are for us, coupled with the more they are perceived to be improved, leads to an increased probability that they will be discarded while they are still fully operational. Some consumers use their mobile phones on average for only about 17.5 months before disposing of them. Phone disposal is linked to the business model and the customer contract for the service.[7]

Suggestions for extending the lifetime of the phone include more modular design to snap in and play new hardware modules such as camera, speaker phone, etc. rather than tightly integrating multiple hardware modules into one changeable unit. The core component in any digital computing device is the CPU. Oliver *et al.* (2007) consider the energy cost in manufacturing CPU ICs, the waste issues in manufacturing them and disposing of them and the social issues of discarding them well before the end of their lifetime. Various reuse schemes are considered. Researchers could standardise embedded processor footprints for a wide range of embedded devices. Instead of reusing a processor in the same device, it could serve a next-generation device with lower performance requirements, applying power-savings techniques like voltage scaling (Section 4.3.4).

ICT systems should be manufactured from eco-friendly materials so if they are discarded, they can be recycled without releasing any toxic substances into the environment. For example, Pioneer Corp have developed a version of the write-once disk for an optical recording system of the Blue-ray AV disk format, whose substrate is made of a natural polymer derived from corn starch.[8]

One of the biggest loads ICT puts on the environment concerns the expansion of the amount of information that we increasingly handle. In theory if information is digitised, it can all be accessed electronically and the main operational impact on the physical environment is energy consumption for ICT devices use. Better designed ICT systems can now support convenient browsing and use of

---

[6] This is true of other hand-held devices such as audio-video remote controls. There seems to be ample scope for reuse to reprogram an old controller to work with a new device, for manufacturers to standardise the remote controller with user-customisable extensions.

[7] Simply by changing the business model of 'giving away' phones with contracts, consumers would use phones for longer.

[8] The disk cannot be eaten as it is coated by a 0.1 mm thick layer of resin that is too hard to bite.

soft copies of information, reducing the need for hard copies, or paper copies of information. For example, car navigation systems can replace the need to print out street plans in order to find how and where to get to a destination. Government, local authority and educational information can all be offered online.

From a technical viewpoint, many devices use common components but in different configurations. Some components seem as if they ought to be easily reusable and reconfigurable, e.g., AC to DC electricity transformers. When a device is replaced or discarded, its working transformer is also discarded rather than being reused. A challenge in reusing power transformers between devices is that there are different AC supplies in different countries and there are device-specific power circuit requirements in terms of current and voltage. A solution is to design a power transformer with reconfigurable current and voltage outputs but this may drive up the cost and the size of the transformer. Current designs require users to manually configure the current and voltage output for each device but if a user does not configure the power transformer to match a device correctly, it can easily destroy the device, e.g., by supplying too high a current. This incorrect use may even cause wider physical environment damage by possibly starting a fire. Hence, designing for reuse adds design complexity and requires safeguards to prevent incorrect and unsafe user configurations.

Sellen and Harper (2001) have used ethnography and cognitive psychology to look at paper media use by individuals and organisations, in particular looking at paper as affordances for interacting with information.[9] There still seems to be a considerable use of paper-based information for several reasons. Some people find paper more tactile to interact with. Paper is considered to support more natural gesture-based inputs such as writing, drawing and painting. Paper is more accessible as we move around. Paper is readily used to distribute copies of information which has a short lifespan of interest to many people such as newspapers and magazines. Electronic copies may be easier to search but paper copies are easier to browse. The pagination in many electronic information display systems does not naturally promote ease of use to continually read text across columns and pages. Paper can be read in a variety of physical environment conditions such as poor lighting and bright natural lightings which limits the use of computer display. Current displays require a lot of energy and can't last without mains power for long meetings or when travelling to and from meetings. It may be socially acceptable to use paper for input and output and formal meetings but sometimes it may not be socially acceptable to use computers.[10]

The use of epaper and ebooks can potentially have a huge impact on the amount of paper used and it also solves some of these limitations of current active ICT displays (Section 2.3.2.2). Blevis (2008) considers the sustainability issues of using epaper and ebooks.

White *et al.* (2003) consider the use of *reverse manufacturing*[11] or *remanufacturing* of computers which in theory focuses on component recovery to remanufacture products and parts for reuse. Complex goods such as computer products are inherently less capable of remanufacture than some other goods because there is less opportunity for reuse. This in part due to more complex

---

[9] The fact that their book only appears to be available in paper form from the publisher, not in electronic form, may seem to confirm their view of the utility of paper versus electronic information.

[10] It is considered by some to be socially acceptable for participants to either just listen and possibly not remember what is said afterwards or to make paper notes at a formal meeting or presentation but not to take notes using an ICT device.

[11] Traditional manufacturing, *forward manufacturing*, moves from high-level abstractions and logical designs to the finished product, formed out of lower-order components. *Reverse manufacturing* is the process that starts with a finished product, an output of a previous manufacturing process, but which seeks to undo this in order to reuse some parts, recycle or even remanufacture a product to make it usable again. Reverse manufacturing can be regarded as a sub-type of reverse engineering.

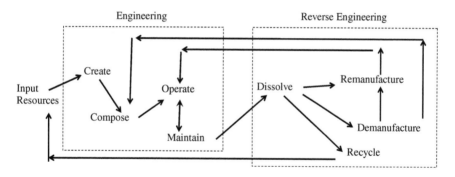

Figure 13.5   The engineering process versus the reverse engineering process

interdependencies and hard-coded links between components. Technology is perceived as being disposable as part of a throw-away culture, partly because it is often cheaper to replace items than repair them. As a result, they are not remanufactured[12] so much as de-manufactured[13] to recover valuable assets wherever and however possible (Figure 13.5). Important aspects and challenges in the acquisition, assessment, disassembly and reprocessing of computer equipment as it moves through this reverse manufacturing process, are as follows. Remanufacturing is more beneficial than recycling in two main ways. Remanufacturing cuts down on the use of energy and resources used for processing because recycled goods are consumed, they are often returned to their original raw material form (consumes energy) and are then used for manufacturing again (consumes energy). *Recycling* largely reduces the raw material extraction and pre-processing of raw materials to be used again. Remanufacturing also differs from recycling, most importantly because it makes a much greater economic contribution per unit of product than does recycling. The cost of labour, energy, and manufacturing operations that are added to the basic cost of raw materials in the manufacture of a product often dominates and determines the cost of the product rather than the cost of the raw materials. Remanufacturing save some of the labour, energy and manufacturing costs, in contrast to recycling which does not.

Note also there is a different challenge in handling the remanufacture and de-manufacture of smart devices embedded in or strewn in physical environments compared to handling self-contained smart devices as the acquisition and disassembly processes are different. In industrialised countries, waste in the form of ICT and other analogue, macro-sized electronic equipment are currently processed separately from other physical world objects because of their material composition. However, as microelectronic components become increasingly embedded in commonly used non-ICT objects, this may make it both ecologically and economically unfeasible to separate these embedded components for special waste treatment. Some studies examine whether the presence of micro-electronics, specifically RFID tags, in traditional non-electronic waste streams will affect the processing of the traditional waste streams. Smart labels induce new material flows into existing recycling loops and impact waste management processes. Potentially valuable materials, e.g.,

---

[12] *Remanufacturing* is the process in which a specific product is disassembled, cleaned, repaired, and then reassembled, to be used again, e.g., printer toner cartridges. Worn out, old, or failed parts may also be replaced with working, new, parts. Worn old parts can be considered for disposal or recycling.

[13] *De-manufacturing* is just the disassembly stage for remanufacturing and reverse manufacturing.

copper or gold, could be irrecoverably lost unless precautionary measures are taken. Potentially dangerous materials could contaminate recycled materials, e.g., aluminium or copper-based RFID antennas and lead solder.[14]

Pister's view,[15] who was the principal investigator on the smart dust project which finished in 2001, was that disposal of smart dust devices represents only a minor concern. For example, if a million grains of smart dust were manufactured, they would take up a litre of volume and if literally thrown in the environment, it would be like throwing mostly a litre's worth of batteries into the environment –although this does not help, that this is not a major environment concern. Also if these millimetre-sized particles were breathed in, possibly this could be dealt with by a coughing reflex.

However, others such as Sutcliffe and Hodgson (2006) have presented a much more systematic analysis of the effects of tiny devices, but from the perspective of nanotechnology rather than micro devices. Some examples of these hazards are that manufactured nanoparticles and nanotubes are likely to be more toxic per unit mass than larger particles of the same chemical in the human body. There is uncertainty on how nanoparticles combine with existing materials and toxins in the environment. Little is known on how they move through the biosphere. Many nano-materials are particularly durable, remaining in the environment long after any product disposal.

As technologists we need to build things in a modular way, to make things easy and cheap to repair. Rather than just replace things, we need to build things to last longer. From the outset, we need to design things to support easy and economical methods of disposal.

13.6 Smart Human–Device Interaction

Smart devices can be used to form a smart personal and pervasive space for their owners that follows them around. There is a need to balance the convenience of automatically tailoring multiple new services to their preferences, versus the loss in privacy in leaving a persistent trail of personalised behaviours which other non-trusted parties could exploit in unwanted ways. Smart devices may be resource-constrained, leading in turn to restricted device behaviours, e.g., oldest or newest messages may be discarded when the local storage is full. Devices may be mobile. In this case, new designs to route incoming messages to mobile hosts need to be formulated. Flexible service discovery is needed to enable devices to discover local access nodes to forward messages over wide areas and to discover and interact with other local services.

Harper *et al.* (2007) in a report called 'Being Human: Human-Computer Interaction in 2020' reflect on changes in smart devices to the current time; how devices are changing lives and society, how changes in smart devices will lead to major transformations in the nature of human computer interaction; how the process to support HCI will need to support general human values such as supporting inclusivity, equality and curiosity; how HCI will change in the future. Some of these aspects are discussed in more detail.

---

[14] The import of aluminium or copper from the smart labels antennas could lead to discolorations or material defects in new glass items. The lead solder used to join the chip with the antenna could, for example, set it free during the smelting process. This is more of a problem in low-tech waste disposal processes, where smart labels are not separated prior to the recycling process. In high-tech waste disposal processes, smart labels can be separated from their host objects, e.g., bottles, by the application of vapour.

[15] See http://robotics.eecs.berkeley.edu/~pister/SmartDust/

Smart devices have changed from the recent past to the current time as follows: from GUIs to gestures, from VDUs to smart fabrics, from stationary box set transceivers to mobile handset-sized transceivers, from simple robots to autonomous machines, from hard disks to digital memories or footprints, from shrink-wrapped (provider produced content) to mash-ups (user produced content) and from sometimes-on phones and answer-phones to always-on.[16]

These developments in smart devices will lead to five main transformations in how humans will interact with devices. First, interface stability has ended as the boundary between computers and humans and between computers and the everyday world becomes more fluid and ecosystems of peer-to-peer ad hoc interaction of devices occur. Second, techno-dependency will increase further because of living in an increasingly technology-reliant world and living with increasingly clever computers. Third, hyper-connectivity[17] will grow further, as we live in a more socially connected world and increasingly become part of a digital crowd. Fourth, short-lived, daily or *ephemeral human memories* will disappear as more information is stored in expanding personalised digital memories and through living in an increasingly monitored world which is embedded with closed circuit video systems. Fifth, new forms of creative engagement such as various kinds of UbiCom and augmented human reasoning will increase.

Three case studies were proposed to highlight trends for HCI in the future. Trading versus trafficking[18] of content uses designs based upon mobile phones to view and share regulated content. Tracking versus surveillance in families uses hidden computer designs such as the Whereabouts Clock to indicate the location of family members (Section 2.2.2.5). The making human memories more tangible scenario uses a digital Shoebox that is wireless connected and that lets users browse through them by running their finger across the top of the box.

13.6.1 More Diverse Human–Device Interaction

The diversity of how devices interact within ICT infrastructures and how they interact with humans past, present and future is summarised in Table 13.2.

Interactions in which humans utilise natural interfaces rather than artificial ones such as keyboards have advantages but they also have limitations. One of the limitations is the expediency and accuracy of selection. The use of gestures to make selections can require good motor skills, effort. It can lack precision and command activation can be long-winded in comparison to the use of mouse commands and keyboard short-cuts. Natural gesture pointing in a 3D space or even 2D space introduces ambiguity in selecting objects. Natural gesture interaction should be seen as complementary rather than as an evolutionary replacement for current HCI.

There are concerns about hidden UI models becoming unattainable, of taking humans out of the loop. These concerns were discussed earlier (Section 1.2.3). It is neither necessary nor desirable that UbiCom systems or human control exist that are mutually exclusive. It is envisaged that multilateral levels of control exist and that UbiCom control and human control are supplementary and that the balance between these depends upon the application requirements.

---

[16] Always-on often means in practice that the first line of response is an automated or so called cold-body response and also that during out-of-office hours messages can be left.

[17] Multiple-modes and multiple channels of communications, mobile phone, VoIP, email, Web, chat etc.

[18] The term *trafficking* was used to depict an exchange where users paid heed to regulations which governed the transfer of the content such as DRM and which could also include concerns about the cultural sensitivity of the content.

Table 13.2 UbiCom Interaction past, present and future (extended from Tesler, 1991)

Decade	1960–70s	1980s	1990s	2000s	2010s
Computer Devices	Mainframe, time-shared, single task computers	Wired MTOS PCs and data servers	Wireless MTOS laptops PDAs,	Wider set of MTOS devices, phone, TV, games console, camera, AV player	Many devices in environments can be interfaced to ICT devices dynamically
Where accessed	Specialist room at work	Desktop	Desktop, Mobile device	Desktop, Mobile device	Mobile devices and environment
What is networked	Terminals Internet started in 1970s	Data IP, non-IP voice and video	Increasing variety or devices networkd	Some integrated AV, radio, data, mobile IP hot-spots	Ubiquitous integrated multimedia networks
Type of hubs that interoperate	Mainframe	Corporate network	PC as hub	Several devices can act as hubs	Virtual, configurable hubs
Internetworks	None	Data internetworks arise	Begin integrated: ICT mobile inter-networks arise	Wider integrated: limited diverse device networks arise	Wider integrated diverse device inter-networks arise
HCI Interface	Physical control interface	Virtual tool interface	Virtual tool interface	Communication Interaction Interface	Hidden, tangible interaction
HCI Technologies	Teleprinter, keyboard, light pen,	WIMPS: keyboard mouse	WIMPS, voice IO, touchscreen,	Gestures, touchscreen voice IO	Organic UI
Mobility Human posture	Not portable Seated	Portable, Seated	Handheld, Standing	Wearable by many	Implants by many
HCI Interaction	Punch and Try, Submit	Text Command	See & Point	Ask & Tell, Delegate	Anticipate, Assist, Delegate
User Tasks	Calculate, Data processing (DP)	Access data, Edit	Present and layout	Remote access, Orchestrate	Remote control, Assist
User types	Specialist	Computer-literate single	Computer-literate singles and groups	Anyone	Anyone

13.6.2 More Versus Less Natural HCI

Although UbiCom systems (artefacts) aim to make some artificial activities seem more natural for humans to interact with, UbiCom inherently does the opposite by changing an activity from being less natural to being more artificial. The nature of what appears to be natural[19] interaction is dynamic, historical, cultural and to an extent personal. The focus is really on making interaction second nature rather than on making it natural. There are several factors that make interaction second nature. First, the degree to which a specific UbiCom system is second nature is rooted in our own individual and societal experience and preferences.[20]

Second, the degree to which a system appears to be second nature depends on the design and the usability of the system. The focus of an UbiCom system is on its use as a tool not on the tool itself. The accessibility and quality of service of a system are important. People expect not to wait, not to have to repeatedly reconfigure technology in order to use it. When technology is introduced, it is often overly complex to use and is often not second nature to use. It often takes several more innovations to make some inventions more second nature to use, e.g., the television.[21] There also exist examples of newer versions of simpler old inventions systems that seem more complex, e.g., clocks. The correct current time of many timing devices needs to be repeatedly manually reset when there is a power failure, the time zone changes, their timing drifts, batteries fail or when nations decide to put clocks forward or back an hour in different seasons. This is reconfiguration is unnecessary. If the clock was an UbiCom system it should automatically synchronise itself to a network timing signal. Raskin[22] (2000) has proposed standardising common functions across applications for the benefit of users.

Third, the second nature of interactions may vary across different functions within a system and across heterogeneous systems using the same function. An example of the former is that operating a mobile phone uses a moderate amount of manual dexterity whereas changing the phone's memory (SIM) card or removing the battery to reboot it when it crashes requires much greater manual dexterity.

There is a trade-off involved in terms of the benefits of going digital versus the disruption to the human experience in changing some traditional way of doing things. Some activities may require transducers or the use of mediated reality support in order to be replaced by digital versions.

13.6.3 Analogue to Digital and Digital Analogues

The physical world itself is not a discrete digital system, although it can sometimes be approximated to one. It exists as an analogue continuum of states in multiple dimensions. In order to sense and interface to the world, analogue to digital[23] conversion is needed.

When we use devices to measure physical world phenomena, it is often not the absolute value measured that is of interest but the user context – the relationship of the current measured values to past values or to some norm. For example, when we weigh ourselves, what we are more interested

---

[19] What we mean by nature is less pure nature and more second nature, i.e., acquired behaviour that is practised for so long it seems innate. Thus travelling by motorised vehicle seems second nature whereas travelling on the back of an animal, although more natural, is considered less normal.

[20] See the earlier comment by Alan Kay about the nature of inventions (Chapter 1).

[21] It took about 40 years of engineering development for a TV that at first required minutes to warm up and display a picture to become a device where the picture is immediately available when you switch it on. The TV also required specialists to install it and to tune and synchronise the receiver in the TV to the transmission carrier frequency for each channel. Today this is automated, so that most people can install a TV.

[22] He is credited with creating the Apple Macintosh User Interface.

[23] *Analogue to digital* is also referred to as *atoms to bits* or *physical to virtual* world.

in, the context, is if have we gained or lost weight and what our weight is compared to the norm. However, weighing scales just give the current weight. Humans must remember these relationships to past values, to the person's norm and to perhaps relate the measurement to time. The weighing scale has no memory, no suitable human–device interface to display the relation of a measurement to a context and no post-processing capability to support this. There are two options here to replace devices so that each has a capability to support users' contexts or to enable non-networked digital devices to be networked and the context-based processing of the situation to occur elsewhere.

Handwriting is considered natural but is manual and is slower and less natural to output information compared to speech. Its text entry speed for the average person is about 30 words per minute, about the same as keyboard entry. It is difficult to learn to operate and difficult to integrate its information into a digital world if we count its transcoding cost into digital form.

Much human interaction remains inaccessible to many humans. Consider the piano and (pipe) organ, these are unique keyboard instruments providing a rich musical experience for players and listeners. However, they are challenging for humans to learn how to play, requiring complex arm and leg coordination, an ability to sight read music, and are static instruments that occupy a significant physical space. This reduces the accessibility for players of these instruments. The accessibility can be increased by creating mobile versions and augmenting the playing using artificial rhythms and beats and allowing music to be downloaded and uploaded to the instrument.[24] The combination of gesture-based user interfaces (Section 5.3.2), near-field electromagnetic sensors incorporated into simpler digital analogues of more complex physical world objects such as very expensive musical instruments, can enable a much less exclusive audience to experience some elements of playing such instruments (Gershenfield, 1999). This may also enable potentially great players to master instruments quicker.

It also apparent that the experience of playing and the music produced changes when using electronic versions compared to using the analogue versions. One of the findings in the Ambient Wood project (Rogers, 2006) noted that there is a danger that the UbiCom systems can end up focusing the student's attention on the technology itself. For example, the goal of the technology in the Ambient Wood Project was perceived by the students to be one of finding all the hidden sounds or images, rather than heightening an awareness of, and triggering a higher-level reflection of, the physical environment.

13.6.4 Form Follows Function

With single function analogue type devices, the functional and physical design can be naturally intertwined so that the physical form follows function, e.g., pen, piano, toaster, etc. It is also possible to mimic analogue form using various digital artefacts such as using flexible continuums of conductive and light emitting materials, organic UIs, or by using ensembles of tiny micro-devices, MEMS, which can self-organise to support a particular form. As assembles can form at the millimetre, micrometer and nano meter scales, we could in theory mimic the form of complex physical objects, such as violins and piano using smart clay type devices. This may enable a much wider audience to experience the interaction with rare physical artefacts such as Stradivarius violins. However, mimicking the physical form using digital artefacts may still make it complex to mimic the equivalent function. An alternative is for digital artefacts to mimic the function but not the form of the equivalent physical world objects, although, this make it less familiar for humans to interact with.

---

[24] For example, a piano has been produced that rolls up and spreads to a metre in length, has a set of 128 tones, 100 rhythms and a MIDI Musical Device Interface.

13.6.5 Forms for Multi-Function Devices

An important design consideration is which form should be used to support multiple functions. The motivation for multiple functions is the interdependence between functions, e.g., previously, cameras recorded but didn't display – now they do both. If a camera displays or outputs internal recorded material, it can just as easily be extended to display multiple remotely recorded content input via a network or local media card. If it has a visual output, it also often needs to output audio. Hence, the device could be extended to support audio players and recorders too, etc. Hence, since the dawn of mobile devices, we often carry around devices with redundant functions, e.g., several devices we carry can store data, can authenticate us, can play audio and can display visual content. The main change is which specific devices can perform which functions. It is also often the case that although devices have a multi-functional capability, users use them for one of the dominant functions, often their original function, for the majority of the time. The assumptions here that each of the different functions can be state of the art or can at least be satisfactory or that the functions are now matured to such a degree so that they can no longer be improved need to be carefully looked at.

Whereas it is sometimes intuitive to design single function devices so that form follows function, e.g., pen, piano, toaster, etc, the form of a multi-function device may be less obvious. The form of a surface could mimic the partial form of a device. Consider a mobile phone which is also a digital camera, one of the two larger planar surfaces could be the phone and the reverse surface could act as the camera. Multi-function devices could be designed to be disassembled into individual components which could be subsequently re-assembled. However, the complexity of multiple functions sharing resources and possibly contending for resources remains.

13.7 Human Intelligence Versus Machine Intelligence

There are different visions of how humans and intelligent systems will coexist in the future. Although, human brains and human ability have not appeared to have changed significantly over a couple of thousand years, humans have developed machines which allows the combined ability of humans using machine as tools to improve pure mental and physical human ability. Russell and Norvig (2003, pp 960–964) discuss the following risks for the future of AI. People may lose their jobs to machines. People may have too little time (as machines take over more and more human tasks) or too much time (as more and more machines become ever more complex, generating ever more inputs for humans to process). People may lose they sense of being unique as humans become more and more subservient to a seamless virtual computer space. People may lose their privacy rights as everything can get logged and replayed and analysed by others. AI systems may result in a loss of accountability by humans as machines make more complex decisions and give more complex advice to guide human behaviour. The success of AI may lead to the end of the human race as machines rather than humans rule the world and can make decisions that rule out humans from the world.

Improved machines are being designed and developed by clever teams of humans. Machines could be designed to improve themselves leading to machines which could replace humans as the most intelligent life-form on earth with humans becoming subservient to machines rather than vice versa. *Emergent intelligence* is intelligence which emerges out of collections of simpler individual behaviours. Vinge (1993) considers a hypothesised point in time in which a *technological singularity*[25] will occur. A technology singularity refers to an intelligence explosion in which machines and software could even slightly surpass human intellect, through being designed to

---

[25] The term *Singularity* is so called as an analogy of the breakdown of modern physics near a gravitational singularity in which drastic changes in society would occur following an intelligence explosion.

self-improve their own designs in ways unseen by their designers. They can iteratively or recursively enhance themselves into more organised and more intelligent systems. Depending on which characteristics of intelligence are chosen, this may have already happened (Figure 13.6). The abilities of human and machines are contrasted in Table 13.3. Modis (2003) considers that if computers can communicate much more information faster than humans, the world could get to state of flux where only machines could perceive certain changes and act on them.

Humans will be accompanied by more micro systems, by more wearable systems, by more direct body, including brain, interfaces (Posthuman model). As humans become more dependent on machines, it may make us become less human. Technology could distance us from nature but it could also allow us to develop richer physical world and mental experiences.

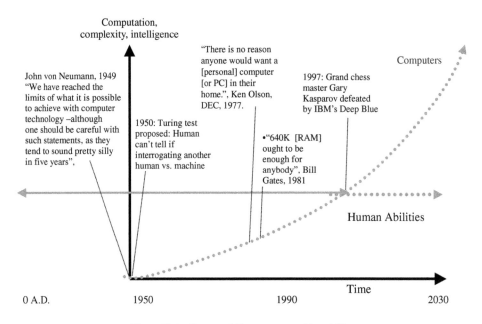

Figure 13.6 Human ability versus machine ability

Table 13.3 Contrasting specific human versus intelligent system behaviours

Humans are better at...	Intelligent machines are better at...
Reasoning inductively, generalising	Reasoning deductively: infer from generalisation
Retrieving details without a priori connection	Detecting prescribed (infrequent) events
Detecting constant patterns in varying situation	Detecting pattern outside human's range
Using experience to adapt decisions	Processing quantitative data in prescribed ways
Selecting from alternatives when failures occur	Performing several activities simultaneously
Remembering and applying principles and strategies	Retrieving details accurately
Acting in novel situations, developing new solutions	Repeating prescribed actions reliably
Focusing on priorities when overload occurs	Maintaining operation under information overload
Communicating richer, softer information	Communicating leaner, harder information, faster

There are also general issues concerning the more active the human environment is made, mediated using smart devices and smart (ICT) environments, the more challenging it can be to make machines understandable so that humans can engage them. In addition, if machines seek to model humans and physical environments so that ICT systems can better adapt to their environment, e.g., through using economic, social and emotional human models, this may lead to even more complex cyclic mutual modelling independencies and difficulty in establishing operational equilibria between multiple active interacting artificial intelligent and physical world systems.

There are two extremes in portraying a future landscape for future human computer interaction. One extreme is where there is an over-reliance on computers and automation which tend to replace the rich and subjective, non-deterministic inter-human interaction and human physical interaction with predominantly ICT system interaction. This has the peril of transferring all the decision-making to computers and use a much less rich, much more deterministic and mechanistic human behaviour and human behaviour which is always mediated by computers and distances human from interacting with the physical and human world. The other extreme is to distrust computers because computers may be open to abuse and to inadvertent errors by specific stakeholder factions in society, e.g., those responsible for surveillance or for administrating personal records, which then cause societal effects, e.g., losing copies of a significant proportion of people's personal records. In addition, some believe that computers have evolved as much as is possible and that the goal of calm computing where computers act in complete harmony with their human users is simply unachievable. Computers are overly constrained in their application which may overload human decision-making.

Several researchers have highlighted the need and benefits in balancing and in combining the best of human and machine intelligence in some kind of symbiotic interdependence between the physical and human environment and ICT systems as environments. Petriu *et al.* (2000) and Fischer (2006) consider how distributed frameworks need to understand the social context of humans and how artefacts and tools can then be better designed and evaluated to empower human beings and to change tasks. Kindberg and Fox (2002) and Rogers (2006) highlight the importance of defining an appropriate balance between human and machine behaviours and the conditions under which each of these is in control of the other. Alterman (2000) discusses the need to balance the design of fully autonomous systems with the benefits of user interaction (see Section 13.4.1) by stating 'The intelligence of a system is indivisible from the people who use it, built it, and designed the task environment in which it runs.' Alternatively, this can be expressed as until clever machines can autonomously design other intelligent machines, intelligent machines will always rely on the human ingenuity of the designer and the operator.

13.7.1 Posthuman: ICT Augments Human Abilities Beyond Being Human

Implants are used to enable less able people to become more normally able but they can also be used to enhance normal abilities. There are many wider societal and moral issues here concerned with the purpose of these implants whether anyone, only some people or everyone can benefit from these, who should be allowed to have them and what new human and social norms could arise as a result of these. In the posthuman model, human beings could become so seamlessly articulated with intelligent machines that the demarcation between what is human and what is human becomes lost (Fukuyama and Rosenthal, 2003).

There are several arguments that human intelligence cannot be modelled on or ultimately subsumed by machine intelligence (Lenoir, 2002). Despite the enormous successes of the past two decades in increasing the computation power of silicon, there are important differences in architecture between the human brain (conceived as a computer) and computing machines, for example there is no equivalent human consciousness in silicon on the horizon, however organic and biological materials and systems can be used as the basis for computers. Second, there is no embodiment - the Cartesian conception of mind as an informational pattern separable from the body is inadequate. The human consciousness is not localised in a set of neural connections in the brain alone, but is highly dependent on the material substrate of the biological body, with emotion and other dimensions as supportive structure. Russell and Norvig (2003, pp. 947–964) also give a discussion in whether machines can actually think, the so called Strong AI, versus whether or not they merely simulate thinking. They discuss the informality of human behaviour together with mathematical and philosophical objections to the idea that machines can actually think.

13.7.2 Blurring of Reality and Mediated Realities

The use of a wide range of versatile and affordable smart sensors, recorders and communicators enables us to document and replay personal memories of many aspects of our personal human and physical world environments which were not possible in the past. Harper *et al.* (2007) have noted through experiences in using a still or photo type wearable camera such as Microsoft SenseCam that our perception of reality and the recent past can change by replaying and (re)experiencing selective views we have recorded of the past. It can make its users reflect and view their social and physical environments in different ways. Mann's WearCam glasses and other devices go a step further enabling its users to record and view video in real-time and to overlay these views with virtual tagged views (Section 2.2.4.5). This enables people to filter, enhance or diminish view of reality in real-time.

The increased use of embedded and implantable devices is changing the nature of and the experiences of being human (Section 13.7.1). Models have been created that autonomously synthesise realistic human motions and possess a broad repertoire of lifelike motor skills which are being deployed as virtual stuntmen in AV films to replace real stuntmen because the stunts performed are considered too dangerous for real stuntmen (Faloutsos *et al.*, 2001). However, we are not yet at the stage where robots and human-like androids are being deployed in everyday life.

Pearson (2000) has made interesting predictions about second-order effects and use of these types of extrasensory reality. Virtualisation may be progressively used to give the illusion of different places and times. Public houses and inns could deploy virtual reality technology to give the illusion of traditions from a bygone era. At home, eWindows, eMirrors, etc, could display four-dimensional views of people, space and time of our choosing, triggering different emotional responses and memories as we view them. As we use more and more mediated electronic communication and virtual interactive environments, we may become unable to differentiate between remote real colleagues and virtual colleagues. Users who frequently use multiple virtual identities when interacting in multiple virtual environments may suffer from cases of multiple-personality syndromes. The use of augmented reality could be used to change the nature of personal and social interaction with family, friends and strangers, changing the experience of these human interactions to make them more exciting, more pleasurable and more tolerable.

13.8 Social Issues: Promise Versus Peril

New technology may be disruptive create new social norms, such as remote calls via mobile phones being accepted within local interactions such as business meetings and social events and more open access increases the need for privacy control of more vulnerable groups such as children. Edwards and Grinter (2001) comment that social benefits are often not envisaged by service providers.

The characteristics of ubiquitous computing are often quoted as being able to support computation anywhere and anytime. However, it is worth considering the exclusivity of the relationships between computation in the digital age and specific social groups. Many markets of goods become saturated before reaching everyone, e.g., in the video game market there is a sharp divide between gamers and non-gamers and this has lead to a stagnant market (Section 5.2.4).

A range of models has been developed by complex systems theoreticians that aim to model social behaviour and processes in terms of relatively simple underlying principles. Algorithmic models enable us to improve our understanding of the components of a complex dynamics in terms of simpler sub-dynamics. (Axelrod, 1997) has noted that although local convergence occurs, this also often generates global polarization because different locales are orthogonal. Hence cultural conventions can limit access to the use of technology to specific social groups. (Leydesdorff, 2002) considers Arthur's model of 'lock-in' to a single but potentially sub-optimal technology. For example, the QWERTY keyboard was invented to prevent keys jamming with the mechanical typewriter but is suboptimal for touch-typing. However, it remains the dominant keyboard layout despite all attempts to replace it. He also considers Kauffman's NK-model of hill-climbing which predicts potentially different sub-optima in a rugged fitness landscape. For example, alternative technologies can only survive in niche markets because of the hyper-selective conditions or niches that provide specific environments in which resources can be mobilised to enable agents to face the competition. From this perspective, adopters of competing technologies are climbing on different hills. In addition to access, it is also worth reflecting on the relations between the individual and personal space of users and technology.

One of the ironies of ubiquitous computing is almost no ubiquitous-computing systems work ubiquitously. Many demonstrations of a smart room or building, work only in their place of development. The creation, operation and management of technology and their associated services are also driven by the economics of usage, by business models and by institutional regulation. The distribution and broadcast of audio-visual content are licensed and regulated regionally – this restricts the 'anywhere anytime' access to within these regions. Licences to access this content must be bought by consumers. For example, UHF and VHF broadcast audio video content is readily accessible only within the range of the regional transmitter. International satellite broadcasts can be restricted by distributing access keys to decode the signals regionally. The DVD, Digital Video Disk is encoded for different regions. However, application gateways can access content locally but then store and forward it outside the regions it was destined for, to support location-free access – *content place-shifting*.

Providers are concerned that people are going to be able to access their product for free because one subscriber can give their access codes away so that others can watch content without paying for it. Manufacturers of place-shift devices say that they observe copyright and digital media laws because their box is a one-to-one transmission device so that owners can't broadcast a show to everyone they know. This is an example of where technological innovation can strain existing business models by driving them to evolve in order to take advantage of new opportunities or to seek to maintain the status quo. Another example is the concern by Internet Service Providers (ISPs) that some application service providers are not paying for the bandwidth needed by central services because they use a decentralised P2P model where clients download applications, acting as servers for further clients.

Systems could seek to orientate their services based upon the social context. Lyon (2001) refers to this as *social sorting* – classifying and categorising populations and persons for risk assessment, and its analogues, and management. The increasing ability of smart environments to analyse and correlate more human interactions and link these to identities raises well-known privacy issues, e.g., RFID tags left on items that accompany you can inform others in public spaces of what items you bought, where you bought them which can be cross-related to how much they cost and hence to income brackets. Gorman *et al.* (2004) discuss the societal issues of UbiCom.

13.8.1 Increased Virtual Social Interaction Versus Local Social Interaction

As computers become more interwoven into the physical world and more accessible, remote access and virtual interactions can take control over the local physical and social space interaction. Multiple different networks are available in many urban and rural places and for many people the network communication devices are by default always on. Whereas in the past single interaction modes were the norm, now multiple synchronous and asynchronous interaction modes are the norm for smart devices. As work and leisure become more flexible, e.g., flexible working locations, flexible working time and as we need to participate in multiple overlapping activities, it may not be so clear-cut to separate work and leisure interaction.

Whereas it was common-place as late as the 1990s to respond to the only other person in the vicinity when they are talking, it is now becoming social etiquette to first consider if people are communicating via an active remote network connection. Although voice call input can be hidden in the personal space, voice output caused by people speaking cannot so easily be hidden, causing possible social contention in physical spaces. Because we now have the whole world at our fingertips and at our beck and call, cooperation and socialization with remote people from a global population who can be selected to share our interests, rather than the smaller population local people who may not, may dominate.

13.8.2 UbiCom Accessible by Everyone

To make systems *accessible* has both a general meaning and a specific meaning in terms of HCI usability criteria. In the general sense, ICT accessibility is the degree to which it can be easily reached or used by as many people as possible. Within the context of HCI, accessible ICT is technology that can be used by people with a wide range of abilities and disabilities.

Within the general sense, making UbiCom accessibility, means taking in account the economic affordability (Section 13.8.4), cultural access[26] as well as the usability of products. Many countries have a sizeable ethnic, diversity, e.g., in the UK this was about 8% in 2001. Internationalism of UIs is well established and seeks to orientate these to the tangible aspects of culture such as the date, address format and currency. Marcus and Gould (2000) discuss earlier work by Hofstede (1977) that attempted to quantify the less tangible, social and psychological culture in terms of a set of five dimensions that was used to rate 53 countries in terms of cultural indices according to the following dimensions.

- *Power-distance*: the extent to which less powerful members expect and accept unequal power distribution within a culture.
- *Collectivism vs. individualism*: Individualism in cultures implies loose ties; everyone is expected to look after oneself or immediate family but no one else. Collectivism implies that people are integrated from birth into strong, cohesive groups that protect them in exchange for unquestioning loyalty.
- *Femininity vs. masculinity*: these refer to gender roles, not physical characteristics: Femininity deals with home and children, people, and tenderness. Masculinity covers assertiveness, competition, and toughness.

---

[26] There are many international national, cultural and religious rules and conventions which govern the use of many products and services. For example, there are rules which govern which sections of the population can drive motor vehicles.

- Uncertainty Avoidance (UA): Cultures vary in their avoidance of uncertainty, creating different rituals and having different values regarding formality, punctuality, etc.
- Long- vs. short-term orientation:a view of life.

As computers are becoming more inextricably embedded into the fabric of society, we share more and more, becoming more closely knitted into a collective and subsequently losing our individuality. Marcus and Gould (2000) have attempted to project how these dimensions may influence the design of user-interface and Web design. For example, High-UA cultures would emphasise: simplicity, with clear metaphors and limited choices; navigation schemes to prevent users getting lost; mental models and help systems to reduce 'user errors and redundant cues to reduce ambiguity. Jones and Marsden (2006) based upon their own experience, however, are sceptical of such a simplistic approach as countries are multi-cultural and because there exist many deeper expressions of culture.

Assuming that the UI is designed to support good HCI usability criteria (Section 5.5), a UI may still not be accessible by all because of user disabilities, the digital divide (those that have access to ICT and those who do not) and heterogeneous cultures. There are several reasons why designing UIs to be accessible by the widest possible group of users is advantageous. The most obvious reason is that it maximises the audience of potential users. While it may not be possible to give an accurate average global figure for the number of disabled people per country, the representation in specific countries can be used as a guide, e.g., in the UK 1 in 6 of the country's population has a disability (PAS, 2006). This seems on the high side but when you also consider the increase in the elderly population, this maybe not so. There is also some evidence that UIs designed to be accessible by disabled may also be used as a model to improve the utility for use by non-disabled users when their ability to interact is constrained, for example, when users need to access devices in poor light, in noisy conditions and while moving. The design of the mobile phone text messaging system has its roots in accessibility design for disabled users. Finally, there is a legal requirement to make systems accessible by everyone unless there are exceptional circumstances.

13.8.3 UbiCom Affordable by Everyone

The use of UbiCom access devices by everyone is dictated to a very large extent by economics and social aspects and in addition to the technology. Norman, (1999) gives a good account of the business and social issues in developing new products for the mass market. In many parts of the world, people still cannot afford the capital outlay to purchase an ICT device or the costs to maintain and operate ICT devices Gupta et al. (2006) have analysed previous approaches to developing ICT infrastructure in emerging economies. For example, multinational corporations set up more than 250 research and development labs in China and India in the years 1997 to 2005. These initially operated as lower-cost subsidiaries to support consumer demand in more developed countries but then these have increasingly broadened their focus to the development of technologies specific to the local economies. As local economies are diverse,[27] it is often much easier for external organizations to reach a new customer base and work through existing distribution channels via value-added intermediaries that better understand the local cultural diversity, for mutual benefit, rather than to create their own competitive channels. Typically,

---

[27] Social, economic and cultural differences include the multiplicity of languages, cultural diversity, literacy rates, price sensitivity, and computer usage.

these intermediaries are large existing entities, government or private, to which customers are already accustomed.

Gupta *et al.* (2006) posit that in order to reach diverse local customers, future research must take an end-to-end perspective in which, R&D must not only consider the front end, e.g., consumer access devices, but also the ICT infrastructure to reach the consumer and the back end, e.g., the enterprise backbone of service provision, to create an economically sustainable, scalable solution.

In order for devices to be used by everyone, low-cost and low maintenance versions of devices by home users on low incomes and in low density rural populations[28] are needed. Strategies to reduce the equipment cost include: using open-source software rather than commercial software,[29] e.g., Linux based digital video recorders; low-cost PCs and laptops for children[30] have been manufactured using low-end, low performance, rather than high-performance, hardware components and through miniaturisation of ICT components (Section 6.4).

It is also necessary to reduce other operational costs, through providing cheaper and multi-service local access networks, e.g., data over TV broadcasts, data over mobile telecoms networks etc, (Section 11.5); through the use of P2P based content sharing (Section 3.2.6), through the use of reduced power consumption (Section 4.3.4) and through reducing the management overhead for users to operate systems (Section 12.3.2). In order for devices to be used in rural areas, computers must be designed to contend with volatile and low power supplies in addition to volatile and lower bandwidth network access (Section 3.3.9). They must be robust to be designed for use in more hostile open living physical environments rather in enclosed living spaces.

13.8.4 Legislation in the Digital World and Digitising Legislation

There are increasing local and global policies, codes of practice and legislation to protect people from being harmed by ICT systems and to protect ICT systems from harming their physical environment. For example, designing UI that are accessible by users with a range of disabilities is a legal requirement. The European Commission Directive 90/270/EEC (Killingley, 1991) obliges employees to take into account five principles when designing, selecting, commissioning and modifying software, and in designing tasks using display screen equipment. These are that systems must be: suitable for the task, easy to use, provide suitable feedback on performance to users, display information in a format and at a pace adapted to operators and must apply principles of ergonomics (Killingley, 1991). Legal claims can be made, for example in the UK under the 1995 Disability Discrimination Act, if websites are not based upon the W3C WAI,

---

[28] For example, with the average population density in India being close to 400 persons/km, the subscriber density ranges from 10,000 phones/km in some urban areas to 5–10 telephones/km in most rural areas. Some sparsely populated pockets may initially require less than 0.5 telephones/km (Jhunjhunwala *et al.*, 1998).

[29] The GNU Project Open Source version of the Unix OS started in 1984 by Richard Stallman as an effort to circumvent AT&T from taking back control of Unix that was previously licensed at low cost to developers and re-releasing it under higher cost commercial licenses. In 1991, Linus Torvalds began developing the Linux kernel based upon GNU Unix. In 1999, embedded computer applications based upon Linux such as TiVo digital set-top boxes to receive and record broadcast video were developed (Barton, 2003).

[30] 'One Laptop Per Child' (OLPC) is an education project, which can be implemented in more several ways, such as the so-called '$100 Laptop'. Many children, especially those in rural parts of developing countries have little access to schools. In parallel with continuing school building programs and teacher education, which is often time-consuming and slow, a parallel method advised by OLPC is to leverage the children themselves by engaging them more directly in their own learning. Although, it may sound implausible to equip the poorest children with connected laptops when richer children may not have them, laptops can be affordable and children are more capable than they are sometimes given credit for. See http://laptop.org/, accessed Nov. 2007.

Web Accessibility Initiative set of guidelines and made usable by disabled people without good cause (PAS 78, 2006).

Because of the complexity and volume of the legal framework and because it is largely oriented to be human-readable and understandable by human legal specialists (there is a human bottleneck), it is sometimes difficult to know exactly which specific legislation applies, whether or not somebody or something is adhering or complying with the legislation and who is liable when legislation is transgressed and what sort of corrective settlement is fair and just. There are two complementary aspects of legislation we consider here. First, as legislation gets so complex, we need to develop automated techniques to show legal compliance. Otherwise, organisations and people may get over burdened with the process of becoming legally compliant. Second, the kinds of legislation needed to protect the rights in society as more innovative UbiCom systems become developed and more widely deployed, need to be considered.

Firms for whom it is paramount to demonstrate regulatory compliance or to advise others about this are increasingly representing legislation in the form of computerised information sets that can be more automatically queried and applied. There is the issue of how to structure and transcode these accurately from legal language into a computation form with the relevant metadata to enhance automated retrieval. The mainstream approach used so far here seems to be to use weak semantic representations of legislation, e.g., to simply represent and store legal rules as natural language text or to partition rules into related facts and store these in a relational database. Legal experts are assisted in being able to more easily search and browse the legislation within a domain. Other approaches being investigated by computer scientists working with legal experts are to model legislation, legal cases and legal *arguments*[31] in much more expressive, semantically richer ways such as *deontic*[32] logic (Abrahams and Eyers, 2007) and the Semantic Web (Uijttenbroek *et al.*, 2007).

It is demanding to engineer a system that can be audited to show legal compliance with the relevant legislation. It is noted that transborder ICT systems are often more complex to engineer because different legislation within the same application domain may apply, within different international and within intra-national boundaries, e.g., ecommerce transactions may have differ-ent taxation liabilities depending on point of manufacture or creation, point of origin of sale, point of completion of sale, point of residence of consumer and point of residence of supplier.

Finally, there are general issues in the development and application of legislation to new types of technology. Generally when any new technology is being researched and developed, society sets up bodies to review and regulate it in order to protect individuals and society. These concerns are enhanced for UbiCom because there is a much greater range of materials, technologies and device sizes becoming more immersed and spread over the physical world and because these are being used in much more unobtrusive ways and being interfaced to humans in new ways. The biggest concerns stem from technical, social and commercial uncertainty about detrimental effects which may or not happen. Of particular concern is the use of technologies that either lead to new forms of soft and hard artificial intelligence supplanting and disrupting human intelligence and in damaging the physical, biological, social and economic environment, e.g., the impact of nanotechnology. In the latter case, voluntary codes of conduct have been proposed.[33]

---

[31] Argumentation is concerned primarily with reaching conclusions through reasoning about claims based on premises. In law, argumentation is used mainly to test the validity of specific types of evidence.

[32] Deontic logic is a logic of obligations and permissions, about what conditions and actions ought to or ought not to be brought about.

[33] The Responsible Nano Code initiative formed by the Royal Society, Insight Investment and the Nanotechnology Industries Association (NIA), http://www.responsiblenanocode.org/, accessed May 2008.

To protect humans from robots and to give robots some rights to protect themselves as sentient beings, a set of the three Laws for Robotics[34] was proposed by Asimov in the 1940s. A robot may not injure a human being or, through inaction, allow a human being to come to harm (First Law). A robot must obey orders given to it by human beings except where such orders would conflict with the First Law (Second Law). A robot must protect its own existence as long as such protection does not conflict with the First or Second Law (Third Law).[35] Robots and software agents are a form of intelligent for which the intelligent system seems reasonably understandable by human (specialists) and systems are well bounded. Emergent technology and intelligence can be hard to legislate to control research and development.

Legislation can be formulated to be too technology specific and thus become too technology restrictive. For example, Gollmann (2005) discusses an interesting point with respect to an early version of P3P which could express only policies about when information was retrieved, e.g., via Web browser cookies. The EU privacy directive requires user consent at the time the data was written because the EU privacy directive expresses privacy concerns relating to databases holding data in a centralised store. However, privacy information can be distributed, e.g., held within a user's system, but which is still accessible by a service provider.

Machines can be designed to incorporate multiple individual human traits such as intelligence, emotions and collective human traits such as social and legislative behaviours. While designing machines to operate legally appears understandable, a further question is whether or not machines should also be designed to be ethical, to support fairness, justice, equity, honesty, trustworthiness and equality. Somebody and something can act legally with or without acting ethically and vice versa. Moor (2006) considers the nature and challenges in supporting *machine ethics*. One way is to constrain the machine's actions to avoid unethical outcomes. It is, however, not clear whether or not machines can be designed to be fully ethical in the sense that they can make explicit ethical judgements and generally are competent enough to reasonably justify them.

13.9 Final Remarks

This book examines in-depth the design and applications of distributed computing, context-awareness, HCI, autonomous systems and intelligent systems all in one volume. It also takes a multi-disciplinary approach to considering the environmental, societal, legal, etc. issues of future computing. To do this, a holistic framework for UbiCom called the Smart DEI (Devices, Environments and Interaction) model has been proposed which is based upon three interlinked system viewpoints:

- UbiCom system properties: distributed, iHCI, context-aware, intelligent, autonomous;
- distinct UbiCom designs and architectures: smart devices, environments and interaction;
- UbiCom system interaction in three distinct environments: ICT or virtual computing, physical world and human world.

---

[34] These laws were proposed as part of a work of a fiction rather than being formulated by legal experts. A new edition of the original book has been published (Asimov (1991).

[35] It is clear that these robot laws are insufficient – it must consider goal conflict between different humans. Consider the case for the foodstuff management scenario (Section 1.1.2.3), what if an owner's robot has the directive to control access to the owner's home, has opened access to the home but at the same time a thief seizes the opportunity to gain entry the owner's home? The application of Asimov's robot laws in this situation is left as an exercise.

No single formal definition for UbiCom systems is given because a diverse range of UbiCom systems is needed. Instead a set of properties is defined which can be combined in different ways to support this diverse range of systems. Five core properties for UbiCom systems: distributed ICT, context-awareness, intrinsic human–computer interaction, autonomous systems and intelligent systems are proposed. UbiCom can be viewed as an application domain in which these different properties are supported. A classification (of over seventy terms) which supports different combinations of UbiCom system properties and sub-properties and which defines synonyms for many different types of ubiquitous computing depending on the application and requirements is given. These properties overlap and are interlinked, e.g., iHCI is interlinked to user context-awareness; autonomous system and anticipatory HCI systems and intelligent systems are interlinked are considered in depth. In terms of types of ubiquitous computing devices, Weiser (1991) proposed three main forms: tabs, pads and boards. This set of forms has been extended to include three additional forms: dust, skins and clay.

Multiple levels of support for these properties are needed, depending on the application and user requirements, ranging from minimal, through basic, moderate, good and full support. It may not be required, or useful or usable in many cases in practice, to support the full set of all of these properties.

Some researchers seem quite pessimistic about the status and rate of progress of ubiquitous computing over the past twenty years or so. This is in part due to the expectation of full support for UbiCom computing generally in any type of system (virtual, physical and human) environment. In contrast to this view, this text contends that substantial progress has been made in achieving the vision of ubiquitous computing. Many specific applications of UbiCom in specific sections and throughout this book argue the case that UbiCom has already succeeded. Many UbiCom systems are in routine use and this use will only increase in the future. Truly, even more exciting times lie ahead for all of us stakeholders involved in the future of computing.

EXERCISES

1. Discuss how the range of six form factors can be used to propose smart environments, and more fluid ensembles of smart devices.

2. Explain some of the key challenges for smart environments based upon intelligible interaction, impromptu interoperability, no (central) system administrator, designing for the home, social effects, reliability, interference in the presence of ambiguity. Consider if these challenges apply equally to smart mobile devices.

3. Discuss the range of smart devices described throughout this text. Predict how likely they are to become mass used and when this will happen.

4. Discuss techniques for lower power usages. Should this be controlled at the operating system level, by remote middleware services, by applications, or by the human user?

5. Compare and contrast the following techniques for lowering energy use: passive electronic components, MEMS, energy harvesting, ultra-capacitors and fuel cells.

6. Define a reverse manufacturing process. Distinguish between recycling, de-manufacturing and remanufacturing. Highlight the benefits of remanufacturing over recycling.

7. Consider how the eco-friendly design and use of micro devices differ from the eco-friendly design and use of macro devices.

EXERCISES (continued)

8. Discuss whether Asimov's robot laws are sufficient Consider the case for the foodstuff management scenario (Section 1.1.2.3), what if an owner's robot has the directive to control access to the owner's home, has opened access to the home but at the same time a thief seizes the opportunity to gain entry into the owner's home? Consider how Asimov's robot laws apply to this situation. Propose how to extend Asimov's robot laws by considering goal conflicts between different humans.

References

Abowd, G.D. and Mynatt, E.D. (2000) Charting past, present, and future research in ubiquitous computing. *ACM Transactions on Computer-Human Interaction (TOCHI)*, 7(1): 29–58.

Abrahams, A. and Eyers, D. (2007) Mapping legal cases to RDF named graphs using a minimal deontic ontology for computer-assisted legal querying. In *Proceedings Workshop on Semantic Web Technology for Law*, pp. 11–20.

Asimov, I. (1991) *I, Robot*. New York: Spectra Books.

Atwood, B., Warneke, B. and Pister, K.S.J. (2001) Smart Dust mote forerunners. In *Proceedings of 14th Annual International Conference on Microelectromechanical Sytsems*, pp. 357–360.

Axelrod, R. (1997) The dissemination of culture: a model with local convergence and global polarization. *Journal of Conflict Resolution*, 41(2): 203–226.

Barton, J. (2003) From server room to living room. *ACM Queue*, 1(5): 20–32.

Beina, E., Steenbruggen, J. and Wagtendonk, A. (2007) Location awareness 2020: a foresight study on location and sensor services. *Vrije Universiteit Technical Report E-07/09 May 2007*. Available from http://www.oracle. com/global/it/mobility/events/ev-mov/2007/euro-beinat.pdf, accessed Feb. 2008.

Blevis, E. (2008) Sustainability implications of organic user interface technologies: an inky problem. *Communications of the ACM*, 51(6): 56–57.

Bullard, G.L., Sierra-Alcazar, H.B., Lee, H.L., *et al* (1989) Operating principles of the ultracapacitor. *IEEE Transactions on Magnetics*, 25(1): 102–106.

Chalasani, S. and Conrad, J.M. (2008) A survey of energy harvesting sources for embedded systems. In *Proceedings of IEEE Southeastcon*, pp. 442–447.

Christensen, C.M. (1997) *The Innovator's Dilemma: When New Technologies Cause Great Firms to Fail: Managing Disruptive Technological Change*. Boston: Harvard Business School Press.

Davies, N. and Gellersen, H-W. (2002) Beyond prototypes: challenges in deploying ubiquitous systems. *IEEE Pervasive Computing*, 1(1): 26–35.

Demers, A.J. (1994) Research issues in ubiquitous computing. In *Proceedings of 12th ACM Symposium on Principles of Distributed computing*, pp. 2–8.

Edwards, W.K. and Grinter, R E. (2001) At home with ubiquitous computing: seven challenges. In *Proceedings of the 3rd International Conference on Ubiquitous Computing*, Atlanta, Georgia, LNCS: 2201, 256–272.

Faloutsos, P., van dePanne, M. and Terzopoulos, D. (2001) The virtual stuntman: dynamic characters with a repertoire of autonomous motor skills. *Computers & Graphics* 25(6): 933–953.

Fischer, G. (2006) Distributed intelligence: extending the power of the unaided, individual human mind. In *Proceedings of Working Conference on Advanced Visual Interfaces*, pp. 7–14.

Fukuyama, F. and Rosenthal, J.H. (2003) *Our Posthuman Future: Consequences of the Biotechnology Revolution*. New York: Picador.

Gershenfield, N. (1999) *When Things Start to Think*. London: Henry Holt & Co, pp. 27–44.

Goldstein, S.C., Campbell, J.D. and Mowry, T.C. (2005) Programmable matter. *Computer*, 38(6): 99–101.

Gollmann, D. (2005) *Computer Security*. Chichester: John Wiley & Sons, Ltd.

Gorman, M.E., Groves, J.F. and Catalano, R.K. (2004) Societal dimensions of nanotechnology. *IEEE Technology and Society*, 23(4): 55–62.

Greenfield, A. (2006) *Everyware: The Dawning Age of Ubiquitous Computing*. Harlow: Pearson Education.

Gupta, A., Ranganathan, P., Sarin, P. and Shah, M. (2006) IT infrastructure in emerging markets: arguing for an end-to-end perspective. *IEEE Pervasive Computing*, 5(2): 24–31.

Harper, R., Randall, D., Smyth, N., *et al* (2008) The past is a different place: they do things differently there. In *Proceedings of the 7th ACM Conference on Designing Interactive Systems* (DIS 2008), pp. 271–280.

Harper, R., Rodden, T., Rogers, Y. and Sellen, A. (eds) (2007) *Being Human: Human-Computer Interaction in the Year 2020*. Technical Report, Microsoft Research Ltd. Available from http://research.microsoft.com/hci2020/downloads/BeingHuman_A4.pdf, retrieved April 2008.

Hofstede, G. (1977) *Cultures and Organisations: Software of the Mind*. New York: McGraw-Hill.

Huang, E.M. and Truong, K.N. (2008) Breaking the disposable technology paradigm: opportunities for sustainable interaction design for mobile phones. In *Proceedings of 26th Annual SIGCHI Conference on Human Factors in Computing Systems*, pp. 323–332.

Jessup, L.M. and Robey, D. (2002) The relevance of social issues in ubiquitous computing environments. *Communications of the ACM*, 45(12): 88–91.

Jhunjhunwala, A., Ramamurthi, B. and Gonzalves, T.A. (1998) The role of technology in telecom expansion in India, *IEEE Communications.*, 36(11): 88–94.

Johanson, B., Fox, A. and Winograd, T. (2002) The interactive workspaces project: experiences with ubiquitous computing rooms. *IEEE Pervasive Computing*, 1(2): 67–74.

Jones, M. and Marsden, G. (2006) *Mobile Interaction Design*. Chichester: John Wiley & Sons, Ltd.

Killingley J. (1991). Directive 90/270/EEC – a job for human factors? (1991). In *Proceedings of 6th BILETA Annual Conference*. Available online from http://www.bileta.ac.uk/pages/Conference%20Papers.aspx, accessed Sept. 2006.

Kurzweil, R. (2001) Promise and peril – the deeply intertwined poles of 21st century technology. *Communications of the ACM*, 44(3): 88–91.

Leydesdorff, L. (2002) The complex dynamics of technological innovation: a comparison of models using cellular automata. *Systems Research and Behavioral Science*, 19(6): 563–575.

Lyon, D. (2001) Facing the future: seeking ethics for everyday surveillance. *Ethics and Information Technology* 3: 171–181.

Marcus, A. and Gould, E.W. (2000) Crosscurrents: cultural dimensions and global Web user-interface design. *ACM Interactions*, 7(4): 32–46.

Marcussen, C.H (2001) Mobile data and m-commerce in Europe – a mobile network operators' revenue perspective, 1999–2003. August 2001. Available online from http://www.crt.dk/UK/Staff/chm/P_CHM.htm, accessed May 2007.

Modis, T. (2003) The limits of complexity and change, *Futurist* (May–June): 26–32.

Moor, J.H. (2006) The nature, importance, and difficulty of machine ethics. *IEEE Intelligent Systems*, 21(4): 18–21.

Norman, D.A. (1999) *The Invisible Computer: Why Good Products Can Fail, the Personal Computer Iis so Complex and Information Appliances are the Solution*. Cambridge, MA: MIT Press,

O'Hara, K., Morris, R., Shadbolt N., *et al.* (2006) Memories for life: a review of the science and technology. *Journal of Royal Society Interface*, 3(8): 351–365.

Oliver, J.Y., Amirtharajah, R., V. Akella, *et al.* (2007) Life cycle aware computing: reusing silicon technology. *IEEE Computer*, 40(12): 56–61.

PAS 78 (2006) Guide to Good Practice in Commissioning Accessible Websites. Available from http://www.drc.org.uk/pas, accessed Sept. 2006.

Patel, D. and Pearson, I.D. (2002) Hype and reality in the future home. *BT Technology Journal*, 20(2): 106–115.

Pearson I.D. (2000) Technology timeline — towards Life in 2020. *BT Technology Journal*, 18(1): 19–31.

Petriu, E.M., Georganas, N.D. Petriu, D.C. *et al.* (2000) Sensor based information appliances. *IEEE Instrumentation & Measurement*, 3(4): 31–35.

Raskin, R. (2000) *The Human Interface*. Reading, MA: Addison-Wesley.

Rogers, Y. (2006) Moving on from Weiser's vision of calm computing: engaging UbiComp experiences. In P. Dourish and A. Friday (eds) *Proceedings of Ubicomp 2006, Lecture Notes in Computing Science*, 4206: 404–421.

Russell, S. and Norvig, P. (2003) *Artificial Intelligence: A Modern Approach*, 2nd edn. Upper Saddle River, NJ: Prentice Hall.

Satyanarayanan, M. (2001) Pervasive computing: vision and challenges. *IEEE Personal Communications*, 8: 10–7.

Sellen, A.J. and Harper, R.H.R. (2001) *The Myth of the Paperless Office*. Cambridge, MA: MIT Press.

Sutcliffe, H. and Hodgson, S. (2006) An uncertain business: the technical, social and commercial challenges presented by nanotechnology. Retrieved from http://www.responsiblenanocode.org/, May 2008.

Taylor, A.S., Harper, R., Swan, L., *et al*. (2007) Homes that make us smart. *Personal and Ubiquitous Computing*, 11(5): 383–393.

Tesler, L.G. (1991) Networked computing in the 1990s. *Scientific American*, 265(3): 86–93.

Uijttenbroek, E.M., Klein, M. and Lodder, A.R. (2007) semantic case law retrieval: findings and challenges. In *Proceedings of Workshop on Semantic Web Technology for Law*, pp. 33–40.

Van der Bush, P. (1945) As we may think. *The Atlantic Monthly*, Vol. 176, 101–108. Reprinted and discussed in *ACM Interactions*, 3(2), Mar 1996: 35–67.

Vinge, V. (1993) *The Coming Technological Singularity: How to Survive in the Post-Human Era*. *VISION-21 Symposium, 1993*. Retrieved from http://www-rohan.sdsu.edu/faculty/vinge/misc/singularity.html, Dec. 2007.

Weiser, M. (1991) The computer for the twenty-first century. *Scientific American*, 265(3): 94–104.

Weiser, M. (1993) Some computer science issues in ubiquitous computing. *Communications of the ACM*, 36(7): 75–84.

White, C.D., Masanet, E., Rosen, C.M., *et al*. (2003) Product recovery with some byte: an overview of management challenges and environmental consequences in reverse manufacturing for the computer industry. *Journal of Cleaner Production*, 11(4): 445-458.

Zhuo, J., Chakrabarti, C., Chang, N., *et al*. (2006) Extending the lifetime of fuel cell based hybrid systems. In *Proceedings of 43rd Annual Conference on Design Automation*, pp. 562–567.

Index

Page numbers in **bold** refer to **footnote** and *figures* are indicated in *italics*.